Santi Cassisi and Maurizio Salaris

Old Stellar Populations

Related Titles

Hoard, D. W. (ed.)

White Dwarf Atmospheres and Circumstellar Environments

244 pages with 84 figures, Hardcover
2011
ISBN: 978-3-527-41031-6

Barnes, R. (ed.)

Formation and Evolution of Exoplanets

320 pages with 136 figures and 5 tables,
Hardcover
2010
ISBN: 978-3-527-40896-2

Wang, Y.

Dark Energy

256 pages with 48 figures and 7 tables,
Hardcover
2010
ISBN: 978-3-527-40941-9

Shore, S. N.

Astrophysical Hydrodynamics
An Introduction

472 pages with 70 figures, Softcover
2007
ISBN: 978-3-527-40669-2

Stahler, S. W., Palla, F.

The Formation of Stars

865 pages with 511 figures and 21 tables,
Softcover
2004
ISBN: 978-3-527-40559-6

Spitzer, L.

Physical Processes in the Interstellar Medium

335 pages with 30 figures, Softcover
1998
ISBN: 978-0-471-29335-4

Shapiro, S. L., Teukolsky, S. A.

Black Holes, White Dwarfs and Neutron Stars
The Physics of Compact Objects

645 pages, Hardcover
1983
ISBN: 978-0-471-87317-4

Santi Cassisi and Maurizio Salaris

Old Stellar Populations

How to Study the Fossil Record of Galaxy Formation

WILEY-VCH

WILEY-VCH Verlag GmbH & Co. KGaA

The Authors

Prof. Santi Cassisi
INAF
Collurania Astronomical Observatory
Via M. Maggini
64100 Teramo
Italy

Prof. Maurizio Salaris
Liverpool John Moores University
12 Quays House
Birkenhead, CH41 1LD
United Kingdom

■ All books published by Wiley-VCH are carefully produced. Nevertheless, authors, editors, and publisher do not warrant the information contained in these books, including this book, to be free of errors. Readers are advised to keep in mind that statements, data, illustrations, procedural details or other items may inadvertently be inaccurate.

Library of Congress Card No.:
applied for

British Library Cataloguing-in-Publication Data:
A catalogue record for this book is available from the British Library.

Bibliographic information published by the Deutsche Nationalbibliothek
The Deutsche Nationalbibliothek lists this publication in the Deutsche Nationalbibliografie; detailed bibliographic data are available on the Internet at http://dnb.d-nb.de.

© 2013 WILEY-VCH Verlag GmbH & Co. KGaA, Boschstr. 12, 69469 Weinheim, Germany

Hardcover ISBN 978-3-527-41076-7
Softcover ISBN 978-3-527-41059-0
ePDF ISBN 978-3-527-66552-5
ePub ISBN 978-3-527-66554-9
mobi ISBN 978-3-527-66553-2
oBook ISBN 978-3-527-66551-8

Cover Design Adam-Design, Weinheim
Composition le-tex publishing services GmbH, Leipzig
Printing and Binding Markono Print Media Pte Ltd, Singapore

Printed in Singapore

Printed on acid-free paper

To my daughter

S. Cassisi

To my mum and the memory of my dad

M. Salaris

Contents

Preface

Without a doubt, one of the main problems in modern astrophysics is understanding how galaxies form and evolve. Within the framework of the currently accepted standard cosmological model, it is possible to perform simulations of the formation and evolution of cosmic structures; the results of these simulations are, however, affected by a number of uncertainties, stemming largely from the extreme difficulty of understanding the gas dynamics/processes that determine how baryonic matter settles into galaxies, cools and collapses to form stars, and how stellar winds and supernova explosions stir, mix and chemically enrich the remaining gas, and control the rate at which new stars form. A detailed calculation of these processes is far beyond the capabilities of modern computers.

Fossil evidence available in nearby galaxies – where "fossil" means that the observables are the result of the past star formation history – in the form of their old stellar populations, is an important part of the research on these open issues. The information encoded in the spectra of these stars contain a detailed record of their origin, hence of the mechanisms of galaxy formation and evolution. Moreover, galaxies in the nearby universe can be studied in much greater detail compared to high-redshift counterparts, and provide a picture of the evolution of cosmic structures complementary to that obtained from high-redshift observations.

This book presents and discusses in detail a wide range of techniques developed in the last decades to extract information about formation and evolution of old stellar populations from their photometric and spectroscopic observations. Our definition of old stellar populations comes from the theory of stellar evolution; a stellar system is denoted as "old" if it harbours in the post-Main Sequence stages – excluding remnants like white dwarfs, neutron stars and black holes – stars with mass below $\approx 2.0 M_\odot$ (low-mass stars). These old populations are characterized by the signature of a well defined red giant branch in the colour–magnitude-diagram or integrated spectrum. In terms of ages, this implies values above ≈ 1 Gyr, meaning that these stars are the fossil record of the development of galaxies over more than 90% of cosmic evolution. We will use a somewhat arbitrary definition of nearby Universe, in the sense that we discuss methods for both resolved and unresolved populations, and their use to study stellar systems at low redshifts, corresponding to look-back times that are a negligible fraction of the age of the Universe.

This work is the "natural" companion and follow-up of our previous volume, "Evolution of stars and stellar populations", that was aimed at introducing undergraduate and postgraduate students to modern stellar evolution studies, and the associated techniques to study stellar populations of all ages, including the basics of "stellar population synthesis techniques". Here, we go several steps further in terms of the level of detail. At the same time, we focus exclusively on old stellar populations. The reader will acquire an in-depth and up-to-date specialistic knowledge of the evolutionary properties of stars hosted by old stellar populations, and how to use their observations to study the evolution of the parent systems. Our book is about "tools and techniques" to study old stellar populations, and the related uncertainties, not about the results in terms of scenarios of galaxy formation; this would require an entire volume on its own. However, we will show some practical applications of the methods, mainly with a view to highlight their reliability.

Chapter 1 briefly introduces the standard Λ Cold Dark Matter cosmology, the problems related to modelling the formation of cosmic structures, and the basic tools for stellar population analyses. Chapter 2 discusses a number of topics related to stellar input physics and model calculations, with particular emphasis on those physical mechanisms that are still uncertain and mostly affect the results presented in the following chapters. The next four chapters present a detailed qualitative and quantitative picture of the life-cycle of low-mass stars, from their formation to the final stage of cooling white dwarfs. Chapters 7–9 present and analyze methods to study simple and composite, resolved and unresolved old stellar populations, based on the stellar evolution previously discussed, providing the reader with the most recent developments in the field. We discuss the recently discovered phenomenon of "multi-populations" in Galactic globular clusters in detail, and its impact on Galactic and extragalactic age-dating techniques. We have also included a large number of references (not necessarily all-encompassing and unavoidably biased by our own familiarity with specific papers) to guide the reader through the recent literature, major milestones and review papers.

We are greatly indebted to Lucio Primo Pacinelli (INAF – Astronomical Observatory of Collurania) and David Hyder (Liverpool John Moores University) for their superb work on many of the figures included in this book, and Katrina Exter for her invaluable help with editing some of the chapters.

Special thanks go to the following colleagues for several figures included in the book as well as for their useful suggestions: F. Allard, H.M. Antia, A. Aparicio, S. Basu, P. Baumann, B. Behr, A. Bellini, D. Bersier, G. Bono, D. Brown, R. Buonanno, E. Carretta, C. Charbonnel, S. Cristallo, E. Dalessandro, P. Eggenberger, J.W. Ferguson, F. Ferraro, G. Fiorentino, C. Gallart, F. Grundahl, S. Hidalgo, A. Irwin, N. Lagarde, A. Marín-Franch, A.F. Marino, G. Michaud, M.M. Miller Bertolami, A.P. Milone, M. Monelli, A. Mucciarelli, S. Percival, L. Piersanti, A.M. Piersimoni, A. Pietrinferni, G. Piotto, J. Richer, L. Sbordone, E. Small, O. Straniero, S. Talon, D. Vandenberg, A. Weiss, P. Wood, and M. Zoccali.

Finally, we wish to warmly thank all of our colleagues and collaborators with whom, over the course of the years, we have investigated and discussed several of the topics addressed in this book.

Teramo   Santi Cassisi
Liverpool   Maurizio Salaris
June 2012

1
Introduction

The so-called Λ Cold Dark Matter (ΛCDM) model is currently the most wide-ly accepted theory for the structure and evolution of the Universe. A combina-tion of experimental data involving the power spectrum of the Cosmic Microwave Background (CMB) radiation temperature fluctuations, the abundance and weak-lensing mass measurements of galaxy clusters, the Type Ia supernovae magni-tude-redshift relation and large scale structure observations has placed strong con-straints on the free parameters of this cosmological model [1].

In the framework of ΛCDM cosmology, the large scale structure of the Universe is assumed to be homogeneous and isotropic – denoted as "cosmological princi-ple" – on scales of the order of 100 Mpc. Therefore, the local inhomogeneities can be treated as perturbations to the general homogeneity of the Universe, and it is represented by the Friedmann–Robertson–Walker metrics. The evolution with time of this model universe is described by

$$\left(\frac{\mathrm{d}R(t)}{\mathrm{d}t}\right)^2 = -kc^2 + \frac{(8\pi G\rho(t) + \Lambda)R(t)^2}{3} \tag{1.1}$$

$$\frac{\mathrm{d}\left(\rho(t)c^2 R(t)^3\right)}{\mathrm{d}t} = -P\frac{\mathrm{d}R(t)^3}{\mathrm{d}t} \tag{1.2}$$

as obtained from the field equations of general relativity. Here, k denotes the spa-tial curvature, $\mathrm{d}t$ the cosmic time separation, ρ the density of matter plus radiation. $R(t)$ is the so-called cosmic scale factor that describes the expansion (or contrac-tion) of the Universe. It has the dimensions of a distance and is dependent on the cosmic time t. The constant k is defined in a way that $k = +1$ for a positive spatial curvature, $k = 0$ for a flat space and $k = -1$ for a negative curvature. The cosmic time is defined by standard clocks comoving with the cosmic fluid. If one sets t to the same value for all clocks when a local property of the cosmic fluid, for example, the average local density of matter, has attained a certain agreed value, by virtue of the cosmological principle, the same value of that property (possibly different from the one at the time of synchronization) has to be measured whenever clocks show the same time.

Equation (1.1) includes the cosmological constant Λ. If Λ were to vary with t, it represents a so-called "dark energy". Current observations are consistent with Λ

Old Stellar Populations, First Edition. S. Cassisi and M. Salaris.
© 2013 WILEY-VCH Verlag GmbH & Co. KGaA. Published 2013 by WILEY-VCH Verlag GmbH & Co. KGaA.

being a constant.[1] The evolution of $R(t)$ is controlled by ρ, the geometry k and Λ. By defining the function $H(t)$ as

$$H(t) \equiv \frac{dR(t)}{dt} \frac{1}{R(t)} \tag{1.3}$$

one can rewrite Eq. (1.1) as

$$H(t)^2 = -\frac{kc^2}{R(t)^2} + \frac{8\pi G\rho(t)}{3} + \frac{\Lambda}{3} \tag{1.4}$$

The value of $H(t)$ determined today is the Hubble constant H_0. It is customary to introduce the critical density $\rho_c \equiv 3H(t)^2/(8\pi G)$ and define the density parameter $\Omega_\rho = \rho/\rho_c$, an equivalent for the cosmological constant $\Omega_\Lambda = \Lambda/(3H(t)^2)$, and the sum $\Omega = \Omega_\rho + \Omega_\Lambda$. With these definitions, Eq. (1.4) becomes

$$(1 - \Omega)H(t)^2 R(t)^2 = -kc^2 \tag{1.5}$$

$\Omega = 1$ gives a flat space, $\Omega > 1$ a positive curvature and $\Omega < 1$ a negative curvature.

At present, ρ appears to be dominated by matter, the contribution from photons being negligible ($\Omega_r = (4.800 \pm 0.014) \times 10^{-5}$). Ω_ρ is split into contributions from baryonic ($\Omega_B = 0.0456 \pm 0.0016$) and dark matter ($\Omega_{DM} = 0.227 \pm 0.014$). The presence of "cold" (with a negligible velocity dispersion) dark matter – called "dark" because it does not interact electromagnetically with the other components of the cosmic fluid – is well-established by observations of galaxy rotation curves, the X-ray emission of the hot ionized gas in clusters of galaxies and gravitational lensing surveys. As for its origin, various candidates predicted by extensions of the Standard Model of particle physics have been proposed, but to date, the question about the nature of dark matter is awaiting a definitive answer.

The dominant contribution to the present energy budget of the Universe is, however, given by Λ, that is, $\Omega_\Lambda = 0.728^{+0.015}_{-0.016}$. This implies that $\Omega_\Lambda + \Omega_B + \Omega_{DM} = 1$ and the space is flat.

As we go backwards in cosmic time, $R(t)$ decreases and the radiation density increases faster than the density of matter. Therefore, there must have been a moment in the cosmic history when the Universe was radiation dominated; the cosmic time t_E of matter-radiation equality is of the order of 10^5 years. Going back to even earlier cosmic times, it is clear that the Universe must have started from a singular state with $R = 0$ and $\rho = \infty$ at $t = 0$. Present estimates of the matter and Λ density coupled to the equations for the evolution of $R(t)$ provide a value of the Hubble constant $H_0 = 70.4^{+1.3}_{-1.4}$ km Mpc^{-1} s^{-1} and an age of the Universe $t_U = 13.75 \pm 0.11$ Gyr.

Now, reversing the arrow of cosmic time, as the Universe expands and cools from the initial singularity during the initial radiation dominated era, a small amount

---

1) An interesting comparison of various points of view on this issue can be found in [2].

of some light elements (mainly D, ^{3}He, ^{4}He, ^{7}Li) is produced when the temperature evolves from $T \sim 10^9$ K down to 3×10^7 K. The amount of light nuclei produced depends on the baryon density and the expansion (i.e. cooling) rate. Calculations of cosmological nucleosynthesis – assuming three neutrino species with
mass much smaller than 1 MeV – provide, for the present estimate of the baryon
density, a primordial He mass fraction $Y_P = 0.2487 \pm 0.0002$, and number abundance ratios $(D/H)_p = (2.52 \pm 0.17) \times 10^{-5}$, $(^3He/H)_P = (1.03 \pm 0.03) \times 10^{-5}$,
$(Li/H)_P = 5.12^{+0.71}_{-0.62} 10^{-10}$ for D, ^{3}He and ^{7}Li, respectively [3, 4].

When the temperature drops below the ionization energy of hydrogen (13.6 eV),
the ionization fraction, however, stays close to one due to the large number of photons over baryons (photons dominate by number, although matter dominates en-
ergetically and therefore gravitationally). The number of photons in the high energy tail of the black body spectrum is high enough to keep the matter fully ionized. Eventually, the temperature and therefore the number density of sufficiently energetic photons drops so low that recombination prevails. It is at this time,
$\sim 10^{5.5}$ years after the singularity (i.e. when $T \sim 4000$ K), that the first atoms
form. The resulting dearth of free electrons has the consequence of reducing the
efficiency of electron scattering, so that matter and radiation decouple. From this
moment on, the temperatures of radiation and matter become different and start to
evolve separately; radiation does not interact any longer with matter and can travel
undisturbed through space. The radiation temperature T_r is reduced according to
$T_r \propto R(t)^{-1}$, and the blackbody spectrum it had at decoupling is preserved. This
blackbody radiation, largely homogeneous and isotropic (because of the cosmological principle) with a temperature T_r of ~ 2.73 K is the theoretical counterpart of
the CMB.

If the Universe was perfectly isotropic and homogeneous, no structures would
have formed with time. However, in the case of inhomogeneities, regions denser
than the background tend to contract and get denser still, inducing a growth of the
initial perturbation. In 1970, Peebles, Yu, Sunyaev and Zel'dovich predicted that
these inhomogeneities had to be imprinted in the CMB as the tiny temperature
fluctuations that have been recently detected (they are of the order of $\Delta T/T \approx$
10^{-5}). Fluctuations of the local density of baryonic matter would have behaved as
sound waves (with their fundamental mode plus overtones) in the cosmic fluid
before recombination, with the photons (to which baryons were tightly coupled
before recombination) providing the restoring force. At recombination, the photons started to travel for the first time unimpeded through space; photons released
from denser, hotter regions were more energetic than photons released from more
rarefied regions. This temperature differences were thus frozen into the CMB at recombination and are detected today. Most importantly, the amplitude and location
of the peaks in the power spectrum of these temperature fluctuations are closely
related to a number of cosmological parameters (for more details see, for example,
the discussion in [5]); in particular, the location of the first peak is mainly related to the geometry of the tridimensional space, whereas the ratio of the heights
of the first to second peak is strongly dependent on Ω_B. Also, the values of the
Hubble constant and of the cosmological constant affect both location and ampli-

tudes of the peaks, albeit with different sensitivities. Observations show that the fluctuations are consistent with a Gaussian random field and the power spectrum is close to scale invariant, that is, $P(k) \propto k^{n_s}$ with $n_s = 0.963 \pm 0.012$ ($n_s = 1$ for a scale-invariant spectrum).

The widely accepted idea about their origin relates to the so-called inflationary paradigm that can, in principle, explain why $\Omega = 1$ and solve the so-called horizon problem, that is, why the CMB across the sky is to a very good approximation isotropic, even though the size of the region causally connected at decoupling corresponds to about only one degree in the sky today.

The central idea envisages a period in the early Universe where a term Λ_{inf} originated by a hypothetical quantum scalar field analogous to the cosmological constant dominates Eq. (1.4). This can be rewritten as

$$H(t)^2 = \frac{\Lambda_{inf}}{3} \tag{1.6}$$

and its solution, assuming a constant Λ_{inf}, is

$$R(t) = R(t_i)e^{\sqrt{\Lambda_{inf}/3}\,t} = R_i e^{H(t)t} \tag{1.7}$$

if t is much larger than the cosmic time $t = t_{inf}$ of the beginning of the Λ_{inf} dominated epoch. Provided this exponential expansion (inflation) is long enough, Ω its driven towards one, irrespective of its initial value. Moreover, during inflation, a very small patch of the Universe can grow to enormous dimensions, so that the isotropy of the CMB temperature we see today, arose from a very small causally connected region that underwent an inflationary growth. An expansion by a factor of $\approx 10^{30}$ solves both the flatness and horizon problem without invoking *ad hoc* initial conditions. The quantum field that originated Λ_{inf} is expected to experience quantum fluctuations that were stretched by the inflation to the scales we see imprinted in the CMB. The general belief is that inflation occurred when the strong force separated from the electroweak one, at about $t = 10^{-35}$ s, and lasted until about $t = 10^{-32}$ s.

1.1
Galaxy Formation

Starting from the ΛCDM cosmological model, one of the main problems in modern astrophysics research is to understand how a galaxy forms and evolves [6, 7]. Here, we will just summarize the main concepts. The basic idea underpinning the currently most accepted paradigm of galaxy formation is that cosmic structures grow through the mechanism of gravitational instability starting from the pattern of density fluctuations imprinted in the CMB. The dark matter component, having no pressure, begins to collapse well before the baryonic matter, and the primordial density fluctuations will grow. The early evolution of this dissipationless (no energy can be lost through electromagnetic interactions) growth is described by the linear perturbation theory. Once the perturbations become non-linear, their evolution is modelled using *N*-body simulations. The final result of the non-linear evolution of a dark matter density fluctuation is the formation of a dark matter halo, an approximately stable state supported by the random motions of the dark matter particles. In standard ΛCDM cosmology, the first haloes form from the smallest scale fluctuations, and are followed by successive merging episodes that produce increasingly more massive structures. The baryons are dragged along by the dark matter that dominates gravity, and the haloes are therefore expected to accrete baryons, so that the individual galaxies we see can be interpreted as the product of the evolution of baryonic matter nested inside a much larger halo of CDM. The efficiency of the accretion process will depend on the depth of halo potential well, and on the pressure of the baryons. Therefore, a complete description of the formation and evolution of galaxies requires a detailed description of the evolution of the dissipative baryons, that is, their accretion, heating, cooling, the associated star formation processes, the chemical evolution, the effect of halo merging on the baryonic component.

Overall, at any given cosmic time, matter is distributed over structures spanning many decades in mass, and growth is driven by merging between haloes of similar mass, by accretion of much less massive haloes and diffuse material, and by destruction by infall onto larger haloes (hence, the name of "hierarchical merging" scenario). A crucial point is that galaxy morphology may be a transient phenomenon and that the different types of galaxies we observe nowadays (the Hubble sequence) reflect the variety of accretion histories. As an example, this scenario envisages that baryons falling smoothly into the potential wells of dark matter haloes produce disks, whereas spheroids (bulges of disk galaxies and isolated elliptical galaxies) are the products of major (i.e. between haloes of similar mass) merger events whereby disks are mixed violently on short timescales, and a burst of star formation depletes substantially the gas content of the merging disks. Smooth accretion of intergalactic gas on spheroids appears then to be able to produce a typical present-day disk galaxy, for example, a rotationally supported disk of gas and young stars and a centrally concentrated bulge system of old stars.

One of the major difficulties in modelling from first principles how galaxies form and evolve is related to the physics that drives the evolution of the baryonic component, that is, gas cooling in dark matter haloes, star formation, chemical evolution and feedback mechanisms that either remove cold gas from a disk or suppress gas cooling. These processes, sometimes labelled "gastrophysics", are still poorly understood in a galactic context and prescriptions that contain free parameters are employed. These parameters are fixed by demanding that models reproduce sets of available observations, typically at low redshifts. A powerful way to provide strong independent constraints on galaxy formation models in general, and on "gastrophysics" in particular, is to determine star formation histories of stellar populations in galaxies using methods from stellar evolution theory.

A wide range of techniques developed in the last decades make use of stellar evolution models to estimate distances, ages, star formation and chemical evolution histories of galaxies. It is these techniques, the related uncertainties and their future developments that we are going to discuss in detail in the following chapters, beginning with their theoretical foundations grounded in stellar evolution theory. We will, at the same time, discuss the use of these methods to determine the star formation and chemical evolution histories of old, local stellar populations.

Although observations at high redshift are certainly a more direct way of looking at the formation of galaxies, these objects are very faint and one can derive a more limited amount of information compared to that obtained from nearby galaxies. Stellar populations in the nearby Universe can be studied in much greater detail and provide a view of galaxy formation and evolution that is complementary to that obtained from high redshift data. Also, observations and analyses of the oldest local stellar populations provide completely independent powerful tests and constraints to the overall cosmological model.

We will use a somewhat arbitrary definition of the nearby Universe, one that often only includes those stellar systems resolvable into individual stars. Here, we will also discuss methods for unresolved stellar populations, and their use to study stellar populations at low redshifts corresponding to look-back times that are a negligible fraction of the age of the Universe.

Our definition of old stellar population comes from stellar evolution theory. We denote as "old" all stellar systems harbouring objects with mass below $\approx 2.0 M_\odot$, that is, those stars that undergo electron degeneracy in their He-core at the end of the main sequence (MS) phase, plus remnants of more massive objects. Observationally, the colour–magnitude-diagram (CMD) of these populations is characterized by the presence of a well-defined red giant branch (RGB) sequence. In terms of ages, our definition of old populations implies values above $\approx 1\,\text{Gyr}$, that is, these stars are the fossil record (fossil in the sense that their observables are the result of the past star formation history) of the development of cosmic structures over more than 90 % of cosmic evolution.

1.2
Decoding the Fossil Records: Photometric and Spectroscopic Diagnostics

The amount of evolutionary information we can gather from a stellar population is encoded in the radiation emitted by its stars. A population that is resolved observationally into its individual constituents is denoted as "resolved", while with an "unresolved" stellar population, we denote a system from which we can only observe the integrated light, that is, the sum of the contribution of all its components.

In the case of resolved populations, one can obtain high-resolution spectra – resolution of the order of $\approx 0.1\,\text{Å}$, most often in the visible wavelength range – of individual stars, that allow the determination of photospheric abundances of several chemical elements using model atmosphere calculations. These abundances provide constraints on the initial chemical composition of the star. Photometric observations through broadband filters (i.e. Johnson–Cousins, Sloan, WFC3, 2MASS) with passbands of order $\approx 10^3\,\text{Å}$, typically spanning the wavelength range from near-ultraviolet (near-UV) to near-infrared (near-IR), are used to produce CMDs. These diagrams are the main tool to constrain ages, star formation and chemical evolution histories – this latter in conjunction with estimates from spectroscopy, if/when available – of resolved stellar populations. Intermediate band photometric filters (i.e. Strömgren, DDO, Walraven) with passbands of order $\approx 10^2$ Å have been devised to derive indices that are particularly sensitive to some specific stellar property, like surface gravity or metallicity (e.g. the indices c_1 and m_1 in the Strömgren system). Narrow band systems – passbands of order ≈ 10 Å – also exist (i.e. Oke, Wing). For example, individual bands in the Wing systems are sensitive to CaH, CN, TiO or VO.[2]

As for unresolved stellar populations, integrated photometry typically in broadband photometric systems provides colour–colour diagrams or – when measurements in several bands are available – a very low-resolution spectral energy distribution (SED) that can be used to constrain ages and metal content of the parent populations. Integrated spectra with resolution of ≈ 1–10 Å can be used to infer ages and, in principle, detailed individual abundances of several chemical species.

Figure 1.1 displays some examples of photometric and spectroscopic data for resolved and unresolved populations.

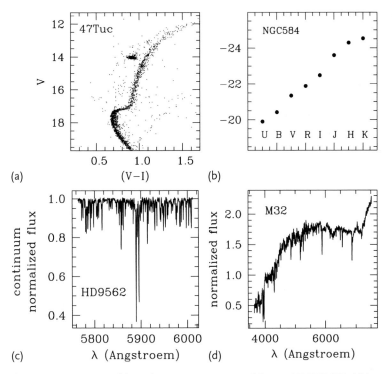

Figure 1.1 (a) A CMD of the Galactic globular cluster 47Tuc [10]. (b) Integrated $UBVRIJHK$ magnitudes of the elliptical galaxy NGC584 [11]. (c) Portion of the spectrum of the star HD9562 [12]. (d) Integrated spectrum of the dwarf elliptical galaxy M32 [13]. The fluxes are normalized to the observed flux at 4000 Å.

2) For an exhaustive review of standard photometric systems and their properties, see [9].

1.3
Decoding the Fossil Record: the Tools

Observed CMDs of resolved populations, SED, colour–colour diagrams, integrated spectra of unresolved populations can be interpreted in terms of evolutionary properties of the parent stars by applying results from stellar evolution theory.

1.3.1
Theoretical Stellar Models, Tracks, Isochrones

The solution of the equations of stellar structure provides a stellar evolution model, that is, the run of physical and chemical quantities from the centre to the photosphere of a star of given initial mass and initial chemical composition, and their evolution with time. It is common practice in stellar astrophysics to specify the initial chemical composition of stellar models by means of the symbols X, Y and Z. They denote the mass fractions of hydrogen, helium and all other elements heavier than helium (called metals, hence Z is also called the metallicity of the star) respectively; these three parameters are related through the normalization $X + Y + Z = 1$. For the metals, the distribution of the individual fractional abundances has to be specified.

This is a convenient choice from the theoretical point of view, though it is not directly related to what is determined from stellar spectroscopy. The helium abundance, for example, cannot be determined for all stars since low-mass objects are generally too cool to show helium spectral lines, and the metal abundances are usually determined differentially with respect to the Sun. The traditional metal abundance indicator is the quantity $[\text{Fe/H}] \equiv \log\left[N(\text{Fe})/N(\text{H})\right]_\star - \log\left[N(\text{Fe})/N(\text{H})\right]_\odot$, that is, the difference of the logarithm of the Fe/H number abundance ratios observed in the atmosphere of the target star and in the solar one. If one assumes that the solar heavy element distribution is universal, the conversion from Z to $[\text{Fe/H}]$ is given by

$$[\text{Fe/H}] = \log\left(\frac{Z}{X}\right)_\star - \log\left(\frac{Z}{X}\right)_\odot \tag{1.8}$$

If one relaxes the assumption of a universal scaled-solar heavy element distribution, the correspondence between $[\text{Fe/H}]$ and X, Y, Z obviously changes because the ratio between the iron abundance and Z is different than in the Sun. In this case, one can still use Eq. (1.8), provided that the left-hand side denotes the ratio of the "total" abundance of metals to hydrogen

$$[\text{M/H}] = \log\left(\frac{Z}{X}\right)_\star - \log\left(\frac{Z}{X}\right)_\odot \tag{1.9}$$

A basic working tool in stellar evolution studies is the Hertzsprung–Russell diagram (HRD), a plot of a star bolometric luminosity versus its effective temperature – both outcomes of stellar evolution calculations – that will be widely used in

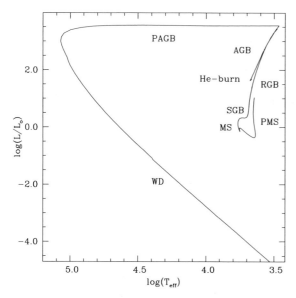

Figure 1.2 Theoretical HRD for the evolution of the Sun. All major evolutionary phases are labelled (see text for details).

the next chapters. The evolution of a stellar model in the HRD is denoted as the "stellar track".

Figure 1.2 displays the complete theoretical HRD, from birth to the final evolutionary stage, for the Sun. All main evolutionary stages of low-mass stars discussed in the next chapters are covered in this diagram. The starting point of the evolution shown in Figure 1.2 is the pre-main sequence (PMS) phase, where the star is convective and evolves vertically in the HRD along its so-called Hayashi track driven by gravitational contraction according to the virial theorem, until a radiative core forms and eventually H-burning ignites in the central regions. At this stage, the evolution slows down, while H is converted into He in the core, during what is denoted as the main sequence (MS) phase. It is the longest evolutionary phase and as a consequence, the number of H-burning stars which can be observed in a given population is much larger than the number of stars in any other phase. When the central H is exhausted, the H-burning shifts to a narrow shell around the pure He-core left over by the central H-burning, and the subgiant branch phase (SGB) starts. During the SGB evolution, the He-core undergoes electron degeneracy, the external layers expand and the photosphere gets cooler, until the model – by now with a very deep convective envelope – reaches its red giant branch (RGB) location, an approximately vertical sequence in the HRD. During SGB and RGB evolution, the He-core grows steadily in mass due to fresh He deposited by the outward moving (in terms of mass layers) H-burning shell, and correspondingly, the surface luminosity increases. Along the RGB, mass loss processes are also efficient, and reduce the mass of the convective envelope.

When the electron degenerate core reaches a mass of about $0.47\,M_\odot$, He-burning ignites in the core (He flash), the electron degeneracy is lifted and the star begins a phase of quiescent central He-burning (labelled as He-burn in Figure 1.2) at a much lower luminosity compared to the end termination of the RGB. The precise value of T_{eff} is determined by the amount of mass left in the envelope, T_{eff} getting higher with decreasing envelope mass (more efficient RGB mass loss). In metal-poor systems like Galactic globular clusters (GCs), the central He-burning stage appears as an extended, more or less horizontal sequence in the CMD at visible wavelengths, and it's called the Horizontal Branch (HB). During this central He-burning phase, the H-burning shell is still active, and when He is exhausted in the core, shell burning provides the energy necessary to maintain hydrostatic equilibrium.

The model now has a C and O core left over by the central He-burning that becomes electron degenerate, marking the beginning of the asymptotic giant branch (AGB) phase. During the AGB evolution, H and He shell burning are never active at the same time, instead they take turns at producing the nuclear energy necessary to maintain hydrostatic equilibrium through a series of so-called thermal pulses (TPs). The AGB sequence practically overlaps with the RGB in the HRD, and the evolution is towards increasing luminosity because of the mass increase of the CO-core due to the He-burning shell. Mass loss is again very effective in reducing the envelope mass. When the envelope mass (reduced by both the growth of the CO-core and mass loss from the photosphere) is reduced to $\approx 0.01\,M_\odot$, the model starts its post AGB (PAGB) phase, moving towards higher T_{eff} at constant luminosity, fixed by the value of the CO-core mass, equal to about $0.54\,M_\odot$, with some residual shell burning.

The PAGB phase terminates when the model reaches its white dwarf (WD) cooling sequence. From this point on, the evolution is towards lower surface luminosities (and T_{eff}), the energy radiated being the free energy of the non-degenerate ions, while hydrostatic equilibrium is maintained by the pressure of the degenerate electrons.

Stellar tracks of different masses and chemical compositions can be combined to predict the HRD of stars harboured by stellar populations with a generic star formation and chemical evolution history. The most elementary stellar population, usually denoted as simple stellar population (SSP), is made of objects born at the same time in a burst of star formation activity of negligible duration, with the same initial chemical composition. Any population with an arbitrarily complex evolutionary history can be reduced to a linear combination of several SSPs.

The theoretical HRD of a SSP is called isochrone. Consider a set of evolutionary tracks of stars with the same initial chemical composition and various initial masses; different points along an individual track correspond to different values of the time t_{tr} and the same initial mass. An isochrone of age t is the line in the HRD that connects the points belonging to the various tracks (one point per track) where $t_{\text{tr}} = t$. This means that when we move along an isochrone, time is constant. However, the value of the initial mass of the star at each point is changing, that is, increasing towards more advanced evolutionary stages. A generic point along

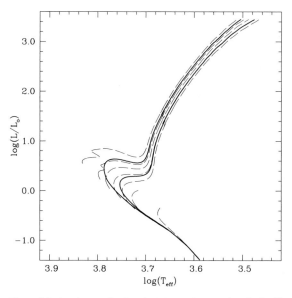

Figure 1.3 Isochrones for 5 and 13 Gyr and solar initial chemical composition (thick solid lines) overimposed on a set of stellar tracks for the same initial chemical composition (thin-dashed lines). We display in order of increasing luminosities along the MS, tracks for 0.6, 0.8, 0.9, 1.0, 1.1, 1.2 and 1.4$M_\odot$.

an isochrone of age t is therefore determined by three quantities: bolometric luminosity, effective temperature and value of the evolving mass. From these three parameters, one can easily derive additional quantities like the radius and surface gravity. It is also straightforward to associate to each point along an isochrone the expected surface chemical abundances – taken from the underlying grid of stellar evolution models – to be compared with spectroscopic measurements.

If mass loss processes are included in the individual stellar tracks – as we have briefly seen before, mass loss is very efficient during the RGB and AGB evolution of low-mass stars – the situation is only slightly more complicated because along each track the total mass is changing with time. The procedure to compute the isochrones is the same, that is, one connects the points of equal age along tracks with various initial masses. However, the value of the mass evolving at a given point along the isochrone is now smaller than the initial mass of the parent track.

Figure 1.3 shows, as an example, two isochrones for ages equal to 5 and 13 Gyr, respectively, and the solar initial chemical composition, that cover the evolutionary phases from the MS to the He flash. Notice the large mass range spanned by the MS phase, whereas the SGB and RGB stages (this is true for all post-MS phases but the final WD stage) are populated by objects with almost the same mass, equal to the value at the termination of the MS (the so-called turn off point – TO). This is a consequence of their much shorter evolutionary timescales, compared to MS lifetimes.

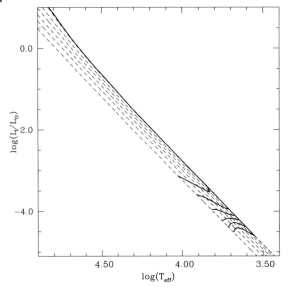

Figure 1.4 Isochrones for 2, 4, 6, 8, 10 and 12 Gyr (solid lines) overimposed on a set of WD cooling tracks with masses between 0.54 and 1.0$M_\odot$ (dashed lines).

Representative isochrones for the final WD stages are shown in Figure 1.4. Along a WD isochrone, it is the sum of the WD cooling age plus the progenitor age at the end of the TP phase that stays constant. Obviously, a fundamental ingredient to calculate WD isochrones is the relationship between the mass of the progenitor on the MS (initial mass) and its final WD mass (initial-final mass relation – IFMR – see Chapter 6 for details). Due to the fast cooling times at high T_{eff}, the bright section of a WD isochrone of fixed age is populated by objects with approximately the same mass, corresponding approximately to the final WD mass of stars that have just left the TP phase. One can also notice that all isochrones in Figure 1.4 share the same location at bright luminosities because empirical and theoretical IFMRs predict essentially the same WD mass ($\sim 0.55 M_\odot$) for all low-mass stars. At the bottom end of the sequence, one recovers the progeny of more massive intermediate-mass stars that have left the TP phase earlier. This explains the sudden increase of T_{eff} (lower radii) since the more massive progenitors produce WDs of increasing mass, and hence lower radius.

Once a generic isochrone of a given age and initial chemical composition is computed, it can be transferred to an observational CMD, that is, the plot of a star magnitude in a given photometric band versus a colour index. To predict magnitudes in a generic photometric system, one needs to assign a spectrum to each point along the isochrone, making use of model atmosphere calculations.

A model atmosphere describes the physical and chemical structure of a stellar atmosphere and the transfer of radiation from the photosphere into interstellar space. It is defined (at least in the plane-parallel approximation) by three quantities: the value of T_{eff}, the surface gravity g and the photospheric chemical composition.

Interpolation among a suitable grid of spectra obtained from model atmosphere calculations provides spectra to assign to each point along the isochrone.

From a given spectrum, it is straightforward to compute bolometric corrections BC_A to a generic filter A from the following equation:[3]

$$BC_A = M_{\text{bol},\odot} - 2.5 \log \left[4\pi (10 p c)^2 \frac{a c T_{\text{eff}}^4}{4 L_\odot} \right] + 2.5 \log \left(\frac{\int_{\lambda_1}^{\lambda_2} F_\lambda S_\lambda \, d\lambda}{\int_{\lambda_1}^{\lambda_2} f_\lambda^0 S_\lambda \, d\lambda} \right) - m_A^0$$

(1.10)

Here, $M_{\text{bol},\odot}$ is the bolometric magnitude of the Sun, F_λ is the model spectrum (i.e. from model atmosphere calculations), S_λ is the response function of the photometric filter (a measure of the efficiency of photon detection within the filter wavelength range) that covers the wavelength range between λ_1 and λ_2, f_λ^0 denotes the reference spectrum that produces a known apparent magnitude m_A^0 (for example, the spectrum of Vega, or a spectrum of constant flux density per unit frequency for the ABmag system. In both cases, the apparent magnitude m_A^0 is usually set to zero). From BC_A, one can then obtain the magnitude M_A from

$$M_A \equiv M_{\text{bol}} - BC_A \qquad (1.11)$$

where $M_{\text{bol}} = M_{\text{bol},\odot} - 2.5 \log(L/L_\odot)$. The value of $L_\odot$ is known, and once the bolometric magnitude of the Sun is fixed (e.g. $M_{\text{bol},\odot} = 4.74$ [14]), what is left in order to convert T_{eff} and the bolometric luminosity L into magnitudes is the choice of f_λ^0 and m_A^0, the so-called "zero points" of the photometric system.

The very popular Johnson–Cousins and the Hubble Space Telescope HST/WFPC2 VEGAmag systems, for example, make use of the star Vega to fix the zero points, assuming that the apparent V magnitude of Vega and all its colour indices are equal to zero. The Sloan system instead uses a zero point flux $f_\nu^0 = 3.63110^{-20} \, \text{erg s}^{-1} \, \text{cm}^{-2} \, \text{Hz}^{-1}$ that gives $m_A^0 = 0$ at all frequencies (see [15] for details about zero points of several photometric systems). Figure 1.5 displays the response function of a few widely used filters;[4] a theoretical low-resolution spectrum of Vega is overimposed, for the sake of comparison.

The transformation from theoretical luminosity and T_{eff} to observed magnitudes and colour indices has been described assuming up to now that F_λ at the stellar surface is obtained from the appropriate theoretical model atmospheres. It is, however, known that current theoretical model atmospheres suffer from at least two main shortcomings:

- Broad-band colours of stars with solar chemical compositions appear to be reasonably reproduced, but many spectral lines predicted by the models are not

3) The terms in the two integrals are formally correct if the calibration of the photometric system is based on energy-amplifier devices; they need to be multiplied by λ when the calibration is based on photon-counting devices like CCDs. The difference between these two types of integrations is usually very small.

4) A complete database of photometric filters can be found at http://ulisse.pd.astro.it/Astro/ADPS/ (accessed 22 January 2013).

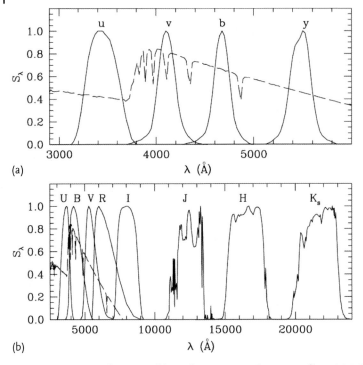

(a)

(b)

Figure 1.5 Response functions of the *uvby* Strömgren photometric filters (a), Johnson–Cousins *UBVRI* and 2MASS *JHK*ₛ (b). Overimposed is a theoretical spectrum of the star Vega (dashed line) [16].

observed in the Sun, and also the relative strength of several lines is not well-reproduced. This affects intermediate- and narrow-band filters in which individual metal lines can significantly affect the bolometric corrections.

- In convective model atmospheres, the energy transport is usually treated with the mixing length theory (see next chapter); this approximation introduces an uncertainty in the predicted spectra, and hence bolometric corrections and colour indices. Recent two- and tridimensional hydrodynamical simulations of stellar model atmospheres are addressing this issue, but they have not produced yet libraries of stellar spectra that cover all the relevant evolutionary phases, mass ranges and chemical compositions.

These shortcomings cause an uncertainty of possibly a few hundredths of a magnitude on the BC_A values. An alternative solution is to use empirical spectra of a sample of nearby stars with independently determined T_{eff}, gravities and chemical composition. A problem with this approach is that stars for which empirical T_{eff} values can be determined are local objects that cover a fairly narrow combined range of chemical compositions, masses and evolutionary phases (reflecting the local population of the Galactic disk) and would not allow a complete modelling of different stellar populations.

1.3.2
Luminosity Functions, Synthetic CMDs

The CMD of an isochrone only provides a morphological counterpart to the observed CMDs of resolved stellar populations, for it does not give any information about the number of objects expected along the different branches of the observational diagrams. As we will see in the next chapters, just using isochrones is enough to constrain ages and initial chemical compositions of resolved SSPs. However, additional information coming from star counts must be accounted for when trying to disentangle the star formation and chemical evolution history of more complex stellar populations.

Given that each point along an isochrone corresponds to a specific evolutionary stage of a model of initial mass M, assuming a stellar initial mass function (IMF) that provides the number of stars dN born with mass between M and $M + dM$, will also provide the number of objects populating a generic interval between two consecutive points along the isochrone. For an IMF of the standard form $dN = C M^{-x} dM$, the normalization constant C can be fixed by specifying either the total mass (M_t) or the total number (N_t) of stars born when the SSP formed. When $x \neq 2$, the value of the constant C is given by

$$C = (2 - x) \frac{M_t}{M_u^{2-x} - M_l^{2-x}} \tag{1.12}$$

where x is the exponent of the IMF, M_u and M_l are, respectively, the upper and lower mass limits of the stellar entire mass spectrum, for example, ~ 0.1 and $\sim 100 M_{\odot}$. If $x = 2$,

$$C = \frac{M_t}{\ln(M_u/M_l)} \tag{1.13}$$

This normalization guarantees that the total mass of stars formed stays constant, independent of the value of x, but the initial number of stars formed changes with changing value of the slope of the IMF. In the case N_t instead of M_t is given, the previous relationships have to be rewritten as

$$C = (1 - x) \frac{N_t}{M_u^{1-x} - M_l^{1-x}} \tag{1.14}$$

if $x \neq 1$ and

$$C = \frac{N_t}{\ln(M_u/M_l)} \tag{1.15}$$

if $x = 1$.

One way to compare observed star number counts with theory is to use luminosity functions (LFs). One can make use of two types of LFs. The differential LF, for example, the run of the observed star number counts $N(M_A^i)$ in the magnitude range between M_A^i and M_A^{i+1}, as a function of M_A^i, or the cumulative LF, that is,

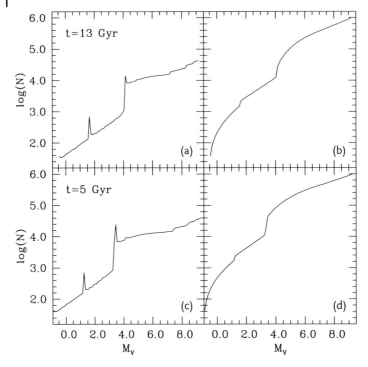

Figure 1.6 Differential (a,c) and cumulative (b,d) LFs for the isochrones displayed in Figure 1.3, computed with a Salpeter IMF and a total population of 10^6 stars.

the run of the number of stars $N(M_A > M_A^i)$ (or $N(M_A < M_A^i)$) with magnitude M_A larger than M_A^i (or lower than M_A^i), as a function of M_A^i itself.

Figure 1.6 displays both the differential and cumulative LF for the two isochrones displayed in Figure 1.3. The normalization constant has been chosen in order to have 10^6 objects along each isochrone. The exponent is $x = -2.35$, the so-called Salpeter IMF [17].

Finally, a synthetic CMD combines both isochrone and the predicted star counts (for a given IMF). Synthetic CMDs can be easily calculated by iterating the following procedure. For a fixed age and initial chemical composition of a SSP (the generalization to complex star formation histories is straightforward), the starting point is to randomly draw a value of the mass M for the synthetic star according to a specified IMF. With this value of M, one interpolates along the appropriate isochrone to determine the magnitude and colour of this synthetic object. Inclusion of photometric errors and unresolved binaries is also straightforward. The magnitude and colour obtained from the isochrone interpolation are modified by adding a value of the photometric error, drawn from a Gaussian (for example) random distribution with σ specified by the data reduction process. A given number of unresolved (non-interacting) binaries or blends of single stars can be included in a similar fashion. To these selected stars (whose mass is denoted as M_{pr}), it is assigned a companion (with the same age and metallicity of M_{pr}) whose mass M_{comp}

is drawn according to

$$M_{comp} = [ran(1 - q_c) + q_c] M_{pr} \tag{1.16}$$

where q_c is the minimum value of the ratio M_{comp}/M_{pr}, to be chosen. The variable ran is a random number with a flat probability distribution between zero and one, so that the previous relationship provides values of M_{comp} with a uniform distribution between M_{pr} and $q_c M_{pr}$. The magnitude of the composite object is evaluated by summing up the fluxes of the components; in a generic band A, the magnitude of an unresolved binary is given by $M_A(bin) = -2.5 \log(10^{-0.4 M_A(1)} + 10^{-0.4 M_A(2)})$, where $M_A(1)$ and $M_A(2)$ are the magnitudes of the two system components M_{pr} and M_{comp}.

It is important to notice that one can also determine theoretical LFs from synthetic CMDs simply by counting the synthetic objects in the appropriate magnitude bins. The main conceptual difference with the "analytical" integration of the IMF along an isochrone described before is that this latter procedure is strictly valid only when the number of stars is formally infinite. The analytical computation implies that all points along an isochrone are smoothly populated by a number of stars that can be equal to just a fraction of unity in case of fast evolutionary phases. However, in real stellar populations (or in synthetic CMDs), the number of objects in a given magnitude bin is either zero or a multiple of unity. When the number of stars harboured by the observed population is not large enough to smoothly sample all evolutionary phases, statistical fluctuations of star counts at a given magnitude will arise, and their extent can be easily evaluated using extensive synthetic CMD simulations.

1.3.3
Stellar Population Synthesis Models

In the case of unresolved stellar systems, photometric and spectroscopic observations can only provide integrated magnitudes, colours and spectra that include the contribution of all the stars belonging to the population. Theoretical predictions of integrated properties of stellar populations are often called "stellar population synthesis" models.

The theoretical counterpart of integrated magnitudes and spectra of SSPs is conceptually easy to determine. Consider a generic CMD of an isochrones of fixed age and initial chemical composition. The integrated magnitude in the photometric band A can be written as the sum of the energy fluxes within the appropriate wavelength range, that is,

$$M_A(t) = \int_{M_l}^{M_u} 10^{-0.4 M_A(M,t)} \Phi(M) \, dM \tag{1.17}$$

where $\Phi(M) \, dM$ is the IMF, M_l is the mass of the lowest mass star in the SSP, M_u is the mass of the highest mass star still "alive" in the SSP (remnants' contribution is negligible), $M_A(M, t)$ is the magnitude of a star of mass M evolving along the isochrone of age t. Integrated colours follow directly from the same equation applied to two different photometric bands.

Equation (1.17) says that the integrated magnitude is the sum of the individual fluxes (in the appropriate wavelength range) of the stars belonging to the SSP; the IMF gives the number of stars formed with a given mass M, and in the assumption of a universal IMF, the effect of age and chemical composition is included in the term $10^{-0.4 M_A(M,t)}$ – for the energy output of a star of mass M and its wavelength distribution depend on both age t and initial chemical composition – and in M_u.

As for the calculation of theoretical integrated spectra, it is easy to realize that the monochromatic integrated flux F_λ received from an unresolved SSP can be written as

$$F_\lambda(t) = \int_{M_l}^{M_u} f_\lambda(M, t) \Phi(M) \, dM \tag{1.18}$$

where $f_\lambda(M, t)$ is the monochromatic flux emitted by a star of mass M evolving along the isochrone.[5] From the integrated F_λ, one can obviously determine the integrated magnitude in the generic photometric band A. In fact, if F_λ^{int} is the total integrated flux obtained by adding up the individual fluxes at the stellar surfaces, the generic integrated absolute magnitude M_A will be equal to

$$M_A = -2.5 \log \left(\frac{\int_{\lambda_1}^{\lambda_2} F_\lambda^{int} S_\lambda \, d\lambda}{\int_{\lambda_1}^{\lambda_2} f_\lambda^0 S_\lambda \, d\lambda} \right) + m_A^0 \tag{1.19}$$

As in the case of the computation of LFs, this analytical formalism to determine integrated magnitudes and fluxes is strictly valid only when the number of stars is formally infinite. When the different evolutionary phases are not smoothly sampled, it is possible to make use of synthetic CMDs to easily determine the statistical fluctuations of the integrated properties. One simply needs to add the monochromatic fluxes – or the fluxes within a given photometric passband – assigned to each of the synthetic stars.

5) It is important to notice that, for example, in the case of theoretical stellar spectra (that give energy per unit area), the fluxes $f_\lambda(M, t)$ need to be rescaled appropriately by accounting for the radii of the models, and eventually to a distance of 10 pc if a prediction of the "absolute" flux scale is needed.

2
Low-Mass Star Physics

A detailed description of stellar physics can be found in several outstanding books (i.e. [18–20] and references therein) and will not be repeated here. The purpose of this section is to summarize the main assumptions behind the derivation of the stellar structure and evolution equations, and the basics of the physical processes that shape the structure and evolution of low-mass stars, for example, the stars that develop an electron degenerate core at the end of the MS. Particular attention will be paid to processes like atomic diffusion, rotation and thermohaline mixing, that are becoming an integral part of modern stellar evolution calculations.

2.1
Basic Equations

To a first order approximation, the evolution of low-mass stars can be described in terms of a temporal sequence of spherical symmetric models in hydrostatic equilibrium, with local thermodynamical equilibrium in each stellar layer, assuming a negligible effect of rotation and magnetic fields, and considering convection as the only efficient chemical element transport mechanism in the stellar interiors.

We denote here as *canonical* models, all evolutionary calculations performed with these assumptions. In the framework of this "standard theory" of stellar evolution, only four differential equations are necessary to fully describe the physical structure of the star and its time-evolution. It is a common procedure to firstly derive these equations by using the distance r from the centre as the independent variable, that is, the so-called "Eulerian" treatment in the sense of classical hydrodynamics, and then transform them into equations that use the local mass value m_r as an independent variable, the so-called "Lagrangian" description. The main advantage of the Lagrangian description is that the radial coordinate r of a given mass element changes in time as a consequence of the contraction/expansion processes that a star experiences during its evolution, and the local value of r is not always associated to the same mass layer. The solution and analysis of the equations is simplified and clearer when using the local mass as independent variable.

The stellar structure equations described below are four differential equations that describe the run of pressure, temperature, luminosity, and radius as a func-

Old Stellar Populations, First Edition. S. Cassisi and M. Salaris.
© 2013 WILEY-VCH Verlag GmbH & Co. KGaA. Published 2013 by WILEY-VCH Verlag GmbH & Co. KGaA.

tion of m_r at a given time t, and their evolution with the time. To these equations, an additional set of equations for the time-evolution of the chemical element abundances has to be added. To solve the final, large system of equations, one needs to include functions such as the equation of state (EOS) and the opacity of the stellar matter, and the rate of energy generation.

Continuity of mass equation: due to spherical symmetry, all quantities describing the stellar structure only depend on the distance r from the centre. By denoting with ρ the value of the matter density at a generic r within the star, the mass contained within a sphere of radius r is given by

$$m_r = \int_0^r 4\pi r'^2 \rho \, dr'$$

Differentiating this equation with respect to the distance from the centre provides

$$\frac{dm_r}{dr} = 4\pi r^2 \rho \qquad (2.1)$$

The "continuity of mass" equation, which is straightforwardly rewritten by using m_r as the independent variable:

$$\frac{dr}{dm_r} = \frac{1}{4\pi r^2 \rho} \qquad (2.2)$$

Hydrostatic equilibrium equation: in hydrostatic equilibrium, all forces acting on a given mass element have to be balanced. In the additional assumption that neither rotation nor magnetic field are present, the only forces acting on a stellar layer are due to its own gravitational field and pressure (P). From these considerations, one immediately derives (see [20] for a detailed derivation of this equation):

$$\frac{dP}{dr} = -\frac{G m_r \rho}{r^2} \qquad (2.3)$$

the "equation of hydrostatic equilibrium". This equation implies that the pressure decreases outwards since the right-hand side is always negative. Equation (2.3) can be rewritten using m_r as independent variable by recalling that $dm_r = 4\pi r^2 \rho \, dr$

$$\frac{dP}{dm_r} = -\frac{G m_r}{4\pi r^4} \qquad (2.4)$$

Conservation of energy equation: the net energy produced in a spherical shell of thickness dr, located at distance r from the centre of the star, can be written as

$$dL_r = 4\pi r^2 \rho \epsilon \, dr$$

where ϵ denotes the coefficient of energy generation per unit time and unit mass. This gives

$$\frac{dL_r}{dr} = 4\pi r^2 \rho \epsilon \qquad (2.5)$$

the "equation of conservation of energy" that in the Lagrangian description becomes

$$\frac{dL_r}{dm_r} = \epsilon \tag{2.6}$$

The coefficient of energy generation ϵ can be expressed as a sum of three terms

$$\epsilon = \epsilon_n + \epsilon_g - \epsilon_\nu$$

ϵ_n is energy per unit time and the unit mass generated by nuclear reactions. It always gives a positive contribution to ϵ; ϵ_g is energy per unit time and unit mass generated by the thermodynamical transformations experienced by stellar matter during the star evolution (expansion, contraction, change of chemical composition). It can be both positive (e.g., in the case of the compression of a stellar layer) and negative (in the case of expansion). The more general expression for ϵ_g is given by

$$-\epsilon_g = \left(\frac{dU}{dv}\right)_{T,\mu} \frac{dv}{dt} + \left(\frac{dU}{dT}\right)_{v,\mu} \frac{dT}{dt} + \left(\frac{dU}{d\mu}\right)_{T,v} \frac{d\mu}{dt} + P\frac{dv}{dt} \tag{2.7}$$

where U is the internal energy per unit mass, $v = 1/\rho$ is the specific volume and μ denotes the mean molecular weight of the stellar matter – that is the ratio between the average mass of a gas particle $\langle m \rangle$ and the atomic mass unit (m_H). The term $(dU/d\mu)_{T,v}(d\mu/dt)$ arises from the variation of U at constant temperature and volume due to the change of chemical abundances. Its contribution to the stellar energy budget is negligible when nuclear reactions are efficient, but is important in the case of WDs, where nuclear burnings are inactive. It is possible to demonstrate [20] that when integrated over the whole stellar structure, ϵ_g is equal to the time-variation of the internal energy plus the gravitational potential energy of the star.

ϵ_ν is energy per unit time and unit mass associated to neutrino production processes (positive by definition) which is effectively subtracted from the stellar energy budget since neutrinos barely interact with the surrounding stellar matter (this explains why ϵ_ν is added to ϵ_g and ϵ_n with a minus sign).

Energy transport equation: inside stars, the energy can be transferred from layer to layer by photons (radiative transport), collisions due to the random thermal motions of particles (conduction), organized, large-scale motions of macroscopic mass elements (convection).

Radiative Transport When the mean free path of the photons l is very small in comparison with the characteristic size of the system, that is, the star radius in our case, the radiative energy transport can be treated in the approximation of a *diffusive process*, which represents a formidable simplification, that can be safely applied to stellar interiors. For example, in the solar interior, $l/R_\odot \approx 10^{-11}$. In this approximation, the radiative energy transport equation can be written as

$$\frac{dT}{dr} = -\frac{3\kappa_{rad}\rho}{4acT^3}F_{rad} \tag{2.8}$$

κ_{rad} is the radiative opacity due to the interactions of photons with the surrounding particles and, in the case of local thermal equilibrium, it coincides with the so-called Rosseland mean opacity; F_{rad} is the flux of energy associated to the outgoing photons.

An immediate consequence of Eq. (2.8) is that whenever there is a temperature gradient, there will always be a radiative flux, although the latter may be smaller than the total energy flux. If the total energy flux is carried by photons, that is, the contribution of the other energy transport mechanisms is vanishing, Eq. (2.8) becomes

$$\frac{dT}{dr} = -\frac{3\kappa_{rad}\rho}{4acT^3}\frac{L_r}{4\pi r^2} \qquad (2.9)$$

This equation takes into account the processes of absorption and re-emission of the photons in the stellar interior through the Rosseland mean opacity κ_{rad}. This treatment of radiative transport is no longer valid when the diffusion approximation breaks down, as it happens in the stellar surface layers, for example, in the solar atmosphere $l/R_\odot \approx 10^{-4}$. A useful parameter to check the validity of the diffusion approximation is the optical depth τ defined as

$$\tau = \int_r^\infty \kappa_{rad}\rho dr$$

This is a measure of the probability that photons interact with the stellar matter before being radiated away. The diffusion approximation for the radiative transport breaks down when τ is lower than ≈ 1–10. This means that one can safely integrate the equations of stellar structure from the centre up to a point where $\tau \sim 1$ using the diffusion approximation for the radiative transport. The layers with $\tau \approx 1$–10 define a sort of "surface" for the stellar model. The layers where τ is below this threshold are called "stellar atmosphere", and their modelling is crucial to both predict the spectrum of the radiation emitted by the star, and provide the outer boundary condition necessary to solve the equations of stellar structure. We will come back to this issue when considering very low-mass stellar structures and white dwarfs.

Electron Conduction   In ordinary stellar matter, the energy transport via conduction is highly inefficient. However, the situation drastically changes after the onset of electron degeneracy in the cores of RGB and AGB stars, and WDs. In these environments, the energy transport associated to electron collisions becomes very efficient and competitive with the radiative energy transport. When electron conduction is efficient, the energy flux transferred to a volume element of unit area and depth dr by an outgoing flux of electrons can be written approximately as

$$F_e \sim -N_e v l \frac{dE}{dr}$$

where N_e is the number of electrons per unit volume, v their average velocity, l the mean free path and E their average kinetic energy. Since $E \propto K_B T$ (K_B is the

Boltzmann constant),

$$F_e \sim -K_B N_e \nu l \frac{dT}{dr}$$

Due to the close similarity with Eq. (2.8); the quantity that multiplies dT/dr can also be written as in Eq. (2.8), introducing the electron conduction opacity – or conductive opacity – κ_e. We will discuss recent updates in the evaluation of the conductive opacity later in this chapter.

In the general case, the total energy flux will be $F = F_{rad} + F_e$, and the equation of the radiative plus conductive energy transport for the stellar interiors becomes

$$\frac{dT}{dr} = -\frac{3\kappa\rho}{4acT^3}\frac{L_r}{4\pi r^2} \tag{2.10}$$

where κ represents the total opacity of the stellar matter, given by the harmonic mean

$$\frac{1}{\kappa} = \frac{1}{\kappa_{rad}} + \frac{1}{\kappa_e}$$

We can now rewrite this equation for the radiative plus conductive transport in the Lagrangian scheme:

$$\frac{dT}{dm_r} = -\frac{3\kappa}{64\pi^2 ac}\frac{L_r}{r^4 T^3} \tag{2.11}$$

It is useful to further transform Eq. (2.11) into a more general expression that can be used for the convective transport. If we denote a generic logarithmic gradient as $\nabla = d\ln(T)/d\ln(P)$ that can also be written as $\nabla = (dT/dP)(P/T)$, then $dT/dm_r = \nabla(T/P)dP/dm_r$. By assuming hydrostatic equilibrium and dividing Eq. (2.11) by Eq. (2.4), we obtain

$$\frac{dT}{dm_r} = -\frac{T}{P}\nabla\frac{Gm_r}{4\pi r^4} \tag{2.12}$$

For the case of radiative plus conductive transport, we denote ∇ with ∇_{rad}, and from Eq. (2.11), we derive

$$\nabla_{rad} = \frac{3}{16\pi acG}\frac{\kappa L_r P}{m_r T^4} \tag{2.13}$$

Equation (2.12) is also suitable to describe the convective energy transport, provided that the appropriate value of ∇ is used, as we will discuss in the following.

Convection One of the basic assumptions of the standard theory is hydrostatic equilibrium. However, in a real star, gas elements are subject to random motions around their equilibrium positions. These fluctuations have a negligible impact if they do not grow with time. Under certain conditions, these small perturbation may grow and trigger large scale macroscopic motions (called convection) involving a significant fraction of the stellar structure. These macroscopic motions can significantly affect the structure and evolution of a star, for they cause a mixing of the stellar matter, and contribute also to the transport energy from hotter to cooler regions within the star.

In order to include convective transport in the stellar model computation, it is necessary to find a criterion for the onset of convection, and a mathematical expression of the temperature gradient in a convective region. To date, the treatment of convection is one of the thorniest problems in stellar astrophysics, for the flow of gas in a stellar convective region is turbulent, for example, the velocity and all other properties of the flow vary in a random and chaotic way, and in addition, convection is a non-local process. An accurate, quantitative and comprehensive description of turbulent convection in the stellar regime is still lacking.

The most widely adopted method for treating convection in stellar evolution models is based on a simple quantitative formalism called the Mixing Length Theory (MLT) [21, 22]. This model is extremely simple; the flow is made of convective elements (bubbles) with a given characteristic size, the same in all dimensions, that move by a fixed mean free path before dissolving. All convective bubbles have the same physical properties at a given distance r from the star centre. Columns of upward and downward moving bubbles are envisaged; upward moving elements start from a given layer, cover a mean free path and then dissolve, releasing their excess heat into the surrounding gas, and are replaced at their starting point by the downward moving elements, that thermalize with the surrounding matter, thus perpetuating the cycle. It is worth noting that the mixing length theory is a "local" theory in the sense that both the criterion for the onset of convection and the evaluation of all relevant physical and chemical quantities are based on the local properties of each specific stellar layer, regardless of the extension of the whole convective region.

The important parameters in this formalism are: the mean free path and the characteristic size of the convective elements which are assumed to be the same for all convective bubbles. Both parameters are assumed to have the same value, denoted with Λ, the so-called "mixing length", equal to a multiple of the local pressure scale height $H_P - \Lambda = \alpha_{ml} H_P$, α_{ml} being a constant to be calibrated – defined as

$$\frac{1}{H_P} = -\frac{1}{P}\frac{dP}{dr} = -\frac{d\ln(P)}{dr}$$

By using the equation of hydrostatic equilibrium, and denoting with g the local acceleration of gravity [$g = (G m_r)/r^2$], the pressure scale height can be rewritten as $H_P = P/(g\rho)$.

In the following, we recall the criterion for the onset of convection [19, 23, 24]. To this purpose, we introduce the adiabatic gradient ∇_{ad}, defined as

$$\nabla_{ad} = \left[\frac{d\ln(T)}{d\ln(P)}\right]_{ad}$$

where the subscript "ad" implies that the derivative has to be computed at constant entropy. One should notice the different meaning of ∇_{rad} and ∇_{ad}; ∇_{rad} refers to a spatial derivative which "connects" the values of P and T in two adjacent mass shell; ∇_{ad} describes the thermodynamical change of a given mass element during the adiabatic transformation it is experiencing. When the values for the two gradients coincide, we refer to an "adiabatic stratification".

Let us also introduce the quantities:

$$\chi_\rho = \left[\frac{d\ln(P)}{d\ln(\rho)}\right]_{T,\mu} \quad , \quad \chi_T = \left[\frac{d\ln(P)}{d\ln(T)}\right]_{\rho,\mu} \quad , \quad \chi_\mu = \left[\frac{d\ln(P)}{d\ln(\mu)}\right]_{\rho,T}$$

and define

$$\nabla_\mu = \frac{d\ln(\mu)}{d\ln(P)}$$

With these definitions, a "local" criterion for the onset of convection, applicable on a layer-by-layer basis, is

$$\nabla_{rad} > \nabla_{ad} \tag{2.14}$$

in regions with uniform chemical stratification, where by definition $\nabla_\mu = 0$ (Schwarzschild criterion).

The situation is more complex in the case of a non-uniform chemical composition. Figure 2.1 displays a qualitative sketch in the $\nabla_\mu - (\nabla_{rad} - \nabla_{ad})$ diagram of the region where instabilities occur, divided into four quadrants. Along the

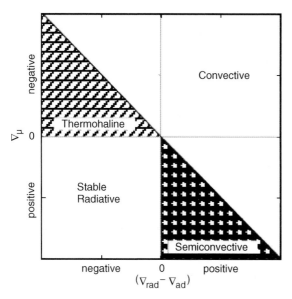

Figure 2.1 Qualitative sketch of the $\nabla_\mu - (\nabla_{rad} - \nabla_{ad})$ stability plane, with different regimes labelled [23].

$\nabla_\mu = 0$ line, one recovers the Schwarzschild criterion. Usually, ∇_μ is positive, for the molecular weight tends to increase (as the pressure) towards the star interior, where heavier elements are synthesized by nuclear reactions. In this case (the lower two quadrants in Figure 2.1), a layer is convective when

$$\nabla_{rad} > \nabla_{ad} - \frac{\chi_\mu}{\chi_T}\nabla_\mu \tag{2.15}$$

This is the so-called Ledoux criterion. Notice how the second term on the right-hand side of Eq. (2.15) usually has a stabilizing effect compared to the case of uniform composition because χ_μ is negative (consider, for example, the equation of state of a perfect gas, with $P \propto (\rho T/\mu)$) while χ_T and ∇_μ are positive. There is potentially an intermediate region in the two lower quadrants of the $\nabla_\mu - (\nabla_{rad} - \nabla_{ad})$ diagram, where a layer can be unstable according to the Schwarzschild criterion but stable according to the Ledoux criterion, labelled as "semi-convective" in the figure. Here, in the case of some thermal dissipation along the motion of gas elements, a small perturbation causes oscillations of increasing amplitude that clearly induce an instability and a large scale motion of matter. A negative ∇_μ (molecular weight inversion) has a destabilizing effect, and indeed no radiative equilibrium is achieved in the two upper quadrants of the $\nabla_\mu - (\nabla_{rad} - \nabla_{ad})$ diagram. There is, however, a region where the layer is formally stable according to the Ledoux criterion, but $\nabla_\mu < 0$. In this situation, a "thermohaline" mixing is efficient and will be discussed later in this chapter. In the laboratory, the instability takes the form of "salt fingers". Given that heat diffuses faster than salt, these fingers sink because they grow increasingly heavier than the environment, until they become turbulent and dissolve. Inside a star, the role of salt is played by a heavier element like helium in a hydrogen-rich medium.

In general, a convectively unstable region mixes matter on very short timescales compared with evolutionary timescales, and the chemical profile in convective layers can always be assumed uniform to a very good approximation (instantaneous mixing approximation). On the other hand, mixing timescales due to semi-convection or thermohaline mixing are expected to be longer, and the approximation of instantaneous mixing is no longer valid.

As already mentioned, these criteria for the onset of convection are local. The boundary of the convective regions are fixed by the layer where ∇_{rad} gets lower than ∇_{ad}, that is the layer where the random motions of the gas stop getting amplified. However, the convective elements have a finite, non-zero velocity at these formal boundaries between the stable and unstable zones, and hence they can penetrate into the stable region beyond, a process usually labelled as "overshooting". The crucial point for stellar evolution calculations is to establish how far these convective bubbles can penetrate the surrounding stable regions. This process is important because it affects the chemical stratification of the models and the evolutionary lifetimes (when additional fuel is engulfed in a convective nuclear burning region) in a way that can be potentially tested by comparisons with observations.

Different approaches based on the concepts of the MLT have been suggested for estimating the overshoot distance λ_{OV} (usually expressed as a multiple of the pressure scale height H_P at the Schwarzschild convective boundary). However, being that the MLT is a local theory, no satisfactory and fully consistent constraints have been obtained, and usually λ_{OV} is calibrated empirically.

After the boundary of convective regions are determined, one needs to determine the appropriate temperature gradient. Obviously, the actual gradient in a convective region, ∇_{conv}, needs to satisfy the condition $\nabla_{conv} > \nabla_{ad}$, that is, it has to be superadiabatic. This requirement can be easily understood by considering a raising gas bubble that dissolves after a length Λ, and releases an amount of heat per unit volume $\delta Q = \rho c_P \delta T$ where c_P is the specific heat per unit mass at constant pressure (pressure equilibrium between the gas inside and outside convective elements is assumed, otherwise the convective bubble would lose its identity) and δT is the temperature difference between bubble and the surrounding gas. The difference δT, neglecting energy losses along the path, is proportional to the difference between the adiabatic gradient associated to the thermodynamical transformation of the gas in the bubble, and the temperature gradient of the environment. If the two gradients were to be the same, no heat would be released and no energy transport is possible. Heat exchange is efficient only if the environment is cooler than the bubble, and hence its temperature gradient (the gradient ∇_{conv} of the convective region) has to be larger than the adiabatic one.

In convective cores, the density is so high that also an extremely small temperature difference between the convective element and the surrounding, that is, a very small difference between the adiabatic gradient and the actual one, is sufficient to guarantee an efficient convective energy transport. In this a case, a good approximation for the actual gradient is: $\nabla_{conv} = \nabla_{ad}$, that is, the degree of superadiabaticity is negligible. In convective envelopes, due to the drastic decrease of the local density, this simplification is no longer valid and one needs more detailed calculations of ∇_{conv} that are usually performed in the framework of the MLT.

Various "flavours" of the MLT formalism can be adopted, depending on the value of three constants a, b, c that enter the following equations of the MLT for the convective velocity v_c, convective energy flux F_c and convective efficiency Γ

$$v_c^2 = \frac{a \Lambda^2 g Q (\nabla_{conv} - \nabla')}{H_P} \tag{2.16}$$

$$F_c = \frac{b \rho v_c c_P T \Lambda (\nabla_{conv} - \nabla')}{H_P} \tag{2.17}$$

$$\Gamma \equiv \frac{\nabla_{conv} - \nabla'}{\nabla' - \nabla_{ad}} = \frac{c_P \rho^2 \Lambda v_c \kappa}{c \sigma T^3} \tag{2.18}$$

where ∇' is the temperature gradient of a rising (or falling) element of matter within the convective region, ∇_{conv} is the average temperature gradient of the matter at a given level within the convective zone and $Q \equiv -(d \ln \rho / d \ln T)_P$. From these expressions for v_c, F_c and Γ, it is possible to obtain a simple algebraic equation whose solution provides the value of ∇ at any point within the convective region. The "classic" MLT flavour adopted in stellar evolution calculations [21] has $a = 1/8$, $b = 1/2$, $c = 24$, and $\alpha_{ml} = \Lambda / H_p$ fixed by some empirical calibration.

When different choices for the constants *a*, *b* and *c* are made, the resulting superadiabatic gradients can essentially be made equivalent by simply rescaling the value of $\alpha_{ml} = \Lambda/H_P$. We will discuss this topic in more detail when describing the empirical calibrations of the MLT.

If convection is efficient such as in the deep stellar interiors, the MLT provide $\nabla \rightarrow \nabla_{ad}$ and velocities of the order of $(1-100)\,\mathrm{m\,s^{-1}}$, that is, many orders of magnitude smaller than the local sound speed. On the contrary, in convective layers close to the surface, the gradient is strongly superadiabatic and velocities are much larger, for example, of the order of $(1-10)\,\mathrm{km\,s^{-1}}$, close to the local sound speed. As for the overshooting region, one can approximate the temperature gradient with ∇_{rad} in the hypothesis that the overshooting gas elements tend to thermal equilibrium with the surrounding medium. In case they are supposed to exchange little heat with the environment, they establish a near adiabatic gradient in this region. In this case, this phenomenon is labelled "convective penetration" rather than overshooting.

Before closing this section, we will briefly mention why convection sets in. Obviously, this must happen either because ∇_{ad} gets very low, or ∇_{rad} becomes very high. This happens under the following conditions:

- The value of ∇_{ad} is fixed by the equation of state of the stellar matter, and its typical value is about 0.40 for a fully ionized monatomic gas. However, in the regions where the most abundant elements such as hydrogen and helium are partially ionized ($T \geq 10^4$ K for H and $T \approx 10^5$ K for He), ∇_{ad} decreases as a consequence of the increase of the number of free particles down to values ~ 0.10. This favours the onset of envelope convection.
- $\nabla_{rad} \propto \kappa_{rad} F$, and therefore large values of ∇_{rad} are attained when the local energy flux is very large. During the core burning stages, if the efficiency of the burning mechanism is a steep function of the temperature (as in the case of the *CNO cycle* for MS stars), the local flux at the centre of the star increases very fast, and hence ∇_{rad} gets very high near the centre and core convection sets in. In region of partial ionizations or when molecules are present, the radiative opacity and ∇_{rad} increase, and this favours the onset of envelope convection.

We can now summarize the stellar structure equations derived before (we drop the index *r* in the notation)

$$\frac{dr}{dm} = \frac{1}{4\pi r^2 \rho} \tag{2.19}$$

$$\frac{dP}{dm} = -\frac{Gm}{4\pi r^4} \tag{2.20}$$

$$\frac{dL}{dm} = \epsilon_n - \epsilon_\nu - c_P \left(\frac{dT}{dt} - \nabla_{ad} \frac{T}{P} \frac{dP}{dt} \right) \tag{2.21}$$

$$\frac{dT}{dm} = -\frac{T}{P} \nabla \frac{Gm}{4\pi r^4} \tag{2.22}$$

where we have rewritten ϵ_g in a more compact form by using basic thermodynamics and dropping the term arising from the variation of μ. To these equations, one needs to add a set of I equations ($s = 1, \ldots, I$) for the change of the mass fraction of the chemical elements considered. In general, an element s is produced by w reactions of the following kind

$$n_h h + n_k k \to n_p s$$

and destroyed by l reactions of the following kind

$$n_d s + n_j j \to n_z z$$

This provides

$$\frac{dX_s}{dt} \frac{1}{A_s} = \sum_z \rho^{n_h + n_k - 1} n_p \frac{X_h^{n_h} X_k^{n_k}}{A_h^{n_h} A_k^{n_k}} \frac{\langle \sigma v \rangle_{hk}}{m_H^{n_h + n_k - 1} n_h! n_k!}$$

$$- \sum_l \rho^{n_d + n_j - 1} n_d \frac{X_s^{n_d} X_j^{n_j}}{A_s^{n_d} A_j^{n_j}} \frac{\langle \sigma v \rangle_{sj}}{m_H^{n_d + n_j - 1} n_d! n_j!} \qquad (2.23)$$

where a generic $\langle \sigma v \rangle_{ij}$ is the cross section for the reaction between two generic nuclei i and j, X_i and X_j the mass fractions of the two elements, A_i and A_j their atomic weights. In the case of convection, the chemical abundances have to be appropriately averaged over the whole convective region in order to determine the corresponding uniform chemical profile produced by the fast convective mixing.

All these equations and the corresponding physical and chemical variables refer to a generic stellar layer located at a mass coordinate m (m runs from zero to the value of the total stellar mass M) and have to be applied to all mass layers from the centre to the stellar surface.

To solve these equations, one needs to know the local values of the ϵ_n (and the related nuclear cross sections) ϵ_ν, the EOS of the stellar matter $P = P(\rho, T)$ and related thermodynamical quantities, the opacities κ and the appropriate temperature gradient ∇ (radiative, adiabatic or superadiabatic). If these auxiliary functions are known, we have at each stellar layer a set of $4 + I$ differential equations for the $4 + I$ variables r, P, T, L, X_s (with $s = 1, \ldots, I$); the independent variables are m and t. In general, one needs to specify the total mass M of the star and its initial chemical composition; then, the equations are solved numerically to determine the structure of the star at the initial instant t_0 and its evolution with time.

Equations (2.19)–(2.22) describe the mechanical structure of the star for a given chemical composition, while Eq. (2.23) describes the chemical structure and its time-evolution. Time derivatives appear explicitly only in Eqs. (2.21) and (2.23), that is, the time-evolution is driven by the energy generation and losses, and the consequent chemical transformations of the stellar matter. It is useful to notice that the timescales associated to the time derivatives in Eq. (2.21) are different from the case of Eq. (2.23). The former derivatives are associated to the rate of change of the gravitational and internal energy (the two contributors to ϵ_g) with time; the corresponding timescale is called Kelvin–Helmholtz timescale and can be estimated from $\tau_{KH} \equiv |\Omega|/L$, where L is the star luminosity and Ω its total gravitational potential energy ($\tau_{KH} \approx 10^7$ years for the Sun).

The time derivatives in Eq. (2.23) are instead associated with nuclear reaction timescales $\tau_n \equiv E_n/L$, where E_n is the nuclear energy reservoir; in the case of the Sun, $\tau_n \approx 10^{10}$ years. In general, $\tau_n \gg \tau_{KH}$ for the whole lifetime of low-mass stars.

The inequality between τ_n and τ_{KH} allows one to decouple Eq. (2.23) from the other four equations. One can therefore solve the mechanical part of the star at a given instant t with a given set of chemical abundances, then apply a time step Δt, solve Eq. (2.23) to determine the new chemical abundances using the value of the physics at the previous instant t, then integrate again the equation of the mechanical structure at time $t + \Delta t$ using these updated chemical abundances and so on. Details about the numerical techniques to solve this system of equations can be found in [20].

2.2
The Thermodynamical Properties of Low-Mass Stars

A detailed discussion about the thermodynamical properties of stellar matter can be found in several textbooks [19, 20]. Here, we will discuss some characteristic thermodynamical properties of low-mass stars.

Stars are made of a mixture of radiation plus gas with several components: electrons, ions, atoms and molecules. The thermodynamical properties of the stellar matter are described by the EOS that determines the fractions of free electrons, neutral and ionized atoms and molecules, their ionization states, pressure P and all other thermodynamic quantities as functions of ρ, T and chemical composition. The chemical composition is often parametrized in terms of μ that can be also expressed in terms of the density and number of particles per unit volume n as

$$\mu = \frac{\rho}{n \cdot m_H}$$

In the case of a fully ionized gas, and denoting, as usual in stellar evolutionary calculations, with X, Y and Z the mass fractions of hydrogen, helium and metals (all element heavier than helium) respectively, one can derive

$$\mu = \frac{1}{2X + \frac{3}{4}Y + \frac{Z}{2}}$$

A major difficulty to derive an accurate EOS is the large range of T and P that has to be covered in order to model the entire structure of stars during their whole lifetime. For example, moving from the centre to the surface in a MS star one first has to deal with completely ionized and then progressively less ionized atoms – the first chemical elements to achieve a state of complete ionization are the two most abundant ones, H and He, until, close to the surface, atoms are neutral and molecules may also form. In more advanced evolutionary phases of low-mass stars, electron degeneracy sets in, and during the WD phase one needs to model also phase transitions from gas to liquid and solid phase.

As a first approximation, one may start with the EOS of a perfect monatomic gas plus radiation, complemented by the EOS of degenerate electrons in electron degenerate cores. For the perfect gas plus radiation, one obtains

$$P = \frac{K_B}{\mu m_H} \rho T + \frac{a T^4}{3} \tag{2.24}$$

$$U = \frac{K_B T}{\mu m_H} \left[\frac{3}{2} + \frac{3(1 - \zeta)}{\zeta} \right] \tag{2.25}$$

$$c_P = \frac{K_B}{\mu m_H} \left[\frac{3}{2} + \frac{3(4 + \zeta)(1 - \zeta)}{\zeta^2} + \frac{4 - 3\zeta}{\zeta^2} \right] \tag{2.26}$$

$$\nabla_{ad} = \frac{1 + \frac{(1-\zeta)(4+\zeta)}{\zeta^2}}{\frac{5}{2} + \frac{4(1-\zeta)(4+\zeta)}{\zeta^2}} \tag{2.27}$$

for the pressure, internal energy per unit mass, specific heat at constant pressure and adiabatic gradient, where $\zeta \equiv P_{gas}/P$.

When $\zeta \to 1$, then $P_{gas} \to P$ and $c_P \to 5/2 \, (K_B/\mu m_H)$, $\nabla_{ad} \to 0.4$. In the case $\zeta \to 0$, then $P \to P_{rad}$, $c_P \to \infty$, and $\nabla_{ad} \to 0.25$.

When the mean distance between particles of mass m is comparable to their de Broglie wavelength $\lambda_{dB} = h/p$ (with h and p being the Planck constant and the particle momentum, respectively), one has to take into account their quantum statistics and the perfect gas approximation breaks down. Assuming that the particles are non-relativistic, their mean kinetic energy is $3/2 K_B T$, and the average momentum p is equal to $(3m K_B T)^{1/2}$. The condition on λ_{dB} becomes

$$\frac{h}{p} = \frac{h}{(3m K_B T)^{1/2}} \ll d$$

One can immediately see that the effect of quantum statistics (degeneracy) become relevant first for the electrons, the lighter particles in the stellar gas.

For non-relativistic electrons with arbitrary degree of degeneracy, by denoting the momentum with p, with η the degeneracy parameter (increasing values of η correspond to increasing levels of degeneracy) and with n_e the electron number density[1]

$$P = \frac{8\pi K_B T}{3h^3} (2m K_B T)^{3/2} F_{3/2}(\eta) \tag{2.28}$$

$$n_e = \frac{4\pi}{3h^3} (2m K_B T)^{3/2} F_{1/2}(\eta) \tag{2.29}$$

---

1) n_e can also be rewritten in terms of ρ and μ_e as $n_e = \rho/(m_H \mu_e)$ with $\mu_e = [\sum_i (X_i Z_i)/A_i]^{-1}$, where the sum is over all elements i with atomic weight A_i and atomic number Z_i.

where $u = p^2/2m K_B T$ and $F_{1/2}(\eta)$, $F_{3/2}(\eta)$ are the Fermi–Dirac functions defined as

$$F_k(\eta) = \int_0^\infty \frac{u^k du}{e^{-\eta+u} + 1}$$

The values of the Fermi–Dirac functions for specified values of the parameter η can be found in tabular form in [25]. It is instructive to rewrite the electron pressure as

$$P = n_e K_B T \left[\frac{2}{3} \frac{F_{3/2}(\eta)}{F_{1/2}(\eta)} \right]$$

closely resembling the relation for a perfect gas, but for the last multiplicative term. The quantity $2/3(F_{3/2}/F_{1/2})$ provides a direct estimate of how much the actual pressure differs from the pressure for a perfect gas of electrons. From the tabulations of the Fermi–Dirac functions, it is immediate to verify that this ratio is of order unity for η values lower than or of the order of 2. For $\eta = 2$, the ratio is equal to ~ 1.5. However, for $\eta = 0$, it is equal to ~ 1.1.

The case of partial electron degeneracy is very important for the evolution of low-mass stars. Low-mass stars evolving along the Red Giant Branch develop a significant level of electron degeneracy inside their helium core, but it is never complete degeneracy. This is shown in Figure 2.2, which displays the values of η within the He core at various stages along the RGB evolution. As detailed in the following sections, the electron degeneracy increases monotonically in the core until the thermal conditions required to ignite the He-burning are achieved. When this occurs, the electron degeneracy starts to be lifted (short-dashed line in the figure).

In the case of complete degeneracy (η approaching ∞),

$$P = \frac{1}{20} \left(\frac{3}{\pi} \right)^{2/3} \frac{h^2}{m_e m_H^{5/3}} \left(\frac{\rho}{\mu_e} \right)^{5/3} = 1.0036 \times 10^{13} \left(\frac{\rho}{\mu_e} \right)^{5/3} \tag{2.30}$$

where the numerical constant is in cgs units. For a mixture of ions following the perfect gas law and non-relativistic degenerate electrons of arbitrary degeneracy parameter, one can estimate the ratio

$$\frac{P_i}{P_i + P_e} \approx 2.5 \frac{\mu_e}{\mu_i} \frac{1}{\eta}$$

where P_i and P_e are the ion and electron pressure, respectively. The higher the degeneracy parameter, the lower the contribution of the ions to the total gas pressure (but the ions provide the main contribution to the density ρ). For a mixture of degenerate electrons and a perfect gas of ions, the electron pressure always dominates, and is strong enough to prevent any substantial contraction of the stellar layers involved (hence, a substantial increase in density), thus averting the onset of quantum effects for the ions. For this same mixture of ions and electrons, one obtains that the internal energy is equal to

$$U = \frac{3}{2} \frac{P}{\rho}$$

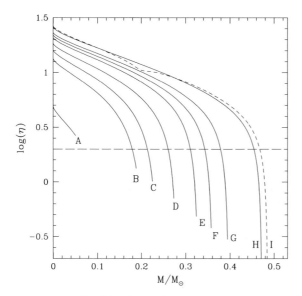

Figure 2.2 Profile of the electron degeneracy parameter η inside the helium core at different stages along the RGB evolution of a $0.8 M_\odot$, $Z = 0.001$, $Y = 0.246$ stellar model. The labels provide a temporal sequence, and correspond to the location in the HRD of the models marked with the same letters in Figure 3.26. The short-dashed line corresponds to the structure at the He-burning ignition, while the long-dashed line marks a value of the η parameter equal to two (see discussion in the text). The local minimum at $M/M_\odot \approx 0.2$ corresponds to the He-ignition.

where $P = P_i + P_e$,

$$c_P \sim \frac{5}{2} \frac{P}{\rho T} \frac{3\mu_e}{2\mu_i \eta}$$

and

$$\nabla_{ad} = 0.4$$

For a fully degenerate relativistic (the speed of the electron approaching the speed of light) electron gas,

$$P = \left(\frac{3}{\pi}\right)^{1/3} \frac{hc}{8 m_H^{4/3}} \left(\frac{\rho}{\mu_e}\right)^{4/3} = 1.2435 \times 10^{15} \left(\frac{\rho}{\mu_e}\right)^{4/3}$$

where the numerical constant is again given in cgs units.

For a mixture of perfect ions and extremely relativistic electrons of arbitrary degeneracy parameter,

$$\frac{P_i}{P_i + P_e} \approx 4 \frac{\mu_e}{\mu_i} \frac{1}{\eta}$$

Moreover,

$$U = \frac{3P}{\rho} - \frac{3 K_B T}{2\mu_i m_H}$$

$$c_P = \frac{4P}{\rho T} \frac{3\mu_e}{\mu_i} \frac{1}{\eta} - \frac{3}{2} \frac{K_B}{\mu_i m_H}$$

$$\nabla_{ad} = 0.5$$

It is useful to have at hand the following simple approximate relation to estimate the value of η in a given stellar layer, that is,

$$\eta \approx 3.017 \times 10^5 \left(\frac{\rho}{\mu_e} \right)^{2/3} T^{-1}$$

It can be also useful to estimate the relative contribution of the degenerate electron to the total pressure P_T in low-mass MS models, in terms of stellar mass M (in solar units) and chemical composition

$$\frac{P_e}{P_T} \sim 0.0114 \times 10^{-6} \frac{T}{(\mu^2 M)^{4/3}}$$

where the stellar mass is in solar unit, T is the central temperature (in K) and μ the mean molecular weight at the centre. From this relation, one derives that for MS stars, the contribution provided by the electron degenerate pressure to the total pressure is important only for very low-mass (VLM) stars with mass $M < 0.15\,M_\odot$.

A first refinement to this approximation to the EOS of the stellar matter must involve the inclusion of ionization, that can be treated using the Saha equation (see, e.g. [19, 20]) in the case of non-degenerate or partially degenerate electrons. This is also not sufficient to accurately model the behaviour of the stellar matter in low-mass stars, because of non-ideal effects that have to be accounted for.

2.2.1
Coulomb Interactions

In a perfect gas, Coulomb interactions are negligible. However, in the stellar matter, atoms and ions cannot be considered completely isolated. As a consequence, the Coulomb interactions of a ion with the neighbouring particles modify its higher quantum states that are the less tightly bound. As shown in Figure 2.3, the actual electrostatic potential acting on electrons bound to a non-isolated ion is obtained by superimposing the electrostatic potential of all the neighbouring particles. As the density increases, the higher quantum levels are progressively more affected, until they are completely destroyed, causing further ionisation, denoted as "pressure ionisation".

One can obtain a rough estimate for the density at the onset of pressure ionization as follows. We assume for simplicity a pure H composition, and denote with d the mean distance between two neighbouring atoms, and a is the orbital radius of the electron in a given bound state. When $a < d/2$, the electron can no longer be bound; by rewriting $a = k^2 a_B$ and $d = (3/4\pi n_H)^{1/3}$, where a_B is the electron Bohr radius, k is the quantum number and n_H is the number density of H atoms, the previous condition becomes

$$k^2 < \left(\frac{3m_H}{4\pi\rho} \right)^{1/3} \frac{1}{2a_B}$$

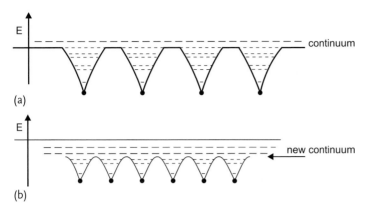

Figure 2.3 Qualitative representation of the pressure ionization process. (a) When the density is not too high, the potential well of the individual atoms (and the electron bound states represented by the thin-dashed lines) is not perturbed by the neighbours. (b) For high enough density, the potential well is perturbed, the continuum energy is lowered and the electron bound states at higher energies disappear.

where n_H has been rewritten in terms of the density. This expression provides an approximate estimate of the principal quantum number of the highest bound state: at the centre of the Sun, $\rho \approx 170\,\mathrm{g\,cm^{-3}}$ and we obtain $k^2 < 0.14$. This implies that hydrogen is ionized. For more complex atoms, the effect of pressure ionization is difficult to model and in general there is no definitive theory for this process, although improvements have been achieved in these last years [26].

The importance of the Coulomb interactions is often parametrized in terms of the parameter Γ defined as (assuming a pure element composition)

$$\Gamma \equiv (Ze)^2/d\,K_B\,T$$

where Z is the charge, and d is the mean distance between ions. The parameter Γ measures the ratio between the electrostatic interactions and the thermal energy of the ions; when the Coulomb interactions are negligible, that is, when the density is sufficiently low at a given temperature, $\Gamma \sim 0$. Values of Γ larger or of the order of unity mean that the Coulomb interactions are very strong; as a order of magnitude one can note that in the Sun $\Gamma \leq 0.05$. When $\Gamma = 1$, the stellar matter behaves like a liquid, and at increasingly larger Γ, the ions form rigid lattice (crystallization). The liquid and solid phase are relevant for the WD evolution, and we will come back to this issue in Chapter 6. It is important to realize that Coulomb interactions do not only affect the level of ionization of the stellar matter discussed above, but also obviously have a direct impact on the thermodynamical properties of the stellar plasma. For example, following [27], it is possible to evaluate the correction (Debye–Hückel correction) to the pressure P_0 of a perfect fully ionized gas as follows

$$P = P_0 \left(1 - \frac{e^3}{3} \frac{\pi^{1/2}}{m_H^{1/2} K_B^{3/2}} \frac{\rho^{1/2}}{T^{3/2}} \mu\,\delta^{3/2} \right)$$

where

$$\delta = \sum_i \left(Z_i^2 + Z_i \right) \frac{X_i}{A_i}$$

where the sum runs over all atomic species i with mass fraction X_i, and where Z_i, A_i are the individual charges and atomic weights. This approximation is valid when $\Gamma < 0.05$.

This Debye–Hückel correction to the ideal EOS always provides a negative contribution to the gas pressure, as a consequence of the interactions between charged particles, and increases with increasing density. In the lowest mass regime, the contribution coming from the Coulomb interaction can be very large: for $M < 0.3 M_\odot$, the Debye–Hückel correction can amount to about 30% or more of the total pressure.

2.2.2
The Exchange Correction

The exchange correction has been often ignored despite the fact that it can be fairly important. The exchange correction is caused by the quantum mechanical properties of the total wave function when electrons (or positrons) are exchanged. Indeed, equivalent quantum mechanical effects exist also for the ions, but these turn out to be negligible because of the much larger mass of the ions with respect to the electrons. Even for the case of no electrostatic repulsion, the anti-symmetrized total wave function of the electrons tends to decrease the spatial overlap of the probability densities of individual electrons. Because of this anti-correlation of the individual electrons, the exchange effect causes the total energy to be decreased in the presence of the electrostatic repulsion, and less pressure is required to confine the gas to a particular volume for a given temperature.

The exchange effect is typically the third-most important non-ideal effect for stellar-interior conditions, after pressure-ionization and the Coulomb effects described before, and accurate treatment is important. As shown in Figure 2.4, for higher temperatures, the exchange correction can be larger than Coulomb effects.

We refer the interested reader to [28] for a detailed discussion of the exchange correction[2].

2.2.3
The Formation of Molecules

The process of formation and dissociation of molecules occurs in the outermost layers of the stellar envelope of cool stars and can have a considerable effect on the general structure of VLM stars – see the discussion in Section 3.2. Clearly, the appearance of molecules introduce an additional degree of difficulty in the evaluation

---

2) An interesting review by A. Irwin can be found at http://freeeos.sourceforge.net/exchange.pdf (accessed 22 January 2013).

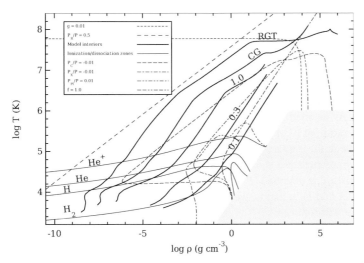

Figure 2.4 Equation of state results and model profiles in the (log ρ–log T) diagram for solar metallicity. The EOS has been calculated using FreeEOS (http://freeeos.sourceforge.net 22 January 2013). The double short-dashed line marks the transition between the two thermal states where electron degeneracy increasingly important. The parameter f is related to the degeneracy parameter η by
$$\eta = \log\left[\left(\sqrt{1+f}-1\right)/\left(\sqrt{1+f}+1\right)\right] + 2\sqrt{1+f}.$$ The parameter g represents the relativity parameter, defined as $g = (K_B T)/\left(m_e c^2\right)\sqrt{1+f}$. The short-dashed line corresponds to the transition to the region where relativistic effects are important. Several lines corresponding to a specific (see the labels) values of the ratio between P_R, P_C, P_X, P_{PI} and the total pressure mark the transition between the different thermal states

where the radiation pressure, the Coulomb effect, the quantum exchange effect and the ionization pressure start to be important. The medium-weight lines labelled H_2, H, He, and He^+ correspond to the midpoints of the hydrogen dissociation and ionization zones, and the midpoints of the helium first and second ionization zones. The thick solid lines labelled by 0.1, 0.3, 1.0, RGT, and CG correspond to model interior loci for Zero Age Main Sequence models of 0.1, 0.3 and 1.0 solar masses, and RGB tip (see Section 3.7) and clump giant (solar metallicity analog of Zero Age Horizontal Branch star – see Section 4.6) $1M_\odot$ models. These stellar model results were calculated using the FreeEOS code. The shaded region corresponds to the high-density, low-temperature computational limit of FreeEOS (courtesy of A. Irwin).

of the thermal properties. This problem is tackled by introducing equations for the molecular dissociation, as many as the considered molecule species, and coupling these equations with the standard Saha equation [20] for estimating the ionization level of the various chemical species.

These equations have the general form

$$\frac{n_{i,0}^{k_i} n_{j,0}^{k_j}}{n_x} = K_{i,j,x}$$

when k_i neutral atoms of type i and k_j neutral atoms of type j combine to form a molecule x. In this expression, $n_{i,0}$ and $n_{j,0}$ are the number densities of the neutral atoms i and j, and n_x is the number density of the molecule x. $K_{i,j,l}$ denotes the dissociation rate and is a function of the temperature T. Analytical formulae for the dissociation rate of several molecules as a function of the temperature can be found in literature.

From the point of view of the definition of the thermal properties of the stellar matter for low-mass stars, the most important molecule is H_2. Its contribution in the lowest mass regime ($M < 0.5 M_\odot$) is relevant in defining the physical structure of these stars. As detailed in Section 3.2, the adiabatic gradient, and hence, the effective temperature of VLM stellar models is strongly affected by the inclusion of the H_2 molecule in the evaluation of the EOS.

Due to its importance, we show here how the fraction of molecular hydrogen in a pure H mixture is obtained [29, 30]. Let us consider three different states for hydrogen: H_2, H and H^+, that is, we are neglecting the possible presence of H_2^+ and H^-. We then introduce

$$N = 2n(H_2) + n(H) + n(H^+) = \frac{\rho X}{m_H}$$

and $x(H_2) = 2n(H_2)/N$, $x(H) = n(H)/N$, and $x(H^+) = n(H^+)/N$. Here, $x(H_2)$, $x(H)$ and $x(H^+)$ represent the fractions of hydrogen atoms in molecular, neutral and ionized states, respectively. Following the approach by [30], we can write

$$\frac{n(H)}{n(H_2)} P(H) = K(H_2)$$

with

$$\log K(H_2) = 12.5335 - 4.9252\theta + 0.0562\theta^2 - 0.0033\theta^3$$

$$\theta = 5040/T$$

with T in K, and $P(H) = n(H) K_B T$ being the pressure associated to the H atoms. One can therefore write

$$\frac{x(H_2)}{x(H)^2} = \frac{2 K_B T N}{K(H_2)} = K_1$$

In addition, from the Saha equations, we can derive the ratio $K_2 = x(H^+)/x(H)$. Recalling that $x(H_2) + x(H) + x(H^+) = 1$, the previous equations can be combined as

$$K_1 x(H)^2 + (K_2 + 1)x(H) - 1 = 0$$

$$\frac{K_1}{K_2^2} x(H^+)^2 + \frac{K_2 + 1}{K_2} x(H^+) - 1 = 0$$

$$x(H_2) = 1 - x(H) - x(H^+)$$

This system of equations for given values of the constants K_1 and K_2 is solved by iteration. For temperatures larger than $\log(T) > 4.5$, the term K_1 becomes small and the fraction of hydrogen atoms bound in molecular hydrogen is also small and can be neglected.

2.2.4
Turbulent Pressure

When discussing the various contribution to the total pressure of the stellar matter, we have thus far neglected to include an additional term, potentially relevant only in presence of convection, the so-called *turbulent pressure*. Indeed, turbulent motions in convective layers can contribute to the total pressure by exchange/transfer of the momentum carried by the convective elements.

In a general case, the total pressure can be written as $P_{tot} = P_{gas} + P_{rad} + P_{turb}$, where the three terms represent the gas, radiation and turbulence pressure, respectively. The two first terms in this sum have to be evaluated according to the adopted EOS, while the turbulent term is defined as

$$P_{turb} = \rho \bar{v}^2$$

where $\bar{v}$ represents the average velocity of the convective bubbles. Due to the lack of a generally accepted turbulence model in stellar conditions, the estimate of this mean velocity, and hence of the turbulent pressure contribution, is a thorny problem. It has been customary to evaluate the mean velocity of the convecting elements by solving the equations of the mixing length theory. A complete description of the derivation of all relevant thermodynamic quantities when including turbulent pressure can be found in [31]; here, we are interested in an order of magnitude estimate of P_{turb} in low-mass stellar models. When treating the convective elements as mass particles, one can write

$$P_{turb} = \frac{2}{3}\left(\frac{1}{2}\rho\bar{v}^2\right) = \frac{1}{3}\rho v_s^2 \left(\frac{\bar{v}}{v_s}\right)^2$$

where v_s is the local sound speed. One can then calculate

$$\frac{P_{turb}}{P} = \frac{1}{3}\Gamma_1\left(\frac{\bar{v}}{v_s}\right)^2$$

with Γ_1, denoting the adiabatic exponent $\Gamma_1 = [(d\ln P)/(d\ln \rho)]_{ad}$, that usually is not much larger than $5/3$. Typical average values for $\bar{v}$ and v_s in a star slightly more massive than the Sun (in order to have a convective core) are $\leq (400-500)$ cm s^{-1} and ~ 550 km s^{-1} so that the ratio $\bar{v}/v_s \sim 10^{-5}$ or less, while in the solar atmosphere $\bar{v} \sim 2$ km s^{-1} and $v_s \sim 20$ km s^{-1}, implying $\bar{v}/v_s \sim 0.1$.

By taking these orders of magnitude into account, the contribution of P_{turb} to the total pressure is actually negligible in the stellar interiors, and is of the order of few hundredths percent or lower in the solar atmosphere. However, when considering more advanced evolutionary stages characterized by extended convective envelopes such as RGB and AGB, the mean velocity of the convective elements can achieve a value of the order of the local sound speed, and hence $P_{turb}/P \approx 0.55$. Obviously, in such cases, turbulent pressure could play a role.

To the best of our knowledge, literature regarding stellar evolutionary computations that include the turbulent pressure is very sparse. However, a few numerical experiments [31] have shown that the structural and evolutionary effects associated to the inclusion of the turbulence term in the pressure balance are, if any, quite marginal. This occurrence is partially due to the fact that, in the assumptions that the MLT provides reasonable estimates of the convective velocities, on average the turbulent pressure amounts at most to about 20–30% of the gas pressure also during the RGB and AGB stage. Based on this, the contribution of P_{turb} is commonly neglected in stellar models computations, although the extremely high accuracy required by the current and future high-precision asteroseismology surveys could make its inclusion necessary in order to obtain a better match between the observed pulsation spectrum and the model predictions.

2.3
Nuclear Energy Production and Nucleosynthesis

In Section 2.1, we introduced ϵ, the energy generation coefficient. As already noted, this term can be expressed as sum of three different contribution: $\epsilon = \epsilon_n + \epsilon_g - \epsilon_\nu$, where the meaning of each term has already been clarified.

A detailed derivation and analysis of both ϵ_n and ϵ_ν terms can be found in several textbooks [18–20, 27] and will not be repeated. Here, we will discuss the nuclear burning processes relevant for interpreting the structural and evolutionary properties, as well as some observations that will be discussed in the following chapters, of low-mass stars.

Let us remember that in a stellar plasma, the kinetic energy of the interacting nuclei is due to their thermal motion. For such a reason, their interaction is referred to as *thermonuclear reaction*. It is possible to demonstrate that for a generic reaction $1 + 2 \rightarrow 3 + 4$, the energy generation coefficient $\epsilon_{n,12}$ can be written as

$$\epsilon_{n,12} = R_{12} \times Q_{12}$$

where R_{12} and Q_{12} are the number of reactions per unit time and unit mass (i.e. the nuclear reaction rate) and the amount of energy released by a single reaction, respectively. The value of Q_{12} can be evaluated from the difference between the sum of the masses of the interacting particles and the sum of the masses of the products. On the other hand, R_{12} is proportional to $\langle \sigma v \rangle_{12}$, the cross section for the reaction between 1 and 2, that is, a measure of the probability that a pair of particles 1 and 2 experiences a reaction.

From the point of view of stellar model computations, the difficulty lies in the estimate of $\langle \sigma v \rangle$. It is a common procedure to define the astrophysical factor, or simply the S-factor, as

$$S(E) = \sigma(E) E e^{2\pi\eta}$$

where E is the energy of the relative motion of the interacting particles, while the parameter η (not to be confused with the electron degeneracy parameter) is proportional to $m^{1/2} Z_1 Z_2 E^{-1/2}$, with m is equal to the reduced mass of the system formed by particle 1 and 2, and Z_i is the electric charge of the i element. With this definition at hand, it is customary to write $\langle \sigma v \rangle$ as

$$\langle \sigma v \rangle = \frac{2^{3/2}}{(m\pi)^{1/2}(K_B T)^{3/2}} \int_0^\infty S(E) e^{-E/K_B T - 2\pi\eta} \, dE$$

that is, the form usually found in the astrophysical literature. It is interesting to discuss in some detail the previous equation by considering the extreme case of a nearly constant S-factor. This case is usually referred to as "non-resonant", and one obtains that $\langle \sigma v \rangle \propto \int_0^\infty e^{-2\pi\eta} e^{-E/K_B T} \, dE$. The factor $e^{-2\pi\eta}$ represents the quantum mechanical probability that a particle with a given energy, tunnels across the barrier due to the Coulomb potential associated with the electric charges of the interacting particle, while the term $e^{-E/K_B T}$ is the high-energy tail of the Maxwell–Boltzmann distribution. The convolution of these two functions produces a strongly peaked curve with a maximum at the so-called "Gamow peak", that denotes the most efficient energy for the nuclear reaction to occur. This means that stars do not burn at high energy where the cross section is very large (for the number of particles with such energies is vanishingly small) neither do they burn at small energies where the number of particles is high because the cross section is vanishingly small. Rather, in a star, nuclear reactions occur at energies where the function $e^{-2\pi\eta} e^{-E/K_B T}$ has a maximum, namely, the *Gamow peak*.

When the trend of the Gamow peak is studied for a given temperature, but for different pairs of interacting particles – different electric charges, hence different Coulomb barriers – one can draw the following conclusions. For increasing electric charges: (i) the Gamow peak shifts to higher energies, that is, larger temperatures are required for thermonuclear reactions to be efficient; (ii) the Gamow peak becomes broader; (iii) most importantly, the area covered the Gamow peak decreases drastically. Roughly speaking, this occurrence means that for a mixture of different nuclei in a plasma at a given temperature, those nuclear reactions with the smallest Coulomb barrier produce most of the energy and the fuel is consumed more slowly. This is of paramount importance for stellar evolution because it provides a natural explanation for the occurrence of well-defined nuclear burning stages.

The case we have just discussed corresponds to a simplified situation. In reality, $S(E)$ can show relevant variations for small changes in the energy, with narrow and/or broad peaks at specific energies. In this case, the nuclear reaction cross section displays so-called "resonances". Generally, in order to evaluate the total rate of a single reaction, several contributions need to be taken into account: narrow and broad resonances, non-resonant processes, sub-threshold resonance and so on [32]. Each individual nuclear reaction represents a special case and the evaluation of the actual reaction cross section is usually complicated. In addition, one has to bear in mind that the Gamow peak, that is, the energy range where one needs to evaluate the S-factor, is usually at least about one order of magnitude lower than the energies at which the nuclear reactions can be studied in the laboratory experiments. When a direct empirical estimate of $S(E)$ is out of current experimental possibilities, one has to rely on – sometimes uncertain – extrapolation in the low-energy regime.

A further important complication arises from the fact that in stars nuclear reactions do not occur in vacuum (as in the case discussed above), but in a plasma where free electrons tend to distribute around the nuclei. The effect is to partially screen the Coulomb potential felt by an approaching particle compared to the case of an isolated nucleus. It is clear that this shielding effect, that is, "electron screening", is very important and has to be properly accounted for when computing the actual rate of a nuclear reaction inside a stellar structure.

We now discuss in some detail the most important nuclear burning processes for low-mass stars. These stars are able to only ignite H- and He-burning. For those stars able to reach the AGB stage, another important nuclear reaction network is that associated to neutron captures (Section 5.3.2).

2.3.1
Hydrogen Burning

The ensemble of hydrogen burning reactions transform four protons into one ^{4}He nucleus, plus two positrons and two neutrinos. The total energy released by this process is $Q = 26.73$ MeV and only a small part of it, about 0.6 MeV, is carried away by the two outgoing neutrinos. This amount of energy is at least almost a factor of 10 larger than in any other set of nuclear reactions occurring in stars. The large efficiency of the nuclear conversion of H into He explains why hydrogen is consumed at a lower rate than any other fuel powering stellar nuclear reactions.

The detailed mechanisms for hydrogen burning were first derived independently in [33, 34]. These works disclosed the existence of two different networks of reactions that can convert efficiently hydrogen into helium and provide the energy needed to fuel MS stars, namely, the *p–p chain* and the *CNO bi-cycle*, that we are now going to review briefly (see [35] for a more detailed presentation). Both burning channels include a number of "secondary elements", that is, elements that are simultaneously destroyed and produced by different nuclear reactions. Following the time-evolution of the abundances of these secondary elements is extremely important for evaluating the relative efficiency of the nuclear burning chains.

2.3.1.1 The *p–p* Chain
At the temperature and density characteristic of the solar interior, the *p–p chain* is the most efficient path for hydrogen burning. The major nuclear reactions involved in the chain are the following:

> **pp I:**
> $$^1\text{H} + {}^1\text{H} \rightarrow {}^2\text{D} + e^+ + \nu_e$$
> $$^2\text{D} + {}^1\text{H} \rightarrow {}^3\text{He} + \gamma$$
> $$^3\text{He} + {}^3\text{He} \rightarrow {}^4\text{He} + {}^1\text{H} + {}^1\text{H}$$

pp II:

$$^3He + {}^4He \rightarrow {}^7Be + \gamma$$

$$^7Be + e^- \rightarrow {}^7Li + \nu_e$$

$$^7Li + {}^1H \rightarrow {}^4He + {}^4He$$

pp III:

$$^3He + {}^4He \rightarrow {}^7Be + \gamma$$

$$^7Be + {}^1H \rightarrow {}^8B + \gamma$$

$$^8B \rightarrow {}^8Be + e^+ + \nu_e$$

$$^8Be \rightarrow {}^4He + {}^4He$$

The *p–p chain* starts with a weak interaction process[3] that has a low cross section ($\approx 10^{-23}$ barn). The *pp I* chain begins to be important only when the temperature is of the order of 5×10^6 K. Below $T \sim 8 \times 10^6$ K, the production of ^3He is more frequent than its destruction, and the abundance of ^3He increases with time. When this temperature is attained, the nuclear reactions $^3He + {}^3He$ and $^3He + {}^4He$ become effective, decreasing the ^3He abundance. The first of these two reactions is strongly favoured ($\sim 80\%$) because the probability of the reaction mediated by the strong interaction is four order of magnitude larger than that of the electromagnetic process, and also because the tunnelling probability through the Coulomb barrier is a factor of 10 larger for the case of the lighter ^3He.

When the burning process of ^3He becomes active, the abundance of this element reaches its equilibrium value – that means that, until the temperature does not vary, the ^3He abundance does no longer change with time. For an increase of T, the equilibrium abundance of ^3He decreases. On the other hand, after the $^3He + {}^4He$ becomes the dominant process (that occurs for $T > 15 \times 10^6$ K), the chain branches again at the ^7Be level due to the competition between electron and proton capture. As an illustrative case, at the thermal conditions characteristic of the solar interiors, the electron capture, although it is a weak process, dominates over the proton capture because of the lack of a Coulomb barrier. When the chain reaches equilibrium, the abundances of all isotopes, namely, ^2D, ^3He, ^7Be, ^8Be, ^7Li and ^8B, remain constant (and small).

Concerning the dependence of the nuclear energy generation of the *p–p chain* as a function of the temperature, on average $\varepsilon_{pp} \propto T^4$, the smallest temperature sensitivity of all nuclear fusion reactions of astrophysical interest.

2.3.1.2 The CNO Bi-cycle

The *CNO bi-cycle* (hereafter the *CNO cycle*) is the combination of two independent cycles: the *CN cycle* and the *NO cycle*. The presence of C, N or O isotopes is necessary for either cycle to begin. Being both produced and destroyed during these cycles, these elements act as catalysts. The reaction networks involved are the fol-

---

3) One has to note that the *p–p chain* can indeed start via $^1H + e^- + {}^1H \rightarrow {}^2D + \nu_e$, the so-called *hep process*. However, its probability is smaller compared to the main reaction ($\sim 0.24\%$).

lowing:

CN cycle

$$^{12}C + {}^1H \rightarrow {}^{13}N + \gamma$$

$$^{13}N \rightarrow {}^{13}C + e^+ + \nu_e$$

$$^{13}C + {}^1H \rightarrow {}^{14}N + \gamma$$

$$^{14}N + {}^1H \rightarrow {}^{15}O + \gamma$$

$$^{15}O \rightarrow {}^{15}N + e^+ + \nu_e$$

$$^{15}N + {}^1H \rightarrow {}^{12}C + {}^4He$$

NO cycle

$$^{15}N + {}^1H \rightarrow {}^{16}O + \gamma$$

$$^{16}O + {}^1H \rightarrow {}^{17}F + \gamma$$

$$^{17}F \rightarrow {}^{17}O + e^+ + \nu_e$$

$$^{17}O + {}^1H \rightarrow {}^{14}N + {}^4He$$

In the *CNO cycle*, any isotope of C, N and O behaves as "secondary elements". This has the important consequence that, in principle, if the temperature is appropriate, the cycle can start almost from any reaction if the involved isotope is present. During the cycle, the isotope is burnt and then produced again. However, the relative abundances of the various isotopes do depend on the relative reaction rates of the nuclear processes involved in the cycle. The whole *CNO cycle* attains the equilibrium configuration for temperatures larger than $\approx 15 \times 10^6$ K.

An important quantity which is also compared with spectroscopic measurements is the carbon isotopic number ratio $^{13}C/^{12}C$, a ratio that is equal to 0.25 when the *CN cycle* attains equilibrium.

The efficiency of the cycle is determined by the reaction with the smallest cross section, which is the $^{14}N(p, \gamma)^{15}O$ reaction, the electromagnetic process with the largest Coulomb repulsion among those involved in the *CNO cycle*. This also has the important implication that the most abundant element in *CNO cycle* processed material is ^{14}N. Another important consequence is that, at odds with the *p–p chain*, the *CNO cycle* is less efficient when it achieves equilibrium: this is because this nuclear reaction acts like a bottleneck where the nuclei involved are blocked up before they can "proceed" through the cycle.

Despite its importance, the cross section of the $^{14}N(p, \gamma)^{15}O$ reaction is still uncertain. It has been recently measured both at the LUNA [36] and at the LENA experiment [37] and the new estimate is lower by a factor of ~ 2 with respect to previous measurements. The new rate causes an increase of the age of old stellar systems by about 0.9 Gyr.

The ratio of the astrophysical *S*-factors for the two reactions $^{15}N(p, {}^4He)^{12}C$ and $^{15}N(p, \gamma)^{16}O$ is about a factor 1000, and the role of *NO cycle* is generally marginal. In stars with mass $\sim 1\,M_\odot$, only the *CN cycle* is contributing to the energy budget. In general, the *NO cycle* only becomes efficient for temperatures larger than $\approx 20\times$

10^6 K. As a rule of thumb, the change of the initial distribution of CNO elements caused by the different burning channels is

- *CN cycle* processed matter $\quad$ −C ↓ N ↑ O ↑
- *CNO cycle* processed matter $\quad$ −C ↓ N ↑ O ↓

where the symbols ↓ and ↑ mean that the final abundance of the element is lower or larger than the initial one, respectively.

The temperature sensitivity of the complete *CNO cycle* is much larger than that of the *p–p chain*, in fact $\varepsilon_{CNO} \propto T^{18}$ at $T \approx 10 \times 10^6$ K. This means that *p–p chain* dominates at low temperatures, $T \leq 15 \times 10^6$ K, that is, in stars with mass lower than $\approx 1.3\,M_\odot$, while the *CNO cycle* dominates at higher temperature, in larger stellar masses.

The very different temperature sensitivities of *p–p chain* and *CNO cycle* have an important consequence: when the *CNO cycle* is the dominating mechanism, the burning process is mostly confined to the central stellar regions. This results in a large central energy flux, an occurrence which favours the presenc e of a central convective region as discussed in Section 2.1.

2.3.1.3 High-Temperature Proton Captures

In addition to the *CNO cycle*, analogous reaction cycles involving heavier elements exist at higher temperatures. In Figure 2.5, we show the *NeNa cycle* and *MgAl cycle*.

The *NeNa cycle* occurs at a temperature $T \sim 40 \times 10^6$ K, while the *MgAl cycle* is efficient above temperatures of the order of $\sim 50 \times 10^6$ K, although to have a significant ^{24}Mg depletion a quite larger temperature, $T \sim (70-80) \times 10^6$ K, is required.

In low-mass stars, during the MS, these two cycles are not efficient because their interiors are not hot enough; during the shell H-burning stage along the RGB, these temperatures (in particular those needed for the *NeNa cycle*) are attained in the H-poor tail of the advancing H-burning shell. In canonical stellar models, their efficiency is so low that the contribution to the nuclear energy budget is negligible, and they are commonly neglected in model computations.

These nuclear burning cycles are very efficient during the MS stage of massive stars, as well as in intermediate-mass stars ($M > 4M_\odot$) during the TPs along the

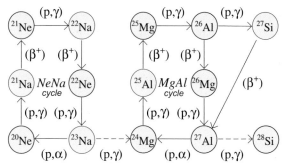

Figure 2.5 Reactions of the NeNa and MgAl high-temperature proton-capture cycles.

AGB, when some H-burning at very high-temperature can occur at the base of the convective envelope, the so-called "hot bottom burning" [38].

In the last decade, the importance of these high-T, H-burning processes has increased because it has been understood that these reactions could be responsible for the existence of well-defined anti-correlations among several pairs of chemical elements, such as O-Na and Mg-Al, that are nowadays routinely observed in Galactic GC stars. We will discuss this topic in detail in Section 7.8.

At high temperatures, larger than the temperatures of the "standard" *CNO cycle*, the following chain $^{15}O(\alpha, \gamma)^{19}Ne(p, \gamma)^{20}Na(e^+, \nu)^{20}Ne$ becomes efficient, producing ^{20}Ne that acts as a catalyst for the *NeNa cycle*. In a sense, this is the sequence of nuclear reactions connecting the *CNO cycle* with the *NeNa cycle*. An analysis of their reaction rates shows that a significant destruction of the ^{20}Ne occurs only when $T \geq 50 \times 10^6$ K. This notwithstanding, even a very small burning of ^{20}Ne is enough to lead to a significant increase of the abundance of the rare ^{21}Ne isotope even at temperature $T < 40 \times 10^6$ K. At higher temperatures, ^{21}Ne is destroyed via the nuclear processes shown in Figure 2.5.

As a result, the ^{21}Ne abundance is larger when H-burning occurs in the temperature range between $\sim 30 \times 10^6$ K and $\sim 35 \times 10^6$ K. The net result of the combined *CNO-NeNa cycles* is to enhance the sodium abundance while oxygen is depleted.

In a typical stellar metal mixture, the most abundant isotopes among those involved in the *MgAl cycle* is ^{24}Mg, whose abundance is unaffected by nuclear burnings, at least for $T < 60 \times 10^6$ K. On the contrary, the transformation of ^{25}Mg into ^{26}Al is very efficient for temperatures larger than about 30×10^6 K. The channel of production of ^{24}Mg via $^{23}Na(p, \gamma)^{24}Mg$, that connects the *NeNa cycle* with the *MgAl cycle* becomes active at $T > 50 \times 10^6$ K, and this affects the whole efficiency of the *MgAl cycle*.

At temperatures larger than about 70×10^6 K, the ^{24}Mg accumulation starts turning into a depletion via proton captures that increase the abundances of ^{25}Mg, ^{26}Al and ^{27}Al. The increase of the abundances of these isotopes is not hampered by their destruction by the proton captures involved in the *MgAl cycle* as a result of their relative slowness with respect to the production reactions. The net effect of the full *MgAl cycle* is to decrease the ^{24}Mg abundance and increase that of ^{25}Mg and aluminium.

We note that reaction rates of many nuclear processes involved in both cycles have a large uncertainty, up to two orders of magnitude or more in some specific temperature ranges, due to the huge number of resonances [39].

We wish to close this section by emphasizing that both the *NeNa cycle* and *MgAl cycle* are nuclear networks that involve the burning of hydrogen into helium. One has to expect that when these cycles are efficient, the produced anti-correlations have to be accompanied by additional helium production. As it will be clear in Section 7.8, this issue is very important in the context of the observational properties of sub-populations in Galactic GCs.

2.3.2
Helium Burning

The most relevant of all helium burning reactions is the so-called *triple alpha reaction* (3α) that corresponds to the production of ^{12}C from the fusion of three ^{4}He nuclei.

Being a triple encounter and very low probability process, this reaction usually occurs in two separate steps:

$$^4\text{He} + {^4\text{He}} \rightarrow {^8\text{Be}}$$
$$^8\text{Be} + {^4\text{He}} \rightarrow {^{12}\text{C}} + \gamma$$

The first reaction is endothermic by about 91.8 keV, and it has a resonance around the fundamental state of ^{8}Be, that is, unstable. In a very short time ($\sim 10^{-16}$ s), ^{8}Be decays back into two α particles. The possibility of the second reaction is therefore extremely low. In this context, ^{8}Be behaves as a secondary element, being involved simultaneously in destruction and production reactions. However, when the interior temperature rises above $\sim 10^8$ K, the probability of the second reaction increases, and there is a significant probability that a nucleus of carbon is produced before ^{8}Be decay, although $T \sim 1.2 \times 10^8$ K is necessary before 3α reactions produce a sizable amount of energy. One also has to note that, given that the reaction rate for the 3α reaction, scales as the square of the density, any increase of the local density affects the efficiency of this process.

The amount of energy released for any ^{12}C nucleus produced is equal to 7.275 MeV, equal to ~ 0.6 MeV per nucleon, that is, about a factor 1/12 of the amount of energy per nucleon released during the H-burning via the *CNO cycle*. This explains, together with the lower luminosity of MS stars, why, for a fixed mass, the core He-burning lifetime is about a factor of 100 shorter than the core H-burning lifetime.

The temperature sensitivity of the 3α reaction efficiency is quite strong: $\epsilon_{3\alpha} \propto T^{40}$ for $T \sim 10^8$ K and $\epsilon_{3\alpha} \propto T^{20}$ for $T \sim 2 \times 10^8$ K. This property has two important consequences: the first is that stars have an extended convective core during the central He-burning stage; the second one is that even a small increase of the local temperature, at a given density, causes a huge increase in the energy release. This property of the He-burning process is one of the reasons for the thermal runaway, named "the Helium flash" occurring at the tip of the RGB in low-mass stars (see Section 3.8).

The other nuclear reactions involved in the He-burning process are the following:

$$^{12}\text{C} + \alpha \rightarrow {^{16}\text{O}} + \gamma$$
$$^{16}\text{O} + \alpha \rightarrow {^{20}\text{Ne}} + \gamma$$
$$^{20}\text{Ne} + \alpha \rightarrow {^{24}\text{Mg}} + \gamma$$
$$^{24}\text{Mg} + \alpha \rightarrow {^{28}\text{Si}} + \gamma$$

Only the first two reactions, together with the 3α reaction, are important. The energy release for each $^{12}C(\alpha, \gamma)^{16}O$ reaction is equal to 7.162 MeV. The other ones are less important from an energetic point of view. Since the rate of the $^{12}C(\alpha, \gamma)^{16}O$ reaction is comparable to that of the 3α reaction, the main outcome of He-burning is the formation of ^{12}C and ^{16}O with traces of ^{20}Ne.

The $^{12}C(\alpha, \gamma)^{16}O$ reaction is one of the key reactions in stellar evolution. Its rate affects the C/O ratio in the stellar core at the end of the central He-burning stage and as a consequence, WD cooling times (see the discussion in Chapter 6). In addition, during the central He-burning phase, when the abundance of He inside the convective core is significantly reduced, α-captures on carbon nuclei becomes strongly competitive with the 3α reactions (that need three α particles) in contributing to the nuclear energy budget, with the immediate consequence that the cross section of this reaction has a strong influence on the core He-burning phase lifetime. In massive stars, this reaction rate affects all the subsequent hydrostatic burning stages and the nature of the remnant left behind after the core collapse.

This reaction has a resonance and a very low cross section ($\sim 10^{-17}$ barn) at low energies, and the nuclear parameters are difficult to measure experimentally or to calculate from theory. A recent analysis [40] has significantly increased the accuracy of the $^{12}C(\alpha, \gamma)^{16}O$ reaction rate, but the estimate should be still affected by an uncertainty of $\sim 35\%$.

When He-burning occurs in a layer with matter previously processed by the *CNO cycle*, for example, in the inter-shell region during the TPs, the nitrogen nuclei in the mixture can activate the following chain of reactions

$$^{14}N(\alpha, \gamma)^{18}F(\beta^+, \nu)^{18}O(\alpha, \gamma)^{22}Ne$$

If the temperature is larger than $T \geq 3.5 \times 10^8$ K, the reaction $^{22}Ne(\alpha, n)^{25}Mg$ can also occur. This reaction is important because it can produce a significant flux of neutrons. Such a high temperature in the inter-shell region during the AGB stage is only attained in stars more massive than $\sim 3M_\odot$. In less massive AGB stars, a He-burning process that can provide an alternative source of neutrons is the reaction $^{13}C(\alpha, n)^{16}O$. This reaction requires a temperature of $\sim 9 \times 10^7$ K that can be easily reached in the inter-shell region. The contribution of the various neutron sources in astrophysical sites and the physical processes at the basis of the nucleosynthesis of neutron capture elements are discussed in Chapter 5.

The availability of a significant flux of neutrons is mandatory in order to synthesize the so-called *s*-elements, that is, those elements beyond the iron peak, such as Sr, Y, Zr, Ba, La, Ce, Pr and Nd, produced through *slow* neutron captures (slow compared to the β decay). Our capability to predict the production of *s*-elements in AGB stars is fundamental in order to properly interpret the spectroscopic observations of the chemical abundances in AGB stars as well as in GC RGB stars in the context of the scenario for sub-populations in these old, Galactic stellar systems.

2.4

Radiative and Conductive Opacity

Whenever energy is transported by radiation, the knowledge of κ_{rad} – that is a measure of the efficiency of all processes that extract photons from the outgoing flux and redistribute them isotropically –, is necessary to compute the thermal structure of the star (see the discussion in Section 2.1). When the diffusion approximation can be applied to the energy transfer process, κ_{rad} is computed according to the so-called Rosseland mean opacity defined as

$$\frac{1}{\kappa_{\text{rad}}} = \frac{\int_0^\infty \frac{1}{\kappa_\nu} \frac{dB_\nu(T)}{dT} \, d\nu}{\int_0^\infty \frac{dB_\nu(T)}{dT} \, d\nu} \tag{2.31}$$

where $B_\nu(T)$ is the Planck function, and κ_ν is the monochromatic flux opacity. Because an harmonic average of the monochromatic opacities, the Rosseland mean is dominated by the frequency intervals where κ_ν is small.

Here, we briefly recall that a photon travelling through matter can undergo electron scattering, or absorption and transfer energy to an electron with three types of interaction: (i) *free–free transition* (*ff*) – a change of the energy level of the free electron; (ii) *bound–free transition* (*bf*) – transition of the electron from a bound to a free state; (iii) *bound–bound transition* (*bb*) – transition of the electron from a lower to a higher energy bound state. The various physical processes are efficient in different thermodynamical regimes.

The electron scattering process is the dominant opacity source in the stellar interiors where atoms are fully ionized. Nucleon scattering is not an important process because the scattering cross section is inversely proportional to the square of the mass of the involved particle. Electron scattering opacity by non-relativistic free electrons, the so-called Thomson scattering, is frequency,[4] temperature and density independent and it is given by $\kappa_s \sim 0.2(1 + X)$ in cgs units. In the case relativistic corrections cannot be ignored (Compton scattering), a more complex expression for the opacity is derived. These relativistic corrections to the Thomson scattering opacity are significant only at temperatures larger than 10^8 K, and they tend to reduce the scattering opacity. There exists an additional scattering process, that is photon scattering by atoms and/or molecule, named "Rayleigh scattering". The cross section for Rayleigh scattering is much smaller than the Thomson cross section, and is proportional to ν^4, that is, the scattering is more efficient at shorter wavelengths.

In the case of *free–free* and *bound–free* transitions, it is still possible to obtain approximate relations for the Rosseland mean opacity, the so-called Kramers law, $\kappa_{\text{bf}} = 4.3 \times 10^{25} Z(1 + X)\rho T^{-7/2}$ and $\kappa_{\text{ff}} = 3.7 \times 10^{22}(X + Y)(1 + X)\rho T^{-7/2}$, for the case of the *bound–free* and *free–free* transitions, respectively (all units are in cgs).

---

4) It is the only grey opacity source.

When *bound–bound* absorption by atoms and/or molecules is important, the evaluation of the κ_ν entering the Rosseland mean require complete line lists of all possible atomic transitions.

In these last 20 years, significant improvements have been achieved in the evaluation of the radiative opacity for the stellar interiors and envelopes, in the range of high (or relatively high) temperatures [41]. As a consequence, currently, estimates

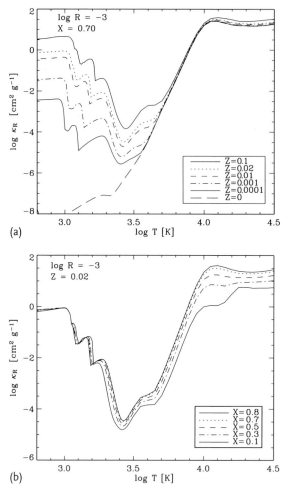

Figure 2.6 (a) Rosseland opacity as a function of temperature for various values of the metallicity, at a fixed value of the H abundance and of the parameter $R = \rho/T_6^3$ where T_6 is the temperature in 10^6 K. (b) As with (a), but for the Rosseland opacity as a function of temperature and various values of the hydrogen abundance at solar metallicity (courtesy of J.W. Ferguson).

of κ_{rad} in this regime[5] are very accurate: a realistic estimate of the κ_{rad} accuracy in the high-temperature regime is of the order of $\sim$ 5%. The evaluation of κ_{rad} in the low-temperature regime is complicated by the formation of molecules at $T < 5000$ K and, when T drops below $\sim$ 1700 K, by grain formation and the associated, and quite complex, absorption processes.

It is the low-T opacities that affect the T_{eff} of RGB models, while the high-T opacities play a major role in determining the mass extension of the convective envelope and the evolutionary timescales [42].

It is instructive to analyze the behaviour of low-T κ_{rad} as a function of various parameters [43]. Figure 2.6a shows the effects of changing the metallicity, while Figure 2.6b discloses the dependence of κ_{rad} on the hydrogen abundance ad fixed metallicity. As a rule of thumb, the opacity is higher when the metallicity increases. In more detail, for $T > 5000$ K, the opacity is dominated by hydrogenic absorption and it is barely affected by composition. In the regime $2500 < T(K) < 5000$, κ_{rad} is still dominated by the hydrogenic absorption mostly associated with the H^- ion and Rayleigh scattering, but the metals make an increasing contribution to the number of free electrons because the H ionization degree is decreasing. When moving towards lower temperatures, the sharp increase occurring at $\log(T) \sim 3.39$ is due to the increasing contribution of the H_2O molecule as an absorber; while at $T \leq 1500$ K, the formation of grains causes an even larger increase in the total absorption.

The dependence of κ_{rad} on the density is shown in Figure 2.7 for a solar chemical composition. As a general rule, when the density increases, the condensation tem-

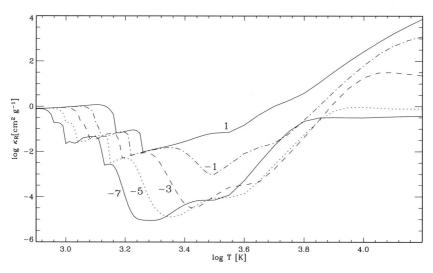

Figure 2.7 Rosseland opacity as a function of temperature for various values of R. All lines correspond to a solar chemical composition (courtesy of J.W. Ferguson).

5) When discussing stellar opacities, it is a customary to assume a boundary between low-temperature and the high-temperature regimes at $T \sim 10\,000 - 12\,000$ K.

perature of grains increases, and this causes a drastic increase in the value of κ_{rad} due to the formation of grains at larger temperature. At $T \sim 5000$ K, Rayleigh scattering dominates at low densities, while the H^- ion is the most important absorber at larger densities. Figure 2.8 shows the opacity calculated when excluding one absorber at a time in two different temperature regimes. This figure emphasizes what has been discussed above. We note how removing molecular sources changes the mean opacity by more than three orders of magnitude at $\log(T) = 3.3$. At even low-

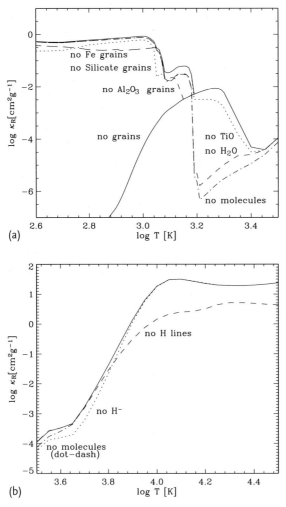

(a)

(b)

Figure 2.8 (b) A partial Rosseland mean opacity for $\log R = -3$ obtained by removing several opacity sources, one at a time from the κ_{rad} computation (see labels). The solid line denotes the total opacity, the dashed-line corresponds to the case with H *bf* and line displays the case without molecules. (a) As for the lower panel, but at lower temperatures (courtesy of J.W. Ferguson). *ff* removed, the dotted-line denotes the case without H^-, and the dash-dotted line

er temperatures, removing all grain species from the opacity has a dramatic effect, that is, at $\log(T) = 2.8$, the difference is more than six orders of magnitude.

2.4.1
Radiative Opacity for C-enhanced Mixtures

It is instructive to discuss the case of low-T radiative opacities for varying carbon and oxygen abundances. As discussed in detail in Section 5.3.1, stellar models with initial mass in range between $\sim 1.2 M_\odot$ and $\sim 4 M_\odot$ experience, during the TPs the so-called *third dredge up*. The outer convective envelope is able to penetrate into the inter-shell region mixed during the previous TP, with the consequence that helium, products of He-burning (essentially carbon) and heavy s elements are dredged-up into the envelope and brought to the surface. As a consequence of re-current *third dredge up* episodes, the stellar surface experiences increasing level of carbon enrichment, that is, increasing values of the ratio C/O. The consequent huge changes in the radiative opacity affect the envelope structure [44] and the model evolutionary and observational properties. Recently, realistic and accurate predictions of κ_{rad} for C/O enhanced mixtures have become available [45, 46].

Due to the large binding energy of the CO molecule, the C/O ratio is more important than the absolute carbon abundance: when the C/O ratio is lower than ~ 1, the major absorbers are TiO, VO, and H_2O molecules. However, when this ratio becomes larger than unity, basically all O is bound in CO with CN, C_2 and SiC and some HCN and C_2H_2 formed from the remaining carbon nuclei.

Figure 2.9 shows the behaviour of κ_{rad} as a function of T for several values of the C/O ratio: one can appreciate the high sensitivity of the radiative opacity around

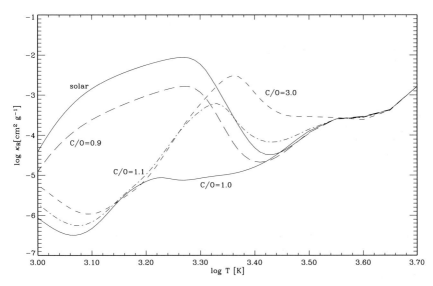

Figure 2.9 Rosseland mean opacity as a function of temperature for solar metallicity, $\log R = -2.0$, and various C/O values (courtesy of J.W. Ferguson).

$C/O \approx 1$, while for $C/O > 3$, the opacity is no longer affected by a further C enhancement. This behaviour can be understood, considering that as the C/O ratio increases, the amount of O available for the molecular H_2O is decreased because the CO molecule is formed preferentially; this causes a decrease of the opacity at temperatures important for H_2O absorption. When the C/O ratio is close to one, CO is the most abundant molecule, a low absorber that causes a decrease of the radiative opacity. Following a further increase of the C/O ratio, the opacity is dominated by CN molecules that produce the characteristic opacity bump at temperature larger than $\log(T) = 3.3$, that is, the temperature range more important for AGB stars.

Let us anticipate here that the use of radiative opacity accounting for the appropriate C/O ratios produces a huge decrease of the effective temperature of AGB models with respect those that do not account for the opacity changes associated with the *third dredge up*.

2.4.2
The H_2 Collision Induced Absorption

An additional source of opacity that is important for specific low-mass objects is the so-called *Collision Induced Absorption* (hereinafter CIA) by H_2 molecules, responsible for much of the opacity in a dense gas at temperature of $\sim 3000\,K$ or lower [47, 48]. Since the H_2 molecule in its ground electronic state has no electric dipole, absorption of photons can take place only via electric quadrupole transition. This is the reason why low-density molecular hydrogen gas is essentially transparent throughout the visible and infrared portion of the spectrum. However, when the density is large enough, each time a collision between two particles occurs, interacting pairs like H_2-H_2, H_2-He or H_2-H, form a sort of "virtual molecule" that, because of its non-zero electric dipole, absorb photons with a probability much higher than that of an isolated H_2 molecule.

The collision-induced absorption coefficient can be formally written as an expansion in terms of powers of the density

$$\kappa_{CIA} = \kappa_1 \rho + \kappa_2 \rho^2 + \kappa_3 \rho^3$$

where κ_n accounts for the absorption by n-body collisions. In the case of H_2, the term κ_1 is equal to zero because there are no allowed dipolar transitions, and $\kappa_2 \rho^2$ is the relevant term, originated from binary collisions. This shows that CIA opacity is proportional to ρ^2, at variance with other opacity sources which are proportional to ρ.

Apart from a relatively large density, low temperatures are also required for the formation of the molecules involved in the CIA process. These conditions on both density and temperature restrict the efficiency of this opacity source to VLM and WD stars. We also note that a decrease of the metallicity favours the effects of CIA absorption because of the decreased contribution coming from other sources associated with the metals, and of the continuum opacity from negative ions (mainly from H^-, where the free electrons mainly come from low-ionization energy metals like Ca, K, Na). An additional reason is due to the fact that, for a given total mass, the lower the metallicity the denser the models.

Before closing this section, we must mention a different approach to determine the mean radiative opacity, the so-called Planck mean opacity κ_P given by

$$\kappa_P = \frac{\int_0^\infty \kappa'_\nu B_\nu(T)\, dT\, d\nu}{\int_0^\infty B_\nu(T)\, dT\, d\nu} \tag{2.32}$$

where κ'_ν is the monochromatic opacity accounting only for absorption processes, excluding scattering. The ratio between the Rosseland and the Planck mean opacity at a given layer provides a criterion for the validity of the diffusion approximation for the radiative transfer. When κ_{rad}/κ_P is equal to ~ 1, this approximation is appropriate, but when this ratio deviates significantly from one, the diffusion approximation can no longer be used.

2.4.3
Conductive Opacity

When the stellar matter is electron degenerate, heat transfer by electrons become an efficient (and in some case dominant) energy transport mechanism. When this occurs, a detailed evaluation of the conductive opacity is crucial. It is useful to briefly recall that, far from degeneracy, electrons cannot efficiently transport energy for their mean free path is orders of magnitude lower than that of photons. However, when electron degeneracy is present, electrons cannot easily exchange momentum with other particles because the quantum states with momentum lower than the Fermi value are occupied. Interactions become very rare and the electron mean free path increases hugely. As discussed in Section 2.1, the total opacity is the harmonic sum of the radiative and conductive opacities. In an electron degenerate environment, κ_e is much smaller than κ_{rad} and $\kappa_{tot} \sim \kappa_e$.

The physical conditions that require a detailed evaluation of the conductive opacity are encountered in the interiors of brown dwarfs, VLM stars with mass $M_{tot} < 0.15\, M_\odot$, in the He-core of low-mass stars during their RGB evolution, in the CO core of asymptotic giant branch stars, as well as in WDs (in both the core and in a portion of the envelope) and envelopes of neutron stars.

Until the mid-1990s, the main sources of electron-conduction opacity tabulations [49–51] suffered of a number of shortcomings:

- The tabulations in [50] included a very limited set of element mixtures, did not take into account relativistic effects, and had significant gaps in the $T–\rho$ coverage;
- The relations in [49, 51] represented a large improvement in their domains of validity over the previous results for the opacity contribution due to electron–ion (ei) scattering. They took into account relativistic effects, more accurate structure factors for an ion liquid, phonon–electron interactions for an ion solid, and can be employed to compute conductive opacities for arbitrary chemical mixtures.

However, these results are only valid when electrons are strongly degenerate, that is, $T \ll T_F$ where T_F is the Fermi temperature given by $T_F = 5.93 \times 10^9 \{[1 + 1.018(Z/A)^{2/3}(\rho/10^6)^{2/3}]^{1/2} - 1\}$ with ρ in cgs units, Z and A being the atomic number and atomic weight of the ions in the plasma, a condition which is not strictly satisfied in low-mass, RGB stars. In addition, they do not take into account important collective plasma effects near the solid–liquid phase transition.

Several of these shortcomings have recently been addressed [52–54]. These new calculations can be used to compute opacities for arbitrary astrophysical mixtures at $T \ll T_F$, and include the contribution from the ei scattering and the electron–electron (ee) scattering in the regime of partial electron degeneracy.

The conductive opacity is related to the thermal conductivity ξ by the equation

$$\kappa_e = \frac{16\sigma T^3}{3\rho\xi} \tag{2.33}$$

where σ is the Stefan–Boltzmann constant. The most practical approach for calculating ξ is the so-called "kinetic method" [55]. By following the elementary theory in which the effective electron scattering rate ν does not depend on the electron

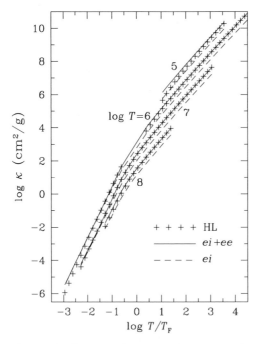

Figure 2.10 Electron conduction opacity ($\kappa = \kappa_{ei} + \kappa_{ee}$) of a pure He mixture as a function of the degree of degeneracy in terms of the ratio T/T_F, following [52]: crosses denote results in [50], while the dashed lines correspond to the opacity in [52], though neglecting the contribution of ee interactions.

velocity, one can write

$$\xi = a \frac{n_e K_B^2 T}{m_e \nu} \tag{2.34}$$

where n_e is the electron number density, m is the electron mass, and $a = 3/2$ for a non-degenerate electron gas or $\pi^2/3$ for strongly degenerate electrons. If the electrons are degenerate and relativistic, the electron mass has to be corrected for the relativistic effects. In a fully ionized plasma, ν is determined by electron–ion (ei) and electron–electron (ee) Coulomb collisions. According to the *Matthiessen rule*, it is possible to assume that the effective frequencies of different kinds of collisions simply add up, that is, $\nu = \nu_{ei} + \nu_{ee}$. Although this approximation is strictly valid only for extremely degenerate electrons, in practice, it gives a good estimate of the conductivity; in fact, one can show that $\nu_{ei} + \nu_{ee} \leq \nu \leq \nu_{ei} + \nu_{ee} + \delta\nu$, where $\delta\nu \ll \min(\nu_{ei}, \nu_{ee})$. The new calculations provide improved determinations of ν_{ei}, ν_{ee} and ξ in the regime of non-degenerate, strongly degenerate and partially degenerate electron gas.

Figure 2.10 shows the dependence of the total conductive opacity as a function of the degree of degeneracy expressed in terms of the ratio T/T_F in a pure He mixture. It is evident that the ee contribution is important for $T/T_F \geq 1$. Figure 2.11 displays the contribution of the ee interaction opacity to the total (ee + ei) conductive opacity in the He core of a low-mass RGB model at different stages of its evolution. This

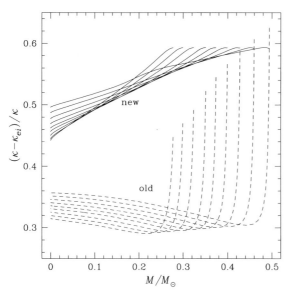

Figure 2.11 Ratio of the ee interaction opacity to the total (ee + ei) conductive opacity throughout the core of a $0.8 M_\odot$ metal-poor RGB model, as a function of the mass coordinate (in solar units). Both the new (solid line, [52]) and old (dashed line, [54]) results employ the same treatment for the ei interaction. Each curve corresponds to a RGB model in a specific evolutionary stage, with He-core mass equal to the value at the end of the curve.

diagram reveals that the contribution coming from the ee interaction is important, amounting to $\sim$ 50% or more of the total conductive opacity.

All of these analytical treatments hold for pure chemical elements. In the case of a mixture of elements, one should take into account the contribution of each ionic specie to the global ei interactions. There are various approaches for solving this problem: for instance, one can treat individually the case of each chemical particle and then sum all specific contributions, or compute the ei contribution assuming an "average" charge number for the mixture of ions present in the stellar matter. At least in the case of RGB low-mass stellar models, one can employ the conductive opacity of a pure He mixture in the electron degenerate core. The inclusion of a realistic fraction of 0.1% (by number) of C or N would cause an increase of κ_e by only about 1%.

2.5
The Atmospheric Structure

A detailed discussion of stellar model atmospheres can be found in [56, 57]. Here, we discuss how the treatment of stellar atmospheres affects the structure of the underlying stellar models. To solve the equations of stellar evolution, one needs the value of the pressure at the surface of the stellar model, defined as the layer where the diffusion approximation for the energy transport starts to be applicable. We denote the corresponding optical depth τ as τ_{ph}. At this layer, the value of the stellar mass is very close to the total mass M since the overlying atmosphere does not contain an appreciable amount of mass, due to the extremely low densities. The problem is to estimate $P(\tau_{ph})$. As it is well known, when the stellar model envelope is in radiative equilibrium, its structure is weakly dependent on $P(\tau_{ph})$, whereas the structure and depth of a convective envelope are affected by the choice of $P(\tau_{ph})$.

The simplest approach routinely adopted in stellar model computations relies on the integration of the atmospheric layers using the following equations

$$\frac{dP}{d\tau} = \frac{g}{\kappa} \tag{2.35}$$

$$T^4 = f\left(T_{eff}, \tau\right) \tag{2.36}$$

plus the equation of state. Here, g denotes the surface gravity of the star.

The first equation is the equation of hydrostatic equilibrium written in the case of constant mass using the optical depth τ as independent variable. We remark that in the definition of optical depth one uses the Rosseland mean opacity that is based on the diffusion approximation. The second equation is a $T(\tau)$ relation that describes the temperature stratification in a "grey" atmosphere as a function of the optical depth for a given T_{eff}. The integration is carried out from $\tau = 0$ (where $T \sim 0$ and $P \sim 0$) to the chosen τ_{ph}. The grey temperature stratification is usually written as

$$T^4 = \frac{3}{4} T_{\text{eff}}^4 \left[\tau + q(\tau) \right]$$

with different possible choices for $q(\tau)$ [57]. The simplest relation for a grey temperature stratification adopts $q(\tau) = 2/3$. This choice is obtained from the Eddington approximation of the grey radiative transfer. When the Eddington $T(\tau)$ relation is adopted, the integration is carried out until $\tau = 2/3$ where $T = T_{\text{eff}}$, and $P(\tau_{\text{ph}})$ is determined. Notice that $\tau = 2/3$ is still lower than that limit for the formal validity of the diffusion approximation.

It is also often employed a grey $T(\tau)$ (that we denote as K66) based on an empirical solar model atmosphere[6] whose form is the following [58]

$$T^4 = \frac{3}{4} T_{\text{eff}}^4 \left(\tau + 1.017 - 0.30 \times e^{-2.54\tau} - 0.291 \times e^{-30\tau} \right)$$

A more rigorous approach is to derive the value of $P(\tau_{\text{ph}})$ from precalculated, detailed, non-grey model atmospheres that are usually plane-parallel and completely defined by the chosen chemical composition, gravity and effective temperature.

When computing stellar models with this approach for the outer boundary conditions, one has to fix the value of τ_{ph} used as the "surface" of the stellar model. A comparison of the radiative flux provided by the exact solution of the energy transport equation, with that obtained with the diffusion approximation, shows that they coincide at optical depths larger than ~ 1 [60]. Values of $\tau \geq 1$ should therefore be adopted for deriving the value of the pressure to be used as boundary condition from model atmosphere computations.

The effect of different choices for $P(\tau_{\text{ph}})$ and τ_{ph} on the model calculations has been extensively investigated [59–61]. Below, we discuss an example that summarizes the main results. As a preamble, we recall that the free parameter α_{ml} entering the MLT has to be calibrated by some empirical constraints when modelling stars with convective envelopes. It is customary to use the solar T_{eff} as a calibrator by computing a Solar Standard Model (SSM, discussed in Section 3.4). Whenever an input entering the model computations is changed, as in the case of a change of the outer boundary conditions, α_{ml} has to be properly recalibrated.

Figure 2.12 shows the run of T and P in the sub-photospheric (in this example, the photosphere is defined as the layer where $T = T_{\text{eff}}$) layers of SSMs computed using various assumptions about the outer boundary conditions. By construction, all thermal stratifications have to coincide at $T = T_{\text{eff}}$, for example, at $T = 5770\,\text{K}$ in the case of the Sun. There are significant differences both in T and P among the various model predictions. The model based on the KS66 grey atmosphere predicts at the photosphere a pressure lower than that of the model based on the MARCS model atmosphere [62]. Figure 2.13 shows the impact of these differences in the sub-atmospheric layers on the location of the evolutionary tracks of the Sun in the

---

6) The solar model atmosphere calculated with this $T(\tau)$ does not reproduce accurately the observed solar flux and limb darkening.

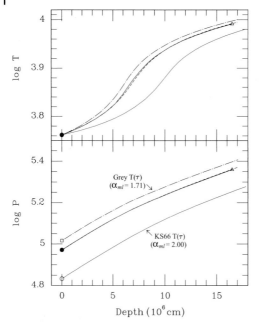

Figure 2.12 Temperature and pressure stratifications as a function of the geometrical depth in the sub-photospheric layers of SSMs computed with different outer boundary conditions. The depth has been set to zero at the photosphere, where all models have exactly the same T_{eff}. The solid line refers to the model obtained once the pressure at the photosphere (defined in this case as the layer where $T = T_{eff}$) has been fixed according to a MARCS atmosphere; the short-dashed line refers to the case of the MARCS atmosphere, but matching the interior structure at $\tau = 100$; the dotted line corresponds to the stratification obtained by using the KS66 grey atmosphere, while the dot-dashed line corresponds to the case with the Eddington approximation (adapted from [59]).

HRD. By construction, all tracks overlap for the present Sun, but the agreement is also very good along the whole MS. On the RGB, the tracks based on the KS66 atmosphere are ~ 100 K hotter than the tracks based on the MARCS atmospheres and the Eddington $T(\tau)$. It is also very important to notice that the two tracks obtained by matching the MARCS atmospheres to the interior structure at different optical depths, $\tau = 1$ (where $T = T_{eff}$) and 100, respectively, do overlap from the Zero Age MS to the RGB. This also occurs at lower metallicities and is proof that the T_{eff} scale for stellar models does not depend on the chosen fitting point (provided the boundary contidion is taken at $\tau \geq \sim 1$).

When comparing low-mass stellar models (based on a solar-calibrated α_{ml}) at lower metallicities, the difference in the T_{eff} scale between models based on both types of grey atmospheres and detailed non-grey models appears negligible [59]. Interestingly, the comparison between theory and observations for the effective temperature scale of approximately solar metallicity RGB stars both in the field and in stellar clusters, suggests that a better agreement does exist when the KS66 $T(\tau)$ relation is employed.

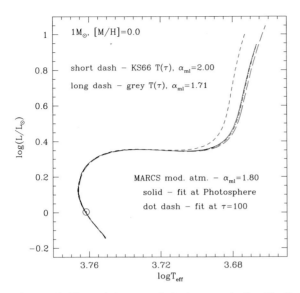

Figure 2.13 The evolutionary tracks in the HRD of various solar standard models computed with various outer boundary conditions. The ⊙ symbol marks the location of the Sun. The T and P stratifications in the sub-photospheric layers at the luminosity and effective temperature of the Sun were shown in Figure 2.12 (courtesy of D. Vandenberg).

However, as discussed in Section 3.2 and Chapter 6, in the case of VLM and WD models, outer boundary conditions from non-grey atmospheres are essential for deriving accurate predictions of their structural and evolutionary properties.

2.6
Mass Loss

One important physical process not included in the definition of standard stellar model is mass loss, although it is routinely included in the computations.[7] The effect of mass loss from the external layers of stars is important for explaining several observations of old stars, even if the observed rates are low and do not affect significantly (apart from specific cases that we will discuss in the next chapters) the structural, chemical and evolutionary properties of low-mass models.

As a general rule, at least in the case of low-mass stars, the efficiency of mass loss increases when T_{eff} decreases and the luminosity increases. This qualitative dependence can be easily understood by recalling that in the atmospheres of cool stars, molecules and grains are present; when the luminosity increases, the surface gravity will correspondingly decrease and the outer stellar layers will be less tightly bound and more easily lost. Therefore, the regions of the HRD more affected are the RGB and AGB. In the case of AGB stars, the general consensus is that the main

---

7) This is accomplished by essentially removing an appropriate number of surface mass shells after each time step.

physical mechanism driving mass loss is radiation pressure on dust particles. For RGB stars, the role of dust and the actual driving mechanisms of their outflows are more uncertain.

The efficiency of mass loss is extremely low in MS low-mass stars. For the Sun, the mass loss rate is of the order of $\sim 10^{-14} M_\odot$/year, and (even though this hypothesis is not backed by any theoretical calculation) an even lower rate is expected in more metal-poor MS objects due to the decreased metal abundance. Because of these low rates, mass loss can be neglected when computing MS models.

2.6.1
The Red Giant Branch Stage

The colour distribution of HB stars in Galactic GCs shows that mass loss affects, possibly with a stochastic behaviour, stars climbing the RGB. However, so far, we lack a reliable numerical recipe to calculate RGB mass loss rates that can be easily included in a numerical evolutionary code.

The majority of studies rely on the Reimers law [63] based on observations of Population I giants. Several different alternative relationships for the mass loss rate, $\dot{M} = dM/dt$, have been proposed, as summarized by [64] and reported in Table 2.1. All of these relationships are based on more or less complicated functions of several variables, usually R or T_{eff}, L and surface gravity. When using these formulas, one should consider the following caveats:

- The original Reimers formula was derived from Pop. I stars with only two RGB stars in the sample, and in principle, it is not suitable for metal-poor RGB stars;
- All these relationships do not consider the effect of those processes such as rotation and magnetic fields that can modify (commonly an enhancement is expected) the mass loss efficiency with respect to the average estimate provided by the given formulae. In fact these formulae only provide the "average" dependence of mass loss on the basic stellar parameters. Star-to-star variations of specific physical properties not included in the formulas (i.e. rotation rates, magnetic fields) can obviously affect the mass loss rates on a star-to-star basis;
- Each one of these relationships provide a different total amount of mass lost at the end of the RGB. This has a large impact on the evolutionary stages that follow.

The mass loss rates given by the relationships listed in Table 2.1 are multiplied by a free parameter, usually denoted as η (again, not to be confused with the electron degeneracy parameter) whose value is chosen to fit empirical constraints like the mean colour of the stellar distribution along the HB of Galactic GCs. When adopting the Reimers law, typical values for η are ~ 0.2–0.4.

In order to overcome these limitations, observations have been devoted to obtain a direct estimate of the mass loss rate in RGB stars belonging to Galactic GCs. Different methods have been envisaged, for example, the analysis of the blue-shifted features of photospheric lines like H_α, or near-infrared photometry, that appears

Table 2.1 Mass loss rate formulas for the RGB and AGB. The mass, luminosity and radius are in solar units, the surface gravity is in cgs units, and the pulsational period (P) is in days. In the Blöcker relation, the $\dot{M}_R$ term represents the mass loss rate provided by the Reimers law.

RGB

Formula name	$\dot{M}\,(M_\odot/\text{yr})$
Reimers	$-5.5 \times 10^{-13}\left[L/(gR)\right]$
Modified Reimers	$-8.5 \times 10^{-10}\left[L/(gR)\right]^{1.4}$
Goldberg	$-1.2 \times 10^{-15}\,R^{3.2}$
Judge–Stencel	$-6.3 \times 10^{-8}g^{-1.6}$
Mullan	$-2.4 \times 10^{-11}\left(g/R^{3/2}\right)^{-0.9}$
VandenBerg	$-3.4 \times 10^{-12}\,L^{1.1}g^{-0.9}$

AGB

Formula name	$\log \dot{M}\,(M_\odot/\text{yr})$
Vassiliadis and Wood	$-11.4 + 0.0123 \cdot P$
Blöcker	$-8.31 - 2.1\log M + 2.7\log L + \log \dot{M}_R$

the most promising technique, having the advantage of sampling an outflowing gas far from the star under scrutiny [65, 66].

The main results of these observational efforts are: (i) mass loss seems to be episodic, and reaches its maximum efficiency near the RGB tip; (ii) the mass loss rates are in the range between $10^{-7}\,M_\odot$/year and $10^{-6}\,M_\odot$/year; (iii) the mass loss efficiency seems to depend very weakly, if at all, on the stellar metallicity. The average of the data can be fit by $\dot{M} \propto (L/gR)^{0.4}$, that is, a dependence on the global variables significantly flatter than for the Reimers formula. This result has the obvious implication that the Reimers law cannot reproduce these RGB mass loss rates.

2.6.2
The Horizontal Branch Stage

The hypothesis that mass loss is efficient during the HB evolution was first suggested with a view that an efficient mass loss when stars cross the RR Lyrae instability strip could explain the HB mass distribution in a more natural way, compared to the standard view of a stochastic mass loss during the RGB stage [67]. After evidence that mass loss within the RR Lyrae strip was incapable of explaining the HB mass dispersion [68], the interest in HB mass loss evaporated.

Owing to the discovery of peculiar chemical patterns at the surface of hot HB stars both in GCs and in the field, interest in this issue is again acquiring momentum. The presence of these chemical peculiarities (see Section 4.7) poses important constraints on the efficiency of the mass loss process in HB stars. It has been shown that the observed chemical patterns can only be explained if mass loss rates are in the range $(10^{-14}-10^{-12})\,M_\odot$/year [69].

To date, the only theoretical analysis [70] devoted to studying the efficiency of HB mass loss in the hypothesis that it is driven by radiation pressure on spectral lines has derived a rate of $\sim 10^{-12}\,M_\odot$/year at solar metallicity, that should decrease by one or two orders of magnitude at metallicities typical of Galactic GCs. In addition, no evidence has been found for the so-called "bi-stability jumps", that is, narrow T_{eff} intervals where the mass loss efficiency suddenly increases by some orders of magnitude, as found in the case of OB supergiants [71].

2.6.3
The Asymptotic Giant Branch Stage

Mass loss is a crucial feature of AGB evolution, for it determines how and when this phase ends. Its efficiency has a dramatic impact on both evolutionary lifetimes and nuclear yields along the AGB evolutionary stage [46]. Two of the most widely adopted formulas in AGB calculations are listed in Table 2.1.

In the case of AGB stars, the mechanism responsible for the observed large mass loss rates has been identified as the interaction between radiation pressure and dust particles. Typical T_{eff} values of AGB stars are around 3000 K, not low enough to allow for condensation of solids in the photospheric layers. Dynamics is commonly assumed to play a key role in the dust formation process: shock waves, caused by stellar pulsation or large-scale convective motions, propagate outwards through the atmosphere and lift gas above the stellar surface, intermittently creating dense, cool layers where dust particles may efficiently form. These dust grains are then accelerated away from the star by radiation pressure, provided that their radiative cross section is high enough, transmitting momentum to the gas through collisions, and dragging it along. This sequence of events is expected to lead to the observed outflows with typical mass loss rates of $(10^{-8}-10^{-4})\,M_\odot$/year and wind velocities in the range between 10 and 50 km/s.

This scenario is made even more complicated by the fact that the surface chemical composition of AGB stars can significantly change as a consequence of the *third dredge up*.

To date, detailed radiation-hydrodynamics models including dust production are only available for carbon-rich chemical compositions in which nearly all oxygen is bound in CO, and the excess carbon gives rise to carbon-based molecules and dust. In this regime, an accurate enough recipe for the mass loss efficiency at solar metallicity as a function of global stellar parameters is [72]

$$\log \dot{M}_{\text{AGB}} = -4.52 + 2.47 \cdot \log \left(10^{-4} \frac{L}{L_\odot} \right)$$
$$- 6.81 \cdot \log \left(\frac{T_{\text{eff}}}{2600\,\text{K}} \right) - 1.95 \cdot \log \left(\frac{M}{M_\odot} \right) \tag{2.37}$$

This formula does not account for any dependence on the actual C/O ratio. Quite recently, it has become clear that what is really important is the amount of condensible carbon, that is, that carbon that is not bound in CO. This depends on both the C/O ratio and metallicity [73]. Therefore, for a given C/O ratio, the amount of condensible carbon in a metal-poor star could be quite similar (or only slightly lower) than that present in a more metal-rich structure, owing to the reduction of O due to the formation of CO molecules.

Since there is no evidence for weaker pulsations in low-metallicity AGB stars, dust-driven AGB mass loss models predict no significant dependence of mass loss rate on the metallicity. The dependence on the pulsation period can be implicitly included in the luminosity term of the previous relation by adopting an appropriate period-luminosity relation for radially pulsating AGB stars.

For the case of oxygen-rich stars, a similar relation based on a fit to empirical data for dust-enshrouded oxygen-rich AGB stars, is the following [74]

$$
\log \dot{M}_{\text{AGB}} = -5.65 + 1.05 \cdot \log \left(10^{-4} \frac{L}{L_\odot} \right)
$$

$$
- 6.3 \cdot \log \left(\frac{T_{\text{eff}}}{3500 \text{ K}} \right) \tag{2.38}
$$

Once again, the dependence on the pulsational properties enters this relation through the luminosity term. This mass loss rate is only applicable to stars with a pulsation period $P > 400$ days. For lower periods, one usually relies on older mass loss recipes as those listed in Table 2.1. Despite the huge complexity of the physical processes at work, a consistent and realistic description of dust-driven mass loss in AGB stars seems within reach.

2.7
Diffusion Processes

Basic physical considerations suggest that, in addition to convective mixing routinely included in stellar evolution computations, additional transport processes are efficient within the stellar interior; they are driven by pressure, temperature and chemical abundance gradients, and by the effect of radiative pressure on the individual ions. These processes are collectively called "diffusion processes". In general, ions are forced to move under the influence of pressure as well as temperature gradients that both tend to move the heavier elements towards the centre of the star, and of concentration gradients that oppose the above processes. We denote these processes as "atomic diffusion". Radiation that has a negligible effect in the Sun pushes the ions towards the surface whenever the radiative acceleration of an individual ion species is larger than the local gravitational acceleration. We denote this process as "radiative levitation".[8] The speed of this diffusive flow depends on the collisions with the surrounding particles as they share the acquired momentum in a random way. It is the extent of these "collision" effects that dictates the

---

8) In the literature, sometimes the term diffusion includes radiative leviation plus the processes discussed in the following subsection.

timescale of element diffusion within the stellar structure once the physical and chemical profiles are specified. The most general treatment for the element transport in a multi-component fluid associated with diffusion can be found in [75].

2.7.1
Atomic Diffusion

Atomic diffusion has been neglected for many years in stellar model computations for at least two reasons: (i) It acts on very long timescales, that is, more than 10^{13} years are necessary for a particle to diffuse from the centre to the surface of the Sun, and the effect on stellar evolution was considered negligible; (ii) The inclusion of atomic diffusion in the calculations requires the computation of both spatial and temporal derivatives that increase the complexity of stellar evolution codes.

Constraints from data about the solar neutrino flux and helioseismology have however forced the inclusion of atomic diffusion in the calculation of the solar-, and more generally, low-mass stellar models [76, 77].

As mentioned before, atomic diffusion (or, briefly, diffusion) in stars is driven by three different gradients:

- *Pressure or gravity gradient*: a pressure gradient favours the concentration of the heavier elements toward the centre. Helium and metals diffuse toward the deeper interiors, whereas hydrogen diffuse outwards. The light electrons would also tend to rise, but are held back by an electric field that counteracts the gravity gradient;
- *Temperature gradient*: a temperature gradient leads to thermal diffusion that tends to concentrate higher charge, and more massive chemical species toward the hottest regions;
- *Chemical composition gradient*: a composition gradient tends to move ions from the high concentration regions to the low-concentration ones, that is, chemical diffusion works in the direction of erasing chemical gradients, and therefore opposes the previous two processes.

A detailed solution of the atomic diffusion equations for a multi-component fluid is available [78], following Burgers formalism [75], together with numerical routines that can be easily handled in stellar evolutionary codes. The temporal evolution of the abundance X_i of a generic element caused by (only) atomic diffusion is obtained from

$$\frac{\partial X_i}{\partial t} = -\frac{\partial}{\partial m} \left(4\pi r^2 \rho X_i \xi_i w_i \right) \tag{2.39}$$

where w_i is the net average (over ionization states; see below) velocity of element i in the centre-of-mass frame of reference. For heuristic purposes, it is convenient to consider a three component fluid made of protons, a small abundance of ions of another kind plus electrons [79]. For this case, w_i of the ions can be written as

$$w_i(r) = D_P(i)\frac{\partial \ln(P)}{\partial r} + D_T(i)\frac{\partial \ln(T)}{\partial r} + D_c(i)\frac{\partial \ln(c_i)}{\partial r} \tag{2.40}$$

where $D_P(i)$, $D_T(i)$ and $D_c(i)$ are the diffusion coefficients due to the presence of pressure, temperature and chemical abundance gradients, respectively, and c_i is the ion number fraction. Of course an appropriate treatment for a multi-component fluid needs to take into account the abundance gradients of all elements to evaluate the velocity contribution due to chemical inhomogeneities.

The diffusion coefficients are obtained assuming that (i) the gas particles have approximate Maxwellian velocity distributions, (ii) the temperatures are the same for all particle species, (iii) the mean thermal velocities are much larger than the diffusion velocities, and (iv) magnetic fields are unimportant. The evaluation of the diffusion coefficients requires the calculation of the resistance coefficients (usually denoted as K, z, z' and z'') under the assumption that collisions are dominated by the classical interaction between two point-charge particles. The most recent and accurate evaluation of the resistance coefficients can be found in [80], that include also the effect of quantum corrections. Although the accuracy of the diffusion constants could be improved considerably when using the correct functions for the resistance coefficients, there still remains the limitation of the Burgers formalism that causes an uncertainty in the diffusion coefficients of order 10%, difficult to reduce further.

In the case of a "simple" binary chemical mixture formed by protons and a single ion species, it is possible to obtain the following approximate relation for the diffusion velocity

$$\langle w \rangle \sim \langle A \rangle \frac{m_i}{K_B T} g$$

where $\langle A \rangle$ is the value of an average diffusion coefficient that can be written as

$$\langle A \rangle \sim 2 \times 10^9 \frac{T^{5/2}}{n_p Z_i^2 \ln \left(\frac{2.73 \times 10^8 T^3}{n_p Z_i^2} \right)}$$

where n_p is the number density of protons, Z_i ion charge, m_i the mass of the ion, and all quantities are in cgs units.

At the centre of the Sun (binary H-He mixture), $\langle A \rangle$ is of the order of $5\, \text{cm}^2\, \text{s}^{-1}$. The timescale for atomic diffusion is given by

$$\tau_0 \sim \frac{R^2}{\langle A \rangle}$$

where R is the characteristic length of the system, that is, the stellar radius in the case of stars. In the case of the Sun, one obtains $\tau_0 \approx 6 \times 10^{13}$ years. For the sake of comparison, the diffusion velocity, coefficient, and the timescale for a particle to diffuse across one pressure scale height for a solar-like star and for a white dwarf are listed in Table 2.2. The diffusion velocity as a function of the distance from the centre in a detailed solar model is displayed in Figure 2.14 for the major chemical species.

During the MS, the evolutionary phase where element diffusion has enough time to produce significant effects (apart from the hot WD phase), the change in the chemical stratification has two main consequences:

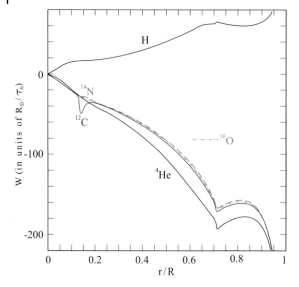

Figure 2.14 Diffusion velocity for H, He and CNO elements as a function of the radial coordinate in a SSM.

Table 2.2 Characteristic quantities related to the efficiency of atomic diffusion at the base of the envelope in a MS star of spectral type A, and in a WD. Time is in years and all other quantities are in cgs units.

	MS	WD
T_{eff} (K)	8500	60 000
Radius	4.9×10^{10}	9.8×10^{8}
Ion–proton diffusion coefficient	2.2×10^{-1}	2.8×10^{-2}
Diffusion velocity	2.1×10^{-10}	3.7×10^{-8}
Local characteristic diffusion time	4.1×10^{11}	5.9×10^{6}

- Due to the increase of the abundance of He in the centre, the amount of H decreases and the core H-burning is shortened;
- Given that the radiative opacity of H is generally larger than that of He, the opacity in the stellar envelope would increase; however, the concomitant reduction in the local abundance of metals would cause a decrease of κ_{rad}. The latter effect counteracts, but only partially, the former one, and the envelope opacity does indeed increase. This net increase of the envelope opacity would tend to increase the core H-burning lifetime; however, the decrease of the available fuel is the dominant process, and the final effect is to shorten the stellar lifetime by $\approx 10\%$. We will discuss this point in more detail in Section 3.9.
- The smaller the convective envelope, the larger the change of the surface abundances due to diffusion from the convective boundary. Given the short

timescales of diffusion compared to convective mixing, convective regions preserve their uniform chemical profile in presence of diffusion. The effect is just a slow change of the absolute values of the individual abundances;

- When the stars move to their RGB phase, the large mass increase of the convective envelope tends to bring back to the surface most of He and metals diffused from the envelope during the MS phase.

2.7.2
Radiative Levitation

When in the stellar layers the radiation field is strong and chemical elements are in a ionization state that favours the interactions with photons, the transfer of momentum from photons to ions is very efficient and the radiative acceleration can become larger than the local acceleration of gravity. This radiative levitation can therefore push chemical elements upwards and cause overabundances in the outer atmospheric layers.

One can treat radiative levitation by considering in the diffusion equations an "effective gravity" defined as $g_{\text{eff}} = g - g_{\text{rad}}^{\text{i}}$, with $g_{\text{rad}}^{\text{i}}$ being the acceleration attributed to a given ion species by the radiation field [81]. One will then face the problem of determining the value of $g_{\text{rad}}^{\text{i}}$ in each stellar layer for *all* chemical species. In a first approximation, the evaluation of $g_{\text{rad}}^{\text{i}}$ amounts to a calculation of the fraction of momentum that the element absorbs from the photon flux

$$g_{\text{rad}}^{\text{i}} = \frac{L_r^{\text{rad}}}{4\pi r^2 c} \frac{\kappa_R}{X_i} \int_0^\infty \frac{\kappa_u^{\text{i}}}{\kappa_u(\text{total})} \mathcal{P}(u)\, du \qquad (2.41)$$

where most symbols have their usual meaning. The quantities $\kappa_u(\text{total})$ and κ_u^{i} are, respectively, the total opacity and the contribution of element i to the total opacity at "frequency" u, with u and $\mathcal{P}(u)$ given by

$$u = \frac{h\nu}{K_B T}$$

and

$$\mathcal{P}(u) = \frac{15}{4\pi^4} u^4 \frac{e^u}{(e^u - 1)^2}$$

The main physical interactions that drive the transfer of momentum from the radiative field to ions are *bound–bound* and *bound–free* transitions.

Equation (2.41) has to be slightly modified in order to take into account that for *bound–free* transition, only a fraction of the momentum of the ionizing photons is transferred to the ion, the residual fraction being transferred to the electron lost by the ion. Detailed numerical computations have shown that the capability of the various chemical species to absorb momentum from the radiative flux strongly depends on their ionization potentials. As a rule of thumb (though the actual behaviour also depends on the thermal state of the stellar layers), elements with ionization potentials in the range between $\sim 10.5\,\text{eV}$ and $\sim 13.6\,\text{eV}$ (i.e. C, P, Cl, Ca) will be pushed up by large radiative forces; those elements with ionization potentials smaller than $\sim 10\,\text{eV}$ or greater than $\sim 18\,\text{eV}$ (i.e. He, Li, Be, B, Ne, Na, Al, K, Ni) do not absorb enough momentum to counter gravity. Those elements with an ionization potential between $\sim 13.6\,\text{eV}$ and $\sim 18\,\text{eV}$ (i.e. Mg, Mn, Fe, Cr) may or or may not be affected by radiative levitation, depending on details of their atomic structure.

For O and Si, the force transferred from the radiation field might be increased by an unusual type of levels, the so-called "autoionization levels". These are energetic levels where two electrons are excited, and the total excitation energy of the two electrons is larger than the energy required to ionize the ground state; then, at the same energy, there exists a continuum of states. As a consequence, there is non-vanishing probability that after the two electrons have been excited, the atom will ionize without collision or absorption of a photon. Most elements are not strongly affected by autoionization levels, while both O and Si have a peculiar behaviour.

The calculations of precise g_{rad}^i involve carrying out the integration over about 10^4 u values for each atomic species [83]. Given that these calculations have to be repeated at each mesh point in the stellar model and at each time-step during the computation of the evolutionary sequence, this explains why there are only a few model sets that include the effect of radiative levitation. One also has to note that the Rosseland mean opacity entering the stellar structure equations as well as in Eq. (2.41) has to also be continuously recalculated. This is, in principle, also true for calculations including only atomic diffusion, to be fully consistent with all composition changes.

Considering again the three component fluid of the previous section, the diffusion velocity of the ions with number fraction c_i can be expressed as

$$w_i = D_c \left(-\frac{dc_i}{dr} + \frac{m_i(g_{\text{rad}}^i - g)}{K_B T} \right)$$

where we have neglected the contribution from temperature and pressure gradients. Strictly speaking, this equation is based on the assumption that D_i does not depend on the excitation state of the ion, and that an equilibrium configuration exists between ionization and recombination.

To describe the migration of an element through the stellar layers, it can be useful to write a single diffusion equation that does not depend on the abundances of its various ions. The mean radiative acceleration of element X can be written as

$$g_{\text{rad}}(X) = \frac{\sum_i n_i D_i g_{\text{rad}}^i}{\sum_i n_i D_i}$$

where D_i and n_i are in this case the radiative diffusion coefficient and number abundance fraction relative to the ionization state i.

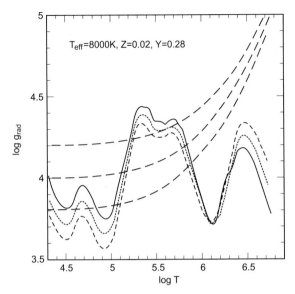

Figure 2.15 Radiative acceleration of Fe as a function of temperature in a stellar envelope with $T_{eff} = 8000$ K, solar chemical composition, and for three surface gravities: $\log(g) = 3.8$ (short-dashed line), 4.0 (dotted line), and 4.2 (solid line). The trend of the local gravities in the envelope is also shown (long-dashed lines) (after [82]).

Figure 2.15 shows the total radiative acceleration of iron in a stellar envelope. When the radiative acceleration acting on the ion is larger than gravity below a convective envelope and the element can diffuse towards the surface. We will discuss the observational evidence of the occurrence of radiative levitation in hot HB stars in Section 4.7.

2.8
Rotation

Spectroscopic observations and helioseismic data show that low-mass stars do rotate. This adds to basic considerations about the angular momentum evolution in a contracting protostellar cloud, and observations of the solar magnetic field that also predict/require stellar rotation.

Rotation has not generally been an ingredient of standard stellar models. This can be largely attributed to the increase in complexity and uncertainty, read free parameters, of models that include rotation, and some estimates that rotation is expected to be at most a perturbation to the structure of low-mass stars. A number of observational data, however, requires mixing mechanisms not present in standard stellar models, and mixing driven by rotation is at least a potential candidate. Also, asteroseismic observations [84, 85] will soon elucidate the role played by rotation in the interiors of stars other than the Sun.

The physical basis for the inclusion of rotation in detailed stellar evolution computations was laid down more than 40 years ago [86, 87] and, more recently, hydrodynamical treatments of the problem have appeared (see, i.e. [88]). A comprehensive, detailed discussion on the various aspects of stellar rotation can be found in [89]. Here, we summarize the main features that are relevant to low-mass stars.

We start by comparing a non-rotating star in radiative equilibrium with its solid-body rotating counterpart. The potential surfaces and the shape of the non-rotating star are spherical. If the star rotates as a solid body, the equipotential surfaces are rotational ellipsoids and two equipotential surfaces will diverge in distance from each other at the equator. This essentially means that gravity on an equipotential surface varies with latitude and that on an equipotential surface the temperature will be hotter at the poles and cooler at the equator (as demonstrated by von Zeipel almost a century ago, the energy flux is proportional to the local g), preventing the star from maintaining hydrostatic equilibrium (von Zeipel paradox). The solution to this paradox is to invoke large scale mass motions that transport energy, the so-called "meridional circulation" that moves material inward from the equator and upwards along the rotational axis towards the poles. The timescale for this mixing process, namely, the "Eddington–Sweet" timescale, was estimated to be

$$t_{ES} \propto \frac{M}{\Omega^2 R^3}$$

where M and R are the stellar mass and radius, and Ω is the angular velocity. This implies that rapidly rotating stars should be well-mixed by this circulation, contradicting observational data. However, one has to take into account that circulation also carries angular momentum. Variation of angular velocity with time due to angular momentum transport, contraction and expansion of the stellar layers and angular momentum loss from mass loss will generate, starting, for example, from solid body rotation, a variation of angular rotation velocity with depth in radiative regions. A "shear" develops between neighbouring layers, leading to instabilities. Given the strong radial stratification in radiative zones and no restoring forces opposing horizontal displacements, horizontal shear is expected to generate a strong turbulence that efficiently suppresses its cause, namely, the differential rotation with latitude [90]. This implies that rotation will be "shellular", with physical and chemical variables constant over isobars, that coincide with equipotential surfaces. This leads to the following set of equations (the stellar structure can still be approximated in 1D) in the case of "shellular" rotation [91–93]

$$\frac{\partial r_\psi}{\partial m_\psi} = \frac{1}{4\pi r_\psi^2 \rho} \tag{2.42}$$

$$\frac{\partial P}{\partial m_\psi} = -\frac{G m_\psi}{4\pi r_\psi^4} f_P \tag{2.43}$$

$$\frac{\partial L_\psi}{\partial m_\psi} = \epsilon - \frac{\partial U}{\partial t} - P \frac{\partial (1/\rho)}{\partial t} \tag{2.44}$$

$$\nabla_{\text{rad}} = \frac{3\kappa}{16\pi a c G} \frac{P}{T^4} \frac{L_\psi}{m_\psi} \frac{f_T}{f_P} \tag{2.45}$$

where

$$f_P = \frac{4\pi r_\psi^4}{G m_\psi S_\psi} \frac{1}{\langle g^{-1}\rangle} \tag{2.46}$$

$$f_T = \left(\frac{4\pi r_\psi^2}{S_\psi}\right)^2 \frac{1}{\langle g\rangle\langle g^{-1}\rangle} \tag{2.47}$$

In this formalism, S_ψ and V_ψ denote the area of a constant pressure surface and the volume enclosed by this surface, respectively. The variable m_ψ is the mass enclosed within this surface, and all thermodynamical and chemical variables are constant on S_ψ. The radial variable r_ψ is determined from $V_\psi = (4\pi/3)r_\psi^3$. The luminosity L_ψ represents the flux of energy through the same S_ψ. Average values of a generic quantity f over an isobar ψ are defined as

$$\langle f\rangle = \frac{1}{S_\psi} \int\limits_{\psi=\text{constant}} f \, d\sigma \tag{2.48}$$

where $d\sigma$ is an infinitesimal element of the equipotential surface ψ. Ignoring rotation induced mixings, the Schwarzschild criterion for convective stability becomes $\nabla_{\text{ad}} < \nabla_{\text{rad}}(f_T/f_P)$ where the gradients entering this relationship are the values determined in the case of no rotation.

Equations (2.42), (2.43), (2.44), and (2.45) practically look identical to the non-rotating case (apart from the factors f_P and f_t), with the difference that now the independent variable m_r is replaced by m_ψ and all quantities are constant on the isobar ψ rather than at radius r. At the non-rotation limit, f_P and f_t converge to unity and the equations are reduced to their non-rotating form.

Once the equations are written, one has to evaluate the surface of a generic isobar from a gravitational potential, like the Roche potential that considers the entire mass M concentrated at the centre of the star (that will be the function of the rotation velocity), and the factors f_T and f_P. In addition, we need a prescription for the transport of angular momentum and chemicals. The major sources of radial element transport associated with rotation are vertical shear (that tends to be inhibited by molecular weight gradients) and the interaction between meridional circulation and horizontal shear. One can treat this transport as a diffusive process, and hence

$$\frac{\partial X_i}{\partial t} = -\frac{\partial}{\partial m}\left[4\pi r^2 \rho \left(D_{\text{sh}} + D_{\text{eff}}\right)\frac{\partial X_i}{\partial r}\right] \tag{2.49}$$

where D_{sh} and D_{eff} are the diffusion coefficients associated with the two aforementioned processes. Evaluations of these coefficients that are proportional to the angular velocity gradient can be found, in [94], and contain at least one free parameter of order unity. It is possible, and sometime convenient, to treat convective mixing in the same approximation of diffusive transport [95] adding inside convective regions an extra term $D_{\text{conv}} = (1/3)v_c\Lambda$ to the sum $D_{\text{sh}} + D_{\text{eff}}$, where v_c is the local convective velocity derived from the mixing length theory, and Λ is the mixing length that contains the free parameter α_{ml}. Within the same formalism, one can also treat semi-convection and overshooting. In an overshooting region, one can employ

$$D_{ov} = D_0 \exp\left(\frac{-2z}{H_\nu}\right)$$

where $D_0 = v_c H_P$, $H_\nu = f H_P$ both evaluated at the Schwarzschild (or Ledoux) boundary, f is a free parameter that fixes the size of the overshooting region (typical values $f \sim 0.02$), that is, larger values correspond to larger overshooting regions, and z is the radial distance from this boundary. In a semi-convection zone, one can employ the coefficient

$$D_{sem} = \frac{\alpha_{sem} K}{6 c_P \rho}\left[\frac{\nabla - \nabla_{ad}}{\nabla_{ad} - \nabla - \nabla_\mu (\chi_\mu / \chi_T)}\right]$$

where K is the thermal conductivity $K = 4ac T^3 / 3\kappa\rho$, and α_{sem} is a free parameter of order 0.1 [96].

In the case where angular momentum in a radiative zone is transported by the same processes responsible for the mixing, the evolution of the angular velocity can be written as

$$\frac{\partial}{\partial t}\left(\rho r^2 \Omega\right) = -\frac{1}{5r^2}\frac{\partial}{\partial r}\left(\rho U r^4 \Omega\right) + \frac{1}{r^2}\frac{\partial}{\partial r}\left(\rho v_\nu r^4 \frac{d\Omega}{dr}\right) \tag{2.50}$$

The first term describes the effect of meridional circulation and contains U, the amplitude of the vertical speed of the meridional circulation given by

$$U = \frac{L}{mg}\left(\frac{P}{c_P \rho T}\right)\frac{1}{\nabla_{ad} - \nabla}\left(E_\Omega + E_\mu\right)$$

where E_Ω depends on the velocity profile, and E_μ on the chemical inhomogeneities along isobars [94]. The second term describes the diffusive transport caused by the vertical shear, and contains the vertical turbulent diffusivity D_v ($v_v \sim D_v$) that also depends on the velocity profile. In the case of convective regions, the transport of angular momentum is much more uncertain, as the interaction between convection and meridional currents is not well-understood. Traditionally, one chooses between the following two limiting cases. If convection inhibits meridional currents, angular momentum is redistributed very efficiently by convection, as in the case of chemical elements, and the result will be solid body rotation, as in the solar convective envelope. If meridional currents dominate, one can hypothesize that this is the case applicable to large, rarefied RGB envelopes, where convective elements may collide elastically rather than inelastically as in the solid body case. It is the specific angular momentum that is expected to be uniform in this case.[9]

9) Three-dimensional hydrodynamics simulations of rotating RGB envelopes, although still with simplifying assumptions, are becoming available [97].

Once the physical description of angular momentum transport and rotational mixing is in place, one still needs (i) to assume an initial rotational profile at the beginning of the pre-MS phase; (ii) to determine the amount of braking and angular momentum loss during the pre-MS phase due to the interaction (driven by magnetic fields) with the circum-stellar disk and magnetic wind braking during the MS; (iii) to include angular momentum loss due to mass loss along the RGB and subsequent evolutionary phases. These processes require the calibration of free parameters that are often poorly constrained by observations [98].

2.9
Internal Gravity Waves

In a stratified fluid, the density changes monotonically with depth and the molecular weight increases in the direction of increasing gravity. When a gas element in this stable stratification is perturbed, the competition between buoyancy and gravity gives rise to an oscillatory motion around the equilibrium position and creates a so-called "internal gravity wave" (see, e.g. [99–101]).

IGWs were originally expected to be an additional source of chemical mixing [102], but it was soon recognized that they contribute to angular momentum transport [103]. As their generation depends on the buoyancy force, they only travel in a radiative stratified region. As for the excitation process of an IGW, they are expected to be generated by the injection of kinetic energy from a turbulent region into an adjacent stable region, as observed both in 2D and 3D simulations of convective mixing [104], for example, by convective overshooting in a stable region and bulk excitation or excitation by Reynolds stresses inside the convection zone [105]. These IGWs penetrate into the radiation zone, transporting angular momentum that is deposited where they are dissipated through heat diffusion by photon exchange, which produces an "attenuation factor" (τ) proportional to the thermal diffusivity and inversely proportional to the IGW frequency ν and amplitude. It is by shaping the internal rotation profile that IGWs indirectly contribute to the mixing of chemical elements.

One can express the solution of the equations describing the propagation of IGWs in a rotating star in terms of Legendre polynomials. At each point within a radiative region, the total angular momentum "luminosity"[10] associated to the IGWs propagation can then be written as

$$\mathcal{L}_J(r) = \sum_{\text{waves}} \mathcal{L}_{J,\nu,\ell,m}(r_c) \exp\left[-\tau(r, \nu, \ell)\right]$$

where ℓ and m represent, respectively, the order and the azimuthal number of the Legendre polynomial.

---

10) The total angular momentum luminosity is defined as the average angular momentum flux transported by IGWs through a surface of radius r.

The deposition of angular momentum is then given by the radial derivative of this quantity. Given that, roughly speaking, we are only considering the radial dependency of IGW transport, all quantities required are evaluated from horizontal averages. The angular momentum evolution only due to IGW transport is given by

$$\rho \frac{d}{dt}\left(r^2 \Omega\right) = \pm \frac{3}{8\pi} \frac{1}{r^2} \frac{\partial \mathcal{L}_J(r)}{\partial r}$$

The $+$ $(-)$ sign in front of the angular momentum luminosity corresponds to waves travelling inward (outward).

Figure 2.16 shows the evolution of the interior rotation profile in a solar model when transport of angular momentum due to both rotation and IGWs has been taken into account. The final internal rotation profile is in good agreement with the helioseismic inferences, showing a near-uniform rotation of the radiative core of the Sun.

Detailed numerical computations have shown that the impact of IGWs of angular momentum redistribution as well as wave-induced mixing of chemical elements could be very important in all evolutionary stages in which an extended convective envelope is present, as it is the case for low-mass stars during the core H-burning stage and more importantly during the RGB phase [106].

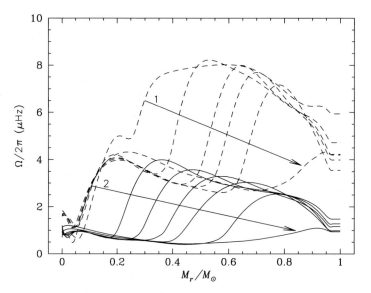

Figure 2.16 Evolution with time of the rotation profile in a SSM when meridional circulation, shear instability and IGW transport are taken into account. The initial equatorial velocity is $50\,\mathrm{km\,s^{-1}}$. The curves correspond to different ages, with time increasing in the direction indicated by the arrows. The dashed lines correspond to ages of 0.2, 0.21, 0.22, 0.23, 0.25, 0.27 Gyr, respectively, while the solid lines refer to ages of 0.5, 0.7, 1.0, 1.5, 3.0 and 4.6 Gyr (courtesy of S. Talon).

2.10
Magnetic Fields

The majority of cool stars show a large number of solar-like activity features: the presence of dark spots, with typical timescales ranging from days to months, with prominences, radio and X-ray emission commonly associated to frequent flaring, coronal mass ejections and winds. All these phenomena are explained as being due to the presence of strong magnetic fields.

The current explanation for the presence of magnetic fields in low-mass stars is the generation within their convective envelope through dynamo processes related to differential rotation, for example, to turbulence and shear at the boundary between the radiative layers and the envelope. This is the so-called "$\alpha\Omega$ dynamo generation" [107].

This scenario predicts a strong correlation between activity and rotation, that is between the magnetic field generation rate and the so-called Rossby number Ro defined as the ratio between the rotational period $P = 2\pi/\Omega$, and the convective turnover time τ_{conv}: $Ro = P/\tau_{conv}$. The Rossby number is a key parameter that measures the efficiency of magnetic field generation: it determines how strongly the Coriolis force affects the path of the convective bubbles. Small values of Ro mean that stars are rotating fast enough to ensure that the Coriolis force largely affects convection.

It has been observed that activity correlates better with Ro than rotation [108, 109]. In cool stars, the average magnetic field, usually defined as $f B_s$, the product of the so-called "filling factor" f that denotes the relative fraction of the stellar surface covered by active regions, and the corresponding value of the average magnetic field strength B_s, increases almost linearly with $1/Ro$ until it saturates at $Ro \approx 0.1$, corresponding to a rotation period of about 2–3 days for solar-like stars.

When analyzing this trend as a function of also the spectral type one finds that this change of the trend of stellar activity with rotation occurs in correspondence of the spectral type M2–M3. It is interesting (see Section 3.2) that the spectral type domain M2–M3 encompasses the mass range where stars are expected to become fully convective $M \sim 0.35 M_\odot$, and that most M dwarfs are characterized by high rotation velocity. This empirical evidence has some important implications: (i) VLM stars (i.e. stars with spectral type later than M2–M3) despite the lack of an interface layer between radiative and convective regions, where dynamo processes could operate according to the standard $\alpha\Omega$ scenario, show strong magnetic fields; (ii) the strength of the magnetic field seems to be uncorrelated with the rotational state. This implies that the physical processes at the origin of the magnetic field in these stars should not be significantly related to rotation.

To solve existing shortcomings of the "$\alpha\Omega$ dynamo" model mainly related to the properties of the magnetic field components generated by this mechanism, a slight modification has been suggested, named the "$\alpha^2\Omega$ dynamo generation" scenario or the advection-dominated dynamo. It is essentially the standard dynamo scenario with the inclusion of a contribution coming from the turbulent velocity field associated to convective motions [110]. A promising scenario for the generation of

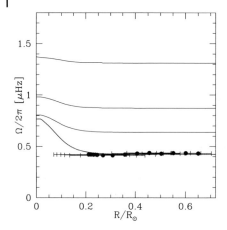

Figure 2.17 Rotation profile of a SSM computed including the contribution provided by magnetic braking at the surface. The various lines represent the rotation profile at (from top to bottom) 0.5, 1, 2 Gyr and at the present solar age of 4.57 Gyr. An initial velocity of 20 km s^{-1} has been adopted. The points with the error bars refer to helioseismic solar rotation measurements [116] (courtesy of P. Eggenberger).

a magnetic field in a fully convective stars, the so-called "α^2 dynamo" model can be sketched. The helicity required to generate the magnetic field is produced as a consequence of the action of the Coriolis force on the convective motions in a rotating, stratified fluid. Although the strength of the magnetic field generated with this process should depend on the rotational rate, the α^2-dynamo should become supercritical, that is, saturate, at large rotational velocities, Rossby numbers lower than about ten, in agreement with the observational scenario [111].

One of the major issues in the treatment of the magnetic field in stellar conditions is related to the stability of the magnetic field itself. This topic has been extensively investigated and is far from settled, given that all analytical solutions for magnetic stability predict that there is no stable configuration [112].

One of the more important evolutionary effects related to the presence of magnetic fields is to induce solid body rotation, due magneto-hydrodynamical processes that oppose differential rotation. The most important of these instabilities for the redistribution of the angular momentum is the so-called "Tayler–Spruit instability" [113, 114]. In the case of the Sun, this instability, together with or instead of IGWs, can contribute appreciably to reduce differential rotation, as displayed in Figure 2.17 [115].

2.11
Thermohaline Mixing

As discussed in detail in Section 3.7, canonical stellar models predict that during the first dredge up along the RGB, the surface chemical composition of low-mass stars is modified due to dredge up into the convective envelope of H-burning processed material. After this mixing episode, canonical models do not predict any further mixing episode along the RGB. Several spectroscopic observations of metal-poor Galactic halo stars provide, however, compelling evidence of an additional mixing process occurring when RGB stars reach the luminosity of the so-called "bump" in the RGB luminosity function. It is observed as a sudden drop of the isotopic ratio $^{12}C/^{13}C$ as well as a decrease of the Li and C abundances, and a slight increase of the N abundance [117]. This "non-canonical" mixing seems to be a universal process as it affects $\sim$ 95% of low-mass stars, regardless of whether they populate the halo field or clusters. Recently, a new mechanism has been proposed as a possible source for this mixing episode: the molecular weight inversion due to the ^{3}He-burning through the $^{3}He(^{3}He, 2p)^{4}He$ nuclear reaction, in the outer wing of the H-burning shell, and the associated thermohaline mixing.

In Galactic halo stars, the main burning mechanism during the MS is the p–p *chain* that, due to its weak dependence on temperature, is efficient also in stellar layers quite far from the star centre. As a consequence, ^{3}He accumulates in a broad zone outside the main energy production region. During the first dredge up, this ^{3}He is mixed within the convective envelope, with the consequence that during the following RGB evolution, the layers above the H-discontinuity left over by the receding convective envelope at its maximum extension will have a uniform ^{3}He, larger than the initial one.

When the shell advances towards the surface during the RGB, in the outer wing above the point of maximum burning efficiency, there is a narrow region where ^{3}He is burnt through the reaction $^{3}He(^{3}He, 2p)^{4}He$. This nuclear reaction is unusual in the sense that two nuclei transform into three and the mean mass per nucleus decreases from three to two. Because the molecular weight is the mean mass per nucleus, this leads to a small local decrease of μ [118]. As long as the H-burning shell moves through layers below the H-discontinuity, that is, the star evolves before the RGB bump, this local negative μ gradient (negative gradient in the sense that locally μ decreases with increasing pressure when moving towards the centre of the star) is negligible because the shell is moving in a region with a much larger positive gradient due to the He profile left over at the end of the MS. However, when the H-burning shell enters the region of the uniform chemical profile beyond the H-discontinuity, the local inversion, that is, the negative gradient in the μ profile of the order of one part in 10^4 becomes much more important. This situation corresponds to the conditions for thermohaline mixing [119–121]. This mixing is usually included in stellar evolution codes as a diffusive process that works in the direction to erase the molecular weight inversion, with diffusion coefficient given by [122]

$$D_{th} = C_{th}\frac{K}{c_P\rho}\left(\frac{\varphi}{\delta}\right)\frac{-\nabla_\mu}{(\nabla_{ad} - \nabla)} \quad \text{for} \quad \nabla_\mu < 0 \tag{2.51}$$

where K denotes the thermal conductivity, $C_{th} = 8/3\pi^2\alpha^2$ with α a free parameter, $\varphi = (\partial \ln \rho/\partial \ln \mu)_{P,T}$, and $\delta = -(\partial \ln \rho/\partial \ln T)_{P,\mu}$. Values of $C_{th} \sim 1000$ are required to reproduce the chemical abundance pattern in halo RGB stars [119, 123]. Evolutionary calculations show that thermohaline mixing extends between the out-

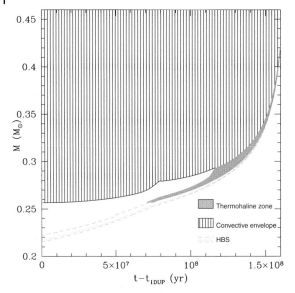

Figure 2.18 A Kippenhahn diagram (location of mass boundaries as a function of time) for a 1.25 $M_\odot$ star computed, including thermohaline mixing (but neglecting any rotation-induced mixing). The hatched area denotes the convective envelope, the dashed lines bracket the shell H-burning region. The thermohaline instability operates within the shaded area. The thermohaline zone develops soon after the H-burning shell reaches the H-discontinuity left over by the first dredge up at $t - t_{1DUP} \approx 7 \times 10^7$ years, and reaches the convective envelope at $t - t_{1DUP} \approx 1.15 \times 10^8$ years (Courtesy of C. Charbonnel and N. Lagarde).

er wing of the H-burning shell and the inner boundary of the convective envelope, merging with the outer convection in a short time ($\sim$ 30 Myr for a model with mass of the order of $1 M_\odot$) as shown in Figure 2.18. Depending on the mixing efficiency, a significant amount of nuclear processed matter in the hotter layers of the H-burning shell can be dredged up to the surface, and reproduce the observations.

However, hydrodynamical simulations predict a different expression for the diffusion coefficient, reasonably reproduced by Eq. (2.51) with $C_{th} \sim 10$ [124, 125]. The expected efficiency is much lower than needed to reproduce RGB spectroscopic data.

We also stress that, indeed, a μ-discontinuity is always produced if a H-burning shell is active inside a chemically homogeneous layer, and hence also during the HB and AGB [126]. Thermohaline mixing brings ^{3}He down from the convective envelope into the burning region of the advancing shell and this leads to a rapid decrease of its surface abundance. Depending on the value of D_{th}, ^{3}He can even be exhausted at the end of the RGB phase as previously discussed. There is empirical evidence showing that some RGB stars can avoid extra-mixing during the RGB. For these stars, the ^{3}He reservoir is intact at He-ignition, and thermohaline mixing can potentially play an important role during the subsequent HB and AGB phases by affecting the surface chemical abundances. Actually, the occurrence in low-mass stars of thermohaline mixing during the AGB would provide an additional mixing mechanism able to connect the outer layers of the H-burning shell with the convective envelope. This process is sometimes denoted as the *cool bottom process*. Its occurrence would explain the observations of some Li-rich stars that appear to be AGB stars, but with no other evident signatures of the occurrence of the *third dredge up* (see Chapter 5).

3
From the Main Sequence to the Tip of the Red Giant Branch

3.1
Overview

In this chapter, we will discuss in detail the structure and evolution of stars with mass lower than about $2.5 M_\odot$ during both the core and shell H-burning phases. The hydrogen burning stage is one of the most important evolutionary phases for all stars, regardless of initial mass and chemical composition. There are several fundamental reasons why this evolutionary stage is so important as:

- It is the longest evolutionary phase and as a consequence the number of H-burning stars observed is much larger than the number of stars in any other phase. H-burning stars are therefore a fundamental component of stellar populations observed in both resolved and unresolved stellar systems;
- The structure and evolution of a star during central and shell H-burning phases determine its evolutionary properties during the following phases;
- The brightness of the termination of core H-burning stage is the fundamental "clock" provided by stellar astrophysics, and its use is the basis of several investigations of the formation and evolution scenario(s) for galaxies;
- The colour of shell H-burning RGB sequences in the CMD is traditionally employed as a proxy of spectroscopic measurements in order to estimate the initial heavy element abundance of stellar systems;
- The brightest portion of the shell H-burning phase provides a standard candle, the brightness of the tip of the RGB;
- Star counts along the faint MS of stellar systems enable the determination of their IMF.

In Chapter 2, we discussed the characteristics of the nuclear mechanisms involved in the H-burning process. During this phase stars evolve with nuclear timescales, orders of magnitude larger than the thermal Kelvin–Helmholtz timescale dictated by the virial theorem. This has the important consequence that after a star attains the H-burning temperature, it is able to "forget" its previous PMS evolution, even before any change in its chemical stratification are induced by nuclear burning. This is why stellar model computations often start from the core H-burning stage by assuming a chemically homogeneous first model.

Old Stellar Populations, First Edition. S. Cassisi and M. Salaris.
© 2013 WILEY-VCH Verlag GmbH & Co. KGaA. Published 2013 by WILEY-VCH Verlag GmbH & Co. KGaA.

When describing the evolution of a star along the MS, we assume that this phase starts with a characteristic configuration denoted as the "Zero Age Main Sequence" (ZAMS). On the ZAMS, the star is in hydrostatic equilibrium, fully supported by H-burning, with all secondary chemical elements involved in the active nuclear processes at their equilibrium abundances, and a negligible fraction of produced He.

3.2
Very Low-Mass Stars

A minimum stellar mass is required to ignite hydrogen, the so-called "H-burning minimum mass" (HBMM). Below this critical value, hydrostatic equilibrium is guaranteed by the electron degeneracy pressure [127]. We will name as very low-mass (VLM) stars those objects whose mass is lower or equal to $\sim 0.5 M_\odot$, and denote objects with mass below HBMM as "brown dwarfs".

VLM stars have been very important in the stellar astrophysics field because before the results from the micro-lensing survey were published, they were considered together with white dwarfs and substellar objects as the main contributors to the dark matter budget in the Galaxy. Nowadays, there is a renewed interest in these stellar structures due to the fact that they are suitable candidates for exoplanets searches.

3.2.1
The Physical Properties

The thermodynamical conditions of VLM stars are extreme. The central temperature of the Sun is of the order of 10^7 K, compared to 4×10^6 K for a VLM star with mass equal to about the HBMM value, for example, $\sim 0.1 M_\odot$ at solar chemical composition. The central density increases from $\sim 100 \, \text{g cm}^{-3}$ for the Sun to $\sim 500 \, \text{g cm}^{-3}$ in a $0.1 M_\odot$ star. The temperature at the photosphere decreases from 6000 to ~ 2500 K, while the density at the same location increases from $\sim 10^{-7}$ to $\sim 10^{-5} \, \text{g cm}^{-3}$.

In VLM stars Γ attains values in the range between 0.1 and 30, meaning that Coulomb interactions among ions are important in the evaluation of the EOS. In Section 2.2.1, we have shown that at large densities (typical of VLM stars), pressure dissociation and ionization have to be accounted for in the EOS, together with formation/dissociation of molecules, mostly H_2. The electron degeneracy parameter η in VLM stars is of the order of unity and increases significantly, thereby decreasing the stellar mass, implying that in the lowest mass tail of the VLM range, there are conditions of partial electron degeneracy.

We can therefore describe a VLM stellar structure as a star where molecular H, atomic helium, and many other molecules (see below) are stable in the atmosphere and in the outer envelope layers, while the stellar interiors are mainly formed by a fully ionied H/He plasma; Coulomb interactions are important in the whole mass regime as well as the pressure ionization/dissociation process, while electron degeneracy increases with decreasing mass and becomes very important for a mass of about $0.12 M_\odot$ (see Figure 2.4).

It is evident that, due to the complex thermodynamical conditions present in these objects, the evaluation of a reliable EOS has been for long time a thorny problem, and only in recent times an accurate EOS suitable for stars in the VLM regime has been computed [26].

Another important problem is related to the complicated task of computing accurate model atmospheres and spectra for these objects that are necessary to provide both bolometric corrections and boundary conditions for the model computations. Difficulties in VLM model atmosphere computations arise, once again, from the peculiar physical properties of these stars. Cool temperatures and high pressures of the outer stellar layers make the opacity evaluations extremely complicated because of the presence of a huge number of molecules. Simply by looking at a stellar spectrum, one easily realize that, around and below $T_{eff} \approx 4000$ K, molecules such as H_2O, CO, VO and TiO are very important. In more detail, TiO and VO control the energy flux in the optical wavelength range, while H_2O and CO are the dominant opacity sources in the infrared window of the spectrum [128]. When the effective temperatures attain values lower than ~ 2500–2800 K, the process of grain condensation becomes extremely important, increasing the difficulty to compute reliable opacities, and hence model atmospheres and spectra.

In addition, as discussed in detail in Section 2.4, VLM stars share with WDs an important opacity source. Due to the large density and pressure in the outer layers, CIA on H_2 molecules becomes a major contributor to the radiative flux absorption. When H_2 CIA is efficient, as for stars with mass lower than $\sim 0.2 M_\odot$, it suppresses the flux longward of $2\,\mu m$ and causes the redistribution of the emergent radiative flux towards shorter wavelengths, that is, for suitable choices of the photometric bands, VLM stars tend to appear "bluer" with decreasing mass [129].

An additional problem is the determination of accurate outer boundary conditions. As discussed in Section 2.5, it is customary to adopt the boundary conditions provided by a grey model atmosphere. In the same section, we have provided evidence showing how this assumption provides results in good agreement with those based on more accurate and sophisticated model atmospheres. However, we have also noted that, strictly speaking, the "grey model atmosphere" approximation is valid only when all the following conditions are fulfilled: (i) presence of an isotropic radiation field, (ii) radiative equilibrium, and (iii) the radiative absorption is independent on the photon frequency. In VLM stars, the strong frequency-dependence of the molecular absorption coefficients yields synthetic spectra which severely depart from a frequency-averaged energy distribution, and so the last condition is not satisfied. More importantly, below ~ 5000 K, molecular hydrogen recombination in the envelope (H + H $\rightarrow$ H$_2$) reduces the adiabatic gradient. This occurrence, coupled with the large radiative opacity, and hence a large radiative gradient, favours convective instabilities in the atmosphere. As a consequence, convection deeply penetrates into the optically thin layers, and the condition of radiative equilibrium

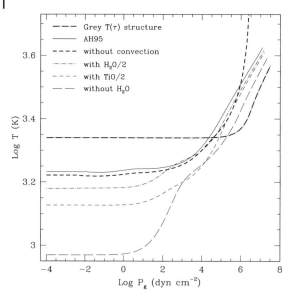

Figure 3.1 The thermal stratification in the atmosphere of a VLM model with $T_{eff} = 2800\,K$, $\log g = 5.0$, and solar chemical composition, as predicted by a grey $T(\tau)$ model and a detailed non-grey model atmosphere [128]. The effects associated to changing the opacity contribution by H_2O and TiO, as well as of completely inhibiting the convection are also shown (courtesy of F. Allard).

is no longer satisfied[1] [130, 131]. Boundary conditions for VLM models therefore need to come from accurate, non-grey model atmospheres [132]. Usually, the pressure from model atmosphere calculations is taken at an the optical depth $\tau \approx 100$. At this depth, the diffusive approximation for the opacities is adequate, although stellar models are almost completely unaffected by any choice of τ larger that unity [133].

Figure 3.1 shows temperature as a function of pressure in the atmospheric layers of a VLM model, as provided by both a grey $T(\tau)$ relation and an accurate, non-grey model atmosphere. For values of the optical depth larger than unity, the thermal stratification provided by a grey atmosphere is cooler and denser than the corresponding predictions provided from non-grey models; the effect is larger with decreasing metallicity as a consequence of the fact that metal-poor atmospheres are denser than more metal-rich ones. Roughly speaking, the difference in the derived outer boundary condition has the consequence that in order to match the internal solution, an atmospheric stratification corresponding to a larger T_{eff} has to be adopted with the grey approximation. One can then expect that this approximation produces VLM models with a hotter T_{eff}. The estimated ΔT_{eff} is $\sim 100\,K$ for a $\sim 0.4 M_\odot$ model.

---

1) This shortcoming is, at least partially, avoided using a grey $T(\tau)$ integrated (in τ) only in layers stable against convection.

Table 3.1 Selected properties of VLM stellar models with $Z = 10^{-3}$. The ages are in years, mass fractions of H and ^{3}He are also listed.

$M/M_\odot$	$\log t_H$	$\log t_{nucl}$	$\log t_{(^3He)}$	X/X_{in}	$^3He^{eq}$	Structure
0.70	10.4	7.5	7.5	1.00	3×10^{-4}	Radiative core
0.60	10.6	7.7	7.7	1.00	6×10^{-4}	Radiative core
0.50	10.9	8.0	8.5	0.99	1×10^{-3}	Radiative core
0.40	11.2	8.2	9.2	0.97	3×10^{-3}	Radiative core
0.30	11.6	8.5	10.2	0.95	7×10^{-3}	Fully convective
0.20	11.9	8.6	10.7	0.94	2×10^{-2}	Fully convective
0.15	12.1	8.7	11.1	0.90	4×10^{-2}	Fully convective

An exclusive (discounting the PMS stage) property of VLM structures is that below a certain mass, they are fully convective. This is due to: (i) the large radiative opacity of both interior and outer layers that strongly increases the radiative gradient ($\nabla_{rad} \propto \kappa$), and (ii) the decrease of the adiabatic gradient in the envelope due to the formation of molecules. The transition mass between fully convective models and structures with a radiative core is around $\sim 0.35\,M_\odot$, the exact value depending on the stellar metallicity (see Table 3.1).

It is also important to note that this convection is largely adiabatic because the large densities of these stars make convective transport quite efficient. This is an advantage in the VLM star modelling with respect to the case of more massive stars because the VLM stellar models do not depend on the value of the free parameter α_{ml} entering in the mixing length theory for the treatment of the superadiabatic layers. On the other hand, the depth of the convective envelope or, for fully convective stars, the thermal stratification of the structure, is affected by the choice of the outer boundary conditions [20]. Therefore, the choice of the outer boundary conditions affects the T_{eff} and more global properties of the VLM stellar models.

As for the H-burning processes, due to the low central temperatures, the important reactions are

$$p + p \rightarrow D + e^+ + \nu_e$$

$$p + D \rightarrow {}^3He + \gamma$$

The process of destruction of ^{3}He, ^{3}He(^{3}He,^{4}He)$^2 p$ is effective only for $T > 6 \times 10^6$ K, meaning that ^{3}He behaves as a pseudo-primary element for long time, and only structures more massive than about $0.15\,M_\odot$ are able to achieve its equilibrium in an Hubble time. This has the important consequence that the definition of the ZAMS for VLM stars is largely meaningless because they spend a huge fraction of their H-burning lifetime without attaining the equilibrium configuration for ^{3}He, as shown in Figure 3.2.

The characteristic timescales for this process are listed in Table 3.1, which also reports the trend with mass of the total H-burning lifetime (t_H), the age of the

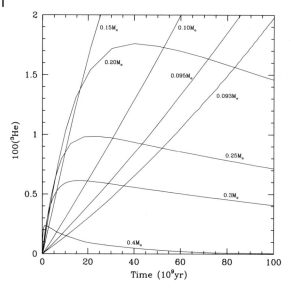

Figure 3.2 The evolution with time of the ^{3}He mass fraction (multiplied by a factor of 100) at the centre of selected VLM models with $Z = 0.0003$ and $Y = 0.23$.

models when the gravitational energy contribution vanishes compared to the nuclear output (t_{nucl}), the stellar age when ^{3}He equilibrium is attained ($t_{^3He}$), the ratio between the central abundance of H when equilibrium is achieved, and the initial one, as well as the abundance of ^{3}He at equilibrium.

3.2.2
Structure and Evolution

Figure 3.3 shows the time behaviour of selected parameters for two VLM stellar models with mass equal to 0.4 and $0.3 M_\odot$, respectively. The $0.4 M_\odot$ model behaves like models with moderately larger masses populating the upper portion of the MS, as it will be discussed in the next section.

The increase of the ^{3}He abundance towards its equilibrium value increases the efficiency of the H-burning[2] and the structure reacts, decreasing both central temperature and density, which start increasing again only when the equilibrium value for ^{3}He has been attained.

On the other hand, fully convective structures such as the $0.3 M_\odot$ model behave quite differently, and the central density keeps decreasing all along the major phase of H-burning. It starts suddenly increasing again only in the very last phases of central H-burning, when the increased abundance of He and the corresponding decrease of radiative opacities induces the formation of a radiative shell that rapidly grows, eventually forming a radiative core.

---

2) We noted in Section 2.3.1 that $p-p$ *chain* is more efficient when it achieves equilibrium.

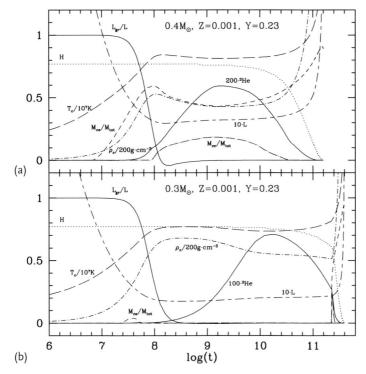

Figure 3.3 The variation with time of the surface bolometric luminosity (L) in solar units, the contribution of gravitational energy to the total luminosity (L_{gr}/L), central density (ρ_c), central temperature (T_c), size of the convective core (M_{cc}) as fraction of the total mass, mass location of the bottom of the convective envelope (M_{ce}), and central abundances by mass of H and ^{3}He for two models at the transition to convective structures: $0.4 M_\odot$ (a) and $0.3 M_\odot$ (b), with $Z = 0.001$ and $Y = 0.23$.

Figure 3.4a displays the effect of age on the HRD of VLM models. One can easily recognize that age plays a negligible role on the HRD location of stars below $\sim 0.5 M_\odot$. It is curious that the $0.3 M_\odot$ model appears the least affected by age, less massive models showing an increasing sensitivity to age. One can understand the reason for this behaviour by considering, as shown in Figure 3.2, that the $0.3 M_\odot$ model at 10 Gyr has already achieved the ^{3}He equilibrium abundance at the centre, and the evolution is governed by the depletion of central H only (with a very long timescale). On the contrary, less massive models keep increasing the central abundance of ^{3}He, an occurrence that affects the *p–p chain* efficiency, readjusting the structure accordingly. This readjustment in the less massive models is amplified by the decreasing electronic degeneracy induced by the decrease of central density.

Figure 3.4b shows the HRD of VLM models with an age of 10 Gyr for selected metallicities. As a consequence of the increased opacity, the sequences become cooler and fainter with increasing heavy element abundance. We wish to empha-

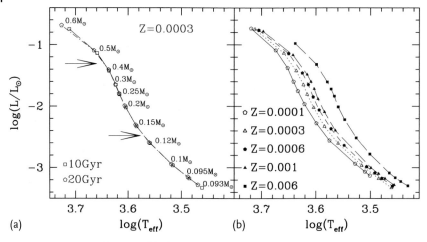

Figure 3.4 (a) HRD of VLM models for two ages and chemical composition: $Z = 0.0003$, $Y = 0.23$. The arrows mark the location of the two bending points discussed in the text. (b) HRD diagram of VLM stars for various metallicities.

size the sinuous shape of the MS of VLM stars that shows two well-defined bending points denoted by the two arrows:

- The brighter point at $T_{eff} \approx 4500$ K corresponds to a mass equal to $\sim 0.45\,M_{\odot}$. The physical process at the origin of this change in the MS slope is molecular hydrogen recombination. With decreasing total mass, H_2 formation becomes efficient in the outer stellar layers. This causes a decrease of the number of free particles in the plasma, and hence a decrease of the adiabatic gradient to $\nabla_{ad} \sim 0.1$ (to be compared with ~ 0.4 valid for a perfect, monatomic gas). The decrease of the adiabatic gradient induced by H_2 recombination produces a flatter temperature gradient hence a larger effective temperature compared to the case without H_2 formation [29]. With decreasing metallicity, this bending point shifts to larger T_{eff} values, that is, larger mass, because the outer layers of metal-poor stars are denser,[3] an occurrence that favours the formation of molecular hydrogen. Given that the location of this point in the HRD is dependent on the thermal properties of the envelope, comparisons with observations represent a formidable benchmark for the reliability of the VLM EOS;
- The faintest bending point is located at $T_{eff} \approx 2800$ K and corresponds to an evolutionary mass of $\sim 0.15\,M_{\odot}$. Its presence is due to the increased level of electron degeneracy. As a consequence, the contribution of the degenerate electron pressure to the total pressure increases, and this will eventually produce the well-known mass–radius relation for fully degenerate objects, that is, $R \propto M^{-1/3}$ (see Chapter 6).

---

3) The lower the metallicity, the lower the opacity; given that $dP/d\tau = g/\kappa$ in the atmosphere, for a fixed gravity, at the same optical depth, the pressure and the density are larger in more metal-poor stars.

One can therefore expect that with decreasing metallicity, this point moves towards larger T_{eff}, and then larger stellar mass. Clearly, the location of this point in the HRD is more affected by the opacity that controls the pressure stratification in the outer layers than by the EOS.

We have already defined the HBMM, the minimum stellar mass that can attain thermal equilibrium supported by H-burning. Detailed numerical computations based on the most accurate available physics for VLM stars have shown that for solar metallicity, the HBMM is equal to $0.075 M_\odot$ [132, 134]. The value of HBMM increases with decreasing metallicity; for instance, it is equal to $0.083 M_\odot$ for $Z = 0.0002$. This is due, as already discussed, to the fact that at fixed total mass, the opacity decreases with decreasing metallicity and the structures become denser, increasing the level of electron degeneracy, a process that opposes the achievement of the thermal conditions required by H-burning ignition.

3.2.3
The Colour–Magnitude and the Mass–Luminosity Diagrams

The location of the MS for VLM stellar models in optical CMDs is shown in Figure 3.5 compared to empirical measurements for both field dwarf stars with well-known parallaxes and a galactic GC. This figure shows that up-to-date VLM stellar models do a good job in matching the observed distribution of metal-poor stars both in the field and in star clusters. However, when considering models at solar chemical composition, there is a serious disagreement between theory and observations. This disagreement is related to shortcomings of the bolometric corrections, probably associated to specific radiative opacity contributions[4] at wavelengths shorter than $\sim 1\,\mu m$.

It is important to note that the morphology of the MS for VLM stars changes significantly when moving from optical wavelengths to near- and far-infrared ones, as shown in Figure 3.6. The most striking evidence is that starting from a few magnitudes below the TO, the MS is almost vertical in these CMDs before suddenly shifting to bluer colours with decreasing stellar mass.

This is related to the competition between the tendency of the star colour to become redder due to the decreasing effective temperature and increasing radiative opacity in the optical, and the increasing efficiency of CIA of H_2 in the infrared that redistribute the flux to shorter wavelengths [135]. This competition leads to quasi-constant colour sequences from a mass of $\sim 0.5 M_\odot$, corresponding to $T_{eff} \sim 4500\,K$, at which the molecular hydrogen recombination starts being efficient, down to $\sim 0.1 M_\odot$. For stellar masses below this limit, CIA becomes the dominant process as a consequence of the larger densities,[5] and this produces the

---

4) There are some indications that this problem could be related to the TiO line list, but there could be the contribution from other molecules as CaOH as well. The same VLM models for solar composition that do not match the data for field star in the solar neighbourhood, do actually fit well data for stars in the Galactic bulge in the near-infrared bands, as shown in Figure 3.6.
5) As discussed in Section 2.4, H_2 CIA is proportional to the square of the density.

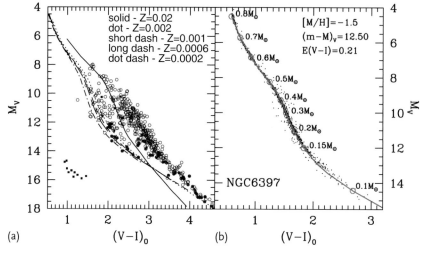

Figure 3.5 (a) The CMD in optical bands for field VLM structures with known parallaxes, with superimposed theoretical VLM models for the labelled values of stellar metallicity. (b) The HST CMD of the MS locus of the Galactic GC NGC 6397 compared with VLM models for a suitable metallicity.

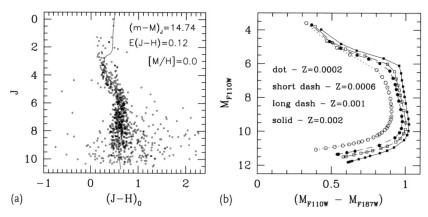

Figure 3.6 (a) The near-infrared CMD of the stellar population in a field of the Galactic bulge compared with model predictions for solar metallicity [136]. (b) CMD in the NIC-MOS *F110W* and *F187W* *HST* filters for stellar models with mass equal or lower than $0.8M_\odot$, and selected metallicities.

blue "excursion" at the very bottom of the MS. This blue excursion becomes more evident with decreasing metallicities due to the larger density that characterizes the atmosphere of metal-poor stars.

Figure 3.7 shows the *mass–luminosity* relations for selected photometric bands ranging from V photometric band to the K band. The *mass–luminosity* relation becomes more and more insensitive to the stellar metallicity when going from the visual to the near-infrared photometric bands, with the *mass–M_K* relation being

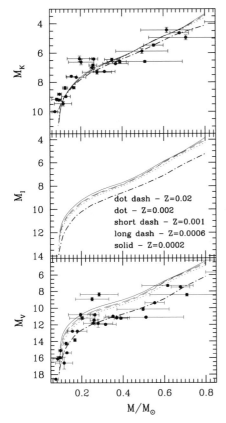

Figure 3.7 The theoretical *mass–luminosity* relation in different photometric bands for various assumptions on the heavy-element abundance, compared to observational data [137].

almost unaffected by metallicity for masses below $\sim 0.4 M_{\odot}$. This is due to the fact that below $T_{\text{eff}} \sim 4000$–4500 K, the opacity in the optical bands, dominated by TiO and VO, increases with metallicity so that the peak of the energy distribution is shifted towards larger wavelengths, in particular to the K band. This process causes a decrease of the flux emitted in the V band and an increase in the K band with increasing metallicity. On the other hand, for a given mass, T_{eff} decreases with increasing metallicity, and the total flux ($F \propto T_{\text{eff}}^4$) decreases. In the K band, these two effects balance each other, and hence the *mass–luminosity* relation in K band is almost independent of the heavy element abundance. This trends holds as long as the H_2 CIA does not largely suppress the flux emitted in the K band as it occurs in the more metal-poor VLM stars with mass just above the HBMM.

The *mass–magnitude* relations also show some inflection points clearly associated to the same physical processes producing the bending points in the HRD. The presence of these points has to be properly taken into account when these *mass–magnitude* relations are used for retrieving the IMF of a stellar population because what is relevant is the first derivative of these relations with respect the stellar mass [138].

3.2.4
The Mass–Radius Relation

Thanks to the recent improvements in the observational facilities, the radii of many VLM stars have been determined accurately. The study of eclipsing binaries, interferometric measurements with the Very Large Telescope Interferometer, and transit observations from micro-lensing surveys, for example, MACHO and OGLE, have provided a large sample of reliable radii and mass measurements for VLM and low-mass stars. Figure 3.8 shows the comparison between theoretical predictions and observed radii for stars in the range of interest: good agreement seems to exist. There is, however, also empirical evidence that stellar models for sub-solar masses underestimates stellar radii, probably for stars with high levels of magnetic activity [139].

3.3
The Main Sequence

All stars leave the PMS stage when they attain the H-burning temperatures. All along the PMS – when neglecting the very small contribution to the energy budget coming from the nuclear burning of light elements such as Li and Be – the main energy source is the release of gravitational energy by contraction according

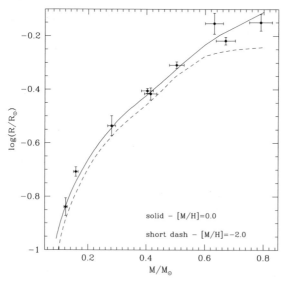

Figure 3.8 The *mass–radius* relation for low-mass and VLM stellar structures for two metallicities [140]. The theoretical data are compared to selected observations [141].

to the virial theorem. When moving to the MS, the contribution to the energy budget coming from nuclear burning increases progressively to become the dominant one. As a consequence, there will be an "intermediate" phase between PMS and ZAMS, during which the star attains the chemical equilibrium of the secondary elements.

Due to the dependence on temperature of the different H-burning processes, in stars with $M \leq 1.3 M_\odot$ (the exact value depending on the initial chemical composition), the main H-burning mechanism is the *p–p chain*. In more massive stars, the dominant H-burning process is the *CNO cycle*.

It is customary to separate MS stars into two distinct groups on the basis of the dominant H-burning process: those stars where the *p–p chain* is the main burning mechanism are denoted as *Low Main Sequence* stars, whereas those structures mainly controlled by the *CNO cycle* are denoted as *Upper Main Sequence* stars.

As a general rule, when electron degeneracy is negligible, a star self regulates its thermonuclear burning rate to maintain hydrostatic equilibrium. Therefore, if, for any reason, the rate of nuclear reactions is larger than needed, the star reacts by expanding, decreasing the temperature and density so that the burning rate decreases and equilibrium is restored. The opposite happens when the nuclear reaction rate is too low.

3.3.1
Low Main Sequence Stellar Models

Let us start by discussing the properties of low MS stars. It is instructive to describe in some detail the approach of these stars to the ZAMS. Near the end of the PMS, due to the lack of ^{3}He nuclei, the dominant reaction is that of ^{3}He production. As the nuclear reactions proceed, the abundance of this element increases, and the number of ^{3}He(^{3}He,^{4}He)2p reactions also increases. This allows the star to complete the *ppI* branch and, for the achieved thermal conditions, to produce more energy per burnt H nucleus.

However, during the previous phase, with only a partially efficient *p–p chain*, the star was forced to reach a higher central temperature and density to meet its energy needs. Now, with a more efficient *p–p chain*, the star slightly decreases the core density and temperature, adjusting the number of nuclear reactions to the exact value required by its energy needs. This situation lasts until ^{3}He equilibrium is achieved. This description provides a useful guideline for interpreting the trend with time of various physical and chemical quantities for a $0.8 M_\odot$ model from the PMS up to central exhaustion of hydrogen, as shown in Figure 3.9.

During the phases preceding ^{3}He equilibrium, a small convective core is present, as shown in Figure 3.9, as a consequence of the fact that for a short time, ^{3}He production is larger in the central portion of the star, and due to the large local energy flux, F ($\nabla_{rad} \propto F$), the central regions become convective. This convective core vanishes as soon as the abundance of ^{3}He is increased through the nuclear burning in surrounding regions, as this redistributes the energy flux over a larger volume, and lowers the local flux in the core.

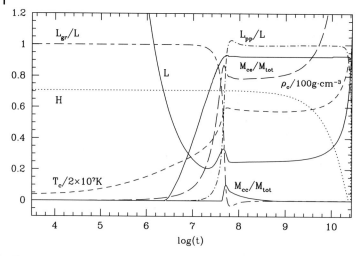

Figure 3.9 As in Figure 3.3, but for a $0.8 M_\odot$ model with solar chemical composition. The time evolution of the fraction of the total luminosity produced by the *p–p chain* (L_{pp}) and the gravitational contribution (L_{gr}) are also shown.

This is the only stage when a low-mass MS star has a convective core. In fact, due to the weak temperature dependence of the *p–p chain* (see Section 2.3.1), H-burning always involves a large fraction of the star. This distribution of the energy flux over a large fraction of the star keeps the radiative gradient lower than the adiabatic one. As a consequence, the H-burning via *p–p chain* always occurs in a radiative core.

To display the large mass extension of the *p–p chain* burning regions, we show in Figure 3.10 the fraction of energy production as a function of the mass coordinate M_r at different stages during the MS, and the associated evolution of the hydrogen chemical profile.

While interiors are in radiative equilibrium, low-mass stars with T_{eff} lower than ~ 8000 K always have a convective envelope associated with the partial H and He ionization regions. Its mass extension increases as the total mass decreases until, as discussed in Section 3.2, for masses below $\sim 0.4 M_\odot$, the stars become fully convective. Convection in the outer envelope is superadiabatic, with the consequence that its efficiency, and hence the radius of the stellar models, depends on the adopted value for the free parameter α_{ml}.

The HRD of low MS stellar models is shown in Figure 3.11. One can notice that along this evolutionary phase, the surface luminosity and effective temperature (and radius) steadily increase. It is instructive to investigate the reasons for this behaviour.

During H-burning, the average number of particles per unit mass decreases, while the mean molecular weight increases in the stellar interiors. Hydrostatic equilibrium can be maintained only by an increase in density and temperature. Given that the nuclear burning efficiency is temperature sensitive, rising central

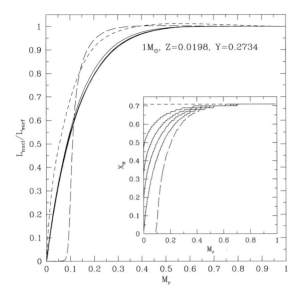

Figure 3.10 The energy produced via nuclear burning per unit time, normalized to the total surface luminosity, in a $1M_\odot$ model with solar chemical composition, at different stages during the MS. The short-dashed line corresponds to the ZAMS, while the long-dashed line corresponds to the end of the core H-burning stage. The inset shows the corresponding hydrogen abundance profiles.

temperatures lead to an increase of the surface luminosity. An additional, although secondary, effect contributing to the stellar luminosity increase is related to the change of the radiative opacity in the stellar interiors because $\kappa_H > \kappa_{He}$.

As for the evolution of the effective temperature, the trend can be explained as follows. Those layers where the molecular weight is increasing most rapidly show a more pronounced tendency to contract; when the molecular weight changes are spread over a large portion of the star, the fraction of the structure that contracts is larger, and the rate of increase of radius relative to the rate of luminosity increase is smaller (note the location of the lines of constant radius in Figure 3.11).

As a consequence of the steady increase of the temperature in the interiors, the *CN cycle* becomes efficient in producing energy. The stage at which this occurs clearly depends on both the total stellar mass, earlier for larger masses, and the chemical composition. As a rule-of-thumb, a larger initial He content favours the *CN cycle* because the interiors are hotter due to the larger molecular weight; the same happens when increasing the metallicity or, more specifically, the abundance of CN elements because obviously the efficiency of the *CN cycle* depends on the abundance of the catalysts. For instance, in metal-poor low MS stars, the *CN cycle* becomes efficient just before the end of the MS. It is worth noting that, in metal-poor, low-mass stars, only the *CN cycle* is activated during the core H-burning stage because the *ON cycle* requires temperatures larger than about 16×10^6 K.

The central H-burning stage ends when the hydrogen is completely exhausted at the centre. This phase roughly corresponds to the hottest point on the tracks

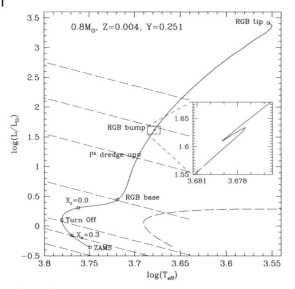

Figure 3.11 HRD evolution of a low MS model from the ZAMS to the RGB tip. The labels mark selected evolutionary stages discussed in the text. The inset shows a enlargement of the RGB bump stage. The long-dashed line corresponds to the evolutionary track of the same mass computed by setting the value of the mixing length parameter to zero (see text for details). The thin-dashed lines represent the loci of constant radius for $R/R_{\odot} = 0.6$, 0.8, 1.0, 2.0, 5.0, 10 and 20, respectively, from bottom to top.

shown in Figure 3.11, the so-called *Turn Off* (TO). It marks the end of the central H-burning phase in low-mass stars, and is the most important observational feature for dating stellar populations (see Section 3.9 and Chapter 7 for a detailed discussion on this issue).

Evolutionary lifetimes along the central H-burning phase are long for low MS stars, ranging from more than 120 Gyr (about an order of magnitude larger than the age of the Universe) for a $0.5\,M_{\odot}$, to ~ 20 Gyr for a $0.8\,M_{\odot}$ and ~ 10 Gyr for a $1\,M_{\odot}$. As one can easily see, the MS lifetime is a strong function of the stellar mass, decreasing as the mass increases: as a first order approximation, one can assume that $t_H \propto M^{-2.5}$.

At central H-exhaustion, the burning already involves a significant portion of the surrounding regions (see Figure 3.10) and the core contracts slightly while H-burning continues in a shell around a He-core. In other words, the transition from core burning to H-burning in a thick shell occurs smoothly and uneventfully.

3.3.2
Upper Main Sequence Stellar Models

We have already emphasized that near the end of the core H-burning stage, low MS stars reach the internal temperatures required to be the *CN cycle* efficient. Upper MS stars have $M > 1.2$–$1.3\,M_{\odot}$ – and are supported by the *CN cycle* already at the

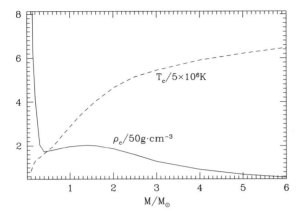

Figure 3.12 Central temperature and density (in cgs unit) as a function of the stellar mass at the beginning of the core H-burning stage, for models with $Z = 0.001$ and $Y = 0.23$.

beginning or in the very early phases of the MS. When further increasing the total stellar mass, the stars also become able to activate the *NO cycle* to complete the full *CNO cycle*. At solar chemical composition, in a $1.5 M_\odot$ model, the *CNO cycle* contributes 70% of the nuclear energy budget at the centre and 50% of the total luminosity, whereas a $1.8 M_\odot$ star is almost completely powered by the *CNO cycle*.

Figure 3.12 shows values of the central T and ρ as a function of the stellar mass at the ZAMS stage. The change in the behaviour of both temperature and density when moving from the low MS to the upper MS regime is explained by the different temperature sensitivity of the *p–p chain* compared to the *CNO cycle*.

The evolutionary tracks of selected upper MS stellar models are shown in Figure 3.13. While the luminosity is steadily increasing as for the case of low MS stars, the effective temperature is almost monotonically decreasing, but at the central H-exhaustion (see below). The burning occurs inside a convective core, and the mass extension of this convective core decreases with time as a consequence of the decreasing interior opacity. In stars slightly more massive than the transition mass between low MS and upper MS stars, the mass extension of the convective core actually increases as central temperature and density increase. This is because the *CN cycle* contribution to energy production increases with respect the *p–p chain* contribution, and energy production becomes more concentrated toward the star centre (see Figure 3.14). Eventually, the size of the convective core attains a maximum and thereafter decreases as the effect of decreasing interior opacity overwhelms the effect of the increasing central concentration of the energy source.

Upper MS stars have a radiative envelope because the regions of partial ionization of H and He are located in atmospheric layers where the density is too low to affect the global thermal properties of the envelope.

Due to the presence of a convective core, hydrogen is homogeneously burnt in the whole convective core, as shown in Figure 3.14, and the end of the central H-burning stage coincides with the exhaustion of the "fuel" in the whole convective zone. At this stage, the star reacts with a fast contraction, the so-called *overall*

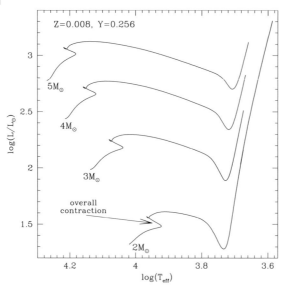

Figure 3.13 HRD for selected upper MS models from the ZAMS to the RGB tip. In the case of the $2M_\odot$ model, the location of the *overall contraction* phase is indicated.

contraction phase, that in the HRD corresponds to a T_{eff} increase before the track moves back towards decreasing effective temperatures. Figure 3.15 displays the evolution with time of selected physical properties of an upper MS stellar model.

A relevant complication in the computation of stellar models for upper MS stars is related to the location of the "true" boundary of the convective region (see Section 2.1). The case for a significant increase of the size of the convective core beyond the Schwarzschild (or Ledoux) boundary has been made many times, from both theoretical and observational points of view. However, we still lack accurate theoretical prescriptions for modelling this phenomenon. This problem is even more crucial when calculating models for more massive stars, with larger convective cores.

In stellar model computations, the extension of the mixed region beyond the Schwarzschild boundary is usually defined in terms of a parameter λ_{OV} that denotes the length – expressed as a fraction of the local pressure scale height H_{P} – mixed by travelling gas elements beyond the Schwarzschild convective boundary.

Regardless of how this "overshooting" is accounted for in evolutionary computations, the effects on the star evolutionary and structural properties can be easily predicted: (i) the star becomes brighter because the increase of the mean molecular weight, related to the conversion of H into He, involves now a larger fraction of the structure, (ii) the core H-burning lifetime increases, as there is more fuel available for nuclear burning, (iii) at the H-exhaustion, the mass size of the He-core is much larger and this will affect the properties of the models during the central He-burning phase; in particular, the star will be brighter, and will spend a shorter time in this stage.

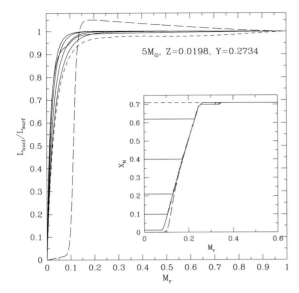

Figure 3.14 As in Figure 3.10, but for an upper MS (intermediate-mass) stellar model (see labels).

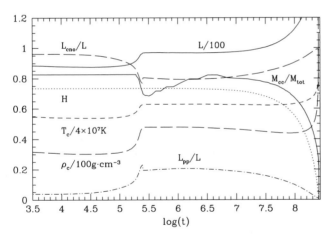

Figure 3.15 The evolution with time of selected physical and chemical quantities for a $3\,M_\odot$ stellar model with $Z = 0.008$ and $Y = 0.256$, from the approach to the ZAMS to the exhaustion of central hydrogen. The ZAMS is reached at $\log(t) \sim 5.4$. The gravitational energy contribution (not shown) is negligible during these evolutionary stages, but for the overall contraction phase, near the upper limit of the time range covered by the diagram.

Another important issue to be addressed is the value of λ_{OV} when stars have small convective cores. In fact, it is well-known that the morphology of the evolutionary tracks and isochrones at the TO depends on how the extension of the mixed regions decreases with decreasing mass.

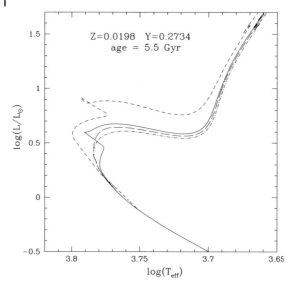

Figure 3.16 Theoretical isochrones (at fixed age and chemical composition) computed with different ways to decrease the core overshooting efficiency with stellar mass: the dot-dashed line corresponds to models without overshooting; the solid line corresponds to models calculated with $\lambda_{OV} = 0.2H_P$ for stars with $M \geq 1.7M_\odot$, and $\lambda_{OV} = (M/M_\odot - 0.9)/4H_P$ for M be-tween $1.7M_\odot$ and $1.1M_\odot$; the short-dashed line corresponds to $\lambda_{OV} = 0.20H_P$ for masses larger than $1.3M_\odot$, linearly decreasing to zero down to $M = 0.95M_\odot$; the long-dashed line corresponds to $\lambda_{OV} = 0.20H_P$ for masses larger than $1.7M_\odot$, $0.15H_P$ at $1.5M_\odot$, down to zero at $1.1M_\odot$ and linearly decreasing in the intermediate range.

When moving deeper inside the star, the pressure scale height steadily increases; this causes a large increase of the size of convective cores in stars whose Schwarzschild convective boundary is fast shrinking, (e.g. for masses below $\sim 1.5M_\odot$) if the overshooting efficiency is kept fixed at a constant fraction of H_P. It is clear the need to decrease λ_{OV} to zero for stars with small convective cores (see [142], [143]). Figure 3.16 shows some isochrones for the same chemical composition and age, obtained using different assumptions about the trend of λ_{OV} with the stellar mass. The change of the isochrone morphology is quite significant and different choices concerning the core overshoot efficiency in the critical mass range $1.1 \leq M/M_\odot \leq 1.5$ mimic different isochrone ages. This occurrence has the obvious consequence that the trend of λ_{OV} with mass, for masses with small convective cores, potentially introduces an additional degree of freedom in stellar evolution models. Even if this problem only affects a restricted range of cluster ages, it has to be taken into account when discussing the uncertainties affecting the comparison between theory and observations.

3.4
The Standard Solar Model

Accurate calculations of models for the Sun have traditionally been one of the fundamental goals of stellar astrophysics for a number of reasons: (i) the Sun is the star for which we can collect the most precise measurements of many fundamental parameters such as the mass, luminosity, radius and age; (ii) the identification of the solar energy sources was a pivotal step towards the solution of the long-standing solar neutrino problem (see below); (iii) it offers an almost unique opportunity for testing the stellar evolution theory and for calibrating some free parameters needed in stellar model computations such as the efficiency of superadiabatic convection; (iv) the study and "interpretation" (via the *inversion technique*) of the spectrum of non-radial solar oscillations represents one of the most important results of asteroseismology (helioseismology in the case of solar studies), and a proof of the importance of this discipline for investigating the properties of stars, as recent asteroseismic surveys such as KEPLER are showing [144].

It is now customary to talk about the *Standard Solar Model* (SSM). *A SSM is a stellar evolution model that best reproduces the observed properties of the Sun.*

Let us start considering a canonical SSM, calculated without including any macroscopic mixing beyond the Schwarzschild boundary of the envelope, nor magnetic fields and rotation, but accounting for atomic diffusion. During the evolution, the initially homogeneous chemical composition is modified as a consequence of nuclear reactions and atomic diffusion. Indeed, the early SSMs were computed by neglecting atomic diffusion because its effect was thought to be negligible. It was later shown that diffusion is actually important for solar modelling and for the predicted solar oscillation spectrum [77, 145].

The basic observational constraints that any SSM must satisfy are the solar mass value $M_\odot = 1.989 \times 10^{33}$ g, the surface luminosity $L_\odot = 3.486 \times 10^{33}$ erg/s, and radius $R_\odot = 6.9599 \times 10^{10}$ cm, at the solar age[6] $t_\odot = 4.57$ Gyr. In addition, the predicted surface chemical composition should be consistent with the observed photospheric composition. Unfortunately, we have no spectroscopical measurements of He in the Sun for the photosphere is not hot enough to produce He lines in the spectrum, and in addition helium is largely lost by the meteorites.

On the other hand, the photospheric ratio $(Z/X)_\odot$ between heavy element and hydrogen abundance can be measured in principle with high accuracy, and has to be reproduced by SSMs [147]. For several years, it has commonly adopted the value $(Z/X)_\odot = 0.0245 \pm 0.001$, though recently this value has been strongly questioned and modified (see Section 3.4.4).

To match the observed properties of the Sun, the parameters involved in the computations of the solar models have to be adjusted. One calculates the evolution of an initially homogeneous solar mass model up to the solar age and compares the model predictions with the empirical constraints. The initial He abundance Y_{ini} is

---

6) The age of the Sun is estimated by dating the oldest meteorites, in the assumption that the formation time of the meteorites and the birth of the Sun are coeval [146].

a free parameter. Once Y_{ini} is fixed, the value of the initial global metallicity Z_{ini} must be also chosen in order to obtain the observed present value of $(Z/X)_{\odot}$. The value of the $\alpha_{ml,\odot}$ also has to be calibrated to match the observed solar radius. In practice, the three parameters: Y_{ini}, Z_{ini} and $\alpha_{ml,\odot}$, are correlated and must be evaluated simultaneously via an iterative procedure. The effects of individual changes of each one of these parameters can be understood easily: the luminosity of the Sun, and more generally of any MS stars, strongly depends on the initial helium content Y; increasing Y, the initial Sun is brighter and a given MS luminosity is reached in a shorter time. Given that the surface abundance ratio (Z/X) is constrained by the observations, Y_{ini} and Z_{ini} cannot be chosen independently: if Y_{ini} increases, Z_{ini} must decrease.

It is clear that any change of the input physics, that is, of the EOS or opacity, would affect the calibration of these parameters. Just to provide typical values for these free parameters in a modern SSM (but with the old estimate for the $(Z/X)_{\odot}$), we report the following values [148]:

$$Y_{ini} = 0.2734 , \quad Z_{ini} = 0.0193 , \quad \alpha_{ml,\odot} = 1.913$$

3.4.1
The Helioseismic Constraints

A new era in investigations of the structure of the Sun was heralded by the advent of accurate measurements of solar oscillation frequencies [149]. The period of these oscillations is in the range between 3 and 15 min, the modes have lifetimes up to several months, and their cyclic frequencies ω have been determined with very high accuracy, typically $\Delta\omega/\omega \sim 10^{-5}$. The solar oscillations are essentially standing sound waves (p-modes) that can be described by spherical harmonics Y_l^m and, for each couple of m, l values, by a radial order n. In the assumption of spherical symmetry, the oscillation frequency only depends on n and l. These oscillations can be considered to a good approximation to be adiabatic, though in the layers near the solar surface.

Helioseismic observations measure the frequencies ω of solar p-modes, and when these measurements are coupled to SSMs, the properties of the solar structure are inferred by means of an "inversion method". In a nutshell, when we wish to estimate the value of a given quantity Q in the Sun, the inversion method works as follows:

- One starts with a solar model with certain values of Q_{mod} and an associated set of oscillation frequencies. In general, these frequencies will be different from the measured values $\omega_{\odot} \pm \Delta\omega_{\odot}$;
- One determines the corrections q to Q_{mod}, to provide model frequencies $\omega_{mod} + \Delta\omega$ that match the observed $\omega_{\odot}$. Expressions for $\Delta\omega = \Delta\omega(q)$ are derived using perturbation theory, where the starting model is used as a 0th order approximation. The correction factors q are then computed, assuming some regularity properties, so that the problem is mathematically well-defined and/or unphysical solutions are avoided;

- The "helioseismic value" $Q_\odot$ is thus determined by adding the starting value and correction: $Q_\odot = Q_{mod} + q$.

Accurate SSMs are always required to deconvolve the information embedded in the solar oscillation spectrum. At the same time, the predicted oscillation spectrum is a critical benchmark for any model of the Sun and stellar evolution models in general.

Helioseismic analyses provide the helium abundance in the photospheric layers Y_{ph} whose value is in the range 0.226–0.260, the location of the base of the convective envelope $R_{bce}/R_\odot = 0.710$–0.716, and the sound speed profile within the Sun. The uncertainties on both Y_{ph} and R_{bce} are due to the uncertainty in the frequency measurements, and more importantly to the errors in the inversion technique as well as to differences in the adopted reference SSMs.

SSMs without atomic diffusion do not match the helioseismic estimate of the photospheric He abundance and the location of the lower boundary of the outer convection zone, and also show a larger disagreement with the inferred sound speed profile. This occurrence has been the strongest evidence of the efficiency of atomic diffusion in the Sun, as radiative levitation has a negligible impact [152].

Figure 3.17 shows the ratio $U = P/\rho$ as a function of the distance from the star centre derived from two SSMs obtained using two different EOSs. The range of values allowed by helioseismic measurements are also displayed [150]. The quantity $U = P/\rho$ can be directly determined by the inversion technique and is related to the adiabatic sound speed c_s by the relation $c_s^2 = \Gamma_1 U$, where Γ_1 is the adiabatic exponent $\Gamma_1 = (d\ln P/d\ln \rho)_{ad}$. Neglecting the atomic diffusion causes a disagree-

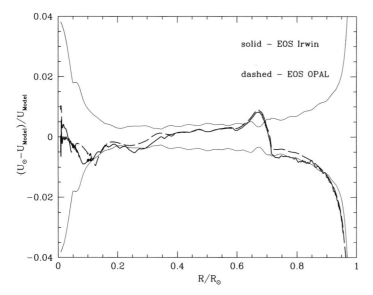

Figure 3.17 Relative differences between the helioseismic solar $U = P/\rho$ profile [150] and the results from two SSMs computed using the OPAL EOS [151] and the FreeEOS (A. Irwin), respectively. Thin solid lines marks the "conservative" error on the helioseismic result.

ment between the helioseismic data and the non-diffusive SSM predictions at the level of 1–1.5% in the solar layers between $r/R_\odot = 0.3$ and ~ 0.7.

3.4.2
Standard Solar Models and the Solar Neutrino Problem: a Solved Conundrum

Standard solar models predict that the temperature at the centre of the Sun is $\sim 15 \times 10^6$ K. Therefore, the *p–p chain* dominates ($\sim 99.6\%$) over the *CNO cycle* ($\sim 0.4\%$). The relative importance of the nuclear reactions involved in the *p–p chain* leads to the result that the majority of the neutrinos produced in the Sun should come from the $p(p, e + \nu_e)$D reaction and be of relatively low energy, while only a minor fraction of the total neutrino flux is expected to come from the ^{7}Be ($\sim 7\%$) and a negligible fraction from the ^{8}B decay ($\sim 0.0075\%$).

The relative contribution to the whole energy budget provided by the various nuclear reactions involved in the *p–p chain* and the associated neutrino fluxes, depends on the rates of the nuclear reactions involved, which are a strong function of density, temperature and chemical composition profile. Therefore, it is evident that the measurements of the neutrino flux coming from the various nuclear branches represent a formidable tool for investigating the deeper solar layers. At the same time, it represents a critical benchmark for SSMs and more generally for stellar evolution models.

The experimental study of the solar neutrinos dates back to more than 40 years ago, namely, to the Davis experiment [153]. This experiment, based on the interaction between an electron neutrino and ^{37}Cl, has a threshold energy of roughly 0.8 MeV and could essentially only detect the neutrinos coming from the ^{8}B decay that constitute a minor fraction of the solar neutrino flux. The main result of the Davis experiment was a neutrino flux equal to roughly one-third of the predicted value.

The discrepancy became more evident thanks to the GALLEX and SAGE experiments [154, 155]. These experiments, designed to detect the bulk of the solar neutrinos produced by the main $p + p$ reaction, confirmed the existence of a severe discrepancy between theoretical and observed fluxes.

This result plunged solar astrophysics into a crisis because these experiments were expected to measure the total number of electron neutrinos emitted by the Sun. This number is a robust prediction of theory because it only relies on the assumption that the solar luminosity comes from the conversion of protons into helium, and not on the details of the internal structure of the Sun or the cross sections of the nuclear reactions involved. If the solar luminosity is powered by the conversion of protons into helium, the total number of electron neutrinos emitted per second by the Sun must be $(2.38 \times 10^{39}/25) \times 2$, that is, the ratio of the energy emitted by the solar surface per unit time and the energy released per each He nucleus produced, multiplied by the number of electron neutrinos per nucleus of He produced. On the other hand, the combined GALEX and SAGE results showed that the measured neutrino flux was just an half of the predicted value.

For several years, many attempts were made to reconcile the predicted neutrino flux with observations. The suggested solutions were based on a change of the physical ingredients used in SSM computations, or on non-canonical physics [156, 157]. The solution arrived when it was shown that neutrinos oscillate among three different species: ν_e, ν_μ and ν_τ. Given that all quoted experiments could only detect the electron neutrinos, they missed the other two species of neutrinos that reach Earth. Recently, the Sudbury Solar Neutrino Observatory (SNO) experiment designed to detect all three neutrino flavours has clearly shown that, indeed, there is a fine agreement between the observed and predicted neutrino fluxes [158].

3.4.3
Non-standard Solar Models

In spite of the success of the class of SSMs described in the previous section, there are observations that cannot be matched without invoking additional physical processes. These observations are the solar internal rotation profile already discussed in Section 2.8, the lithium depletion in the solar photosphere, and the evidence that helioseismology suggests a smoother He abundance gradient below the bottom of the convective zone.

The photospheric lithium abundance is about 160 times lower than that measured in meteorites: [Li] $= 1.05 \pm 0.10\,\text{dex}$[7] [159]. This difference between the current solar and protosolar values is not predicted by standard stellar evolution models. One also has to note that the lithium abundances of solar-like stars in the solar neighbourhood spread over more than two orders of magnitude, a much larger range than for other elements.

This wide range of observed [Li] in nearby solar-like stars is most likely due to a correlation between [Li] and the star age and mass. We recall that Li is easily destroyed by proton capture reactions in stellar interiors. If lithium is transported between the outer convection zone and deeper regions with temperatures high enough for Li destruction, the photospheric abundance will decrease with time. Atomic diffusion probably contributes to the lowering of the surface Li abundance throughout the MS stage. This would explain why the photospheric solar abundance is much smaller than the meteoric one. We expect an enhanced lithium depletion in stars with larger MS convection zones as well as in stars with a higher degree of differential rotation between the radiative core and the convective envelope (see discussion in Section 2.8). The reason is that Li is only depleted as it moves to deeper and therefore hotter regions of a star, where the temperature is high enough ($T \sim 2.5 \times 10^6$ K) for proton captures to become efficient.

Whatever the physical mechanism responsible for this extra-mixing, it has to not only satisfy the helioseismic constraints, but also the observational evidence that the ratio ^{3}He/^{4}He does not seem to have increased from the early stages of

---

7) We define a generic chemical abundance $[X]$ as $[X] = \log(n_X/n_H) + 12$, where n_X and n_H are the number densities of element X and hydrogen, respectively.

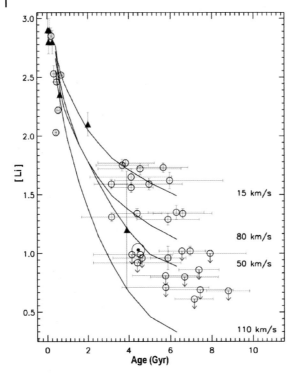

Figure 3.18 [Li] versus age for solar twins and in open clusters with solar metallicity [160]. The ⊙ point shows the location of the Sun. Solid lines represent the values predicted by models that include both rotation and IGW [161] for various assumptions about initial rotational velocities (see labels) (courtesy of P. Baumann).

the solar evolution until now, putting a strong constraint on the efficiency of any mechanism envisaged to explain the surface Li depletion.

In recent times, several mechanisms have been suggested to explain this Li depletion: convective overshoot, rotation-induced mixing, internal gravity waves. Each one of these processes can satisfy an observational constraint, but fails in reproducing the others. However, it seems that models accounting for both the presence of rotation-induced mixing and internal gravity waves are able to reproduce both the Li depletion in the Sun and the spread in the abundance of this element among solar-like stars[8] (see Figure 3.18).

Concerning the solar rotational profile, we have already shown in Sections 2.8 and 2.9 that rotational induced mixing(s) and internal gravity waves can significantly contribute to the redistribution of angular momentum in the Sun, but a contribution to this process coming from the braking induced by the presence of a magnetic

8) Recently, it has been suggested that the presence of a planet around a star could affect the Li depletion process, in the sense that planet-host stars could have experienced more Li depletion than stars without planets. However, this claim is not supported by analysis based on larger stellar sample including stars both with and without planets [160].

field also seems to be necessary to achieve the observed flat rotation profile in the core of the Sun.

3.4.4
The Metallicity of the Sun and the "Solar Abundance Problem"

Modern solar models including atomic diffusion reproduce the sound speed profile determined from seismic inversions to within $\sim 0.4\%$, as well as the seismically-inferred depth of the convection zone and the helium abundance in the convective envelope. This situation has changed due to recent new analyses of the solar spectrum that have revised downward the solar heavy element abundance, particularly carbon, nitrogen, oxygen and neon, the more abundant metals that contribute to the opacity just below the convective zone [162].

The first revision (denoted as AGS05 [162]) determined a decrease of $\sim 35\%$ in the C abundance, of $\sim 27.5\%$, in the N abundance, while O and Ne are decreased by 48%, and by 74%, respectively, compared to the "reference" metal mixture used in SSMs ([147] denoted as GS98). The abundances of elements from Na to Ca were lower by 12 to 25%, and the iron abundance was lowered by 12%. In the case of the GS98 abundances, the ratio of metals to hydrogen mass fraction $(Z/X)_\odot = 0.023$, and the global metallicity is $Z \sim 0.018$, while for AGS05, $(Z/X)_\odot = 0.0165$, and $Z \sim 0.0122$.

SSMs computed with the new AGS05 composition are in disagreement with helioseismic determinations of the convective envelope boundary and surface He abundance. In addition, the predicted sound speed profile is in worse agreement with helioseismic results compared to models with the GS98 mixture. These discrepancies have raised the so-called "solar abundance problem".

The main cause for the revision of the solar abundances is the use of a 3D model atmosphere (in place of classical 1D models) and improved non-LTE corrections. The main effect of these reduced metal abundances is a reduced opacity in the radiative interior, and considerable effort has been devoted to modify the physics inputs of SSMs (i.e. atomic diffusion efficiency, radiative opacities). There is, however, little physical justification for the modifications required to match the helioseismic constraints [163–165]. A more recent analysis ([166] denoted as Caf11) with independent 3D model atmospheres has determined an oxygen abundance intermediate between the GS98 and AGS05 values, while a reanalysis of the AGS05 results has revised the O abundance ([167] denoted as GASS10) that is now consistent with Caf11 results within the 1σ uncertainties.

Table 3.2 lists the solar heavy element abundances for these four mixtures, whilst Figure 3.19 displays the predicted SSM sound speed profiles, and Table 3.3 compares the predicted positions of the convective envelope boundary and envelope He abundances with the helioseismic constraint.

The best agreement is for the GS98 abundances, although the Caf11 results produce comparable SSMs in spite of the lower metallicity. The agreement between SSMs based on the Caf11 solar mixture and helioseismic constraints can be further improved by increasing $\sim 40\%$ the neon abundance in the solar mixture.

Table 3.2 Recent estimates of the solar heavy element abundances (see text for more details). Abundances are expressed as $[X] = \log(n_X/n_H) + 12$. The abundances given in parenthesis have been obtained from [168].

Element	GS98	AGS05	GASS10	Caf11
		log ϵ		
C	8.52	8.39	8.43	8.50
N	7.92	7.78	7.83	7.86
O	8.83	8.66	8.69	8.76
Ne	8.08	7.84	7.93	(8.05)
Na	6.32	6.27	6.27	
Mg	7.58	7.53	7.53	(7.54)
Al	6.49	6.43	6.43	
Si	7.56	7.51	7.51	(7.52)
S	7.20	7.16	7.15	
Ar	6.40	6.18	6.40	(6.50)
Ca	6.35	6.29	6.29	
Cr	5.69	5.63	5.64	
Mn	5.53	5.47	5.48	
Fe	7.50	7.45	7.45	7.52
Ni	6.25	6.19	6.20	

Table 3.3 Comparison between helioseismic estimates and theoretical predictions from SSMs with different solar heavy elements mixtures about the position of the lower boundary of the convective envelope, the photospheric- and the initial He abundance [163].

Solar mixture	Z/X	$R_{bce}/R_\odot$	Y_{ce}	Y_{in}
GS98	0.0230	0.7154	0.2464	0.2768
AGS05	0.0165	0.7272	0.2296	0.2601
GASS10	0.0181	0.7225	0.2363	0.2666
Caf11	0.0209	0.7166	0.2425	0.2725
			Helioseismology	
		0.7133 ± 0.0005	0.2485 ± 0.0035	$0.273 \pm 0.006 \pm 0.002$

3.5
Blue Stragglers

Blue Straggler (BS) stars were first discovered in the Galactic GC M3 [169]. Nowadays, they are found virtually in all stellar systems, for example, the Galactic field, open and globular clusters as well as Local Group dwarf galaxies. BS stars are brighter and bluer than the MS TO of the bulk stellar population, and appear linger or straggle in their evolution, hence their name *Blue Stragglers*. The evidence that

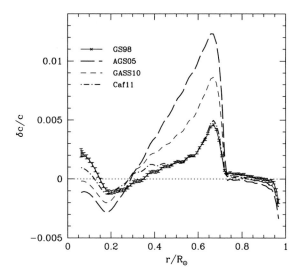

Figure 3.19 Relative difference between the helioseismic sound speed and SSM predictions based on the various solar heavy elements mixtures discussed in the text (courtesy of H.M. Antia and S. Basu).

BSs are found in different environments suggests that their production is efficient regardless of the properties of the stellar environment.

Their origin has to be related to binary system evolution and collisional processes involving single stars and/or binaries (or a combination of these scenarios). A thorough analysis of the origin and evolution of BSs can help elucidate how stars in binary systems evolve, as well as how the dynamical processes efficient in dense stellar systems can affect the evolution of isolated, non-interacting stars [170].

The BS observational and theoretical scenario has significantly changed in the last years. For many decades after their early discovery, these stars were only detected in the field, in open clusters, and in the outer regions of GCs; an evidence that generated the idea that low-density environments were their natural habitats. However, this was just an observational bias and, starting from the early 90s, mainly thanks to the *HST*, high-resolution studies discovered BSs also in the highly-crowded central regions of GCs.

A powerful tool has been the use of ultraviolet (UV) photometric bands. Whereas optical CMDs of old stellar populations are dominated by the cool stellar component, in the UV bands, both MS, SGB, and RGB stars are fainter, and BSs appear among the brightest objects, defining a narrow, nearly vertical sequence, as shown in Figure 3.20 [171].

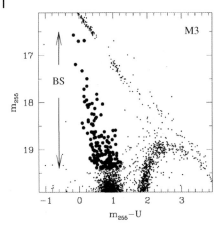

Figure 3.20 The CMD in UV filters for the GC M3. The cluster BS population is represented with big filled circles (courtesy of F. Ferraro).

3.5.1
Formation Channels

On the basis of their observational properties as well as some direct measurements, BSs are more massive, up to a factor $\sim$ 2, than stars currently evolving at the TO [172]. This implies that any evolutionary channel devised to explain the BS origin has to provide a mechanism for increasing the original mass of these stars.

There is a general consensus about the idea that BSs are the products of stellar mergers between two or more low-mass stars. However, the physical mechanism triggering the merging event is still under debate. So far, there are two proposed scenarios:

- Coalescence[9] of a binary system [173]. For a binary system to coalesce, Roche lobe overflow must occur that triggers a mass transfer from the outer envelope of an evolved donor onto the envelope of its companion. Since this mass transfer can only occur when the radius of the primary component becomes larger than the radius of its Roche lobe, this process depends on the evolutionary status of the donor. It is common to refer to BSs originating via this process as *primordial* BSS, where the term "primordial" means that this process is not affected by the dynamical evolution of the parent stellar system.

  These primordial BSs are mainly formed in relatively wide binaries [174]. Rather than having the two stars merged via angular momentum loss, in these systems, a BS is formed when the primary evolves off the MS and fills its Roche lobe. Mass transfer onto the secondary (still on the MS) may then produce a BS. Obviously, very tight binaries, that is, a binary system with a very short separation, can also merge via a common envelope phase, but this is expected to only produce a small fraction of the BS population, at least in the Galactic field.

9) Stars share a common envelope or have completely merged.

- Collision or merger in high-density environments as GC cores.[10] Two different channels can be envisaged: (i) mass transfer in head-on collisions between two single stars, and (ii) mass transfer occurring in low-velocity encounters with an impact parameter corresponding to nearly grazing collisions.

  The second mechanism would act in two separate steps: (i) the two stars first become gravitationally bound into a binary system due to the energy dissipation during the encounter. (ii) Once the binary system is formed, it coalesces violently [175]. One has to note that even in the densest GC cores, the cross section for these process is extremely small, and in addition, the constraints on the initial relative velocity and impact parameter are extremely restrictive. A possible way-out for this problem is to invoke a dynamical interaction between a pre-existing (primordial) binary system whose cross section for dynamical interaction is larger than for single star interactions, and a third, unbound, star. This kind of interaction would produce a reduction of the separation (hardening) of the two binary components drastically accelerating the coalescence process. Regardless of the fact that the process requires the interaction between single stars or a single star and a binary, BSs formed as a consequence of dynamical processes are usually called *collisional* or *dynamical* BSs.

In principle, it is possible that both scenarios are at work simultaneously, but with a relative efficiency strongly dependent on the properties of the parent stellar system. For instance, one can easily realize that almost all BS stars observed in the Galactic field should be derived from the evolution of primordial binary systems.

Concerning the effects that the cluster environment can have on primordial binaries, one can make some speculations. Binaries could simply be broken up by close stellar encounters; however, this occurrence is unlikely, as most binaries are hard, for example, their binding energy is larger than the kinetic energy of any incoming star. Another possibility is that exchange encounters could alter the system, for instance, if a neutron star is exchanged into the system, this could form a low-mass X-ray binary. This process can surely occur and could explain the presence of at least a fraction of the low-mass X-ray binaries observed in Galactic GCs. An additional occurrence that could be important in the context of BS formation is that in high collision rate GCs, exchange encounters can efficiently produce binaries containing more-massive (with respect the initial components) MS primaries [176]. Once they evolve, mass transfer onto the secondaries produce BSs, but they will have been formed earlier with respect to those present in clusters having low collision rates (where the primary masses are lower). By today, most of the BSs formed in this way in high collision rate GC have evolved and then it would not be longer

---

10) In post-core-collapse GCs, the stellar density in the central regions can be larger than $\sim 10^5$ stars/pc^3. The core collapse is a catastrophic dynamical process consisting of the runaway contraction of the core of a star cluster. Binary-binary and binary-single collisions are thought to halt (or delay) the collapse of the core, thus avoiding infinite central densities. A common core-collapse observational signature is a steep cusp in the projected star density profile. Observational evidence suggests that $\sim 15\%$ of the GC population in the Galaxy has experienced this evolutionary stage.

observable. This occurrence would reduce the effective number of primordial BSs in high-density GCs with respect low collisional rate stellar systems.

Let us now discuss the expected differences in the evolutionary and chemical properties of BSs originated by different formation channels. This is crucial because the final structure and chemical composition profiles determine the observational properties and evolutionary tracks of the merger products.

In the case of coalescence of primordial binary systems, we start considering tight binaries that enter in a common envelope phase, a situation resembling that of full-contact W Ursae Majoris (W UMa) stars. In this case, the objects evolve with a mass transfer from the original secondary to the primary. The secondary component can be completely absorbed or alternatively the two stellar objects can coalesce in a prompt event. The chemical profile of the final stellar outcome is difficult to predict. One can envisage that in the case of a shallow contact, a small portion of the nuclear processed matter belonging to the secondary component could be mixed throughout the envelope of the final merged product. In the case of a rapid coalescence event, a large fraction of the nuclear processed matter in the core of the secondary star could remain "trapped" in the core of the final merged stellar object. Clearly, one expects that all matter belonging to the primary component remains trapped in the core of the new-born BS star. As a conclusion, in the case of primordial BSs, the surface may be enriched with matter processed by H-burning, but the surface He-enrichment should be limited. In addition, since only the less massive stars contribute to the surface He enhancement in the merger product, binaries with nearly equal mass should have more He enhancement than those with more extreme mass ratios.

In the case of a wide binary system, the mass transfer is modulated by the primary star filling its Roche lobe, and one can expect that once the primary has lost a significant amount of mass, the transfer stops. Also in this scenario, one can envisage that the newborn BS should display a chemical surface composition not extremely different from the original one.

For the case of collisional BS stars, it is essential to perform accurate, but very time-consuming, hydrodynamical simulations. The first investigations of this type concluded that in the case of head-on collisions, the mixing efficiency of the matter coming from the two interacting stars would be lower than that of grazing impact [177]. If two stars near the end of the central H-burning stage, with significantly He-enriched cores, collide, the He in the interiors would be more thoroughly mixed throughout the remnant in a grazing collision than in a head-on collision. In the same way, these simulations predicted that for unequal mass collisions, the more massive stars are thoroughly mixed, while the less massive star would settle in the core of the more massive one. This implies a larger envelope abundance of He and a mixing of H into the core of the remnant, with the consequence that the BS lifetime would be longer. It has been therefore commonly assumed that as a consequence of the induced larger envelope He abundance, collisional BSs would be bluer that the BSs formed by binary merger process, and for a given total mass collisional BSs should be brighter than primordial BSs.

However, more recent hydrodynamical simulations have disclosed a different scenario [178–180]:

- Significant mixing between core and envelope does not occur in stellar interactions involving equal-mass stars, regardless of whether the interaction is a head-on or a grazing collision, or a collisional induced binary merger. Grazing collisions and binary mergers result in slightly more mixing of helium into the envelope than head-on collisions. Grazing collisions and induced binary mergers produce remnants that appear to be extremely similar. In addition, it has been found that stars with less dense cores experience a larger mixing, but not enough to alter the subsequent evolution of the remnant;
- The amount of mass lost during the collision is a negligible fraction of the total mass of the remnant. A significant mass loss could, in principle, occur via the formation of a disk around the remnant; but if this is the case, the final colour of the remnants would be too red, in disagreement with observations;
- A still-unsettled issue is that related to the possible occurrence of a global mixing just after the remnant is formed. In fact, the input of kinetic energy from the interaction could drive an expansion of the remnant [181]. If this expansion is large, the star is driven to its Hayashi track where the convective envelope mixes much of the interior. In the case of a fully convective structure, the chemical abundances in the core would be mixed with the matter in the envelope and the result would be a chemically homogeneous stellar structure.

  Current simulations suggest that this mixing process involving the whole remnant structure should not occur for two reasons: (i) the injection of kinetic energy during the collision decreases the temperature gradient in the core and this helps quenching core convection, and (ii) the temperature in the interiors is large enough that matter is fully ionized and electron scattering is the dominant opacity source. This means that the radiative opacity in the core after the collision is low enough to disfavour the onset of convection.

  In addition, since the core is enriched in He, the scattering opacity is even lower than in PMS stars with the same physical conditions but a "normal" He content. If the remnant is massive enough to maintain hot enough temperatures during the expansion phase, the high He abundance in the core would prevent a surface convective region from penetrating into the central regions;
- The difference between "mixed" and "unmixed" remnants is substantial both in the morphology of the evolutionary tracks, and in the evolutionary lifetime as a BS star. Clearly, unmixed models have a shorter lifetime due to the lower central H abundance. As for the morphology of the evolutionary tracks, unmixed models start their core H-burning stage on the red side of the reference ZAMS, in contrast with mixed models.

  The most massive mixed models are able to reproduce the brightness and colour of the brightest and bluer BSs observed in Galactic GCs. However, they fail to reproduce both the reddest BSs as well as the observed spread in the BS colour. The most massive, unmixed models are able to match the brightness of the brighter BSs, but are in disagreement with the observed colour of the bluest

BSs, as well as the BS colour scatter. Less massive models for both the mixed and unmixed case show similar evolutionary tracks because of small amounts of He present in the core of the progenitor stars, with the tendency of unmixed models to be a shifted to the red with respect to the reference ZAMS;

- Empirical evidence suggests that real BSs spend a significant fraction of their lifetime away from the reference ZAMS. This supports the predictions of unmixed models. However, one has to bear in mind that these models cannot reproduce all the features of the BS distribution in the CMD; if unmixed models are the best candidates, the assumptions that collisional BSS have to be bluer and brighter than their primordial counterparts should be significantly revisited.

Another important issue related to BS formation involves their rotational velocity. In the case of BS stars originated by collisions, the merger product is expected to rotate rapidly, with a total angular momentum about a factor of 10 larger than low-mass MS stars, that is, of the order of $10^{51}\,\mathrm{g\,cm^2\,s^{-1}}$.

To date, we have few accurate measurements of rotation rates of BSs in clusters. Rotational velocities have been obtained for one BS in 47 Tuc ($v \sin i = 155 \pm 55\,\mathrm{km\,s^{-1}}$), one object in M3 ($v \sin i = 200 \pm 50\,\mathrm{km\,s^{-1}}$) and a few objects in the open cluster M67, with values ranging in the interval between 10 and 120 km s^{-1} [172, 182]. Despite the lack of direct empirical measurements, both the observed colour distributions and star counts of BSs in clusters show that they are not significantly affected by rotation-induced mixing as they would be in the case of large rotation rates.

We therefore face the problem of understanding how BS stars lose angular momentum in their pre-BS stage. It has been early suggested that this could occur via a magnetically induced stellar wind, but some analyses do not support this scenario due to the lack of a long-lived convective envelopes during the pre-BS stage that enable dynamo processes to generate the required magnetic field. Other scenarios can be envisaged to explain the angular momentum loss from an initially rapidly rotating BS. This could be achieved, for example, by angular momentum transfer to a circum-stellar disk, ejecta from the collision, or a nearby companion possibly captured during a binary interaction. Indeed, angular momentum transfer to a circum-stellar disk appears as an extremely efficient mechanism for slowing the rotation of stars as they contract to the MS [183]. This mechanism does require that stars have a convective envelope for the generation of a magnetic field, but its strength does not have to be as high as that needed for a magnetic-driven wind mechanism. The transfer of angular momentum to a nearby companion has the advantage that a convective envelope is not required and demands that many BSs (also those originating via a collision) were initially in a binary system, as it is indeed observed.

3.5.2
Specific Frequency and Radial Distribution

BSs in different environments could have formed via different evolutionary channels, as those previously discussed [184]. One could expect that the BS stars in
low density environments are produced by the coalescence of primordial binaries,
while in high-density GCs they arise mostly from dynamical interactions between
single stars and, more probably, between a single star and a binary system. However, there are strong observational findings supporting the idea that BS stars arising
from both formation mechanisms could actually coexist within the same cluster.

It is customary to define the BS star "specific frequency", F_{BSS}, as the ratio between the number of BSs in a given region of the cluster, to the number of "normal" single stars in the same region, typically ether RGB or HB stars. Extended
BS surveys have disclosed the existence of a large spread of F_{BSS} values in Galactic
GCs, as shown in Figure 3.21 [185, 186]. The parameter F_{BSS} has been extensively
used to disentangle the contributions of the various BS formation channels in the
various environments. The same Figure 3.21 shows the trend of the BS frequency
as a function of fundamental cluster properties as the total visual magnitude, the
central density and the stellar collision rate.[11]

It is interesting to note that the leftmost point in Figure 3.21a represents the BS
frequency in the Galactic field, $F_{BSS} \sim 4$, that is, about one order of magnitude
larger than the average specific frequency in Galactic GCs [174].

The data show a correlation between F_{BSS} and the integrated absolute magnitude M_V. Since the integrated luminosity is a proxy of the total clusters mass, this
correlation reveals that the faintest (less massive) clusters have a BS specific frequency up to a factor ~ 20 larger than that for the brightest (more massive) clusters. The BS specific frequency seems also to depend on the cluster central density
although the correlation is much less strong than for the cluster luminosity, and
it does not correlate with the collisional rate. One can also notice that the post-
core-collapsed GCs do not show any peculiar (or specific) trend. These features are
somehow puzzling. In fact, one could naively expect to observe more BSs in those
GCs where the probability of collision is higher, while the largest BS frequency is
indeed observed in the Galactic field.

The results seem to point out that there is not a single, dominant, evolutionary
channel for the BS formation. It could be that, in high-density environments, the
fraction of collisional BSs indeed increases but this occurrence is balanced by the
concomitant destruction of primordial binaries whose evolution could produce BS
stars. This scenario would consistently explain the larger frequency of BSs in the
field, where the collisional channel is ruled out. Clearly, the problem of the BS
formation in clusters is not yet a settled issue, but what is becoming more and
more evident is that the presence of primordial binaries, and how the evolution

---

11) The rate $\Gamma_\star$ of stellar collisions per year can be expressed as $\Gamma_\star = \rho^2 r_c^3 / \sigma$, where ρ is the central
cluster density in unity of solar luminosity per cubic parsec, r_c is the core radius in parsec, and σ
is the central velocity dispersion in km s^{-1}.

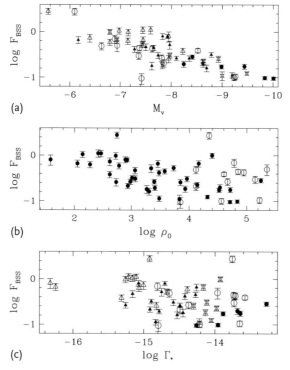

(a)

(b)

(c)

Figure 3.21 The relative frequency of BSs with respect to HB stars in Galactic GCs, as a function of the integrated absolute magnitude of the cluster (a), the central density (b), the collision rate (c). Different symbols are used for GCs with different central densities: $\log \rho_0 < 2.8$: open trian-gles; $2.8 < \log \rho_0 < 3.6$: filled triangles; $3.6 < \log \rho_0 < 4.4$ crosses; $\log \rho_0 > 4.4$: filled circles. The open circles in each panels mark the position of post core-collapse GCs. The leftmost point in (a) marks the value of the BS specific frequency in the Galactic field.

of their properties is modified by dynamical interactions in high-density stellar environments, are crucial ingredients [187].

It has been suggested that the BS Luminosity Function (LF) could show features connected with the BS formation process. This idea arises because early theoretical BS models predicted that collisional BSs should be bluer and brighter than primordial BSs as a consequence of their larger surface He abundance. Observationally, the results are not clear cut because data seem to support both the occurrence of a brighter peak in the BS LF of more luminous GCs compared to the less luminous ones and the lack of any brightness difference among GCs [186, 187].

On the other hand, a robust result is obtained from the analysis of the radial distribution of BSs within a given cluster. There is general consensus that the BS radial distribution in GCs is bimodal, as shown in Figure 3.22. A peak in the BS frequency is present at a few core radii (r_c), followed by a decrease to a minimum, and an upturn in the external regions. This trend is confirmed in almost all GCs. It was suggested early on that this bimodality could be the signature of the two

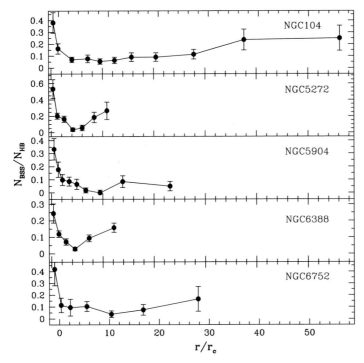

Figure 3.22 Radial distributions of the specific frequency N_{BS}/N_{HB}, as observed in five Galactic GCs. The minimum in the star counts coincides in all clusters, but for NGC 6388, with the location of the *radius of avoidance* (courtesy of E. Dalessandro).

formation mechanisms acting simultaneously in the same cluster; the external BSs would essentially be originated from the evolution of primordial binaries, while the central BSs would be generated by stellar collisions.

An alternative scenario is based on the idea that almost all BSs were formed in the core by dynamical collisions and then ejected to the outer regions by the recoil of the interactions [188]. Those BSs that get kicked out to a few core radii would rapidly drift back to the centre due to mass segregation, leading to the central BS concentration and a paucity of BSs in the intermediate regions. More energetic recoils would push the BSs to larger distances and since these stars require a longer time to drift back towards the core, their presence would account for the larger specific frequency in the GC outskirts.

Recent dynamical simulations show that the observed bimodality can only be explained by accounting for a significant fraction of BSs originated from primordial binary systems [189]. In particular, a large fraction, between 20 and 40%, of the global BS population has to be formed in the cluster outskirts, where primordial binaries can evolve without any dynamical interaction, and experience mass transfer processes. It is worth noting that (but for a few notable exceptions as

NGC 6388[12] [190], the minimum in the BS radial distribution coincides always with the so-called *radius of avoidance*, the radius within which all stars with mass larger than $\sim 1.2\,M_\odot$ have already sunk to the cluster core because of dynamical friction effects.

3.5.3
Pulsating Blue Stragglers: SX Phoenicis Stars

A significant fraction of BSs shows luminosity variations associated with the star motion within a binary system (eclipsing binaries) and intrinsic oscillations (pulsations) in single stars. The binaries observed within the BS population may well result from only partially completed binary mergers that have pushed one of the components into the BS domain. They might also result from the complete coalescence of two out of the three stars in a single star, that is, a binary encounter.

SX Phoenicis (SX Phe) pulsations arise from the fortuitous coincidence that the region of the HRD populated by BSs overlaps with the instability strip extension (the same of RR Lyrae and classical Cepheids at larger luminosity) that includes δ Scuti stars in the field. The SX Phe stars represent the low-metallicity counterparts of δ Scuti variables.

However, one has to bear in mind that there are also many BSs located well inside the pulsational instability strip that do not show any evidence of oscillations. The reason for this behaviour is still unknown. Figure 3.23 shows the BS population in the GC 47 Tuc, the location of the instability strip boundaries as well as the position of the discovered SX Phe stars in this cluster. The typical pulsation period is in the range ~ 0.0035 days to ~ 0.06 days, while the amplitude in the optical V-band ranges from ~ 0.10 mag to ~ 0.5 mag [191, 192].

The existence of oscillations in some BSs is of interest because the periods provide direct constraints on their structure. The case of double pulsator SX Phe, for example, BSs pulsating simultaneously in two different pulsational modes (usually the fundamental and first overtone), is particularly important. Using the Petersen diagram, that displays the ratio between the two pulsational periods versus the fundamental mode period, in comparison with model predictions, it is possible to accurately constrain the stellar mass and luminosity simultaneously [193]. Figure 3.24 shows an application of this method to the two double mode pulsating BSs in 47 Tuc.

The inferred BS masses, $(1.35 \pm 0.1)\,M_\odot$ and $(1.6 \pm 0.2)\,M_\odot$ respectively, are well above the mass of TO stars in 47 Tuc, and are consistent with the masses inferred from their location in the CMD, as shown in Figure 3.23.

---

12) The peculiar radial distribution found in this GC is explained as a consequence of the fact that the cluster may not be dynamically relaxed, at least the more central regions. A possible explanation could be the presence of an intermediate-mass black-hole at the centre of the cluster.

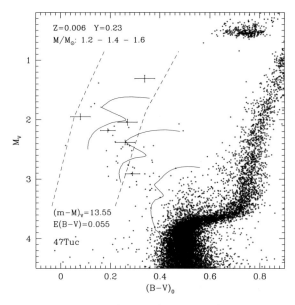

Figure 3.23 The CMD of 47 Tuc. The position of the pulsating BSs is marked by the points with the error bars. The location of the blue and red edge (dashed lines) of the δ Scuti instability strip and the evolutionary tracks for selected stellar models (see labels) are also shown.

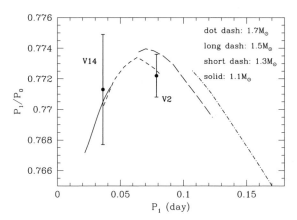

Figure 3.24 Comparison in the Petersen diagram between theoretical period ratios and observational data for two pulsating BSs in 47 Tuc, showing multi-periodicity [191]. The pulsation models have been computed for

$Z = 0.006$ and $Y = 0.23$, and luminosities and effective temperatures determined from stellar evolution models for the labelled values of the mass.

3.6
Subgiant Branch Evolution

At the central exhaustion of H, the star remains with a He-core that does not produce nuclear energy because temperatures are too low for the ignition of He. The nuclear contribution to the energy budget comes from H-burning in a shell surrounding the He-core.

When the central abundance of hydrogen drops to zero in low MS stars, the region of active nuclear burning smoothly shifts above the He-core, where H is still abundant. This is not the same as in upper MS stars, where the H-burning shell ignites after the *overall-contraction* phase.

Once H-burning settles in a shell, low- and upper MS stars have a similar structure. Their following evolution will depend on the mass of the He-core, which in turn depends on the initial mass. It has been shown a long time ago by Schönberg and Chandrasekhar that for a given stellar mass, there is an upper limit, namely, the so-called *Schönberg–Chandrasekhar limit*, for the mass of an isothermal He-core made of an ideal gas to remain in hydrostatic equilibrium. The ratio between the He-core mass (M_{core}) and the total stellar mass (M_{tot}) has to be lower than the upper limit ($M_{core}/M_{tot})_{SC}$ given by

$$\left(\frac{M_{core}}{M_{tot}}\right)_{SC} = 0.37 \left(\frac{\mu_{env}}{\mu_{core}}\right)^2$$

where μ_{env} and μ_{core} are the mean molecular weights in the stellar envelope and in the core, respectively. At the exhaustion of central hydrogen in a star with solar chemical composition $\mu_{env} \sim 0.6$, while $\mu_{core} \sim 1.3$ and $(M_{core}/M_{tot})_{SC} \sim 0.08$.

When the ratio (M_{core}/M_{tot}) exceeds this limit, the core cannot support the weight of the outer envelope and is forced to contract. As a consequence of this contraction and the associated gravitational energy release, the core heats up and a temperature gradient builds up in the core. When this occurs, the core is no longer isothermal and the Schönberg–Chandrasekhar limit no longer applies.

In stars more massive than about $3 M_\odot$, He-core mass exceeds the Schönberg–Chandrasekhar limit at the end of the MS stage, or in the early phases of shell H-burning. The contraction induces a strong heating of the He-core that quickly reaches a temperature of $\sim 10^8$ K, which is required for He-burning fusion reaction. For stars with mass in the range between ~ 3 and $\sim 2.3 M_\odot$, the Schönberg–Chandrasekhar limit is reached later during the shell H-burning stage. In less massive stars, the isothermal core becomes degenerate while (M_{core}/M_{tot}) is lower than the Schönberg–Chandrasekhar threshold. In these conditions, the limit no longer holds; this is due to the evidence that the tendency of the He-core to contract under the weight of the envelope is balanced by the contribution to the total pressure provided by the degenerate electrons that support the structure against gravity without the need for additional contraction.

The mass threshold for stars developing a significant level of electron degeneracy in their He-core during the H-burning shell stage is a very important quantity for each given stellar population. It is customary to define this transition mass with the acronym M_{HeF}. The value of M_{HeF} determines the transition between low-mass stars, that is, those igniting He-burning under conditions of partial degeneracy, and intermediate-mass stars. This mass limit is of the order of $2.3 M_\odot$ for a solar chemical composition, and decreases with decreasing Z, as discussed in Section 3.7.1.

After the MS, there is a transition phase during which all stars move from the blue side of the HRD to the red one; this is the subgiant branch (SGB) phase (see Figure 3.11). During the SGB, the star is supported by H-burning occurring in the shell via the *CNO cycle*. As a general rule, whenever H-burning is efficient in a shell, the dominant nuclear process is always the *CNO cycle*. Due to the huge dependence of the nuclear efficiency on temperature, the mass thickness of the burning shell is very small. Right at the end of the MS, the shell is still thick ($\sim 0.2 M_\odot$) because it is still mainly driven by the *p–p chain*; as the star evolves from the TO to the RGB, it becomes progressively narrower because H is quickly exhausted in the inner parts of the shell and the temperature drops fast in its outer layers. As a consequence, when the star reaches the base of the RGB, the thickness of the shell is equal to about $10^{-3} M_\odot$, and will be further reduced by about one order of magnitude at the end of the RGB.

During the SGB, regardless of the possible presence of a convective envelope during the MS, an extended outer convective region appears. This occurrence is a consequence of the huge envelope expansion, the temperature drop in the outer layers, and the ensuing increase of the radiative opacity that decreases the efficiency of the radiative energy transport. During the whole SGB and early RGB stage, the outer convective zone deepens steadily.

Before closing this section, we note that for stars with non-degenerate He-cores, the evolutionary rate during the SGB phase is roughly the thermal Kelvin–Helmholtz timescale, that is, of the order of ~ 12 Myr for a $3 M_\odot$ and ~ 1 Myr for a $6 M_\odot$. Therefore, this evolutionary phase is so short that the chance of observing objects in this phase is very small. This leads to the so-called *Hertzsprung gap* in the CMD of intermediate-age stellar systems. On the contrary, for low-mass stars, the SGB lifetime is of the order of 1800 Myr for a $0.8 M_\odot$ and of 250 Myr for a $1.5 M_\odot$ with chemical composition $Z = 0.001$ and $Y = 0.246$.

3.7
Red Giant Branch Evolution

During the early phases of shell H-burning, the He-core of low-mass stars attains partial electron degeneracy (see Figure 3.25).

Along the RGB, the He-core mass does monotonically increase due to the conversion of H into He produced by the H-burning shell. Due to the release of gravitational energy by slow contraction (the gas is only partially electron degenerate), it tends to heat up. Electrons are able to transport energy efficiently by conduction, and this favours the persistence of an isothermal core temperature profile. At the same time, with increasing densities, the neutrino production in the central regions of the core becomes progressively more efficient (see [20]). This favours cooling of the innermost layers of the core. Neutrino energy losses eventually overcome the energy gain via gravitational energy, and this causes the appearance of a local maximum (T_{max}) in the temperature, away from the star centre. The trend is shown in Figure 3.26 for a stellar model at various stages along the RGB.

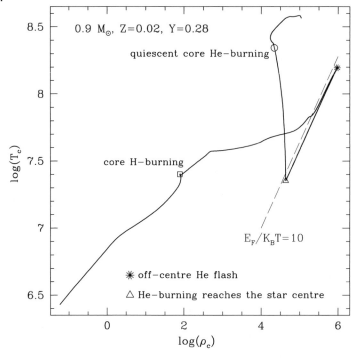

Figure 3.25 Evolution of temperature and density at the centre of a low-mass stellar model from the pre-MS stage to the core He-burning phase, through the He-flash at the tip of the RGB. The dashed line (corresponding to Fermi energy E_F equal to $10K_BT$) can be assumed as a boundary between the electron degenerate and non-degenerate regions. Points corresponding to the ignition of the off-centre He flash, the arrival of the He-burning front at the star centre, and the onset of qui-escent core He-burning are clearly marked (see Section 3.9.1). We note that the evolution of the central conditions approaching quiescent He-burning is slightly different from what usually shown (see i.e. Figure 5.12 in the textbook [20]) because the finer details of the evolution through the He flash until electron degeneracy is lifted from the entire core are usually not discussed in detail, especially in textbooks aimed at giving a general introduction to the subject (courtesy of L. Piersanti).

The key stellar parameters that control the location of T_{max} are the metallicity, or more appropriately the abundance of CNO elements, and the initial mass of the star. The higher the abundance of CNO elements, the faster the conversion of H into He (due to the larger efficiency of the *CNO cycle*), the larger the growth-rate of the He-core, the stronger the heating of the degenerate core, and the closer T_{max} to the star centre. Also, the higher the initial mass of the star, the lower the degree of electron degeneracy (because the core is hotter) and density in the He-core, the lower the efficiency of neutrino production, and again, as a consequence, T_{max} is located nearer to the star centre.

As previously described, while the H-burning shell affects the thermal evolution of the He-core, the growth of the core does in turn affect the evolution of the shell. The steady heating of the core raises the temperature at its outer boundary. This results in a sustained positive feedback: the H-burning shell increases the He-core

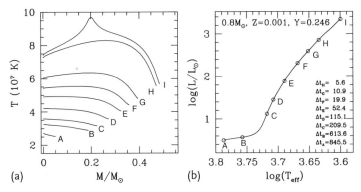

Figure 3.26 (a) The temperature stratification of the He-core at different stages along the RGB evolution of a $0.8 M_\odot$ stellar model with $Z = 0.001$ and $Y = 0.246$. The hottest sequence corresponds to the He-ignition at the tip of the RGB. (b) HRD of the corresponding stellar models. For each model, the time (in Myr) needed to reach He-ignition, is also listed. The structures shown in (a) are the same that are employed in Figure 2.2.

mass by He-accretion, and the core heats up. This heating leads to an increase of the temperature and density in the shell that increases the efficiency of H-burning, and hence the conversion rate of H into He, and the heating of the core. This feedback produces a strong correlation between the stellar surface luminosity and the He-core mass, the so-called *He-core mass–luminosity* relation for RGB stars, that has only a mild dependence on the initial chemical composition (see Section 3.9).

An additional reason for the existence of the *He-core mass–luminosity* relation is related to the fact that the stellar core is compact. As a consequence, the quantity $|dP/dM_r| \propto m/r^4$ at its outer boundary is very large; this means that within the thin shell, the pressure drops by several orders of magnitude. This implies that the H-burning shell does not "feel" the presence of the outer envelope and its properties only depend on the properties of the He-core.[13]

The RGB evolution ends when the off-centre maximum temperature reaches $\sim 10^8$ K, which is required to ignite He-burning via triple-α captures at the tip of the RGB (the brightest point along the evolutionary track shown in Figure 3.11). This process has the characteristics of a thermonuclear runaway and is called the *He Flash*.

He-burning ignition occurs when the He-core mass achieves a value of $M_{cHe} \sim 0.48$–$0.49 M_\odot$, almost independent of the total stellar mass. This is because low-mass stars attain similar levels of electron degeneracy, and M_{cHe} needs to reach similar values for the release of gravitational energy via contraction to be able to overcome neutrino energy losses. When the total mass increases above $\sim 1.5 M_\odot$, the level of electron degeneracy decreases and the thermal conditions for the He-burning ignition are achieved at a lower core mass. For a given total stellar mass, the value of M_{cHe} at the RGB tip depends on the initial chemical composition; it

---

13) This also means that the RGB evolution is not affected by mass loss processes, unless the efficiency is so huge to peel off almost the whole stellar envelope (see Section 4.4).

decreases when increasing the metallicity and/or the initial He. This behaviour can be easily understood when remembering that an increase of the metal content and/or of the He abundance increases the efficiency of shell H-burning in the shell; the growth rate of M_{cHe} is therefore larger, and the thermal conditions for He-burning ignition are achieved earlier.

When discussing the SGB phase, we have emphasized that as a consequence of the huge envelope expansion and consequent cooling, an extended, both in mass and radius, outer convective region develops and penetrates deeper inside the structure during the early RGB stage. This outer convection zone brings to the stellar surface matter partially processed by H-burning nuclear processes during the MS.[14] The largest variation of the surface chemical composition is achieved at the maximum penetration of the convective envelope. This process is known as the *first dredge up* and occurs in the early RGB evolution as shown by Figure 3.11.

The maximum penetration of the convective envelope at the *first dredge up* increases when decreasing the stellar mass and/or increasing the metallicity and/or decreasing the initial He abundance. The latter dependence on the chemical composition can be easily understood when considering the impact of a change of metals and/or He on the stellar opacity, while the dependence on the total mass is due to the fact that decreasing the total mass the RGB location becomes cooler, and this increases the opacity of the stellar envelope.

The *first dredge up* has two remarkable effects on the evolution of low-mass stars:

- A change of the initial surface chemical composition. In more detail, for a $\sim$ $1\,M_\odot$ star with initial solar chemical composition, the surface He mass fraction is increased by $\sim 8\%$, that is, ~ 0.02 (see Figure 3.27 for the case of metal-poor stellar structures) while the ^{3}He abundance increases by one order of magnitude. The ^{12}C/^{13}C ratio drops from the solar value of ~ 90 to ~ 25–30, while the ^{14}N/^{15}N ratio increases by a factor of 2, and oxygen isotopes are not significantly affected because during the core H-burning stage the *NO cycle* was barely active. In more massive stars also, the surface oxygen isotopic ratio is affected by the *first dredge up*. The surface abundance of lithium is drastically reduced due to the mixing of the outer layers with inner regions where Li has already been burnt. The efficiency of the dredge up increases moving from $0.5\,M_\odot$ to $\sim 1.5\,M_\odot$ stars (see 3.27). This occurrence is due to the fact that models in this mass regime are fully supported by the *p–p chain*; with increasing mass, this burning process has to extend to a larger fraction of the stellar interiors, and even if the outer convection is less deep when the total mass increases, the amount of He dredged-up the surface increases. More massive stars are supported by the *CNO cycle* that is more centrally confined; this explains the rapid drop of the dredge up efficiency when the stellar mass is further increased;
- The creation of a chemical discontinuity. Right after the *first dredge up*, the lower boundary of the outer convection zone is forced to move outwards in mass

---

14) One has to remember that, due to its weak dependence on temperature, the *p–p chain* involves a large fraction of the star.

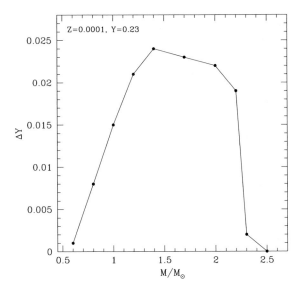

Figure 3.27 The behaviour of the mass fraction of He brought to the surface during the *first dredge up*, as a function of the total stellar mass.

as a consequence of the advancing H-burning shell. As the convective boundary recedes, a chemical composition discontinuity is left over in the structure, marking the point of the maximum penetration of the outer convective region (see Figure 3.28). The presence of this chemical discontinuity produces a relevant observational feature in the evolution of RGB stars.

When the advancing H-burning shell approaches this composition discontinuity, the surface luminosity temporarily drops before starting again to monotonically increase (see the inset in Figure 3.11) as soon as the shell has crossed the discontinuity and enters the overlying layers with uniform chemical composition, as shown in Figure 3.28. This behaviour of the luminosity has been traditionally explained as due to the variation of the H-burning efficiency caused by the abrupt change in the mean molecular weight at the composition discontinuity. This is because the H-burning efficiency is strongly dependent on the mean molecular weight: $L_H \propto \mu^7$. However, the luminosity starts to drop before the H-burning shell actually reaches the composition discontinuity [194]. The luminosity drop may also be due to the increase in the opacity just above the advancing shell due to the larger H abundance on the outer side of the discontinuity.

This peculiar behaviour of the time evolution of the surface luminosity has the important consequence that a low-mass star crosses the same luminosity interval three times – the time spent during this phase is about the 20% of the total RGB lifetime. In an old stellar system, one can verify an overabundance of stars at a specific position along the RGB that appears as a peak in the star count distribution as a function of luminosity (see Section 3.9): this is the so-called *RGB bump*.

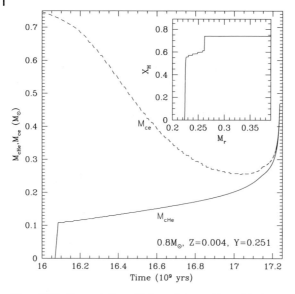

Figure 3.28 The evolution with time of the position of the bottom of the convective enve-lope (M_{ce}) and of the upper boundary of the He core during the RGB stage of a low-mass mod-el (see labels). The inset shows the corresponding hydrogen profile after the *first dredge up*.

The luminosity of the RGB bump is fainter, the deeper the H abundance dis-continuity. Therefore, the RGB bump brightness has the same dependency on the chemical composition and the stellar mass as the lower boundary of the convective envelope at its maximum penetration.

The presence of an extended convective envelope is one of the distinctive prop-erties of RGB stars. Indeed, the morphology of this evolutionary sequence in the HRD is strictly related to the presence of convection in the outer stellar layers. Giv-en that the nuclear energy released in the shell increases during the RGB phase, once on the RGB a star is forced to expand during its evolution to dissipate the surplus of nuclear burning energy by means of the work made against gravity. In the meantime, the stellar effective temperature decrease is bound by the existence of a lower limit to the actual gradient in a convective zone, the adiabatic value, that can be overcome only in the outer layers. This can be easily demonstrated by computing a "toy model" where the efficiency of convection is *by hand* set to zero and a radiative gradient forced on the whole envelope. The result of this ex-periment is shown in Figure 3.11. Given that a RGB star simultaneously attains large luminosity and large radius, one can notice that a stellar model where con-vection is turned off does *not* become a proper RGB star. Apart this evidence, the exact physical reason(s) why a star inflates to red giant dimensions is (are) not fully understood and various authors provide different, sometimes also contradictory, interpretations [195, 196].

3.7.1
The Red Giant Branch Transition

We have already noted that for any given chemical composition, a minimum mass exists M_{HeF} that is able to ignite helium in non-degenerate conditions. For masses below this critical value, the RGB stage has a longer lifetime and the He-burning ignition occurs via a mildly violent He flash. The transition mass M_{HeF} between low- and intermediate-mass stars has relevant implications on the observational properties of a stellar population [197]. It is customary to use the designation *RGB phase transition* to describe the changes that occur in both the morphology of the CMD and the integrated spectral energy distribution of a stellar population, as stars with mass lower than M_{HeF} begin to populate the RGB.

Figure 3.29 shows the trend with the total mass of selected evolutionary quantities at the He-burning ignition. One can easily note that in the low-mass regime, due to the large, and similar, level of electron degeneracy in the core, all stars reach a similar value of the He-core mass (M_{cHe}) and, as expected from the *He core mass–luminosity* relation, a similar brightness at the RGB tip. In the higher mass regime shown in the same figure, where degeneracy is not important, M_{cHe} increases al-

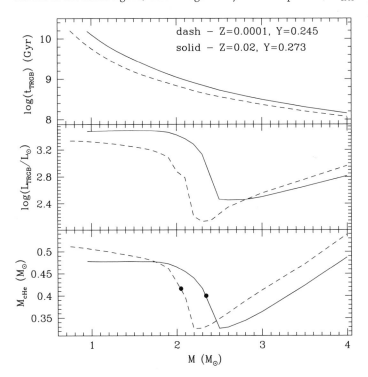

Figure 3.29 Evolutionary lifetime, surface luminosity and He-core mass at the RGB tip as a function of the initial mass, for two different chemical compositions. The full dots mark the values of the transition mass M_{HeF}.

most linearly with increasing mass as a consequence of the increasing mass size of the convective core during the previous core H-burning stage. In the regime corresponding to the *RGB phase transition*, M_{cHe} decreases drastically with increasing stellar mass. This quasi-discontinuity in M_{cHe} occurs over a mass range of the order of $0.15\,M_\odot$, almost independent of the chemical composition. The trend of both the RGB tip brightness and the stellar age at the He-ignition can be easily interpreted when accounting for the behaviour of M_{cHe}. Since the value of M_{cHe} affects both the luminosity and evolutionary lifetime of the evolutionary phases that follows, such discontinuity of the trend of M_{cHe} with mass affects the observational properties of core and shell He-burning stages.

Another interesting feature emerging from Figure 3.29 is that the minimum value of M_{cHe} does not significantly depend on the chemical composition (whereas the total mass that achieves this minimum value for the He-core mass, and its age, depend on the initial chemical composition). The impact of this property will be discussed in Section 4.6 when referring to the observational properties of red clump, He-burning stars.

As for the dependence of M_{HeF} on the metallicity, the general trend is a decrease with decreasing heavy element content. M_{HeF} is equal to $\sim 2.3\,M_\odot$ at solar metallicity, and $\sim 1.2\,M_\odot$ for extremely metal-poor models. This behaviour can be understood as follows. The decrease with decreasing metallicity down to $Z \sim 10^{-4}$ is due to the fact that in this regime, the lower the metal content, the larger the stellar luminosity and the convective core during the MS stage, and hence the larger the He-core and the earlier the He-ignition. When the metallicity decreases further, the monotonic decrease of M_{cHe} is not due to the size of the convective core (because extremely metal-poor stars have smaller, if any, convective cores, but to the deficiency of CNO elements that force these stars to burn H via the *p–p chain* at larger temperatures, an occurrence that causes a decrease of the electron degeneracy in the core and, in turn, an earlier He-burning ignition.

An important consequence is that the *RGB phase transition* occurs at an age that strongly depends on the metallicity of the stellar population. At solar chemical composition the transition occurs at an age of $\sim 630\,\mathrm{Myr}$, that increases to $\sim 4.5\,\mathrm{Gyr}$ at the lowest metallicity.

3.8
The Helium Flash

Following the discussion given in the previous sections, it is now clear that all stars less massive than M_{HeF} ignite He-burning in conditions of electron degeneracy. At He-ignition, the density and temperature at the centre of the star are $\sim 10^6\,\mathrm{g\,cm^{-3}}$ and $\sim 8 \times 10^7\,\mathrm{K}$, respectively. Due to the discussed evidence that the temperature in the He-core has an off-centre maximum, the He-burning ignition occurs in a shell located at about $M_r \sim 0.2$ (see the hottest profile in Figure 3.26a).

The matter in the He-core is in a state of "partial, non-relativistic, degeneracy". Under these conditions, the gas pressure is only a function of density. When He-

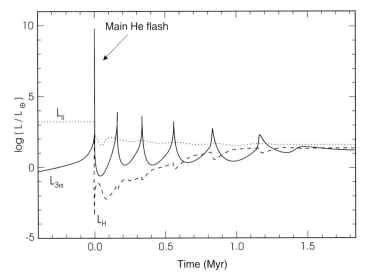

Figure 3.30 Time evolution of the surface luminosity (L_S), the 3α-reaction luminosity ($L_{3\alpha}$), and the H-burning luminosity (L_H) during the onset of He-burning at the tip of the RGB. The zero of the time-axis corresponds to the start of the main He flash.

burning is ignited in a non-degenerate layer, the fresh energy input causes a temperature increase that induces a pressure increase. To preserve hydrostatic equilibrium, the layer does expand, reducing the local temperature; in other terms, the expansion moderates the burning rate. In a degenerate layer, the pressure is insensitive to temperature changes, and an expansion does not occur soon after the temperature rise at the nuclear reaction ignition. As a consequence, the nuclear burning rate increases dramatically following the large temperature increase, and a thermal runaway occurs, the so-called *He flash*. For a detailed analysis of the evolution of the model structure during the *He flash*, we refer to [198, 199].

About $10^{10} L_{\odot}$ are released in a few seconds, as shown by Figure 3.30, but the surface luminosity does not increase at the flash. Instead, it starts to decrease soon after the ignition. The strength of the *He flash* is moderated by two factors. First, as the temperature increases, the level of electron degeneracy decreases and the structure can expand; the second important factor is that as soon as the energy flux in the core increases, convection sets in, carrying out efficiently the energy produced in the inner layers.

Standard stellar model computations have shown that during the *He flash*, convection reaches layers very close, both in radius and mass, to the lower boundary of the convective envelope, but the huge discontinuity in pressure and entropy at the position of the H-burning shell that separates the outer convective envelope from internal He flash-induced mixing zone prevents the possibility of any mixing with the H-rich envelope.

However, this process could occur in models igniting the *He flash* under conditions of extreme electron degeneracy, that is, while cooling along the WD sequence, an occurrence that could explain the formation of *Blue Hook* stars (see Section 4.4) as well as in extremely metal-poor, Population III RGB stars at the He ignition [200–202].

Although the local increase of the temperature removes the electron degeneracy at the point where He-burning ignites, the inner regions of the core remain degenerate following the main flash. This occurs because during the main flash, there is insufficient time for the heat to diffuse inward. The degeneracy of the inner regions is subsequently removed through a series of much lower amplitude secondary flashes, until He-burning eventually reaches the star centre.

The evolution of temperature and density at the star centre is shown in Figure 3.25; the drastic temperature and density decrease occurring right after the ignition of the main *He flash* is due to the nearly adiabatic expansion of the inner core. A detailed analysis of the models discloses that the temperature and density decrease is not monotonic, for during the recurrent secondary flashes the star centre moves back and forth in the diagram, along a sequence of almost constant $E_F / K_B T$ ratio. This shows that the level of electron degeneracy in the centre is barely affected by the occurrence of the main *He flash* and the secondary ones, until the burning reaches the inner layers of the core. These "oscillations" of the central thermal conditions – whose amplitude decreases during the secondary flashes – are due to the alternate expansion and compression of the inner core caused by the the secondary flashes. When the He-burning region sets in the centre of the star – corresponding to the local minimum of the central temperature in Figure 3.25 – electron degeneracy is lifted from the core, and quiescent He-burning commences in the convective inner core.

The nuclear burning occurring during the flash increases the carbon mass fraction by about 0.03–0.05, and the corresponding evolutionary path in the HRD is shown in Figure 3.31. The time spent from the start of the main flash to the beginning of the core He-burning phase (zero age HB) lasts $\sim 10^6$ years, and the probability to observe a star in this phase is extremely small.

Recently, it has been discussed the possibility to directly probe the occurrence of the flash by using asteroseismic data for RGB stars [203]. This idea relates to the evidence that the rapid contraction of the stellar envelope soon after the flash ignition would improve the coupling of the p-mode acoustic oscillations to the core g-modes, making the detection of $l = 1$ mixed modes possible during the development of the flash event.

3.8.1
3D-Simulations of the Helium Flash

Despite the numerical difficulties and extremely time-consuming computations, detailed hydrodynamical simulations of the flash have been performed in the last years [204–206]. The main results obtained by these 3D-simulations can be summarized as follows:

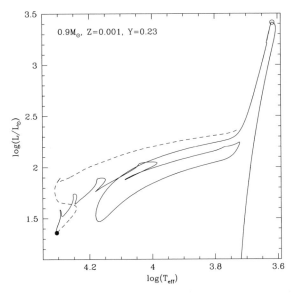

Figure 3.31 HRD of a stellar model from the RGB to the core He-burning stage, showing the evolution through the main and secondary He flashes. The open and full circle mark the He-burning ignition at the RGB tip and the ZAHB location, respectively. The portion of the evolutionary track displayed as a dashed-line corresponds to the core and shell He-burning phases. The stellar model has been computed including mass loss along the RGB according to the Reimers law with the free parameter η set to 1.1 (courtesy of L. Piersanti).

- During the main *He flash*, the flash-induced mixing zone is very turbulent. It is characterized by convective velocities that are roughly four times higher than those predicted by the MLT, and the width of the convection zone grows on a dynamical timescale;
- The flash-induced convection zone grows on a dynamical timescale due to efficient turbulent convection. This growth can lead to injection of H (from the envelope) into the He-core in all stellar models regardless of their metallicity, whereas standard stellar models predict that this occurrence is possible only for extremely metal-poor stars. The nuclear burning caused by the ingestion of H leads to a split of the convection zone into two parts separated by a radiative zone [202]. The 3D simulations predict that this double convection zone should disappear quite soon because the convective motions occurring inside these convective shells decays very fast. The final outcome of this process is still under debate because the results seem to depend significantly on the numerical details of the computations;
- The interiors of stellar structures as predicted by 3D models seem to retain more or less spherical symmetry;
- The 3D simulations predict turbulent motions not only at the outer edge of the convection zone, but also at the inner boundary of the convective shell powered by the huge energy release associated to the *He flash*. As a consequence, the convection zone also extends below the point where the He-burning ignition

occurred. This result seems to indicate that the extension of the convective zone induced by the flash is not accurately predicted by stellar models. If a rapid growth towards the stellar interiors indeed occurs, the main flash will never be followed by subsequent secondary flashes because convection will lift quite efficiently (in ~ 1 month) the electron degeneracy in the whole core.

3.9
Theoretical Isochrones for the Hydrogen Burning Stages

One of the main properties of an isochrone is that earlier evolutionary phases, in particular, the MS, display a much larger range of masses than later phases. For instance, in the case of a 10 Gyr, $Z = 0.001$ isochrone the MS covers a mass range from the minimum possible stellar mass (~ $0.08 M_\odot$) to $0.837 M_\odot$, the mass of the model at TO; at the base of the RGB one finds a mass of $0.863 M_\odot$, while at the RGB tip the mass is $0.872 M_\odot$. Therefore, the mass evolving in post-MS phases is approximately constant and very close to the TO mass.

This happens because beyond the TO, there is a rapid increase of the evolutionary speed and therefore the stars evolving in post-MS stages may be regarded as having started out with the same initial mass to within a small percentage. As a consequence, the morphology of an isochrone of any given age along the RGB and following stages, approximately coincides with the evolutionary track of a model with mass similar to that evolving at the isochrone TO.

Figure 3.32 shows selected isochrones with the same initial chemical composition but different ages. One can easily note that the luminosity of the TO decreases with increasing age, as a consequence of the decrease of the mass of the star evolving at the TO. The TO brightness does not depend linearly on age when considering the whole age interval displayed in Figure 3.32. However, when restricting only to ages larger or of the order of 6 Gyr, the relation age-TO luminosity is almost linear, $d \log(L/L_\odot)/dt \approx -4.1 \times 10^{-2}$ (dex/Gyr). This property makes the TO the most important clock provided by the theory of stellar evolution for age-dating resolved stellar systems. Figure 3.33 shows some isochrones in an optical CMD for selected ages and fixed chemical composition.

Another interesting feature is the significant change of the SGB slope when changing age; the slope increases with increasing age.

The effective temperature (hence colour) difference between the TO and the base of the RGB increases with decreasing age because the TO gets hotter whilst the RGB $T_{\rm eff}$ is almost unchanged.

Also, decreasing the age the RGB bump brightness increases, and the bump disappears for ages lower than an age limit equal to ~ 0.6 Gyr at $Z = 0.01$, as this age limit depends on the initial chemical composition. This is a consequence of the fact that models more massive than about $2.2 M_\odot$ do not experience the *first dredge up* during the shell H-burning phase.

Another important feature is that the bolometric luminosity of the RGB tip does not depend on the isochrone age, at least for age larger than ~ 4 Gyr. However,

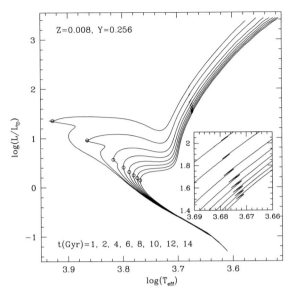

Figure 3.32 HRD of some selected isochrones with the same chemical composition and various ages (see labels). Open dots mark the location of the TO along the MS, while the inset shows the RGB bump region.

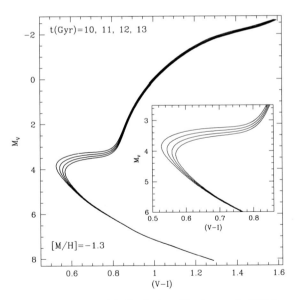

Figure 3.33 An optical CMD of selected isochrones for $[M/H] = -1.3$. The inset shows the TO and SGB portion of the isochrones.

it changes significantly for younger ages. This occurrence is what one obviously expects on the basis of the previous discussion on the *RGB phase transition*.

3.9.1
The Effect of Chemical Composition

Figure 3.34 shows the optical CMDs of selected 12 Gyr H-burning isochrones of various metallicities, spanning the whole range of Galactic metal-poor and intermediate metallicity systems. Figure 3.34 reveals some interesting features:

- When decreasing the metallicity, isochrones become brighter and bluer; this behaviour is the consequence of the global reduction of the radiative opacity when the heavy element abundance decreases. It is well-established that metals contribute to the radiative opacity in two well-defined temperature regimes, around 10^5–10^6 K and below ~ 4000 K, namely. Opacity computations show that in the range between 10^5 and 10^6 K, the continuum absorption is mainly controlled by the abundances of C, N, O and Ne, with the largest contribution coming from the most abundant of these elements, that is, oxygen; concerning the line opacity all elements provide more or less the same contribution. In the low-temperature regime, the main "indirect" contributors to the radiative opacity are Mg, Si and Fe, because of their low first-ionization potential and non-negligible abundances, and act as *donors* of the free electrons required for forming H^-, one of the most important opacity sources at low temperatures, especially in the

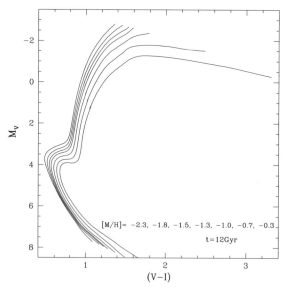

Figure 3.34 An optical CMD of 12 Gyr isochrones for various assumptions about the metallicity [M/H] (see labels, metallicity increases, moving from left to right).

metal-poor regime (TiO and H_2O molecules are important especially at metallicities around solar and for T_{eff} below $\sim$ 3000 K).

Detailed evolutionary computations have shown that the MS is essentially affected by changes of the high-temperature radiative opacity (especially at low metallicities) while the TO and SGB brightness are mainly affected by the change of the *CNO cycle* efficiency, induced by a variation of the sum of the CNO elements. On the other hand, both the location and slope of the RGB depend mainly on the low-temperature opacities and are affected by changes of the abundance of low first-ionization potential elements such as Mg, Si, and Fe [42];

- The SGB becomes steeper when the metallicity decreases;
- Both the location and slope of the RGB disclose a remarkable dependence on the metallicity, becoming redder and more tilted with increasing Z. Indeed, the RGB shows the largest sensitivity to the heavy element abundance. The RGB colour has therefore been employed as a photometric metallicity (iron) indicator by comparing the location of the target RGB sequence with the RGBs of selected GCs with well-known (from high-resolution spectroscopy) metallicity, or alternatively isochrone RGBs of various metallicities. An application of the first approach is shown in Figure 3.35, where the colour of bulge RGB stars is compared with the location of empirical RGBs with known metallicity;
- There are two additional features along the RGB, namely, the RGB bump and the brightness of the RGB tip, whose metallicity dependence deserves to be discussed. The RGB bump becomes fainter when the metallicity increases, as a consequence of the larger envelope radiative opacity that extends the convective envelope deeper into the structure at the *first dredge up*.

Depending on the adopted photometric filters, for metallicity larger than $[M/H] \sim -1.0$, the RGB tip is no longer the brightest point along the sequence. In the HRD, the RGB tip is *always* the brightest point along the RGB, but not in CMDs, due to the metallicity dependence of the bolometric corrections. For the Johnson V-band, in the high-metallicity regime, the effect of the bolometric corrections does monotonically increase along the RGB as a consequence of the formation in the stellar atmosphere of molecules (in particular, TiO and H_2O). This effect even overcomes the monotonic increase of the bolometric luminosity and produces a bending of the RGB in the optical CMD. The appearance of this bending moves towards lower metallicities with using shorter wavelength filters, while in the near- and far-infrared observational plane the RGB tip is the brightest point in the whole metallicity range explored.

3.9.1.1 The Heavy Element Mixture

It is well-known that old, metal-poor, stars are characterized by a distribution of metals different from the Sun. Metal-poor stars display an overabundance of the so-called α-elements, O, Ne, Mg, Si, S and Ca, with respect to Fe compared to the solar metal mixture (i.e. $[\alpha/Fe > 0]$).

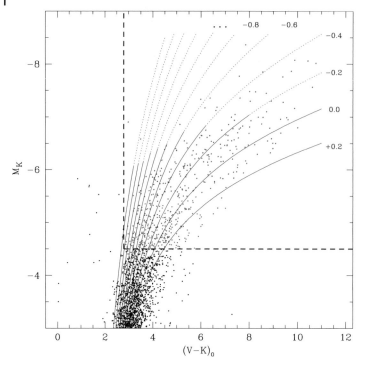

Figure 3.35 The optical-near infrared CMD of a stellar field in the Galactic bulge. The solid/dotted lines corresponds to fiducial RGB sequences for various [Fe/H] (as labelled) obtained from an empirical calibration based on the RGB of Galactic GCs with known metallicity. The two dashed lines show the portion of the CMD more appropriate to estimate the heavy element abundance.

It has been demonstrated that in the metal-poor regime, for globular cluster-like ages, if the following condition on the mass fraction of metals is satisfied,

$$\left[\frac{X_C + X_N + X_O + X_{Ne}}{X_{Mg} + X_{Si} + X_S + X_{Ca} + X_{Fe}} \right] \sim 0 \tag{3.1}$$

Isochrones computed for an α-enhanced mixture ($[\alpha/\text{Fe}] > 0$) at a given [Fe/H] are well-mimicked in the HRD by scaled-solar ones with a total metallicity provided by

$$[M/H] = [\text{Fe}/H] + \log(0.638 f_\alpha + 0.362) \tag{3.2}$$

where $f_\alpha = 10^{[\alpha/\text{Fe}]}$ [42].

When mimicking the CMD of an α-enhanced isochrone with a scaled-solar one, one has to appropriately choose the independent variable for the interpolation among the tables of bolometric corrections. There are two natural choices for this; the first one is to determine the scaled-solar bolometric corrections at the same [M/H] of the α-enhanced isochrone. The second possibility is to consider scaled-solar transformations with the same [Fe/H] of the α-enhanced models; this second

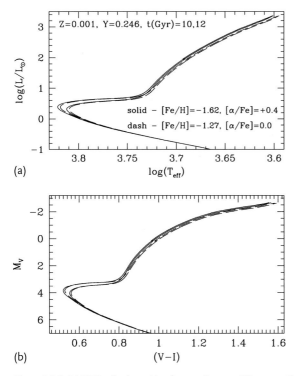

Figure 3.36 (a) HRD of selected isochrones for two different α-element enhancements and the same global metallicity. (b) $M_V-(V-I)$ CMD of the same isochrones.

option provides an overall better agreement with the self-consistent α-enhanced isochrone [207].

Figure 3.36 shows a comparison both in the HRD and in an optical CMD between scaled-solar and α-enhanced isochrones with the same $[M/H]$. The two sets of isochrones are in good agreement in both the HRD and CMD; in more detail, the brightness is the same in all evolutionary stages, and also effective temperature and colour are very similar, the scaled-solar isochrones being slightly cooler.

The situation changes significantly when the abundance of CNO elements is enhanced (or depleted) with respect to a reference mixture, and Eq. (3.1) is not satisfied. The effect is shown in Figure 3.37, where metal-poor α-enhanced isochrones are compared with an isochrone computed by accounting for an enhancement of about a factor of 2 in the sum C + N + O, but the same iron content.[15]

Given that the CNO element abundance is increased and the iron content is the same, the total metallicity of the CNO-enhanced isochrone is slightly larger.

---

15) Indeed, in the case of the CNO-enhanced mixture, the models in the figure also include light element anti-correlations, like the O-Na anti-correlation, observed in Galactic GCs (see the discussion in Section 7.8). However, the impact of these light elements on both opacity and the nuclear burning efficiency is negligible. Also, the small difference in the initial He abundances does not play a major role in the comparison.

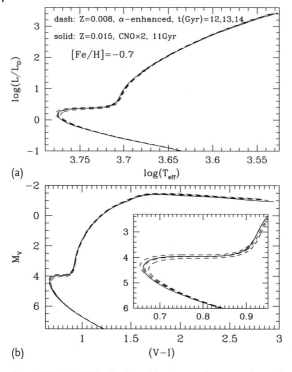

(a)

(b)

Figure 3.37 (a) HRD of selected isochrones for an α-enhanced mixture and $Y = 0.256$, and one isochrone with a CNO-enhanced mixture, $Y = 0.266$, but the same [Fe/H] of the α-enhanced isochrones. (b) As for (a), but in the M_V-$(V-I)$ CMD.

This occurrence explains why the CNO-enhanced isochrone is slightly redder with respect to the other isochrones. The CNO-enhanced isochrone for an age of 11 Gyr is almost perfectly matched at the TO and SGB regions by a 13 Gyr α-enhanced isochrone, for example, when fixing the TO brightness there is a $\sim$ 2 Gyr age offset between the two sets of isochrones. When accounting for the small difference in He content between the two sets of isochrones, the TO age offset due to the CNO elements enhancement is slightly decreased to about $\sim$ 1.2–1.5 Gyr.

The two sets of isochrones almost show the same colour along the other evolutionary stages in optical CMDs. At bluer wavelengths the differences in the heavy element mixture produce remarkable differences in the spectra. These properties will be extensively discussed in the section devoted to the multiple stellar populations in Galactic GCs.

A CNO-enhanced mixture affects two other important features along the RGB, that is, the RGB bump and the RGB tip brightness. The RGB bump brightness is decreased by about $\Delta \log(L/L_\odot) \approx 0.12$, with respect to "normal" α-enhanced stellar models, when the CNO sum is doubled, keeping all other element unchanged. The RGB tip brightness is increased by about $\Delta \log(L/L_\odot) \approx 0.02$, despite the fact that the He-core mass at the He-burning ignition is decreased by $\sim 0.005\,M_\odot$.

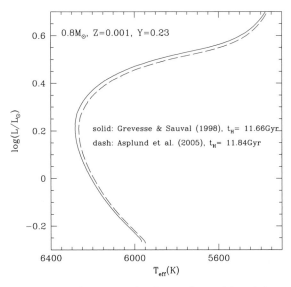

Figure 3.38 Evolutionary tracks of two stellar models with the same initial mass and chemical composition, but for two different assumptions about the scaled-solar heavy elements mixtures, namely, GS98 and AGS05. The MS lifetime of the two models is also displayed.

As for the effect of varying the abundance of *single* elements one at a time at fixed [Fe/H], luminosities and T_{eff} of the TO region are affected, consistently with the results reported above, by variations of C, N and O (TO luminosity and T_{eff} decrease with increasing abundances). However, the RGB T_{eff} is affected by variations of mainly Mg and Si, for example, T_{eff} decreases with increasing abundances [208].

Before closing this section, we wish to comment briefly about varying the solar heavy element distribution. We have already discussed this issue in Section 3.4.4 in the context of the SSM. When considering metal-poor, low-mass stars as those currently evolving in Galactic GCs, the effect of using a different solar mixture is not as relevant as in the case of solar metallicity stellar models, as shown for a selected model in Figure 3.38.

3.9.1.2 The Helium Content

Figure 3.39 shows the effect of a change of the initial He abundance on the morphology of H-burning phase isochrones. An increase of He at fixed Z decreases the radiative opacity because of the corresponding decrease of H to preserve $X + Y + Z = 1$, and κ_{He} is lower than κ_H; moreover, an increase of He causes an increase of the mean molecular weight μ, and since the H-burning efficiency strongly depends on μ ($L_H \propto \mu^7$), the efficiency of the nuclear burning is affected.

The combination of these effects makes the He-enhanced evolutionary tracks hotter and brighter, whilst the H-burning lifetime decreases. The net effect on the isochrones is that a He-enhanced isochrone for a given age is hotter both along the MS and the RGB, *but* the TO is slightly fainter as a consequence of the faster

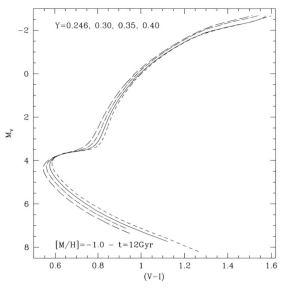

Figure 3.39 Optical CMD of isochrones with the same age and metallicity, but various assumptions about the initial He abundance. The value of He increases when moving from right to left.

evolutionary lifetimes during the MS. The SGB location is not significantly affected by a change of the He abundance. The increase of the evolutionary rate has the consequence that the mass of the star "evolving" at the TO at a given age decreases with increasing He abundance. This has a major impact on the morphology of the HB in He-enhanced stellar populations discussed in the next chapter.

An increase of the initial He abundance also affects the RGB bump brightness. At a fixed metallicity and age, the RGB bump will appear more luminous as a consequence of the more external location of the H-discontinuity produced by the *first dredge up*. This is obviously due to the decreased envelope radiative opacity. In addition, the extension of the luminosity drop that characterizes the bump, decreases with increasing He, because the larger the initial He content, the lower the H abundance discontinuity. As a consequence, there is an upper limit for the initial He content – the exact value depending on metallicity and age – for the occurrence of the RGB bump (see Figure 3.40).

Given that an increase of the initial He makes the stellar structures hotter, the level of electron degeneracy in the He-core during the RGB is decreased; as a consequence, the He flash is attained at a lower core mass and fainter RGB tip brightness.

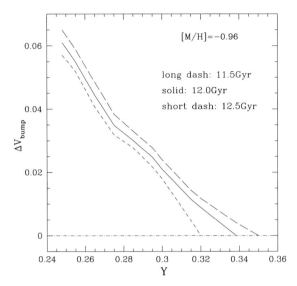

Figure 3.40 The luminosity drop in the Johnson V-band corresponding to the RGB bump for a given metallicity and various assumptions about the initial He abundance and age.

3.9.2
The Luminosity Function

Theoretical isochrones not only provide information about the position of stars of different masses in the HRD and CMDs, but they also allow one to predict the relative number of stars along the various evolutionary stages.

Given that to each point along an isochrone, a value of the actual (different than the initial one, if mass loss is efficient) mass of the star evolving at that HRD (or CMD) location can be assigned. The adoption of an IMF that determines the number of stars dN born with mass between M and $M + dM$, provides the number of objects populating the interval between two consecutive points along the isochrone.

When using an IMF of the form $dN = C M^{-x} dM$, the normalization constant can be constrained by specifying either the total mass (M_{tot}) or the total number (N_{tot}) of stars in the given SSP (see Chapter 1). In case the exponent x in the power law is equal to 2.35, the adopted IMF corresponds to the so-called Salpeter IMF.

When considering a generic photometric band, for instance, the standard Johnson V-band, the number of stars dN between magnitudes M_V and $M_V + \Delta M_V$ along an isochrone is given by

$$dN = \frac{dN}{dM} \frac{dM}{dM_V} dM_V = C M^{-x} \frac{dM}{dM_V} dM_V$$

where the derivative dM/dM_V is estimated along the isochrone.

Star counts N as a function of magnitude along the isochrone provide the so-called *differential luminosity function*, hereinafter simply the *luminosity function* (LF, see also Chapter 1). When computing the LF, one must not take account the actual mass along the isochrone that could be different from the initial one as a consequence of mass loss, but the initial mass. In fact, it is the initial mass that determines the evolutionary timescales and the number of stars predicted by the IMF.

Figure 3.41 shows the CMD of a Galactic GC with the corresponding LF from the lower MS to the RGB tip. Notice the correspondence between features of the LF and the corresponding evolutionary stages in the CMD.

3.9.2.1 The Shape of the Luminosity Function

The LF of the RGB is a simple straight line on a magnitude-log(N) plane, except in correspondence with the RGB bump. The slope of this line allows a major test of the evolutionary rate along the RGB that is virtually independent of the isochrone age or metallicity. This linearity is a direct consequence of the *He-core mass–luminosity* relation for RGB stars. The increase of M_{cHe} in a given time step Δt can be written as $\Delta M_{cHe} = m_{He} L \Delta t$, where m_{He} is the mass of He produced per unit of energy released by the H-burning shell. Remembering that $L \propto M_{cHe}^{\alpha}$, then

$$\Delta t \equiv \frac{1}{m_{He}} \frac{\Delta M_{cHe}}{L} \propto \frac{1}{m_{He}\,\alpha} L^{\frac{1}{\alpha}-1} \frac{\Delta L}{L}$$

When considering small time steps, for example, small changes of the luminosity, and defining the quantity $\tau = dt/d\log L$, that is, the time spent in a small luminosity interval centred at L, one can derive

$$\log(\tau) = \frac{1-\alpha}{\alpha} \log(L) + \text{constant} = -\frac{1-\alpha}{2.5\alpha} M_{bol} + \text{constant}$$

Similar relations can be obtained for any photometric passband. Since the number of stars in each luminosity interval is proportional to the time spent by stars in that luminosity bin, the logarithm of the star counts along the RGB also clearly scales linearly with the magnitude. As already noted, the M_{cHe}–L relation slightly depends on the initial chemical composition and the initial stellar mass (for models that develop electron degeneracy in their He core), and as a consequence, the slope of the RGB LF does not significantly depend on chemical composition and age.

The RGB LF exhibits the characteristic RGB bump, discussed before; the level of agreement between empirical and theoretical values of the magnitude of the RGB bump will be discussed in the following sections.

The SGB portion of the LF is very interesting, for it is the feature most sensitive to the parameters of the stellar population such as its age. Due to the rapid increase in the number of stars towards the TO, and the fact that the SGB is almost horizontal in the CMD, the SGB LF appears as a steep drop in the star counts, the so-called SGB *break*, with a *peak* just above the break (see the location of these features in Figure 3.41). The magnitude level of these features mainly depends on the isochrone age and metallicity. One can also note that the TO is not located at the break of the SGB, but ~ 0.5 magnitudes fainter.

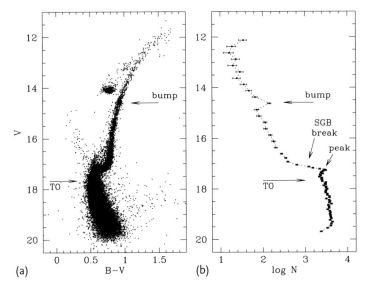

Figure 3.41 CMD of 47 Tuc (a) and the corresponding LF (b) obtained by discarding the core and shell He-burning phases.

3.9.2.2 The Dependence of the Luminosity Function on the Stellar Population Properties

Figure 3.42 shows the dependence of the LF on metallicity and the IMF exponent. For a given age, the RGB bump and the location of both the LF SGB *break* and *peak* become fainter with increasing metallicity. This directly arises from the dependence of the isochrone morphology on the metallicity. The slope of the RGB LF is largely independent of the metallicity as discussed. The shape of the LF for the lower MS is affected by a change of the exponent of the IMF, while the post-MS portion of the LF is unchanged because, contrary to the MS, the value of the initial mass of objects evolving along the RGB is approximately constant.

The position of the SGB break is a strong function of the age. To highlight this property of the LF, Figure 3.43 compares theoretical LFs for ages of 8, 10, 12 and 14 Gyr at fixed chemical composition. The LFs have been normalized to have the same RGB levels. The rate at which the SGB *break* moves towards lower luminosities is essentially independent of chemical composition, but does decrease somewhat with increasing age. For stellar populations older than $\sim$ 12 Gyr, this rate is approximately $\sim$ 0.07 mag/Gyr. In this respect, the age sensitivity of the SGB *break* is quite similar to that of the TO brightness; the only possible advantage of using the LF is that the magnitude of the *break* is easier to determine observationally, while the TO magnitude measured on a CMD could be affected by an uncertainty of the order of a few tenths of a magnitude. The position of the SGB *break* at a given age is also a function of chemical composition. This is shown in Figure 3.43b. The dependence of the SGB *break* on the metallicity is stronger than that on age, and increases with increasing metallicity. There is also a dependence of the *break*

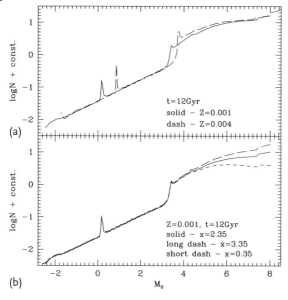

Figure 3.42 (a) Theoretical LFs for the same age and two different metallicities for a Salpeter IMF. (b) As for (a), but for various values of the exponent of the IMF power law, at fixed age and metallicity. In both panels, the LFs have been normalized to the same star counts at $M_V = 2.0$.

position on the helium content: an increase of 0.01 in the initial He abundance makes the SGB *break* ~ 0.10 mag fainter.

Finally, the LF SGB *peak* is much less prominent at the older age. However, for a given age, it becomes more and more evident with increasing metallicity.

3.9.3
The Calibration of Superadiabatic Convection

As previously discussed, the temperature gradient along the bulk of the convective envelope can be approximated by the adiabatic value. However, in the layers close to the stellar surface, the convective gradient becomes strongly superadiabatic. To determine the actual gradient in these outer layers, the mixing length theory (MLT) is almost universally used, as discussed in Section 2.1. The MLT contains several free parameters whose values affect the predicted $T_{\rm eff}$ of the stellar models. Most of these parameters are fixed beforehand and the only one left to be calibrated is $\alpha_{\rm ml}$, the ratio of the mixing length to the pressure scale height $H_{\rm P}$, that provides the scale length of the convective motions. As a general rule, an increase of $\alpha_{\rm ml}$ corresponds to an increase of the convective transport efficiency and an increase of the stellar model $T_{\rm eff}$.

As discussed in Section 3.4, the value of $\alpha_{\rm ml}$ is usually calibrated with a SSM, and this solar-calibrated value ($\alpha_{\rm ml,\odot}$) is then used for computing models of stars very different from the Sun (e.g. metal-poor RGB and main-sequence stars of various

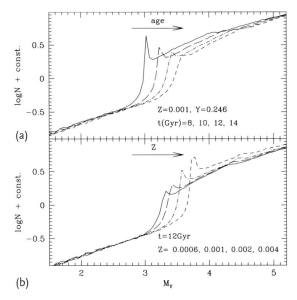

Figure 3.43 (a) LFs around the TO for the same metallicity and various ages. (b) LFs at a fixed age and various values of the metallicity. In all cases, a Salpeter IMF has been adopted and the LFs have been normalized to the same star counts at $M_V = 2.0$.

masses). It is clear that, even if a physical input employed in the stellar model computations is not accurate, it is possible to mask this shortcoming, at least from the point of view of the predicted T_{eff}, by simply recalibrating α_{ml} on the Sun. This guarantees that the models always correctly predict the T_{eff} of, at least, solar-type stars. However, since the extension of the superadiabatic layers is larger in RGB stars, theoretical RGB models are much more sensitive to α_{ml} than MS ones. Therefore, it is not safe to assume a priori that the solar-calibrated value of α_{ml} is also adequate for RGB stars of various metallicities.

The strong dependence of the RGB model T_{eff} scale on the treatment of superadiabatic convection offers an independent method for calibrating α_{ml}; one can compare empirical estimates of T_{eff} at a fixed absolute magnitude for the RGB of Galactic GCs with theoretical models of the appropriate chemical composition and various assumptions on α_{ml}. Figure 3.44 shows a comparison between the empirical RGB T_{eff} for a sample of Galactic GCs and model predictions based on a solar-calibrated α_{ml} [209]. One can easily note that the solar-calibrated α_{ml} also provides a satisfactory match to the effective temperatures of RGB metal-poor stars.

It is important to emphasize that there is in principle no reason why α_{ml} should be kept constant when considering stars of different masses and/or chemical composition and/or at different evolutionary stages; even within the same star, α_{ml} might in principle vary from layer to layer, although stellar model calculations usually keep α_{ml} constant throughout the convective zone.

It is also worth mentioning that observational uncertainties can affect the MLT calibration based on the empirical RGB T_{eff} scale, though the effect is probably

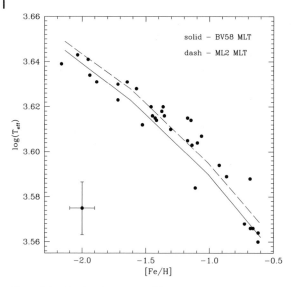

Figure 3.44 Empirical estimate of the average RGB T_{eff} at $M_{bol} = -3$ for a sample of Galactic GCs, compared to theoretical predictions alternatively using the solar-calibrated BV58 and ML2 flavours of the MLT. The typical observational error bars are displayed in the lower left corner.

small. For instance, to determine the RGB T_{eff} at a fixed absolute magnitude of a given GC, one has to rely on GC distance estimates because the RGB is roughly vertical, an uncertainty of $\sim \pm 0.2$ mag in the individual cluster distance modulus does not affect at all the MLT calibration. The same outcome applies for the presence of random uncertainties on the cluster reddening and metallicity.

To give an estimate of the sensitivity of the calibrated α_{ml} value to systematic errors on the temperature scale, chemical composition, and adopted distances, we note that systematic changes of the empirical RGB T_{eff} by ~ 70 K, or cluster $[M/H]$ by 0.2 dex, or adopted cluster distance moduli by 0.25 mag would cause a ~ 0.1 variation of the calibrated α_{ml}.

3.9.3.1 The Mixing Length Flavour

There are three additional free parameters entering the MLT that are generally fixed a priori, before the calibration of α_{ml} is performed. These free parameters were denoted as a, b, c, in Chapter 2.

Depending on the values adopted for these parameters, one can define different MLT "flavours". The "classical" formulation of the MLT, nowadays almost universally used in the computations of stellar evolution models, is the so-called BV58 "flavour" of the MLT [21]. Another widely used flavour of the MLT is mainly employed in calculations of model atmospheres, spectra and bolometric corrections for WD stars, and is denoted as ML2, with its own specific choice of a, b, and c different from the case of BV58 [210]. The values of these three free parameters for the ML2 and BV58 flavours are listed in Table 3.4.

By using the relations for v_c, F_c and Γ provided in Chapter 2, it is possible to obtain a simple algebraic equation whose solution provides the value of ∇ at a given value of r in the stellar structure. A widely used implementation of the BV58 flavour of the MLT gives this equation for the effective temperature gradient [18]

$$\zeta^{\frac{1}{3}} + B\zeta^{\frac{2}{3}} + a_0 B^2 \zeta - a_0 B^2 = 0 \tag{3.3}$$

where ζ is defined as $\zeta \equiv (\nabla_r - \nabla)/(\nabla_r - \nabla_{ad})$. Once the value of ζ is computed, the knowledge of ∇_r (radiative gradient) and ∇_{ad} (adiabatic gradient) immediately provides the actual gradient ∇. The quantities B and a_0 entering Eq. (3.3) are obtained through the following relationships:

$$\Gamma = A(\nabla - \nabla')^{\frac{1}{2}}$$

$$\nabla_r - \nabla = a_0 A(\nabla - \nabla')^{\frac{3}{2}}$$

$$B \equiv \left[\frac{A^2}{a_0}(\nabla_r - \nabla_{ad}) \right]^{\frac{1}{3}}$$

To adopt Eq. (3.3) with the ML2 choices of the constants a, b and c, the quantities a_0, A, and B calculated with the BV58 choices have to be transformed as follows:

$$a_0(\text{ML2}) = \frac{8}{3} a_0(\text{BV58})$$

$$A(\text{ML2}) = 3\sqrt{2} A(\text{BV58})$$

$$B(\text{ML2}) = \left(\frac{27}{4} \right)^{\frac{1}{3}} B(\text{BV58})$$

Once the BV58 and the ML2 flavours of the MLT are accounted for in a stellar evolutionary code, the first step is to obtain a SSM in order to evaluate the appropriated value of α_{ml} in the two cases.

The solar-calibrated values of α_{ml} are, respectively, $\alpha_{ml,\odot}(\text{BV58}) = 2.01$ and $\alpha_{ml,\odot}(\text{ML2}) = 0.63$ with the models in [209]. The value obtained from $\alpha_{ml,\odot}(\text{ML2})$ is similar to that adopted for WD model atmospheres in order to obtain an overall consistency between temperature estimates from both the UV and the optical spectrum, observed photometry, gravitational redshift mass estimates and trigonometric parallax [211]. The position of the bottom of the solar convective envelope and surface He abundance are the same in both ML2 and BV58 calibrated models.

Table 3.4 Values of the free parameters (besides α_{ml}) in the BV58 and ML2 MLT flavours.

MLT flavour	a	b	c
BV58	$\frac{1}{8}$	$\frac{1}{2}$	24
ML2	1	2	16

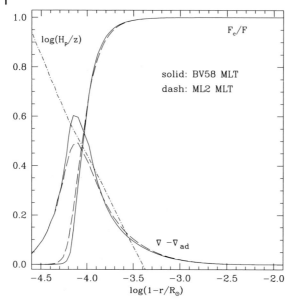

Figure 3.45 The trend of the superadiabaticity $(\nabla - \nabla_{ad})$ and of the ratio of the convective to the total energy flux as a function of the radial location in the outer layers of the solar convection zone as obtained by using the BV58 and ML2 flavours of the MLT. The dashed dotted line displays the ratio between the local pressure scale height and the geometrical distance from the top of the convective region.

Figure 3.45 shows the trend of some physical variables within the solar convective envelope, as obtained by using alternatively the BV58 and ML2 "flavours". The two approaches provide overall similar results in the superadiabatic region. The same figure also displays the ratio between the local pressure scale height and the geometrical distance from the top of the convective envelope (z). Around the peak of the superadiabatic region, the local value of H_P is ~ 3 times the distance from the surface. This highlights a well-known inconsistency when using the BV58 calibrated values of α_{ml}: with $\alpha_{ml,\odot}(BV58) = 2.01$, the mixing length l in the superadiabatic region is about 6–7 times longer than the distance z from the surface. Interestingly, the use of the ML2 partially alleviates, but does not fully solve this problem.

Figure 3.44 shows the effective temperature scale of RGB stellar models obtained with the ML2 flavour of the MLT and the solar-calibrated value of α_{ml} (dashed line). The models appear ~ 50 K hotter than BV58 results, but both model T_{eff} scales appear consistent with empirical results within current uncertainties.

All of these findings support the conclusion that different MLT choices are largely equivalent, but not exactly coincident, once α_{ml} is calibrated with a SSM.

3.9.4

The RGB Luminosity Function Bump

In the previous sections, we have discussed the physical reasons for the bump in the RGB LF of old stellar populations. The RGB bump was first predicted by theory, and only later was its existence confirmed by observations of the Galactic GC 47 Tuc [212, 213].

Given that the bump brightness depends on the maximum depth attained by the convective envelope and the chemical profile above the advancing H-burning shell, the comparison between predicted and observed luminosities provides valuable information about the internal structure of low-mass stars. The RGB bump has therefore been the subject of several theoretical and observational investigations [214–216]. Nowadays, the RGB bump brightness has been accurately measured not only in Galactic GCs, but also in Local Group dwarf galaxies and the Galactic bulge [217, 218].

The observational parameter routinely adopted to compare observations with theory is the quantity $\Delta V_{HB}^{Bump} = V_{Bump} - V_{HB}$, that is, the V-magnitude (or filters similar to Johnson V) difference between the RGB bump and the HB at the RR Lyrae instability strip. This approach has the advantage of being formally independent of distance, reddening, and uncertainties of photometric zero points.

Figure 3.46 shows a comparison between the most recent measurements of this parameter for both a large sample of Galactic GCs and Local Group dwarfs, and theoretical predictions (for the Local Group dwarfs, the predictions were obtained from synthetic CMDs that match the observed complex populations, see Chapter 8). This comparison discloses a discrepancy, at the level of ~ 0.20 mag or possibly larger, for GCs with $[M/H]$ below ~ -1.5, the predicted ΔV_{HB}^{Bump} values being larger than observed. Obviously, one cannot discriminate whether this is due to too bright theoretical bump luminosities, or underluminous HB models (or a combination of both effects). Due to the strong dependence of ΔV_{HB}^{Bump} parameter on metallicity, a quantitative assessment of the actual discrepancy between theory and observation depends on the adopted metallicity scale. It is interesting to note that, unexpectedly (for Local Group dwarfs host complex stellar populations), the agreement between theory and observations seems to improve when comparing model predictions with empirical measurements of the ΔV_{HB}^{Bump} parameter in Local Group dwarfs.

The main drawback of using the ΔV_{HB}^{Bump} parameter as a diagnostic of the RGB bump luminosity is that uncertainties in the estimate of the HB level for GCs with blue HB morphologies and in theoretical predictions of the HB luminosities (see Chapter 7 for a detailed discussion on both issues) make very difficult the interpretation of discrepancies between theory and observations. Also, ΔV_{HB}^{Bump} depends on the age due to the dependence on the stellar population age of the RGB bump brightness (the HB level is largely unaffected by age for old stellar populations, as discussed in the following chapter). One can, in principle, find a GC age that brings model predictions in agreement with the observations. However, the ages necessary to match the observed ΔV_{HB}^{Bump} for metal-poor clusters need to be well

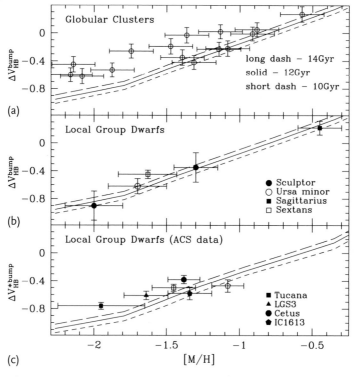

Figure 3.46 (a) Comparison between theoretical values of ΔV_{HB}^{Bump} as a function of metallicity for various ages, and empirical data for a sample of Galactic GCs. (b) As with (a), but for a sample of Local Group dwarf galaxies. (c) As with (b), but using the magnitude $M_{V*} = (M_{F475W} + M_{F814W})/2$.

above 14 Gyr, the age of the Universe according to the commonly accepted cosmological parameters.

A complementary avenue is offered by the magnitude difference between the TO and RGB bump, $\Delta V_{TO}^{Bump} = V_{TO} - V_{bump}$, which bypasses the HB.

On the observational side, an accurate estimate of the TO brightness requires high-quality photometric datasets. Figure 3.47 shows this kind of comparison for a sample of GCs observed with the ACS detector (in the F606W filter) on board HST. The apparent TO magnitudes have been converted to absolute magnitudes using a MS-fitting distance scale, and these $M_{F606W}(TO)$ values are compared in Figure 3.47a with α-enhanced theoretical predictions for various cluster age assumptions [219]. From this comparison, one can obtain an estimate of the absolute age of the GCs in the sample. The level of agreement between theory and observations for the quantity $\Delta M_{F606W}^{TO\text{-}Bump}$ (the equivalent of ΔV_{TO}^{Bump}) is shown in Figure 3.47b. The age obtained from $M_{F606W}(TO)$ is clearly not consistent with that obtained from the $\Delta M_{F606W}^{TO\text{-}Bump}$ parameter (younger ages by on average $\sim$ 3–4 Gyr are obtained from the latter). This is better outlined in Figure 3.47c that shows

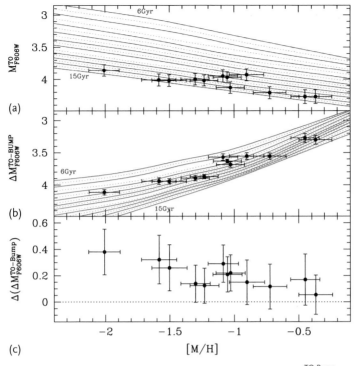

Figure 3.47 (a) Absolute M_{F606W} magnitude of the TO as a function of [M/H] for a sample of Galactic GCs. The theoretical predictions are provided by α-enhanced isochrones for ages between 6 and 15 Gyr, in steps of 0.5 Gyr, shown as solid and dashed lines. (c) As for the (a), but for $\Delta M_{F606W}^{TO\text{-}Bump}$. (c) The difference between the values of $\Delta M_{F606W}^{TO\text{-}Bump}$ predicted by the stellar models for the cluster age estimated by using the absolute TO magnitude, and empirical measurements, as a function of [M/H].

the differences between the observed values of $\Delta M_{F606W}^{TO\text{-}Bump}$ and those predicted by stellar models for the cluster age estimated from $M_{F606W}(TO)$.

This plot clearly shows that the expected $\Delta M_{F606W}^{TO\text{-}Bump}$ values are systematically larger than observed: the mean difference being of the order of ~ 0.20 mag. Only unreasonably young TO ages (obtained by "stretching" the GC distances) can solve this discrepancy without invoking inaccuracies of the models. Given that the observed TO magnitude is by definition matched by the theoretical isochrones in order to determine the TO age, this discrepancy implies that the absolute magnitude of the RGB bump in the models is too bright, for example, RGB models predict a too shallow maximum depth of the envelope at the *first dredge up*. This also explains the similar discrepancy found for the ΔV_{HB}^{Bump} parameter.

On the basis of the most recent updates in the input physics, it does not appear realistic to modify the maximum depth of the convective envelope at the dredge up by reasonable changes in the adopted physics (as for instance, the radiative opacity). The most plausible solution of this discrepancy involves some amount of convective overshoot, beyond the Schwarzschild convective boundary by about $\sim 0.25 H_P$.

3.9.5
The Tip of the Red Giant Branch

In Section 3.8, we noted that He-burning ignition terminates the RGB evolution at the so-called RGB tip. It has been also shown (see Section 3.7.1) that for each given chemical composition, the luminosity of the RGB tip does not depend on the stellar population age, for ages older than $\sim$ 4–5 Gyr.

Figure 3.48a shows that the bolometric luminosity of the RGB tip for an age $\sim$ 12 Gyr monotonically increases from the very low-metallicity regime up to about the solar chemical composition, despite of the decrease of the He-core mass at the He ignotion shown in Figure 3.48b. This can be understood by considering that the efficiency of the H-burning shell increases with increasing heavy element abundance. In other terms, this is due to the fact that the $M_{cHe}-L$ relation depends, as a second order effect, also on the chemical composition. On the other hand, the increase of M_{bol} (TRGB) for super-solar metallicities is a consequence of the fact

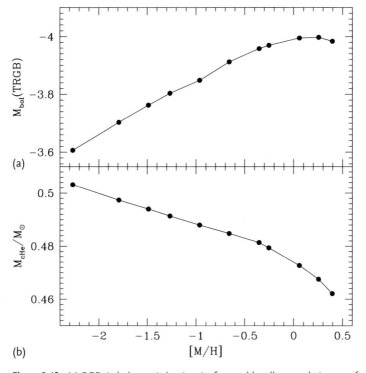

Figure 3.48 (a) RGB tip bolometric luminosity for an old stellar population as a function of [M/H], as predicted by evolutionary stellar models. (b) As for (a), but for the He-core mass at the same stage.

that the effect associated with the huge increase of the radiative opacity overcomes the increased efficiency of the H-burning shell.

The theoretical predictions of the RGB tip (TRGB) brightness in various photometric bands for a $\sim$ 12 Gyr stellar population – indeed these results hold for any age larger than about 4 Gyr – are shown in Figure 3.49. In the I-Cousins band, the brightness of the TRGB is largely independent on the metallicity for $[M/H] < -0.9$; the total variation of M_I (TRGB) is lower than 0.10 mag. This is due to the fact that in this metallicity range, $M_{\rm bol}$ (TRGB) and the I-band bolometric corrections have a similar dependence on $[M/H]$, and their variation when $[M/H]$ changes keeps M_I (TRGB) roughly constant. This important property of the RGB tip brightness in the I-band is at the basis of its use as a distance indicator. At the same time, when considering metallicity larger than ~ -0.9 dex, M_I (TRGB) shows a non-negligible dependence on the metallicity.

In the near-infrared (NIR) bands, the RGB tip brightness displays an almost perfectly linear dependence on the heavy element content in the whole explored metallicity range. The slope $d M_{\rm NIR}^{\rm TRGB}/d[M/H] \approx -0.28$, -0.49, and -0.62 mag/dex for the J, H and K-band respectively.

The theoretical predictions for the RGB tip brightness in several photometric bands appear in good agreement with empirical measurements in ω Cen and 47 Tuc, with distances determined from eclipsing binary systems. Evolutionary

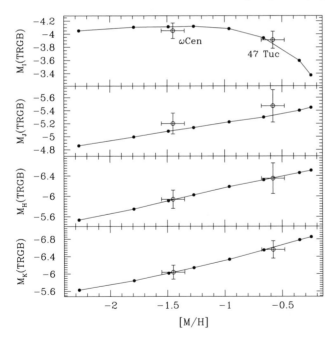

Figure 3.49 The magnitude of the RGB tip in several photometric filters as predicted by the same models shown in Figure 3.48. The empirical measurements of the RGB tip in ω Cen and 47 Tuc are also shown.

computations also show that RGB tip brightness at fixed Z is largely independent on the exact heavy element distribution, that is, scaled-solar or α-enhanced.

3.9.6
The Effect of Atomic Diffusion and Rotation

We wish to now discuss how the inclusion of atomic diffusion in model computations affects the isochrones morphology and the calibration of the TO brightness as a function of the age. To gain insights on the effect of diffusion, let us first discuss its impact on the evolution of models representative of stars currently evolving in an old star cluster, that is, a $0.8\,M_\odot$ star with $Z = 0.0006$ and 0.01.

Figure 3.50 shows the evolutionary tracks computed under various assumptions about the efficiency of atomic diffusion: (i) no diffusion, (ii) diffusion of He but not metals, (iii) the diffusion of both He and metals, but without accounting for the effect of diffusion of metals on the radiative opacity, (iv) a self-consistent case where *all* effects due to diffusion of all chemical species are taken into account. Table 3.5 lists the selected properties of these calculations.

Figure 3.50 clearly reveals how the evolutionary track morphology is affected by the inclusion of diffusion:

- When only He diffusion is taken into account, the track is redder, with a lower TO luminosity that is also reached earlier, in comparison with the no-diffusion model. As a consequence of He diffusing inward, the envelope H abundance has to increase and this causes an increase of the radiative opacity that makes the model cooler and fainter. The corresponding reduction of the abundance of hydrogen in the interiors decreases the amount of fuel available for the nuclear burning and the central H-exhaustion is attained earlier;
- The track corresponding to He diffusion matches perfectly that obtained by taking into account the diffusion of both He and metals, but for the effect of metal diffusion on the stellar opacity (the two lines overlap almost perfectly). The explanation is that, in low-mass metal-poor stars, the contribution of the *CNO cycle*, whose efficiency is affected by any change of the abundances of CNO el-

Table 3.5 Selected properties at the TO of a $0.8\,M_\odot$ model for two initial chemical compositions, with and without including atomic diffusion. Y_{surf} denotes the surface helium abundance.

Scenario	$\log(L_{TO}/L_\odot)$	$\log(T_{eff}^{TO})$	t_{TO} (Gyr)	$\Delta \log(L_{TO})$	$\Delta t_{TO}/t_{TO}$	Y_{surf}
		$Z = 0.0006$, $Y = 0.246$				
No-diffusion	0.3165	3.8193	11.152			0.246
Diffusion	0.2710	3.8087	10.410	−0.045	−6.6%	0.115
		$Z = 0.01$, $Y = 0.259$				
No-diffusion	−0.0142	3.7529	18.471			0.259
Diffusion	−0.0448	3.7464	16.775	−0.031	−9.2%	0.180

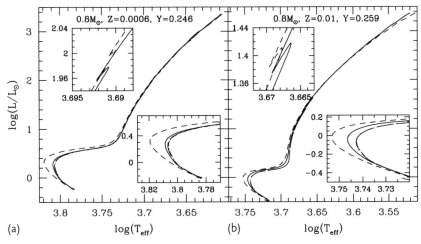

Figure 3.50 Evolutionary tracks of a $0.8\,M_\odot$ stellar model for two different initial chemical compositions and various assumptions about the efficiency of atomic diffusion, for example, no-diffusion (short-dashed line), diffusion (solid line), only He diffusion (dot-dashed line), diffusion of both He and metals, but without including the effect of metal diffusion on the radiative opacity (long-dashed line). The insets enlarge the TO and RGB bump stage.

ements, to the nuclear energy budget is negligible. Therefore, the changes in the CNO abundances induced by diffusion do not affect the nuclear burning efficiency and the evolution of this model;

- The diffusion of heavy elements plays a role, however, through the impact on the radiative opacity. When the heavy element diffusion is calculated self-consistently, it partially counteracts the effect of He diffusion;
- The effects of diffusion are larger at lower metallicities.[16] When increasing the metallicity, the extension of the outer convection zone increases, increasing the surface reservoir of He and metals that diffuse less efficiently towards the centre. For the same reason, at a given metallicity, when decreasing the total mass below $\sim 0.8\,M_\odot$, the effect of diffusion decreases as a consequence of the progressive deepening of external convection. On the other hand, when increasing the total stellar mass well above the quoted value, the effects of diffusion start to decrease as a consequence of the shorter evolutionary lifetimes, as diffusion in MS stars acts on long timescales.

Figure 3.51 compares two sets of isochrones computed with and without atomic diffusion. At a given initial metallicity and age, isochrones with diffusion have a cooler and fainter TO. As rule of thumb, diffusion reduces the age of old stel-

---

16) For the calculations in Table 3.5, the effect of diffusion is larger in the more metal-rich model due to its longer MS lifetime; this allows diffusion to work for a longer time, compensating the effect associated of the more extended outer convective zone. When comparing models with similar core H-burning lifetimes, atomic diffusion effects are, indeed, larger at lower metallicities.

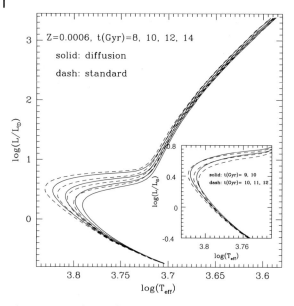

Figure 3.51 Isochrones for $Z = 0.0006$ and selected ages (see labels), computed with and without atomic diffusion. The inset enlarges the CMD around the TO and SGB regions.

lar systems by $\sim 1\,\mathrm{Gyr}$. For the very same reasons discussed before, this effect is slightly larger (lower) for the most metal-poor(-rich) stellar populations.

As for radiative levitation, its effect on the isochrone morphology is generally negligible. However, the efficiency of radiative levitation, in addition to atomic diffusion, needs to be properly accounted for in order to provide accurate predictions of the surface chemical evolution of certain elements; in fact, it has a major impact on the MS surface chemical abundances of specific elements, like Fe, at low metallicities [220, 221].

The impact of atomic diffusion (and levitation) on RGB models is generally negligible because the evolutionary timescales are too short. During the early RGB, due to deepening of the outer convection zone, most of the helium and heavy elements that were diffused toward the interiors during the MS are dredged back to the surface. As a consequence, after the *first dredge up*, the original (initial) chemical abundances are largely restored inside the convective envelope. The first direct consequence is that RGB effective temperatures are essentially the same as for models without diffusion, as shown by Figures 3.50 and 3.51.

This notwithstanding, the efficiency of atomic diffusion during the MS has the following effects on the RGB evolution:

- After the *first dredge up*, the He abundance in the envelope is slightly lower. As an example, in a $0.8 M_\odot$ model with $Z = 0.0006$ and initial He abundance $Y = 0.246$, the post dredge up surface He abundance is lower by about $\Delta Y = 0.008$ with respect to the same model, but computed without accounting for the occurrence of atomic diffusion. This will cause a slight decrease of the brightness of the stellar models during the following core He-burning phase;

- As a consequence of the decrease of the envelope He abundance and the corresponding increase of the radiative opacity, the RGB bump brightness is decreased, as shown by Figure 3.50 [222, 223]. The decrease of the V-band brightness of the RGB bump is ~ 0.07 mag (the exact value depending on metallicity and age). Since the V-magnitude of the zero age HB is decreased by about 0.02 mag by the inclusion of atomic diffusion, this effect partially compensates the change of the RGB bump brightness, and the $\Delta V_{\mathrm{HB}}^{\mathrm{Bump}}$ parameter discussed in Section 3.9.4 is increased by only about 0.05 mag;

- The He-core mass M_{cHe} at the RGB tip is increased by about 0.003–$0.004\,M_\odot$ for $Z \leq 0.001$, slightly decreasing when increasing the metallicity [223, 224]. The increase of the He-core mass at the RGB tip of diffusive models should be a consequence of the decrease of the He abundance in the envelope that changes the mean molecular weight, hence the efficiency of the H-burning process in the shell ($L_{\mathrm{H}} \propto \mu^7$);

  Despite the increase of M_{cHe}, the zero age HB luminosity is, however, lower. As it will be discussed in Chapter 4, this shows that the reduction of the envelope He abundance overcomes the effect associated to the He-core mass increase;

- Due to the He-core mass increase, the luminosity of the tip of the RGB increases: for a $0.8\,M_\odot$ model, $\Delta \log(L_{\mathrm{tip}}/L_\odot) \sim 0.003$ dex at $Z = 0.0002$, and ~ 0.01 dex at $Z = 0.002$.

Along the RGB, stellar models that consistently account for atomic diffusion and levitation predict almost exactly the same properties of models computed by accounting "only" for atomic diffusion. Radiative levitation can be indeed effective in selectively modifying the concentration of metals in the interiors, but the evolutionary effects are very small, if any.

Finally, the diffusion of helium in the layers between the bottom of the convective envelope and the advancing H-burning shell can produce a mean molecular weight inversion larger by a factor of ~ 4 than the μ inversion produced by the ^{3}He-burning discussed in Section 2.11. He-settling, therefore, could be just as important as ^{3}He-burning in generating mixing instability between the H-burning shell and the outer convection zone. The μ inversion induced by atomic diffusion should occur slightly earlier during the evolution, but while that associated to ^{3}He-burning is occurring on a nuclear timescale, the He-settling is governed by the (longer) atomic diffusion timescale.

In Section 2.8, we discussed the role of rotation as an important "source" of mixing instabilities that are able to affect the surface chemical abundances. It is important to assess whether the presence of rotation (at realistic rates) could alter the morphology of low-mass evolutionary tracks and isochrones. Early calculations suggested that rapid differential rotation in the stellar interiors might strongly affect the structural properties and evolutionary lifetimes of stellar models [225, 226]. Nevertheless, these early rotating stellar models had large physical limitations due

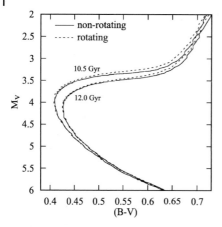

Figure 3.52 Optical CMD of isochrones for $Z = 0.001$ and selected ages (see labels), computed with and without the inclusion of rotation.

to the fact that angular momentum redistribution and rotationally induced mixings were ignored.

More recent comparisons with models computed accounting for angular momentum redistribution and losses, mixing instabilities driven by differential rotation and meridional circulation show that, within the explored range of initial angular momentum distributions, the HRD of low-mass stars are largely unaffected by rotation [227, 228]. Also, the evolutionary rate of the rotating models is marginally affected: for a given stellar mass and chemical composition, the age difference at the central H-exhaustion is of the order of 0.10 Gyr (i.e. $\sim 1\%$). Therefore, within $\sim 1\%$, stellar rotation does not affect age estimates of old stellar systems from their TO brightness (see Figure 3.52). The RGB location at fixed magnitude, for in-

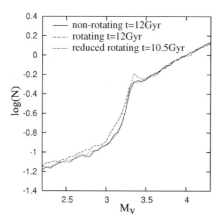

Figure 3.53 LFs for 12 Gyr, $Z = 0.001$ isochrones, with and without rotation. The LF obtained from a 10.5 Gyr old isochrone, computed from rotating stellar models with a reduced value of the initial angular momentum, is also shown. All LFs have been normalized to the same counts at $M_V = 4$, and are based on the Salpeter IMF.

stance, at $M_V = 2$, of the isochrones including rotation is bluer by $(B-V) \approx 0.01$ and 0.007 mag for 10.5 and 12.0 Gyr, respectively.

Figure 3.53 shows the comparison between theoretical LFs obtained from the same isochrones of the previous figure (employing the same IMF in all cases). For a given age, rotation reduces the SGB slope and increases the number count at the base of the RGB by $\sim 11\%$ in the case of the 12 Gyr LF.[17] The shape of the RGB bump is not changed by rotation, only its location is shifted to brighter luminosities by $\Delta M_V = 0.1$ mag is the LF corresponding to the 10.5 Gyr old isochrone.

More controversial is the impact of stellar rotation on the structural properties at He-ignition, and in particular on the He-core mass at the He flash. According to the early computations, an increase of $\Delta M_{cHe} \approx 0.013 M_\odot$ should be expected for an initial rotational velocity of 2×10^{-4} rads^{-1} [226]. However, some indications [228] were provided that, due to the quoted limitations in the models provided by [226], the actual increase of M_{cHe} should be lower than the value early predicted. Nevertheless, the recent evolutionary computations by [227], based on reasonable assumptions about the initial angular momentum value, angular momentum redistribution and losses, predict a huge increase of the He-core mass by about $0.048 M_\odot$. This large He-core increase would lead to a zero age HB luminosity brighter by $\Delta M_V \approx 0.33$ mag, an increase that seems to be ruled out by current empirical estimates of the HB luminosity in GCs whose distances have been obtained with independent methods.[18] It is clear that further investigations on this issue are needed.

17) These results are in agreement with [229] in spite of a different treatment of rotation in the model computations.

18) The initial angular momentum distribution in the calculations performed by [227] has been derived from empirical measurements of the rotation rate in solar metallicity stars. Numerical experiments show that when reducing this initial value by $\sim 60\%$ (that produces SGB models with surface velocities still consistent with empirical estimates of the rotation rate in low-metallicity stars), the He-core mass would increase by $\sim 0.016 M_\odot$, corresponding to an increase of the zero age HB brightness by about 0.12 mag, in better agreement with current uncertainty on independent estimates of this quantity.

4
Horizontal Branch and Red Clump

4.1
Overview

In the previous chapter, we discussed the structural and evolutionary properties of low-mass stars during the main core and shell H-burning stages, as well as some related evolutionary features such as the MS TO and the RGB tip, which are extremely useful when measuring the properties of old stellar populations such as their age and distance. In this chapter, we describe the main structural and evolutionary characteristics of low-mass stars during the following evolutionary stage, that is, the core He-burning phase. The subsequent shell He-burning phase is discussed in the next chapter.

The nuclear reactions involved in the He-burning process were discussed in Section 2.3.2 together with the most important uncertainties affecting the available rates for these nuclear reactions. We remind you that the He-burning mechanism is the most efficient burning process after the H-burning process, and as a consequence, the core He-burning lifetime is the second longest after the core H-burning lifetime.

The central He-burning stage in low-mass stars is very important for many reasons:

- Its evolutionary lifetime is strongly affected by the efficiency of the mixing processes occurring in the stellar interiors. The comparison between theoretical predictions on evolutionary lifetimes and star counts in stellar systems such as star clusters, therefore provides a formidable tool in the investigation of the convective processes in these stars;
- The luminosity of low-mass, core He-burning stars is one of the most important distance indicators for Population II stellar systems, and it also plays a fundamental role in the age dating of old stellar systems such as Galactic GCs;
- RR Lyrae stars, one of the most important class of variables, are low-mass, core He-burning stars. These pulsating stars are a fundamental standard candle, and the analysis of their pulsational properties provides a valuable tool for studying the structural and evolutionary properties of low-mass stars in an advanced evolutionary stage;

Old Stellar Populations, First Edition. S. Cassisi and M. Salaris.
© 2013 WILEY-VCH Verlag GmbH & Co. KGaA. Published 2013 by WILEY-VCH Verlag GmbH & Co. KGaA.

- Indeed, other classes of important variable stars, such as Population II Cepheids and Anomalous Cepheids, are also associated with the evolution of core He-burning, low-mass stars;
- Star counts for the core He-burning stage provide the most important tool for estimating the initial helium abundance of old stellar systems, the so-called *R parameter*;
- For many decades, the colour distribution of core He-burning stars in the CMD of old stellar systems, and its correlation with the main properties of the stellar population such as metallicity, age, and so on, is one of the longstanding problems in stellar astrophysics: the so-called *Second Parameter Problem*.

Since in the optical photometric bands, the distribution of core He-burning stars in the CMD of old stellar systems corresponds to a largely constant luminosity sequence, that is, almost horizontal (at least in its reddest portion), it is customary to refer to the central He-burning stage of low-mass stars as the "Horizontal Branch" (HB). In the following, we will adopt this definition.

4.2
Mixing Processes

Due to the large sensitivity of the 3α reaction on temperature ($\epsilon_{3\alpha} \propto T^{40}$ at $T \sim 10^8$ K), core He-burning stars, regardless of their total mass, have an extended convective core until He-exhaustion at the center.

The development of convective instabilities at the base of these extended mixings in low-mass stars during the HB stage is discussed in detail in textbooks such as [20] and many scientific papers [230–233]. Such a detailed discussion will not be repeated here. Here, we summarize the physical phenomena causing the various mixing processes and highlight their impact on the structural and observational properties of core He-burning stars and of their progeny. However, before discussing these mixing processes, we will summarize the physical properties in the interiors of low-mass stars in this evolutionary stage.

4.2.1
The Physical Conditions in the Convective Core

At the beginning of the central He-burning stage, the physical conditions at the centre of the star are: $\log(T_c) \sim 8.06$, $\log(\rho_c) \sim 4.2$, and a chemical composition that is almost a pure He mixture – there is a mass fraction of $\sim 5\%$ of carbon, produced by He-burning occurring during the He flash at the tip of the RGB and during the very short relaxation phase of the star just after the He-ignition.

In these conditions, the radiative opacity is dominated by electron scattering, which is independent of core composition, and the opacity is increased thanks to a significant contribution from *free–free* absorption. The relative weights of these contributions to the Rosseland mean opacity vary somewhat during the evolution of the star, because the carbon and oxygen fractions increase as a consequence of the nuclear burning and the temperature rise. An approximate expression for *free–free* absorption is provided by Kramers's formula, $\kappa_{ff} \propto \sum_i (X_i Z_i^2 / A_i) \rho T^{-7/2}$, where X_i is the mass fraction of the i element with atomic number Z_i and weight A_i. As a consequence of the conversion of helium into carbon, the contribution of the *free–free* opacity to the total opacity increases with the time.[1]

At the beginning of this evolutionary stage, the thermodynamical properties of the stellar plasma closely resemble those of a perfect gas, with extremely small corrections from the ideal, related to radiation pressure and electron degeneracy. However, during the core He-burning stage, because of the increase of the CO fraction, the Coulomb interactions also increase. In fact, the Coulomb parameter Γ (see the definition in Section 2.2.1) can be expressed as a function of the average atomic number and weight of the stellar plasma as

$$\Gamma \propto \frac{\langle Z \rangle^2}{\langle A \rangle^{\frac{1}{3}}} \frac{\rho^{\frac{1}{3}}}{T}$$

When considering pure gases, the composition dependent term $\langle Z \rangle^2 \langle A \rangle^{-1/3}$ is close to 1 for H, about 2.5 for He and ~ 16 and ~ 26 for C and O. During the core He-burning stage, because of the increase of the C and O abundance, it can happen that Γ becomes of the order of unity. As a consequence, within the convective core of low-mass He-burning stars, Coulomb interactions can be more important than in the earlier RGB stage, despite the higher temperatures. The changes in the thermodynamical properties induced by non-ideal effects obviously also affect the properties of the convective core.

4.2.2
The Expansion of the Convective Core

Due to He-burning, the C abundance builds up inside the convective core, and the opacity increases, producing a discontinuity in the radiative and the chemical gradients at the convective core boundary (see Figure 6.4 in [20]). With time, ∇_{rad} will increase just inside the convective core edge until it exceeds the adiabatic gradient ∇_{ad}. It is obvious that this situation cannot be stable and has to give rise to a convective boundary instability. It is widely recognized that the layer where the Schwarzschild stability criterion is satisfied does not represent the true edge of a convective region since the velocity of the convective elements do not fall to zero there. Thus, in a real word, some amount of convective overshoot at the canonical boundary of the convective core occurs.

Owing to the presence of the chemical discontinuity previously mentioned, and the increasing opacity inside the convective core, to allow the occurrence of convective overshoot inside a radiative shell surrounding the central convective region causes a local increase of the opacity and, in turn, a convective boundary instability.

---

1) As an order of magnitude, one can estimate that κ_{ff} can contribute to about the 25% of the total opacity.

As a result, a *self-driving mechanism* for the extension of the convective core occurs: any radiative shell which is mixed with matter coming from the central convection zone as a consequence of the convective core overshoot will definitely belong to the convective core.[2] This physical behaviour is fully supported by detailed computations of HB stellar models. Due to this process, at a certain moment at the edge of the convective core, the radiative gradient becomes equal to the adiabatic one, and the core stops increasing in mass.

Because of the mass growth of the convective core, this early stage of the core He-burning phase is often referred to as the "overshooting phase". However, the extension of the convective core is indeed a self-driving mechanism; induced by the drastic change in the chemical composition and hence in the opacity stratification, both caused by the He-burning. So, although the mechanism by which the convective core boundary moves outward with time is probably some degree of convective overshoot or penetration, it would be more appropriate to refer to this early stage as "convective core expansion".

4.2.3
Induced Semi-convection

It is evident that as a consequence of the burning occurring in the interior a sharp composition discontinuity appears at the transition between the convective core and the external radiative layers. This situation is unstable [235] and it requires that the star develops a semi-convective zone of varying chemical composition just exterior of the fully convective core. The need for the formation of a semi-convective region is also implied by the behaviour of the radiative gradient, which instead of decreasing monotonically when moving from the stellar centre towards the outer layers, exhibits a minimum and then turns upwards in the neighbourhood of the convective core boundary. This trend forces the star to readjust as follows: because of overshoot, the edge of the convective zone will shift outwards, causing a He-enrichment within the convective core and thus globally lowering ∇_{rad}. When the minimum of ∇_{rad} becomes equal to ∇_{ad}, further mixing of stellar material across the minimum radiative gradient point will be shut-off. The outer boundary of the fully convective core will then be located at this minimum point. However, outside the convective core, there will still exist a separate region which is convectively unstable because $\nabla_{rad} > \nabla_{ad}$. Overshoot between this region and the outer layers will continue until the increasing He abundance in this region has established convective neutrality (by lowering the ∇_{rad} to ∇_{ad}). In this way, a semi-convective region is created. A qualitative visual description of this process can be found in Figure 6.6 of [20].

Figure 4.1 shows the profile of the helium abundance in the fully convective and semi-convective regions in a low-mass star at various stages of the central He-

---

2) As will be discussed in Section 4.7, it has been recently suggested [234] that the role of overshoot could be significantly reduced because of atomic diffusion of carbon at the boundary between the canonical convective core and the outer radiative zone.

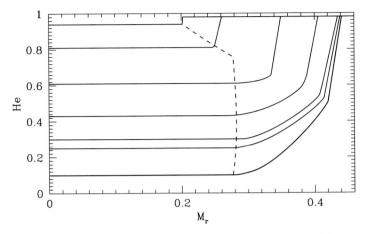

Figure 4.1 The helium abundance profile inside the convective core and the semi-convective zone at various levels of core He depletion for a $0.64\,M_\odot$ model with $Z = 0.001$ and $Y = 0.246$. The dashed line represents the outer boundary of the canonical convective core.

burning phase: one can easily note the core convective expansion stage and the development of the semi-convective zone.

The primary effect of both convective core expansion and semi-convection on the interior structure is an increase of the mass size of the He-depleted region as shown in Figure 4.1. This increase in the available He fuel causes an increase of the core He-burning lifetime t_{HB} (roughly speaking by a factor of 2) with respect to model predictions that do not account for both mixing processes. This occurrence also means that when the star eventually approaches the next evolutionary stage, namely, the Asymptotic Giant Branch (AGB), the size of the carbon-oxygen (CO) core is also larger and this affects the AGB evolutionary lifetime. For a given total mass, the AGB lifetime depends on the mass size of the CO core: the larger the CO core mass, the shorter the AGB lifetime is because there is less He fuel available for this evolutionary stage. In passing, we note that this dependence provides a tool for checking on the accuracy of this mixing process scenario for low-mass, He-burning stars: the comparison between theory and observations concerning the ratio between the number of AGB stars and those in the core He-burning stage, the so-called R_2 parameter can constrain the numerical algorithms adopted in stellar model computations to account for semi-convection. So far, the available analysis fully supports the present theoretical framework, which predicts a value for R_2 in the range 0.12–0.15, while the empirical value in Galactic GCs is 0.15 ± 0.02.

Concerning the morphology of the evolutionary tracks in the HRD: the presence of a semi-convective zone forces the models to perform a more extended blue loop during the core He-burning stage with respect to models not accounting for this mixing process.

Another important implication of the occurrence of these mixings is that the CO abundance profile in the He-exhausted core strongly depends on the adopted numerical method for modeling them (see [236]). This is an important issue because,

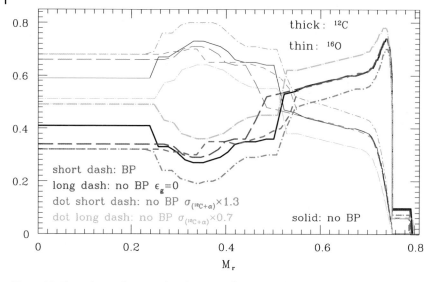

Figure 4.2 The carbon and oxygen abundance profiles in the same stellar structure considered in Figure 4.1 at the beginning of the thermal pulses stage along the AGB. The various lines correspond to different assumptions about the treatment of the breathing pulse process: *no BP* refers to HB models where the occurrence of the breathing pulses has been inhibited by imposing that the core He abundance never has to increase, *no BP* $\epsilon_g = 0$ refers to a model where the occurrence of these convective instabilities has been avoided by putting equal to zero the gravitational energy release when the central abundance of He is below 0.1; two additional models computed by alternatively increasing or decreasing by about 30% the rate for the nuclear reaction $^{12}C(\alpha, \gamma)^{16}O$ with respect the standard value are also shown (see text for more details).

as discussed in Chapter 6, the WD cooling time depends on this CO chemical profile at the beginning of the WD cooling phase. Figure 4.2 shows the interior C and O profiles at the beginning of the thermal pulses stage along the AGB for the same stellar models whose He abundance profiles were shown in Figure 4.1, but for various assumptions in the method adopted for inhibiting the convective instabilities arising at the end of the central He-burning phase (which will be discussed in the following section) and the efficiency of $^{12}C(\alpha, \gamma)^{16}O$ nuclear reaction.

4.2.4
The Breathing Pulses

Detailed evolutionary computations (see [237] and reference therein) have highlighted that when the core He abundance has been reduced by nuclear burning to about $Y \sim 0.10$, a convective instability can affect the convective core. As the core He abundance drops below this limit, α-captures by ^{12}C nuclei tend to overcome ^{12}C production by 3α reactions because of the paucity of α particles, thus He-burning becomes mainly a $^{12}C + \alpha$ production of oxygen, whose opacity is even larger than that of ^{12}C. On the basis of the same physical mechanism previously described, this occurrence causes an increase in the size of the semi-convective region and, in turn, fresh helium is transferred into the core, which is now nearly He-depleted. As shown by [232], even a small amount of He driven into the core becomes very important in comparison with the vanishing amount of He otherwise present in the core. This increase of the He abundance enhances the rate of energy production by He-burning, and thus the luminosity increases, driving an increase in the radiative gradient. As a consequence, a phase of an enlarged convection zone is started: the so-called *breathing pulse*. After this breathing pulse, the star readjusts itself, to quietly burn the fresh He in the core. Numerical simulations show that a few (usually three) major breathing pulses are expected before the complete exhaustion of He in the core.

From the point of view of the evolutionary properties, the main effects induced by the occurrence of the breathing pulses are the following: (i) the star performs, in a very short time, a loop on the HRD at each pulse (see Figure 4.6a); (ii) the He-burning lifetime is increased by more than 20%; (iii) the mass of the CO-core at the He-exhaustion is increased. This last effect implies a huge reduction of the AGB lifetime, and indeed the ratio between the core He-burning lifetime and the AGB lifetime is decreased to ~ 0.08, well below the range of values allowed by observational constraints (see [238]).

This evidence seems to suggest that the efficiency of the breathing pulse phenomenon is very low and for such a reason, their occurrence is inhibited in most of the modern numerical stellar evolution codes. Various algorithms for avoiding the occurrence of breathing pulses in models computations have been proposed in the literature: for instance, by preventing any increase of the He abundance in the convective core in the late stages of the central He-burning phase. An alternative method, suggested by [233], to suppress these convective instabilities consists in neglecting the gravitational energy contribution during the central He-exhaustion stage. Although both approaches succeed in inhibiting the occurrence of the breathing pulses, the choice of one or another method affects the final evolutionary results: for instance, the second approach produces core He-burning lifetime[3] longer by about 5% and, maybe more importantly, a different CO profile stratification is predicted with respect to stellar models based on the first approach, as shown in Figure 4.2.

4.3
The Horizontal Branch Evolution

As already mentioned, the evolutionary sequence in the HRD corresponding to the central He-burning stage in low-mass stars is named the *Horizontal Branch* (HB). From a theoretical point of view, an important definition is that corresponding to the *Zero Age Horizontal Branch* (ZAHB): a ZAHB stellar model corresponds

---

3) This is because models based on this approach show a larger semi-convective zone during the late core He-burning phase.

to a structure in equilibrium, burning He in a chemically homogeneous core, and with a chemical abundance profile in the H-burning shell resembling the chemical abundance profile of the shell when the He flash occurred at the RGB tip, this being that the CNO elements are in equilibrium in the H-burning shell. The requirement that all secondary elements involved in the H-burning are at their equilibrium configuration is because it is customary to initiate the core He-burning evolutionary sequence from an equilibrium model, with the *CNO-cycle* in the H-burning shell out of equilibrium. This approach avoids the need to compute the evolution of the stars from the RGB tip up to the beginning of the core He-burning stage.[4]

The fundamental properties of a ZAHB star are completely determined by four parameters: the He-core mass, M_{cHe}, the total mass, M, the abundance of He, and the metallicity in the envelope. The value of M_{cHe} depends on the initial chemical composition and on the total mass of the star. However, in Section 3.7.1, it was shown that in the low-mass regime, that is, $M \leq 1.4 M_\odot$, M_{cHe} is only weakly dependent on the stellar mass. This is because all low-mass stars develop a quite similar level of electron degeneracy in the core during the RGB stage, and so the He-core mass has to achieve a similar value in order to attain the thermal conditions required for the He-ignition. The He abundance in the envelope of ZAHB stars is slightly larger than the initial value by $\Delta Y \sim 0.02$–0.04 because of the *first dredge up*. The total stellar mass at the He-ignition can be different than the initial one because of mass loss occurring during the RGB evolution.

A core He-burning star is supported in its equilibrium configuration by two nuclear burning sources: He-burning occurring inside the convective core and H-burning in a shell surrounding the He-core. For each fixed mass of the He-core, the efficiency of the H-burning shell is mainly controlled by the mass of the envelope: the larger the mass of the envelope, the hotter the H-burning shell, and thus the more efficient is the H-burning process.

For a fixed chemical composition of the envelope and He-core mass (which is equivalent to a fixed total mass and initial chemical composition of the RGB progenitor), ZAHB stellar models of different total masses, that is, different envelope masses, define an almost horizontal locus in the HRD. Those with the lowest mass envelopes are located in the hottest part of the sequence, the ZAHB location moving towards cooler T_{eff} with increasing envelope mass. The range of effective temperature values spanned by these ZAHB models is very large: from $\sim 35\,000$ K for the stars with a negligible envelope mass ($M_{env} \sim 10^{-4} M_\odot$) to ~ 4000 K for more massive envelopes ($M_{env} \sim 0.4 M_\odot$). This means that the T_{eff}, and hence the colour of ZAHB models is an highly non-linear function of the residual envelope mass (see Figure 4.3). Due to the dependence of the shell H-burning efficiency on the envelope mass, the ZAHB luminosity increases when moving from the hot side of the ZAHB (structures with the H-burning shell switched-off) to the cooler side.

---

4) The He-burning occurring during the He flash is taken into account, by considering that a mass fraction of $\sim 5\%$ of C is produced during this process.

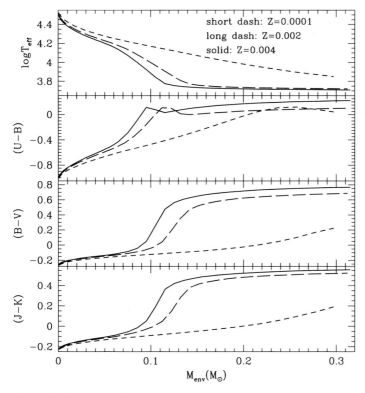

Figure 4.3 The trend of effective temperature and photometric colours from the ultra-violet to the near-infrared as a function of the envelope mass of ZAHB models for selected metallicities.

Observations of core He-burning stars in old stellar systems such as Galactic GCs fully support these theoretical predictions. In the canonical evolutionary scenario (but, see also Section 7.8 for some recent updates on this topic) the mass spread among HB stars is explained by considering that RGB stars lose, *on average*, about 0.3–0.4 $M_\odot$ due to mass loss, with some dispersion around this mean value because mass loss is an intrinsically stochastic phenomenon.

As has been discussed, HB stars are supported by two nuclear burning sources: during the ZAHB stage, the surface luminosity is primarily controlled by the He-core mass, and as a second order effect by the envelope mass via its effect on the shell H-burning efficiency. As a rule of thumb, the larger the He-core mass, the brighter the ZAHB structure is. This has in important implication: since the value of M_{cHe} does not depend on the RGB progenitor mass for masses below $\sim 1.4 M_\odot$, that is, for stellar ages larger than 4–5 Gyr, the ZAHB and, more generally, the HB brightness, is one of the most important standard candles for Population II stellar systems (see Section 7.2.3).

The dependence of the observable ZAHB properties on the initial chemical composition such as metallicity, initial He content and heavy elements chemical dis-

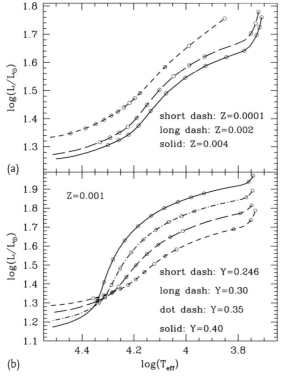

Figure 4.4 The dependence of the ZAHB location in the HRD on the metallicity and initial He content. Along each ZAHB, points mark the location of the same stellar masses, from left to right: 0.51, 0.52, 0.53, 0.54, 0.55, 0.56, 0.57, 0.58, 0.59, 0.60, 0.65, 0.70 and $0.80M_\odot$; only for the ZAHB loci corresponding to values of the He abundance $Y > 0.30$ the $0.8M_\odot$ models are not plotted because the RGB progenitor has a mass equal to $0.70M_\odot$.

tribution, is shown in Figures 4.4 and 4.5. One can summarize the results as follows:

- *Metallicity*: any increase of the global metallicity makes the ZAHB locus fainter. This is due to the combined effect of the reduction in the He-core mass at the RGB tip (see Figure 3.48) and the increase of the envelope radiative opacity. At the same time, for a given total mass of the ZAHB star, its HRD location becomes redder as a consequence of the larger opacity of the envelope;
- *Initial helium abundance*: generally speaking, when increasing the initial He abundance for a given RGB progenitor, the value of M_{cHe} decreases. This tends to make the ZAHB locus fainter as it occurs on the hot side of the sequence; at the same time, if the H-burning shell is efficient, the larger envelope He abundance increases the efficiency of the H-burning.[5] The second effect overcomes

---

5) We remind the reader that $L_H \propto \mu^7$.

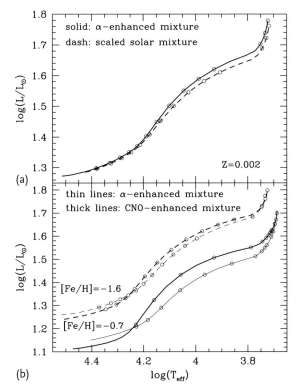

Figure 4.5 (a) The dependence of the ZAHB location in the HRD on the heavy element mixture for a fixed global metallicity. Along each ZAHB, the points mark the same mass, from left to right: 0.50, 0.51, 0.52, 0.53, 0.54, 0.55, 0.56, 0.57, 0.58, 0.59, 0.60, 0.65, 0.70 and $0.80 M_\odot$. (b) As for (a), but the ZAHB loci correspond to models computed with an α-enhanced mixture with $[\alpha/Fe] = 0.4$ and CNO-enhanced (by a factor of ~ 2 with respect the reference mixture) for two values of [Fe/H] (see labels).

the first one for ZAHB stars with a massive enough envelope and so the brightness of the ZAHB increases with the He content. One also has to note that for a given total mass, the larger the envelope He content, the hotter the ZAHB location;

- *Heavy element mixture*: when comparing α-enhanced ZAHB models with similar models with a scaled-solar heavy element distribution with the same global metallicity Z, a very good agreement is found: the luminosity difference at maximum is only of the order of $\Delta \log(L/L_\odot) \sim 0.02$, while the difference in T_{eff} is negligible. This is a proof that the "rescaling law" discussed in Section 3.9.1 is largely valid also for core He-burning structures.

On the other hand, the situation is different when comparing α-enhanced ZAHB models with those computed by accounting for an enhancement of the sum of CNO elements. We have performed this comparison, keeping the [Fe/H] value fixed, so that the global metallicity is different between the two sets of ZAHB models. Despite the lower value of M_{cHe} and the larger Z, the CNO-enhanced

ZAHB models are brighter than the α-enhanced ones. This is probably due to the fact that in the CNO-enhanced models the H-burning via the CNO cycle is more efficient as a consequence of the increased $(C + N + O)$ sum. The brightness difference between the ZAHBs at $\log T_{eff} = 3.83$ is $\Delta \log(L/L_\odot) \approx 0.05$, that is ≈ 0.12 mag in the optical photometric bands, for models with $[Fe/H] = -0.7$. This luminosity difference decreases with decreasing metallicity, as can be appreciated by considering the data for ZAHB models with $[Fe/H] = -1.6$ shown in Figure 4.5.

4.3.1
The Morphology of the HB Evolutionary Tracks

Previously, we emphasized that in a ZAHB model, the brightness is mainly controlled by the He-core mass-size while the T_{eff} is mainly controlled by the mass of the residual envelope. In some sense, during the central He-burning stage, the rule governing the path of the evolutionary track in the HRD is reversed with respect to the ZAHB stage. In fact, the main characteristic of HB evolution is that the efficiency of the H-burning shell decreases monotonically as the efficiency of the central He-burning steadily increases. As long as the luminosity produced by the H-burning shell is larger than a fixed amount, corresponding to $\sim 20\%$ of the total surface luminosity, the star evolves towards larger effective temperatures. When L_H drops below this limit, the evolutionary path reverses towards the red side of the HRD, as shown in Figure 4.6. As a consequence, the stars perform a loop in the HRD, the effective temperature extension of this loop strongly depends on the parameters affecting the H-burning shell efficiency, mainly on the envelope mass (or the ratio, q, between the envelope mass and the total stellar mass), and the envelope He abundance. Note that gravitation provides a negligible contribution to the whole energy budget during the core He-burning stage. However, at the central He-exhaustion due to the lack of He nuclei, the energy released via nuclear burning is not sufficient to keep the star in equilibrium and so the stellar core rapidly contracts; at this stage, the gravitational energy contributes largely to the energy budget of the star.

As can be seen from Figure 4.6b, at the end of the HB stage, the surface luminosity temporarily shows a small drop, followed by a monotonic rise: in old stellar systems, this causes an increase of the number of stars in the magnitude interval in the CMD corresponding to this luminosity drop. This observational feature is known as the *AGB clump*; this will be discussed in Section 5.2.

As for the central thermal properties: the central temperature increases monotonically along the whole HB stage, and only after the exhaustion of He at the centre, will the core start cooling via cooling processes associated with the increasing level of electron degeneracy; the central density decreases during the major part of the core He-burning phase; it begins to increase when the central He abundance is ~ 0.30. This occurs because, due to the reduction in the number of He nuclei available for the burning, the He-burning efficiency decreases and the star

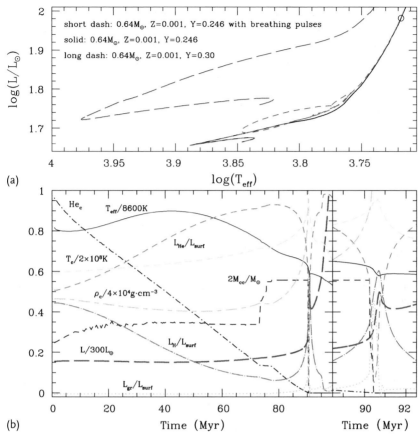

(a)

(b)

Figure 4.6 (a) The morphology of the evolutionary track for an HB structure with mass equal to $0.64\,M_\odot$ and chemical composition: $Z = 0.001$ and $Y = 0.246$. The evolutionary track corresponding to the same model in which the occurrence of the breathing pulses has not been inhibited, is also shown. The long-dash line represents the evolution in the HRD of a similar HB structure without the breathing pulses but with an initial He abundance equal to $Y = 0.30$. The open circle marks the location of the core He-exhaustion. (b) The time evolution of several physical quantities corresponding to the $0.64\,M_\odot$, $Z = 0.001$, $Y = 0.246$, without breathing pulses, model shown in (a). The symbols have their usual meaning and the physical quantities have been normalized as it has been labeled. The small panel on the right side shows the same quantities, but only during the stage of central He-exhaustion.

reacts slightly contracting; the central density increase partially offsets this process because the triple alpha energy generation rate is proportional to ρ^2.

4.3.2
The Core He-burning Lifetime

For a low-mass star, the core He-burning lifetime is of the order of 100 Myr. In fact, the actual HB lifetime (t_{HB}) depends significantly on the initial chemical composition and the residual mass of the envelope. In particular, for a given chemical composition, the smaller the envelope mass,[6] the longer the HB lifetime. In more detail, HB stars whose ZAHB location is hotter than $T_{eff} \approx 20\,000$ K, have a t_{HB} larger than $\sim 20\%$ of the corresponding evolutionary lifetime of stars populating the cool side of the HB locus (see Figure 4.7). This is because these stars are less massive and, hence, less luminous than cool (red) HB stars. This effect has to be properly taken into account when comparing theory with observations of star counts along the HB, to correctly measure the properties of a stellar population such as the initial helium abundance (see the discussion about the *R parameter* in Section 7.6).

As for the dependence on the initial chemical composition, the data shown in Figure 4.7 show that when increasing the metallicity for a given total mass of the HB star, the HB lifetime increases. This occurs because metal-rich HB stars are fainter than their metal-poor counterparts due to the combined effect of the reduction of the He-core mass at the He flash and the larger envelope opacity.

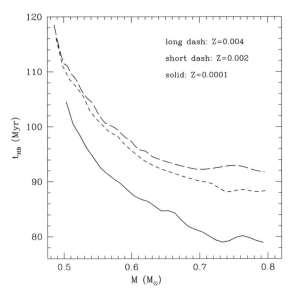

Figure 4.7 The core He-burning lifetime as a function of the total mass for HB models with different metallicities.

---

6) Since we are considering low-mass stars, we are in the regime for which it can be assumed that the He-core mass at the RGB tip is almost independent of the total mass of the RGB progenitor.

As already discussed, an increase of the initial He abundance causes a decrease of the He-core mass at He-ignition at the RGB tip, but despite this, He-enhanced HB models, massive enough to have an efficient H-burning shell, are brighter than HB models with a lower initial He content; whereas those populating the hottest portion of the HB are fainter (see Figure 4.4). As a consequence, the hot HB, He-enhanced stars have slightly longer HB lifetimes than their less He-rich counterparts (the difference is less than about 5% when moving from $Y = 0.246$ to $Y = 0.30$). The stars populating the red side of the HB also have slightly larger values of t_{HB}, despite being brighter. For instance, for the two HB models with $Y = 0.246$ and 0.30, whose evolutionary tracks are shown in Figure 4.6, the relative variation in the core He-burning lifetime is only about 2%. The longer core He-burning lifetime of these red HB stars is due to the large contribution to the energy budged supplied by the H-burning shell in He-rich stellar models.

4.3.3
Synthetic Horizontal Branches

For a given initial chemical composition, the HRD and CMD location of HB models only depends on the envelope mass of the stellar structure because in progenitors with mass well below the RGB transition, the He-core mass at the ZAHB does not depend on the initial mass. In analyses of the CMD distribution of HB stars in old simple stellar populations, one usually assumes that the envelope mass at the ZAHB is the single parameters that determines the CMD evolution of the models.

Perhaps the best approach to interpret the observed distribution of HB stars in a CMD relies on the computation of synthetic HBs. A synthetic HB is typically a Monte Carlo simulation of the expected CMD (or HRD) distribution of core He-burning stars in a simple stellar population, as determined from a grid of evolutionary HB models, for a given initial chemical composition, rate of formation of HB stars, and their mass distribution[7] [239].

Let us assume that the initial chemical composition of the HB stars has been chosen. The synthetic HB calculation starts by first selecting the mass of the HB stars (the probabilistic mass distribution has to be specified as an input) and then determining the age t_{HB} since its arrival on the ZAHB. It is usually (and reasonably) assumed that stars are being fed onto the HB at a constant rate; this implies that t_{HB} can be determined by considering a uniform probability distribution for t_{HB}, whose zero point is set at the ZAHB location.[8] Once the stellar mass and

---

7) Due to the presence of multiple populations firmly established in the last decade (see Section 7.8), it is actually possible that along the HB of most, if not all, Galactic GCs, subpopulations of HB stars with different initial chemical compositions do exist. This can be modelled in synthetic HB computations by including additional parameters in the simulations.

8) To correctly account for the variation of the HB lifetime as a function of mass, the maximum age used in the Monte Carlo simulation has to be at least equal to the maximum HB lifetime among the available tracks. If the randomly selected t_{HB} for the selected mass is larger than its HB lifetime, the object will not appear in the synthetic CMD, meaning that it has evolved to the following AGB and WD phases.

t_{HB} are specified, interpolations in mass among the available evolutionary tracks and interpolation in time along the interpolated track have to be performed.

A standard assumption for the mass distribution is a Gaussian of the form:

$$P(M) \propto \exp - \left(\frac{\langle M \rangle - M}{\sigma(M)} \right)^2$$

where $\langle M \rangle$ is the mean HB mass, and $\sigma(M)$ is the mass spread parameter. If M_{RGB} is the initial mass of the RGB progenitor, the quantity $\Delta M = M_{RGB} - \langle M \rangle$ is the mean mass lost along the RGB and $\sigma(M)$ is the mean dispersion around this value. This is derived from the assumption that mass loss along the RGB is a stochastic process, with a specified unimodal distribution.

From a historical point of view, this parametrization of the HB mass distribution was suggested by the fact that it allows one to roughly reproduce the mean colour of the observed HB distribution in several Galactic GCs but it has no real physical

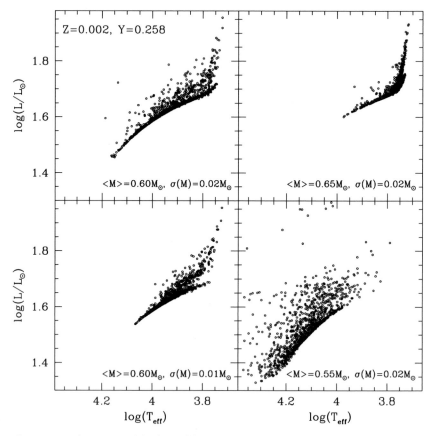

Figure 4.8 Synthetic HB models obtained for various assumptions about the value of the mean mass $\langle M \rangle$ and $\sigma(M)$, but for a fixed initial metallicity and the same initial mass of the RGB progenitor.

justification. A uniform mass distribution could be assumed as well, and indeed in some cases combinations of two (or three) Gaussian distributions or a Gaussian plus a uniform mass distribution have been also adopted to reproduce extended blue tails in the HB distribution and/or multi-modal colour distributions.

Obviously, the values of $\langle M \rangle$ and $\sigma(M)$ are free parameters in the synthetic HB simulations, although there are some constraints on $\langle M \rangle$; in fact, it cannot be larger than the initial mass of the RGB progenitor (for a given age of the population), or lower than the He-core mass at He-ignition. The effects of changing the mean mass and/or the mass spread are shown in Figure 4.8. When decreasing $\langle M \rangle$ the HRD of the synthetic HB becomes bluer and slightly fainter, because of the increasing number of bluer (less massive) HB stars; on the other hand, a reduction of $\sigma(M)$ at fixed $\langle M \rangle$, decreases the mass dispersion and the colour dispersion along the HB.

Due to the strong dependence of the bolometric corrections on T_{eff}, the morphology of the synthetic HB in the CMD depends on the adopted photometric filters, as is shown in Figure 4.9. The bluest stars along the HB become fainter when moving from the UV filters to the near-infrared ones, whereas the opposite occurs to the coolest, red HB stars. To compare synthetic HBs with observations, it is necessary to "perturb" the magnitudes of each synthetic star by a random value to mimic the

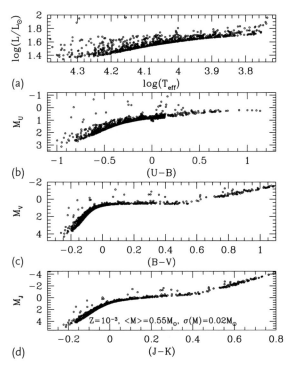

Figure 4.9 Synthetic HB simulation as obtained for a fixed initial chemical composition and values for $\langle M \rangle$ and $\sigma(M)$ (see labels in (d)), in the HRD and in various photometric planes.

effect of photometric errors following a (usually) Gaussian error distribution with dispersion (σ) provided by the photometric analysis.

Synthetic HBs are also a fundamental tool for retrieving important information on the properties of variable stars associated to the core He-burning stage, such as the RR Lyrae. It is straightforward to convert T_{eff}, luminosity and mass for each synthetic stars into, that is, pulsation period and amplitude estimates, by using predictions on the boundaries of the instability strip and analytical relations between pulsational and evolutionary properties (see Section 4.5).

4.3.4
The Second Parameter Problem

At the beginning of this chapter, we discussed the dependence of the basic properties of core He-burning, low-mass stars on the initial chemical composition, and we noted that, for a given mass, a more metal-rich ZAHB model will be fainter and redder in the HRD compared a more metal-poor model. Using appropriate synthetic HBs, one can easily predict the change in the global HB morphology induced by a variation of fundamental properties of a stellar population, such as its metallicity.

As a general rule, once the mean mass lost during the RGB, the mass dispersion around this value, and the age of the stellar population (hence, the initial mass of the RGB progenitor) are fixed, any increase of the global metallicity Z forces the HB star to lower T_{eff}, and hence redder colour. The physical explanation for this behaviour is related to the following combined effects associated with the metallicity increase: (i) for any given initial mass of the RGB progenitor, the He-core mass at the He-ignition will decrease, and so for a fixed mass of the ZAHB star, the envelope mass is larger in more metal-rich stars, this causes the H-burning in the shell to be more efficient and the ZAHB location (and indeed the global HB evolutionary track) become(s) cooler; (ii) the radiative opacity in the envelope increases; (iii) at a fixed age for the stellar population, the evolutionary mass at the RGB tip in a metal-rich isochrone is larger than that of a metal-poor isochrone as a consequence of longer MS lifetimes of metal-rich stars.

In the previous discussion, we assumed that the amount of mass lost by RGB stars with different metallicities is on average the same. However, when assuming that mass loss occurs at a rate governed by the Reimers law (see Section 2.6 for a discussion on the shortcomings associated with the use of this mass loss law), more metal-rich RGB stars will lose more mass than metal-poor RGB stars, as a consequence of their cooler T_{eff} and brighter luminosity at the RGB tip. However, the previously quoted structural and evolutionary effects in metal-rich HB and RGB stars largely overcome the effect of a higher mass loss rate during the RGB stage; so theoretical predictions about the general trend of the HB morphology with the metallicity are robust. As a consequence, metallicity is the main, "the first", parameter controlling the HB morphology.

The importance of metallicity in governing the HB morphology was early discovered thanks to accurate observations of Galactic GCs [240] and it was rapidly

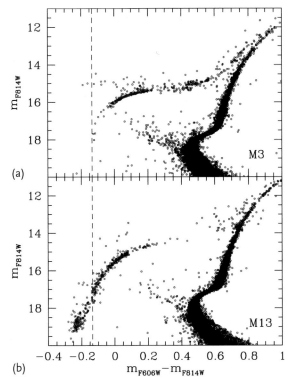

Figure 4.10 The CMDs of the two Galactic GCs, (a) M3 and (b) M13, in the F814W and F606W HST filters. The dashed line marks the colour of the hottest stars along the HB of M3.

confirmed by the first reliable HB models [241]. However, soon after this important theoretical achievement, it was pointed out [242, 243] that the correlation between colour/effective temperature and metallicity had several exceptions: there are several pairs of Galactic GCs with exactly the same metallicity but markedly different HB morphology. This raised the still unsettled problem in stellar astrophysics that has become known as the *second parameter problem*.

Figure 4.10 shows the comparison between the HB morphologies of two Galactic GCs (M3 and M13) presenting one of the best examples of the second parameter at work. These two GCs have a similar iron content, [Fe/H] ~ -1.50, but strikingly different HB morphology: in particular, M13 has a tail of HB stars much bluer than the hottest HB stars in M3.

Before continuing the discussion on this issue, it is useful to remember that the HB morphology is commonly characterized by adopting the parameter:

$$HB_{type} = \frac{N_B - N_R}{N_B + N_V + N_R}$$

where N_B, N_R and N_V denote the number of stars at the blue side and red side of the RR Lyrae instability strip, and within it. The parameter HB_{type} is degenerate for clusters populated only at the red or blue side of the instability strip; in these cases, different colour distributions give the same value of this parameter (as shown in Figure 9.16 of [20]). For M3 and M13, the parameter HB_{type} is equal to 0.08 and 0.98, respectively.

Since the discovery of this discrepancy between the expected (on the basis of the GC metallicity) HB morphology and the observed morphology, several explanations have been suggested (see [244] and references therein). In the following, we summarize the most important candidates for the role of the second parameter:

- *Cluster age*: if metallicity is kept fixed and the age of the stellar population is increased, the HB becomes bluer due to the lower average mass evolving along the HB. In fact, mass loss along the RGB is practically independent of age because low-mass stars have quite similar evolutionary lifetimes, T_{eff} and brightness during the RGB stage. So in principle, an age difference between two GCs affected by the second parameter problem could be a viable solution for explaining the striking difference in the HB morphology. The problem is that the age difference, required for explaining the different HB morphology among various GCs with the same metallicity, seems to be ruled out by accurate GC relative age estimates [245];
- *Initial He abundance*: an increase of the initial He abundance tends to make the evolutionary path in the HRD of a HB star hotter (bluer), a consequence of their more extended blue loop. However, in addition to this effect one has to bear in mind that He-enhanced stellar models, during both the core and shell H-burning stages, have shorter evolutionary lifetimes. This means that, for a given age, a He-enhanced isochrone involves an evolving mass lower than that of a "normal" He isochrone. This occurrence, in turn, makes bluer the HB morphology for any fixed average mass loss efficiency along the RGB.

  The problem with this scenario was, until a few years ago, that there were convincing estimates of the initial He abundance in the Galactic GC population showing an almost constant He abundance, in very good agreement with the cosmological expectation, that is, $Y_{GC} \approx 0.245$. However, recent empirical findings concerning the multi-population phenomenon in Galactic GCs are changing this view, and there is striking evidence for the presence of various stellar sub-populations characterized by different initial He contents within the same GC. As a consequence, the role of He as the second parameter is becoming more plausible;[9]

Besides these second parameter candidates, many more have been suggested during these last decades, such as enhanced mass loss during the RGB stage as a consequence of dynamical stellar interaction in high-density GCs, and peculiar CNO elements abundance, and so on; but these appear to play a marginal role, if any, in comparison with the previously mentioned candidates.

Nowadays, the emerging complex view of stellar populations in Galactic GCs clearly shows that it is misleading to refer to the problem of the HB morphology

---

9) This argument will be discussed in more detail in Section 7.8.

as only the problem of the *second parameter*. It is becoming clear that two or more parameters (such as age, helium, and so on), together with the metallicity, play a relevant role, with different efficiencies in the various clusters, in driving the observed HB morphology (see Section 7.8).

4.4
Extremely Hot Horizontal Branch and "Blue Hook" Stars

We now know that, for a fixed chemical composition, the luminosity of the ZA-HB of low-mass stars is mainly governed by the He-core mass, while the T_{eff} is controlled only by the envelope mass. In the canonical stellar evolutionary framework, the formation of extremely hot (blue) HB (EHB) stars, that is, those stars with $T_{eff} > 18\,000\,K$ and envelope mass less than $\sim 10^{-2}\,M_\odot$ can only occur if their RGB progenitor experienced a very efficient mass loss. Although, on general grounds this is possible, by investigating the envelope mass range corresponding to EHB stellar models (see Figure 4.3) one discovers that it is very narrow and depends in a non-linear fashion on the metallicity. This points to some shortcomings in this evolutionary scenario for the formation of EHB stars:

- It is evident that the mass loss efficiency needs to be extremely well fine-tuned, such that a RGB star loses a large fraction of its original envelope but it stops the process just short of losing the entire envelope;
- Since EHB stars are present in stellar environments with a wide range of metallicity, it could be expected that the quoted mass loss "fine tuning" would require some significant adjustment with metallicity;
- It is commonly assumed that mass loss does not appreciably affect the evolutionary properties of RGB stars [246], and for a "not-too-large" mass loss efficiency[10] this is an accurate approximation. However, when considering very high mass loss efficiencies, one would expect that the structural and evolutionary properties of the stars are actually affected by the mass loss process.

Let us now discuss the impact of a high mass loss efficiency on the evolution of low-mass RGB stars. Since it is customary in stellar model computations to adopt Reimers mass loss law, in the following we specify the mass loss efficiency on the basis of the value adopted for the free parameter η in this mass loss prescription. For "canonical" values of η ($\eta \sim 0.4$), the structural and evolutionary properties of low-mass RGB stars are not affected at all. Evolutionary computations show that stars start to react to the mass loss only when the envelope mass, M_{env}, drops below $\sim 0.1\,M_\odot$.

---

10) When accounting for mass loss during the RGB stage, it is customary in stellar model computations to use the Reimers law. The use of a value of ~ 0.3–0.4 for the free parameter η present in this mass loss recipe allows one to reproduce the colour distribution of the bulk of the HB stars in Galactic GCs.

When $M_{env} \sim 0.06\,M_\odot$ for stars with metallicity[11] $Z = 2 \times 10^{-4}$, the T_{eff} starts to increase and the star definitively ends its normal RGB evolution. However, the star remains near the RGB, still losing mass, and so reducing its envelope mass. When $M_{env} \sim 0.007\,M_\odot$, the star finally leaves the RGB, though before experiencing the He flash, and achieves an effective temperature $\log T_{eff} \approx 3.80$, the outer residual envelope shrinks rapidly, and the star moves rapidly to the hot side of the HRD. These peculiar stars are the so-called *RGB stragglers* [247].

Depending on the mass of the He-core developed during the RGB stage, that is, on the thermal conditions attained by the He-core before the star leaves the RGB, there are two possible evolutionary routes for *RGB stragglers*:

- The most obvious fate for these stars is that they reach the WD cooling sequence and start dying quietly as He-core WDs [248]. Before [247], it was commonly believed that this was the only evolutionary scenario for the RGB stragglers progeny. This scenario has an important implication: the smallest envelope mass allowing a star to experience the He flash at the RGB tip sets the minimum envelope mass, and hence the maximum T_{eff} of HB stars;
- It has been shown that, despite the fact that RGB stragglers are cooling down along the He WD cooling sequence, for selected choices of the free parameter η, is still possible for them to experience a He flash [247]. This can be understood by considering that, although the residual mass of the H-rich layers is extremely small, the H-burning shell can contribute to a slight increase in the mass of the He-core. In addition, the outer layers of the He-core, which are in conditions of not high electron degeneracy, can continue to contract.
  The release of thermal energy associated with He-core contraction is eventually able to warm the core and hence to allow He-burning ignition. Because this delayed He flash occurs along the WD cooling sequence, that is, at very large values of T_{eff}, this sub-class of RGB stragglers are named *hot He-flashers* [249]. These stars, as they approach the core He-burning stage, populate the hottest portion of the HB sequence because of their extremely small H-rich envelope.

The existence of this second evolutionary channel for producing EHB stars has an important consequence (see the discussion in [249]): the range of values for the free parameter, η, suitable for producing EHB stars is almost constant, regardless of the metallicity, and, more importantly, a fine-tuning no larger than that required for producing cooler HB structures is needed.

The evolutionary computations of low-mass RGB stars that include a high value for the efficiency of the mass loss have shown that the structural properties and, in particular, the He-core mass of RGB stragglers when they leave the RGB, depend on the mass loss efficiency: the larger the mass loss efficiency, the lower the He-core mass. From the data in Figure 4.11, it is clear that for η values lower than about 0.65, the He-core mass at the coolest point along the RGB is barely affected by the occurrence of the mass loss. However, for larger values of η, M_{cHe} drops by about $0.01\,M_\odot$ with respect to the standard prediction. So stellar models predict that the

---

11) This critical value for the envelope mass is only very slightly dependent on metallicity.

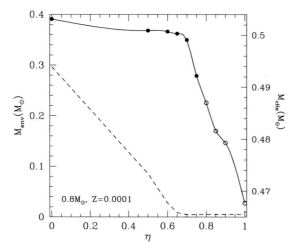

Figure 4.11 The He-core mass (solid line) and the residual mass of the envelope (dashed line) at the RGB tip for models experiencing the He flash at the RGB tip, or the same quantities at $\log T_{eff} = 3.80$ for RGB stragglers, for various values of the mass loss efficiency, specified by the value of the free-parameter η entering in the Reimers law. The full dots represent those stellar models able to ignite He at the RGB tip or as hot He-flashers, the empty dots mark those models that cool down along the WD sequence without igniting He-burning.

He-burning ignition in hot He-flashers has to occur in a He-core whose mass is slightly reduced with respect canonical models, thus making the corresponding ZAHB models slightly fainter in comparison with ZAHB models for EHB stars computed under the assumption of a constant He-core mass.

Detailed analysis of the evolutionary properties of hot He-flashers [200, 201, 247, 249–251] have shown that within this scenario, there are significant differences in the evolutionary outcomes. This has important consequences on the modelled properties of the core He-burning progeny. The hot He-flashers have to be divided in two sub-classes, the *early He-flashers* and the *late He-flashers*, depending on the location of the star along the WD cooling sequence when the He flash occurs:

- *Early He-flashers* (EHeF): these stars are able to achieve the thermal conditions for He-ignition along the brighter portion of the WD cooling sequence, at $\log(L/L_{\odot}) > 1.8$–2.0. The development of the He flash follows the canonical prescriptions, and their progeny have a ZAHB location corresponding to the hottest point along the ZAHB locus predicted by the canonical evolutionary framework;
- *Late He-flashers* (LHeF): these stars experience the He flash when they have already cooled down the WD sequence. He-burning ignition occurs under conditions of strong electron degeneracy, and with the H-burning shell almost completely extinguished. What distinguishes the behaviour of EHeF stars from the LHeF stars is that in the LHeF stars, the convective shell developing in the He-core (above the point where the He-burning starts) is able to reach the H-rich

envelope because of the huge energy release associated with the He flash. This leads to the ingestion of protons from the outer envelope into the very hot interior of the He-core; a process known as *He flash-induced mixing*.

The critical quantity at the basis of the dichotomy in the He flash development between EHeF and LHeF stars is the H-burning efficiency in the shell: if the H-burning is efficient, the local entropy is so large to act as a barrier, preventing the outward penetration of the He flash induced mixing into the H-rich envelope. This corresponds to the normal situation in canonical stellar models which ignite He at the RGB tip. However, when the H-burning in the shell is not very efficient, as happens when the envelope mass is very small and the burning efficiency in the shell drops as the RGB straggler cools down the WD cooling sequence, the entropy barrier is not high enough to prevent the He flash-induced mixing process. In passing, we note that such behaviour is very similar to that characteristic of extremely metal-poor (Population III) stars during the He flash at the tip of the RGB (see [202] and references therein).

Figure 4.12 shows evolutionary paths in the HRD for various assumptions about the mass loss efficiency along the RGB, corresponding to the various evolutionary channels discussed here, that is stars igniting He at the RGB tip, EHeF and LHeF objects. Figure 4.13 shows the development of the He flash induced mixing in the same LHeF model of Figure 4.12. This figure reveals that soon (∼ 5 months) after the He flash has attained its maximum, the He flash-induced mixing zone reaches

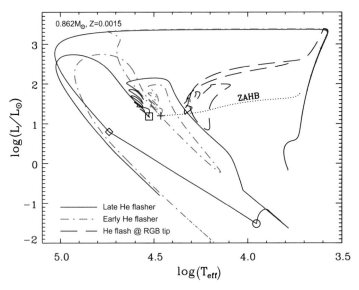

Figure 4.12 The path in the HRD of evolutionary tracks corresponding to models with the same RGB progenitor (see the label) but with three different efficiencies for the mass loss during the RGB stage. The various evolutionary tracks correspond to: a star able to ignite He at the tip of the RGB, an early hot He-flasher, and a late hot He-flasher. For the latter model, the diamond, the circle and the square mark the occurrence of the He flash, the flash-induced mixing, and the ZAHB location, respectively.

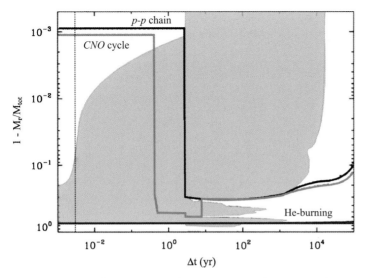

Figure 4.13 The evolution with time of the location of the convective region boundaries from the beginning of the He flash for the LHeF star shown in Figure 4.12. The temporal behaviour of the location of the maximum energy release via *p–p chain*, *CNO cycle* and He-burning is also indicated. The dotted line marks the time when He flash luminosity achieves its maximum.

the lower boundary of the H-rich envelope. As soon as the convective shell starts ingesting protons in the hotter internal layers, this hydrogen is burnt via *CNO cycle* at an extremely high rate (due to the large temperatures in the He-core, and the large amount of carbon produced by 3α-reactions).

Despite the huge amount of energy released by the H-burning, the star does not expand to an RGB configuration because of the extremely small size of the residual envelope. Due to the highly efficient H-burning, the He flash-induced mixing zone initially splits in two, and then in three, distinct convective shells: the He-burning shell and two shells associated with *p–p chain* and *CNO cycle* burning processes.

An important consequence of the He flash-induced mixing is that since the upper boundary of the convective zone moves further outwards during this process, it can eventually reach the outer layers of the stars; so a huge amount of material processed by both H- and He-burning is dredged up to the surface. The surface of LHeF stars should thus be strongly enhanced in C and He.[12]

These huge variations in the surface chemical abundances affect the opacity of the outer layers and, therefore, also the predicted emergent radiative flux. Figure 4.12 suggests that the ZAHB location of LHeF objects is expected to be ~ 4000–5000 K hotter than the ZAHB location of canonical EHB stars, and slightly fainter as a consequence of the smaller He-core at the He flash. A deeper study [200] on the predicted spectra of stars with surface chemical abundances expected for LHeF

12) Evolutionary simulations [201, 251] predict that, depending on the properties of the star, such as its total mass and initial metallicity, when the He flash occurs, the final surface abundances by mass of H, He and C should be: $\sim (2-4) \times 10^{-4}$, $\sim (0.94-0.96)$, and $\sim (0.03-0.04)$, respectively.

stars shows that: in a stellar atmosphere with a normal H abundance, the H opacity shortward of ∼ 910 Å redistributes the flux from the far-UV photometric bands to longer wavelengths; in a He-enhanced stellar atmosphere, the H-opacity source is largely depressed and a relatively larger fraction of the flux is emitted in the far-UV. When both He and C are enhanced, as seems the case for LHeF stars, the opacity contribution coming from C partially restores the opacity at short wavelengths, and so the resulting spectra are redder and fainter in the far-UV with respect to HB stars with a normal surface He abundance.

In conclusion, due to the combined effects of their characteristic structural properties and spectral peculiarities associated with the huge enhancement in the He and C abundances, the progeny of LHeF stars populates the extremely hot portion of the HB, and is eventually separated by a gap in T_{eff} from the hottest point of the canonical ZAHB locus. They should also appear significantly fainter in the far-UV photometric bands than canonical stellar models.

Accurate photometric observations in the far-UV of hot HB stars in the two Galactic GCs ω Cen [252] and NGC 2808 [200] have revealed the presence of a significant group of EHB stars, underluminous by about ∼ 0.7 mag in the far-UV bands, and forming a blue hook-like feature in the far-UV CMD (see Figure 4.14); for this reason, these peculiar stars are now known as *Blue Hook* stars.

Accurate spectroscopic measurements of effective temperature, gravity and surface chemical abundances in a large sample of Blue Hook stars in Galactic GCs nicely confirm the theoretical predictions about the properties of the progeny of LHeF objects ([253, 254] and references therein). However, empirical measurements find a surface abundance of H larger than predicted by models. This could be readily explained by considering the outward diffusion of H, and the concomitant gravitational settling of He, in the atmospheres of Blue Hook stars. Indeed,

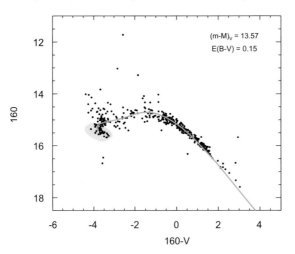

Figure 4.14 The far-UV CMD of the Galactic GC ω Centauri [252]. The shaded region indicates the region of the CMD where Blue Hook stars are located. The solid line corresponds to a theoretical ZAHB locus.

the net effects of these physical processes has to be considered together with mass loss during the HB stage (see [251]). In fact, the final surface chemical patterns are the result of the convolution of various effects associated with the different physical processes occurring once an LHeF star settles on the ZAHB.

We note that the He flash-induced mixing scenario is important not only for explaining the presence of Blue Hook stars in some Galactic GCs, but also because it provides a viable evolutionary channel for explaining the observational properties of hot subdwarfs in the Galactic field. These hot subdwarfs are considered the counterparts of hot HB stars in GCs, and they are spectroscopically classified as sdB and sdO[13] [255]. Many sdO stars appear to be largely He-enhanced, and this He-enhancement is always accompanied by an enhanced C abundance. For the sdB subdwarfs, only a small fraction ($\sim$ 5%) shows a He-enhancement, but again, the He-enhanced sdB are also C-enhanced.

The chemical peculiarities observed in sdO and sdB stars could be evidence that they are the equivalent in the field of the Blue Hook stars found in Galactic GCs. Although there is, in general, a good agreement between the empirical evidence and the theoretical predictions on the LHeF scenario, the finer details of the observed chemical patterns can be fully explained by only taking into account additional non-canonical processes such as gravitational settling, radiative levitation and mass loss [251].

It is important to consider how it is possible to achieve the large mass loss efficiencies required to drastically reduce the mass of the stellar envelope during the RGB stage. Indeed, it is quite improbable that such a large mass loss efficiency can be achieved in the context of a "normal" mass loss associated with stellar winds during the RGB stage. There are only two viable channels for attaining the required reduction of the envelope mass in RGB stars: (i) dynamical interaction between single stars and/or a single star and a binary system (this has a larger interaction cross section than a single star) in the crowded stellar environments of the nuclei of GCs, (ii) mass loss as a consequence of overflow of the Roche lobe in binary systems. Both scenarios have their own shortcoming and advantages from the theoretical and the observational sides.[14]

4.5
Low-Mass Pulsating Stars in the Core He-burning Stage

In the HRD, there are several specific regions where if a star crosses them during its evolution, the star will experience radial oscillations. As a consequence of these pulsations, the luminosity and the radius of the stars change periodically with time,

---

13) Stars with $T_{eff} < 30\,000$ K are commonly classified as sdB, while stars with $T_{eff} > 40\,000$ K are commonly considered sdO. The objects whose T_{eff} is within the range from 30 000 to 40 000 K are classified as sdO, sdB or sdOB.

14) We note that the occurrence of multi-populations in Galactic GCs, in some cases also including He-enhanced sub-populations, seems to offer an alternative possible scenario for the formation of Blue Hook stars (see Section 7.8).

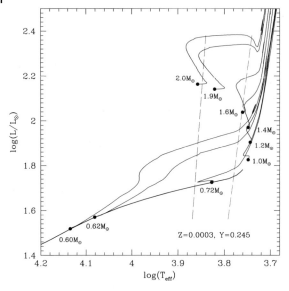

Figure 4.15 The location in the HRD of evolutionary models for He-burning stars corresponding to various kinds of pulsators such as BL Her, RR Lyrae and Anomalous Cepheids. The two dashed lines show the approximate location of blue and red edge of the instability strip.

that is, the stars become radial variable stars. These peculiar regions of the HRD are called *instability strips*.[15] We are interested here in the instability strip associated with the He-burning stage, and, in particular, in the low-mass, variables stars populating this strip, such as RR Lyrae, Population II Cepheids and Anomalous Cepheids. This instability strip is also known as the *Cepheid instability strip* because it also hosts (at brighter magnitudes) the most important Population I variables, the Cepheids. In this work, we are interested in the fainter part of this instability strip, corresponding to smaller pulsational periods, which is usually called the *RR Lyrae instability strip*. The relative position in the HRD of the various classes of pulsating stars with which we are faced in this section is shown in Figure 4.15.

Population II variables, and in particular RR Lyrae stars, are fundamental distance indicators. They have been used extensively for measuring the distance to stellar systems within the Local Group. Different methods for using these variables as standard candles have been designed, each with its own advantages and drawbacks, and some of them will be discussed in Chapter 7.

However, one has to bear in mind that radial pulsators are also an important diagnostic tool of the internal structural properties of stars. In fact, it has been known since the fundamental analysis performed by Ritter in 1879 that the pulsational period of a homogeneous sphere experiencing adiabatic radial pulsations is

---

15) A detailed discussion of the various instability strips can be found in the textbook from [256] and several review papers such as [257] and [258].

connected to its mean stellar density by the relation:

$$P\sqrt{\rho} = Q$$

where Q is the pulsational constant, which depends slightly on the mass of the variable.

The existence of this relation, connecting an observable, namely, the pulsational period, with a structural property, that is, the mean stellar density, provides a plain evidence of the importance of stellar pulsations to stellar evolution theory: radial pulsations analysis provides a formidable tool for investigating stars, allowing a quantitative test of the reliability of theoretical evolution predictions independently of the comparison with photometric data in the CMD.

The main theory of pulsating stars was developed by Eddington [259, 260] who showed the physical reasons why some stars pulsate while others do not. In principle, any star experiencing a transient perturbation in its interior or in the external regions may experience radial pulsations. However, as demonstrated by Eddington, these induced pulsations would be damped out very quickly, on a timescale of few thousand years. For a given stellar mass, the radius of a star is roughly fixed by the energy flow through the star. In order to have stable radial pulsations, this energy flux, and in turn the radius, has to vary in a periodic fashion.

It was recognized early on that the necessary modulation of the outward energy flux would be achieved if the opacity at some suitable level in the stellar envelope would increase during the phase of compression (during which the envelope becomes hotter) and decrease during the expansion phase, thus releasing the energy absorbed during compression. Detailed analysis showed that the He- and H-ionization zones located near the surface, for instance, in RR Lyrae stars, the total mass above the base of the He-ionization zone is only $\approx 10^{-7} M_\odot$, are responsible for driving the pulsations, during which the stellar luminosity for a RR Lyrae star can vary by a factor of 2 and the radius can change by $\approx 20\%$.

A fundamental role in the developing/quenching of the radial pulsation is played by convective instabilities in the outer stellar layers: as a rule of thumb, when convection becomes an efficient energy transport mechanism in comparison with radiative transport, it significantly contributes to the quenching of the pulsations. As we will discuss more in detail, the existence of a finite width of the instability strip is due to the efficiency of outer convection and to the shift of the location of the H and He partial ionization zones in the stellar envelope as a consequence of a change of T_{eff}.

4.5.1
RR Lyrae Stars

Since the early twentieth century, thanks to the extensive work of Bailey [261, 262], it is known that all GC stars populating a well-defined region of the HB with $(B-V)$ colour in the range ~ 0.2–0.4 mag, are variable stars with regular light curves and periods from ~ 0.2 to ~ 0.8 days. They were called RR Lyrae stars, a name taken from RR Lyrae itself, a field variable star found by Fleming in 1899.

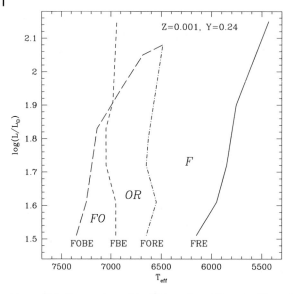

Figure 4.16 The topology of the RR Lyrae instability strip. The various lines mark the location of First Overtone blue edge (FOBE), the Fundamental blue edge (FBE), the red edge of the First Overtone zone (FORE), and the red boundary of the Fundamental mode region (FRE).

By analyzing the distribution of data in the *Period-Pulsational Amplitude (A)* diagram, the so-called Bailey diagram (see Figure 6.14 in [20]) – [261] found that these pulsators can be divided in two distinct classes: RR_{ab}, with asymmetric light curves and large amplitudes, of the order of 1 mag in the *B* band, decreasing for longer periods; and RR_c, with short periods and symmetric, smaller amplitude light curves. On the basis of the accurate theoretical investigations, it is now known that the former class of variables are stars radially pulsating in the fundamental mode, whereas those of the latter class are pulsating in the first overtone mode. The two types of RR Lyrae are not uniformly distributed inside the instability strip. This is because the pulsational modes are stable in two different regions of the HRD. This can be seen in Figure 4.16, which shows the complex topology of the RR Lyrae instability strip.

4.5.1.1 The Instability Strip

The hottest boundary, the *blue edge*, of the instability strip is located at an effective temperature of $\sim$ 7200 K at the ZAHB luminosity level: a value that slightly decreases with increasing stellar luminosity. The reason for the existence of a blue edge, and of the change of its location with luminosity, has to do with the fact that, for a given mass and luminosity, as the surface temperature increases, the mass-size of the stellar layers above the ionization zones decreases, and, in turn, the contribution of these ionization zones to the pulsational driving decreases. Moving from the blue edge to the red side of the HRD, there is a small region within which only the first-overtone mode is stable (the so-called *FO zone*) and on the right of this on the HRD, both modes can be stable (the *OR region*). Here, a star can pulsate as a fundamental, as a first-overtone, or as a double-mode variable. Many RR Lyrae have been found to pulsate in both modes: these are called *double pulsators* or RR_d variables. The RR Lyrae pulsating simultaneously in both modes are very important for constraining evolutionary models for low-mass, core He-burning stars. In fact, a comparison between theory and observations in the $P_1/P_0 - P_0$ plane (where P_0 and P_1 are the fundamental mode, and the first overtone pulsational period), the so-called Petersen diagram (see Figure 6.15 in [20]), is a fundamental tool for measuring the mass of these pulsating stars.

Moving further to the red, a region appears where only the fundamental mode is stable (the *F zone*). The red side of the instability strip is delimited by the so-called *red edge*, located at $T_{eff} \sim 5900$ K, whose existence is related to the presence of convection in the stellar envelope; this being an efficient energy transport mechanism, it quenches the pulsational process. At larger luminosities, no first overtone pulsators are expected, only pulsations in the fundamental mode are possible.

The width and the topology of the RR Lyrae instability strip is almost independent of the metallicity in the whole metallicity range of Galactic GCs. Only when considering very metal-rich stellar populations, with metallicity $Z \geq 0.02$, one can note a global shift of the instability strip towards lower luminosities, and a reduction of the strip width by about 200–300 K due to the shift of *F zone* red edge towards larger T_{eff} then found for metal-poor stellar populations.

4.5.1.2 The Pulsational-Period Relation $P = f(M, L, T_{eff})$

As already mentioned, the pulsation period of an RR_{ab} is connected to the main stellar structural parameters such as mass, luminosity and effective temperature: $P = f(M, L, T_{eff})$ [263, 264]. An updated relation for fundamental mode pulsators valid in the metallicity range $10^{-4} \leq Z \leq 0.006$ is that obtained by [265]:

$$\log(P_0) = 11.038 + 0.833 \log\left(\frac{L}{L_\odot}\right) - 0.651 \log\left(\frac{M}{M_\odot}\right)$$
$$- 3.350 \log(T_{eff}) + 0.008 \log(Z)$$

that can be also written in the form

$$\log(P_0) = 11.038 + 0.833 A - 3.350 \log(T_{eff}) + 0.008 \log(Z)$$

where A is equal to $\log(L/L_\odot) - 0.781 \log(M/M_\odot)$, and depends on the mass-to-luminosity ratio of the star. A similar relation can be also derived for first-overtone variables.

These relations are a fundamental tool in investigating the link between pulsations and stellar evolution. Note that, for each fixed T_{eff}, the range of allowed masses has little influence on the periods. So, in principle it is possible to use a pulsation period to constrain the HB luminosity [266], while it has been shown that the mass-luminosity parameter, A, is a sensitive function of the original He abundance (as discussed in Section 7.6); therefore, the pulsational properties of

RR Lyrae stars can also be linked to a parameter of cosmological relevance such as the initial He abundance.

When combining $P = f(M, L, T_{eff})$ relations for fundamental and first overtone variables, one finds

$$\log P_1 \approx \log P_0 - 0.130$$

This relation has often been used to remove the complication of the two mode periods, dealing in the analysis only with fundamental and "fundamentalized" periods, the last one obtained by implementing the empirical first overtone period according to the previous relation.

4.5.1.3 The Oosterhoff Dichotomy

In the field of RR Lyrae stars, a very important issue is the so-called Oosterhoff dichotomy in the properties of RR_{ab} stars in Galactic GCs. The problem was raised as early as 1939 by Oosterhoff [267], who discovered that the mean periods of RR_{ab} populating different clusters are clumped around two distinct values: $\langle P_{ab} \rangle = 0.55$ and $\langle P_{ab} \rangle = 0.65$ days for the GCs belonging to the so-called Oosterhoff I (Oo I) group and Oosterhoff II (Oo II) group respectively; where the Oo I GCs (which include M3, M5) are more metal-rich than the Oo II GCs (which include M15, M68).

The Oosterhoff dichotomy has been a source of great interest for decades, especially after it was shown that our Galaxy is the only known example to exhibit such a dichotomy. The RR Lyrae in neighbouring dwarf galaxies have values of $\langle P_{ab} \rangle$ different to that of either the Oo I or Oo II groups in the Galaxy. Therefore, an understanding of the physical reasons at the root of the Oosterhoff dichotomy is essential to obtaining a deeper insight into the formation and evolution of the globular cluster system, as well as the formation history of the Galaxy as a whole.

The leading theoretical explanation for this phenomenon involves a dichotomy in the "transition period" between the RR_{ab} and RR_c variables, reflecting a difference in effective temperature at the transition point, that is, when a variable star eventually switches its pulsation mode from fundamental to first overtone or vice versa. This explanation, named the *hysteresis effect* [268], is based on the idea that a variable star eventually entering in the *OR region* of the instability strip, continues pulsating in its pulsation mode: it would continue pulsating in the fundamental mode if coming from the *F zone* or in the first overtone if it was a RR_c variable.

The *hysteresis effect* would cause a delay in mode switching (from, for example, RR_{ab} to RR_c and vice versa) and would occur at different temperatures, depending on the direction of evolution. According to this scenario, for the Oo I intermediate metallicity clusters, most of the HB stars begin their lives in or near the instability strip. The RR Lyrae become hotter and their periods become shorter as they evolve, and eventually the RR_{ab} stars switch their mode to become first-overtone variables. In the Oo II metal-poor clusters, the progenitors of RR Lyrae stars begin their core He-burning stage on the hotter (blue) side of the instability strip and then evolve to lower temperatures and longer periods as they traverse the instability strip at luminosities appreciably higher (giving rise to a longer pulsation period) than the

ZAHB. Thus, a luminosity difference contributes to the hysteresis effect, in addition to the direction of evolution.

This scenario has the advantage of explaining the larger fraction of RR_c variables in Oo II clusters [269]. Although we still lack a sound physical explanation for the *hysteresis mechanism*, it is recognized that the Oosterhoff dichotomy in Galactic GCs can be properly explained when accounting for the *hysteresis mechanism* and for the canonical dependence of pulsational and evolutionary properties of stars on the metallicity.

4.5.1.4 The Rate of Pulsational-Period Change

Within the legacy left by Eddington (1918), one of the most simple and weighty insights connecting stellar evolution and pulsation was the evaluation of period changes due to evolutionary effects. It is evident from the functional relation connecting the pulsation period to the stellar luminosity and effective temperature, that during the evolution of a star within the instability strip, its pulsational period has to change. Period changes can provide useful clues about the variation of the density stratification in a star caused by the evolution, and can also be connected to the evolutionary rate in the HRD. Even though the measurement of this quantity is, in principle, simple, reliable estimates are often lacking due to systematic errors and to spurious effects introduced by the use of different sets of sometime very old photometric data.[16]

On the basis of the $P = f(M, L, T_{eff})$ relationship between the period and the main evolutionary properties of the stars, for fixed total mass, the period increases with increasing luminosity and/or decreasing T_{eff}. Thus, depending on the evolutionary path of the star inside the instability strip, one can obtain different solutions for the behaviour of the period change, $\dot{P} = dP/dt$, as a function of time: a monotonic increase of the period for stars that cross the strip moving towards the red side of the HRD during their off-ZAHB evolutionary phases (see Figure 4.17), or a non-monotonic behaviour for those stars which experience a blue loop inside the instability strip. In any case, it is obvious that the faster the evolutionary rate, the larger the rate of period change. This is the reason why the data in Figure 4.17 show that type II Cepheids have rates of period change as a function of time that are, on average, 2 orders of magnitude larger compared to RR Lyrae stars.

On the observational side, it is well-established that both fundamental and first-overtone variables tend to have positive rates of period change: usually in the range (0.0–0.1) days/Myr. While the trend towards an increasing pulsational period is supported by evolutionary computations, the measured value is about an order of magnitude larger than model predictions. Observations also show that there is an excess of large period changes among RR_{ab} stars compared to RR_c stars. A plausible explanation for larger period changes among fundamental pulsators when compared with first overtone pulsators is explained by the topology of the instability strip. The region where the fundamental mode is stable is a factor of 3–4

---

16) Needless to say that in order to obtain a firm estimate of the rate of period change,
$\beta = \dot{P} = dP/dt$, a large temporal baseline in the RR Lyrae observations is mandatory.

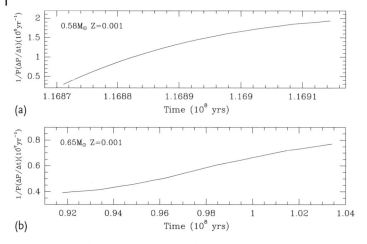

Figure 4.17 (a) The behaviour with time of the pulsation period change for a typical Pop. II Cepheid while the star evolves within the instability strip. (b) As for (a), but for a typical RR Lyrae star in a metal-poor stellar system.

larger (in T_{eff}) than that for first overtone mode. This implies that there is a higher probability of detecting larger rates of period change among fundamental mode variables.

Thus far, there is no sound explanation for the disagreement between theory and observations concerning $\dot{P}$: it could be due to an incorrect theoretical prediction of the evolutionary rate in the HRD and/or to the presence of random observational errors.

4.5.2
Population II Cepheids

Those Cepheid-like variables that do not clearly belong to the Pop. I Classical Cepheid class have had a confused nomenclature over the years. They have been known as Pop. II Cepheids, even though many do not show the chemical abundances and/or kinematics of halo stars. On the other hand, some of the "Cepheid-like" variables in Galactic GCs have been called BL Her or W Vir stars, after the prototypes in the field, on the basis of similarities in their light curves and periods. However, these field stars often have the chemical composition and kinematic properties of Pop. I objects. To add to the confusion, various other names, which attempt to imply an evolutionary status, have also been used, such as "Above Horizontal Branch" stars (AHB).

Therefore, we wish to at first put some order to the nomenclature following the most up-to-date literature. We name *BL Herculis* (BL Her) variables the Pop. II Cepheids with period lower than ∼ 10 days, the *W Virginis* (W Vir) pulsators are those showing a pulsational period between 10 and ∼ 25 days, and the *RV Tauri* (RV Tau) are the variables with period between 25 and ∼ 150 days (e.g. the mini-

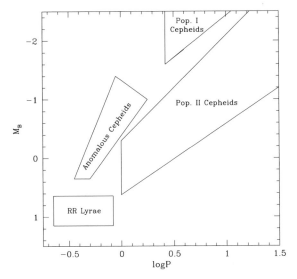

Figure 4.18 Qualitative representation of the pulsation period-mean B-band magnitude for the various classes of Pop. II variables discussed in the text.

mum period of long period variables). For the two first classes of Pop. II Cepheids, the pulsational amplitude in the B-band (A_B) is in the range ~ 0.5 to ~ 1.8 mag, while RV Tau variables have blue amplitudes of the order of or larger than 2.5 mag. Note that, unlike for RR Lyrae stars, there is no clear distinction between the fundamental mode and first-overtone mode (if any) pulsators among the type II Cepheids.

Also, unlike the RR Lyrae stars (see the discussion in Section 7.2.3), Pop. II Cepheids (as the more famous classical Cepheids) show a well-established *Period–Luminosity* relation in the classical Johnson–Cousin photometric bands; this is shown in Figure 4.18.

Besides Pop. II Cepheids discovered in the Galactic field, these pulsators are mainly found in Galactic GCs with a blue HB morphology and moderate to extreme metal deficiencies.

4.5.2.1 The Evolutionary Status of Population II Cepheids

The evolutionary channels producing the various types of Pop. II Cepheids have been extensively investigated over many years; a detailed review can be found in [270].

The BL Her objects, that is, the variables characterized by the shorter periods and fainter mean brightness among the Pop. II Cepheids, are associated with hot, low-mass stars that began their main core He-burning stage on the blue side of the RR Lyrae instability strip, and then crossed the instability strip during their post-HB evolution when moving from the blue to the red side of the HRD. Figure 4.19 shows the evolutionary paths in the HRD of BL Her progenitors for various metallicity values. Since the characteristic mean luminosity during this crossing increas-

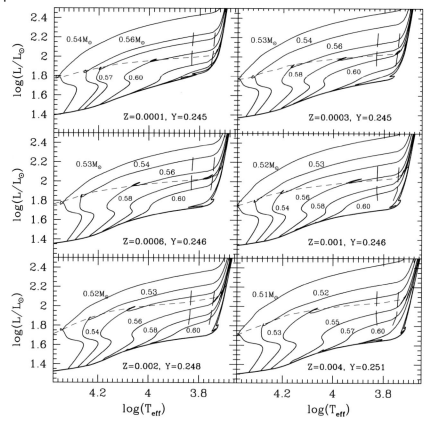

Figure 4.19 Evolutionary tracks for selected HB stars crossing the type II Cepheid instability strip at different luminosity levels, with various initial chemical compositions. The ZAHB and the locus (short-dashed line) corresponding to the central He-exhaustion are also plotted. The boundaries (long-dashed lines) of the instability strip are also shown.

es with decreasing total mass, these variables span a range of periods of several days (see Figure 4.20); the crossing time of the instability strip depends on the evolutionary mass, a typical value being $\sim 80 \times 10^4$ years.

The common view is that the W Vir variables stem from the post-HB evolution of low-mass stars which approach the AGB evolutionary stage with a not too massive envelope. It is instructive to describe the evolution of these stars after central He-exhaustion. When the fuel supply in the core is over, He-burning occurs in a shell which progressively narrows, and its efficiency increases. Depending on the residual mass of the H-rich envelope, such increases of the energy release from the He-burning shell, causes the overlaying layers to expand and, as a consequence, the energy supply from the H-burning shell begins to decline (see Figure 4.6d). As a result of the fall in L_H, the surface luminosity drops while the stellar envelope contracts, and the star moves towards the blue side of the HRD. This blueward path is halted and reversed as the H-burning shell begins to regain its strength as a con-

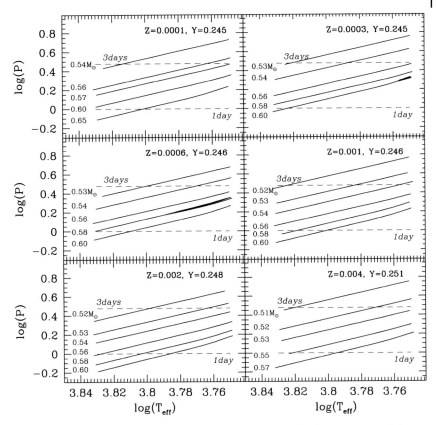

Figure 4.20 The *Period-T*$_{eff}$ diagram for selected stellar models (also shown in Figure 4.19) spending a significant fraction of their evolutionary lifetime within the BL Her instability strip during their off-ZAHB evolution, for various initial chemical compositions.

sequence of the warming associated with gravitational energy release. At odds with the situation occurring during the thermal pulsing stage along the AGB, where the re-ignition of the H-burning shell coincides with the switch-off of He-burning shell (see Chapter 5), in this case, the luminosities of both shells rise together. Because of this behaviour, the stars perform *blueward noses* in the HRD, thus again entering the instability strip. The typical timescale of these blueward noses is of few Myr. These loops in the HRD can be recursive as a consequence of a mutual readjustment of the burning efficiency in the two shells; an occurrence allowing the star to enter the instability strip many times.

From a structural point of view, the crucial parameters governing the occurrence and the extension of these blueward noses are: the residual mass of the envelope at the end of the core He-burning stage and the abundance of He in the envelope. If the envelope mass is too small, that is, less than $\sim 10^{-2} M_\odot$, the star at central He-exhaustion does not move towards the AGB, but behaves as an *AGB-manqué*, evolving almost vertically in the HRD before approaching the WD cooling

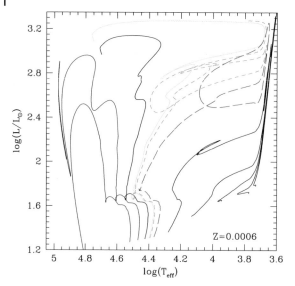

Figure 4.21 The HRD of models with different evolutionary patterns during their post-HB evolution. When starting from the left, the first three stellar models, with masses 0.5045, 0.510 and 0.515 $M_\odot$, behave as *AGB-manqué* stars; the models with masses 0.519, 0.520 and 0.525 $M_\odot$ behave as *post-Early AGB* stars; the models with masses in the range between 0.55–0.8 $M_\odot$ climb the AGB after the central He-exhaustion and experience the thermal pulse stage.

sequence; this is shown in Figure 4.21. If the envelope is too massive, the star, once it reaches the AGB, climbs the AGB without experiencing any blueward nose before reaching the thermal pulsing phase. For a star to experience the blue loop in the HRD, its envelope mass at this stage has to be a value larger than the minimum required to reach the AGB, but less than a critical value M_{bn}. The value of M_{bn} is $\sim (0.05-0.06) M_\odot$ for He-enhanced stars with an initial helium content of $Y = 0.30$, but its value decreases significantly with decreasing He abundance, being of the order of $(0.01-0.02) M_\odot$ for $Y \sim 0.25$.

It is now commonly accepted that RV Tau stars are shell He-burning stars which are experiencing thermal pulses, with a very low envelope mass, but sufficient enough to keep the more external H-burning shell active. The extremely low mass of the envelope can be achieved via the combination of efficient mass loss and H-burning occurring during the AGB stage. It is also possible that some RV Tau objects are stars leaving the AGB and crossing the instability strip during their blueward motion towards the WD cooling sequence.

It is clear that, in any given stellar system, the range of total stellar masses able to produce type II Cepheids is very small when compared with the mass range of classical Cepheids. In fact, the type II Cepheid mass range is limited by the requirement that the envelope mass has to be low enough to allow the star to perform a blue loop in the HRD. This is at the root of the fundamental difference in the way the strip is populated by Pop. I Cepheids and by Pop. II Cepheids: the main reason for the wide luminosity range associated with classical Cepheids is the wide mass range of these intermediate-mass pulsators, while for both BL Her and W Vir stars, a small range in mass (of few tenths of a solar mass) is sufficient to produce the observed range in luminosity.

Before closing this section, we remark that detailed pulsational models show that Pop. II Cepheids, at least for the explored cases concerning BL Her and W Vir objects, obey a pulsational-period relation, linking the period with fundamental evolutionary parameters such as mass, luminosity and T_{eff} [271] as is for RR Lyrae stars. For fundamental BL Her pulsators, this relation is

$$\log(P) = 11.579 + 0.89 \log \left(\frac{L}{L_\odot} \right) - 0.89 \log \left(\frac{M}{M_\odot} \right) - 3.54 \log(T_{eff})$$

Needless to say, this relation (as for the RR Lyrae) is very important for investigating the evolutionary properties of type II Cepheids observed in stellar clusters, as one can combine pulsational and evolutionary model prescriptions. These model computations show that there is, in the HRD, an upper luminosity limit above which the *FO region* inside the instability strip disappears, and only fundamental variables are expected.

4.5.3
Anomalous Cepheids

Anomalous Cepheids (ACs) are radial pulsators with periods in the range ~ 0.5 to ~ 2 days. As for Pop. II Cepheids, ACs are brighter than RR Lyrae and are core He-burning stars. However, at any given metallicity, ACs are more massive than RR Lyrae: their typical mass is in the range $\sim 1.2-1.8 M_\odot$. The reason they are called "anomalous" is related to the evidence that they do not follow the same *Period–Luminosity* relation of type II Cepheids: at any period, ACs are brighter than Pop. II Cepheids. The location of ACs in the period-Wesenheit $(V, V-I)$ plane[17] with respect to other classes of variables is shown in Figure 4.22. Indeed, ACs show pulsational properties corresponding to both fundamental and first-overtone pulsators, but the first-overtone ACs closely match the properties of short period Classical Cepheids, which sometimes makes their identification as "genuine" ACs very problematic.

The pulsational-period relation for fundamental mode ACs [272] is

$$\log(P_0) = 10.88 + 0.82 \log \left(\frac{L}{L_\odot} \right) - 0.62 \log \left(\frac{M}{M_\odot} \right) - 3.31 \log(T_{eff})$$

while, for a given luminosity, effective temperature and mass, the relation between the first overtone period and the fundamental one is $\log P_1 = \log P_0 - 0.13$.

---

17) The Wesenheit index is defined, in this case, as $W(V, V-I) = V - 2.54 \times (V-I)$. The Wesenheit index can also be defined using different photometric filters, and its main advantage is that it is a reddening-free index. This is because the multiplicative factor to the colour index is chosen to cancel the effect of the extinction on the magnitude and the colour.

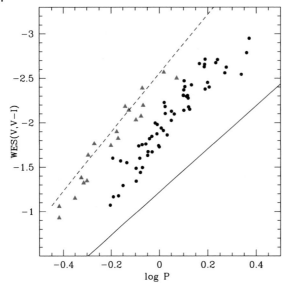

Figure 4.22 The *Period–Wesenheit* diagram for a large sample of Anomalous Cepheids in the Large Magellanic Cloud. The triangles correspond to first-overtone pulsators, while the circles to fundamental mode ACs. The dashed line represents the theoretical expectation from pulsational models for Classical Cepheids, while the solid line shows the same prediction but for BL Her variables (courtesy of G. Fiorentino and M. Monelli).

4.5.3.1 The Evolutionary Scenario for AC Stars

An important piece of information for establishing the evolutionary channel(s) producing ACs is the fact that these pulsators are quite abundant in the majority of the Local Group dwarf galaxies, but they are almost absent in other metal-poor, very old stellar systems such as Galactic GCs.[18] This evidence, together with analysis of their pulsational properties, suggests that ACs are more massive than RR Lyrae stars; they should only be present in stellar systems with a Star Formation History "extending" to intermediate ages, or alternatively they could be the final result of exchange of mass in binary systems.

On theoretical grounds, it is well-known that, for ages lower than $\sim$ 4–5 Gyr, the He-core mass at He-ignition at the RGB tip decreases with decreasing age (i.e. increasing mass of the RGB progenitor), as discussed in Section 3.7.1. This has the important consequence that the He-core mass at the beginning of the core He-burning stage cannot be considered any longer to be independent of the mass of the RGB progenitor, as one commonly assumes for older stellar systems.

Evolutionary computations show that when the value of M_{cHe} decreases, the ZA-HB brightness should also decrease. Actually, this is only true when considering

---

18) Only one AC has been found in a Galactic GC: V19 in NGC 5466 [273]. A detailed spectroscopic study of this star provided a mass estimate of $(1.66 \pm 0.5) M_\odot$, a small rotational velocity, and iron, s-process, α-element abundances supporting its cluster membership, and some evidence for a very long-period binary orbit.

a mass distribution along the ZAHB appropriate for low-mass RGB progenitors.[19] In fact, evolutionary computations show an interesting behaviour: for a fixed value of M_{cHe}, when first increasing the envelope mass, the T_{eff} of the ZAHB models decreases monotonically, but when continuing to increase the envelope mass, a minimum effective temperature[20] is reached, after which the ZAHB T_{eff} as well as the ZAHB brightness starts increasing. This behaviour causes the appearance of an upper branch along the ZAHB [274]. This upper branch can be present only in a stellar system with age ranging from $\sim (1–2)$ Gyr to $\sim (4–5)$ Gyr, the exact value depending on the chemical composition. In fact, older stellar systems have an evolving mass at the RGB tip that is not massive enough to allow ZAHB stars to populate the upper ZAHB locus; while younger stellar systems would populate the instability strip with stars too massive and so behaving as Classical Cepheids. This scenario is outlined in Figure 4.23 for a fixed metallicity and various RGB progenitors, while Figure 4.24 shows the dependence on the metallicity.

Depending on the metallicity, the initial mass of the RGB progenitor, and the efficiency of mass loss during the previous evolutionary stage, this "upper branch" of the ZAHB can enter the instability strip as early suggested by [275, 276]. Metallicity is the critical issue: when the metal content is larger than $Z \sim 4 \times 10^{-4}$, the minimum effective temperature of the ZAHB is too low to allow the more massive stars to enter in the pulsation strip. For this reason, AC pulsators stemming from this evolutionary channel are not expected in intermediate-metallicity and metal-rich stellar environments.

These theoretical predictions nicely explain the existence, in metal-poor stellar systems with an intermediate-age population (such as Local Group dwarf galaxies), of variable stars, brighter and more massive than the RR Lyrae with the pulsational properties observed for ACs. These upper branch pulsators, despite their larger mass and brightness, have core He-burning lifetimes similar to the evolutionary lifetime of RR Lyrae stars (i.e. $\sim 10^8$ years); a consequence of the fact that in such massive HB stars, H-burning in the shell significantly contributes to the whole energy budget, together with core He-burning.

We mentioned an alternative scenario for the formation of ACs, related to the binary evolution and in particular to the exchange of mass between the components of a binary system. Clearly, this would be the only possible scenario explaining the presence of ACs in very old stellar environments such as V19 in the GC NGC 5466.

Binary evolution and stellar collisions can produce stellar products whose mass is larger (up to twice as large) than the mass currently evolving at the cluster MS TO; this is supported by the presence of large sample of Blue Stragglers found in the majority of Galactic GCs. The progeny of these BSs during the core He-burning stage populates the red side of the HB, and if the cluster metallicity and

19) It is clear that as a consequence of mass loss occurring during the RGB stage, the maximum allowed mass for a ZAHB star has to coincide with the initial mass of the RG progenitor; so it is of the order of $\sim 0.8–1.2 M_{\odot}$ for old stellar systems.

20) This minimum effective temperature is log $T_{eff} \sim 3.72$ at $Z = 0.0004$ and ~ 3.74 for $Z = 0.0001$, and corresponds to a total mass of the ZAHB star of $\sim 1.0 – 1.2 M_{\odot}$.

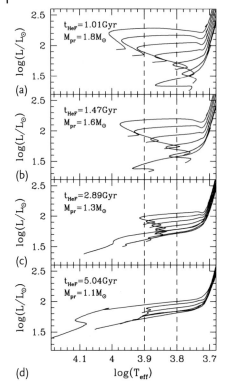

Figure 4.23 The location in the HRD of selected He-burning models for $Z = 10^{-5}$ and various assumptions about the RGB progenitor (see labels). The age of each selected RGB progenitor at He-ignition at the RGB tip is listed. In each panel, the more massive star has the same mass as the RGB progenitor and the step in mass is $0.2M_\odot$ in (a) and (b), and $0.1M_\odot$ in (c) and (d). The approximate location of the instability strip is also shown as dashed lines.

their masses are adequate, they can populate the upper HB branch previously described, producing ACs [277]. It has been shown that binary systems suitable for producing ACs can survive long enough in a dense stellar environment, such as a GC, only if the density of the stars is sufficiently low (as indeed it is the case of NGC 5466) [278]. This would help a lot in explaining (together with the requirement for an appropriate metallicity) the absence of ACs in most Galactic GCs. Nevertheless, there is a problem with this binary scenario that is related to the observed paucity, if any, of ACs in the Large Magellanic Cloud, which is a low-density stellar environment, although the average metallicity of this galaxy is too large to allow the formation of ACs.

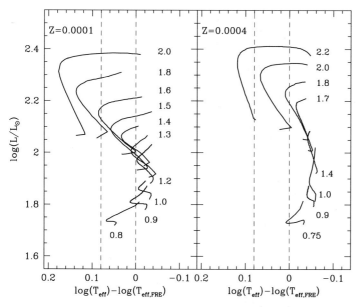

Figure 4.24 The location in the HRD of selected core He-burning models with various assumptions about the RGB progenitor (see labels) and for $Z = 10^{-4}$ and 4×10^{-4}. The effective temperature of the models have been scaled to the T_{eff} of the red edge of the instability strip. The location of the boundaries of the instability strip are also shown (dashed lines).

4.5.4
Extremely Hot HB Pulsators: EC 14026

There is an additional, important, class of pulsators among He-burning, low-mass stars, that is, characteristic of the extremely hot HB stellar population and their progeny; this class of pulsators is known as *EC 14026* after the discovery of its prototype in the Galactic field [279]. The existence of this class of pulsators was predicted on theoretical grounds by [280] before being discovered observationally.

The pulsational properties and physical mechanism causing the pulsations in *EC 14026* objects are very different to those of the other classes of variable stars already discussed. Taking as a reference the properties of the *EC 14026* prototype, these objects have a very small-amplitude and very rapid brightness variations: the main stable period is ~ 144 s and the amplitude is ~ 0.024 mag, but there is clear evidence supporting the presence of multiple mode pulsations. These objects lie in a specific region of the $T_{\text{eff}}-\log g$ plane, clustered about $T_{\text{eff}} \sim 35\,000$ K and $\log g \sim 5.9$.

Theoretical investigations have shown that low-order radial and non-radial modes can be present in these objects (as is actually observed), and have demonstrated that the pulsation phenomenon is driven by an opacity bump mechanism due to a local enhancement of the metal (mostly iron) abundance in the envelope [281]. A local metallicity value $Z \geq 0.04$ is required to drive the pulsations.

The presence of this local heavy element enhancement can be physically explained as a consequence of radiative levitation in the envelope of these extremely hot structures (see Section 4.7 for a discussion on this topic). Theoretical predictions on the instability strip associated with this class of pulsators reveal that it is located between ~ 29 000 and ~ 36 500 K.

The red edge of the instability strip is linked to the need for a "minimum" T_{eff} value such that radiative levitation is efficient enough to produce the required metallicity enhancement. When moving towards larger T_{eff} values, the radiative levitation efficiency increases, but the iron opacity bump moves towards the stellar surface, and when it is too high in the atmosphere, it contains a very small amount of mass and it can not continue to contribute to the pulsation driving.

Thanks to extensive observational surveys, these *EC 14026* variables have been found among the field hot subdwarfs of spectral type B; no firm identification of any object representative of this class of pulsators has been collected so far in old stellar systems, but this could be due to current shortcomings in the photometric observations of extremely hot HB stars in GCs [282].

4.6
Red Clump Stars

We have discussed in Section 3.7.1 the trend of the He-core mass at the He-ignition as a function of the initial mass, showing (see Figure 3.29) that M_{cHe} is almost constant for masses $(0.2-0.3)\,M_\odot$ lower than the transition mass M_{HeF}; as a consequence, the ZAHB luminosity of the progeny of these objects is almost constant. For masses larger than M_{HeF}, the He-core mass monotonically decreases down to a minimum value ~ $0.32\,M_\odot$ that is weakly dependent on the metallicity (see Figures 3.29 and 4.25a) and corresponds to an initial mass $[M(M_{cHe}^{min})]$ in the range ~ 2 to ~ $2.5\,M_\odot$ (from metal-poor to solar metallicity composition). Therefore, one expects a minimum in the ZAHB luminosity for masses equal to $M(M_{cHe}^{min})$.

When moving to more massive stars igniting He in non-degenerate cores, the core mass at ignition steadily increases with the total mass, following the monotonic increase of the convective core during the MS. Due to the combined effect of M_{cHe} increase and increased efficiency of the H-burning shell (see discussion in Section 4.5.3), the ZAHB location of these models is brighter than the minimum value corresponding to $M(M_{cHe}^{min})$.

Figure 4.25 shows the behaviour with the total initial mass of the relevant evolutionary parameters. Figure 4.25c displays the core He-burning lifetime, t_{He}, as a function of the initial mass; t_{He} is almost constant for stars less massive than M_{HeF} because of similar He-core mass on the ZAHB and similar luminosity, but increases around the transition to non-degenerate RGB cores, attaining a maximum for $M(M_{cHe}^{min})$, a factor ~ 2–3 larger than the value for lower masses. This is because the lower the initial He-core mass at the ZAHB stage, the larger the contribution to the energy budget provided by the H-burning shell, and in addition the lower the brightness of most of the core He-burning stage (see Figure 4.25).

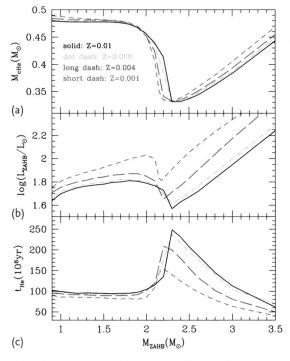

Figure 4.25 The trend with the total initial mass of the He-core mass and of the surface luminosity at the beginning of the central He-burning stage for selected metallicities. (c) shows the same trend, but for the evolutionary lifetime during the core He-burning stage.

Figure 4.26 shows the evolutionary tracks in the mass regime under scrutiny, during the core He-burning stage. When excluding models significantly less massive than $M_{HeF} - M < (0.7-0.8) M_\odot$ –, all tracks have a similar morphology, and span a "limited" interval in luminosity and effective temperature. For instance, at solar metallicity, the models cover a range of ~ 500 K in T_{eff} and $\sim (0.5-0.6)$ mag in bolometric luminosity. For any given metallicity, the location of the ZAHB for masses between $\sim 0.8 M_\odot$ and $M(M_{cHe}^{min})$ defines a kind of hook in the HRD, whose bluest and faintest termination[21] corresponds exactly to $M(M_{cHe}^{min})$. With increasing stellar mass the ZAHB location becomes progressively brighter at almost constant T_{eff} (i.e. at almost constant colour).

So far, we have neglected the occurrence of mass loss during the RGB stage. If we relax this assumption, the theoretical framework is not significantly modified; mass loss (for realistic rates) along the RGB can appreciably affect models with initial mass below about $1.2 M_\odot$ due to their longer RGB lifetime and brighter maximum luminosities. For more massive stars, RGB mass loss can only slightly

---

21) The detection of the faintest point along this blue hook would provide a strong constraint on the properties of models with mass around M_{HeF}, as well as also on the value of M_{HeF} for a given stellar population.

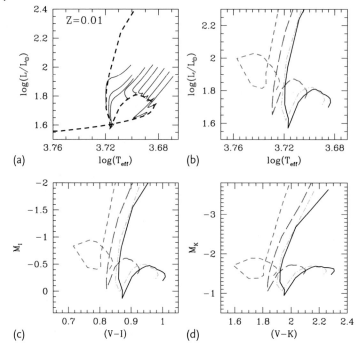

Figure 4.26 (a) The HRD location of evolutionary tracks for core He-burning stars with mass equal to that of the RG progenitor: 0.7, 1.1, 1.3, 1.5, 1.7, 1.9, 2.1, 2.3, 2.4, 2.6$M_\odot$, and with metallicity $Z = 0.01$. For each track, only the central He-burning stage is shown; the dashed line corresponds to the ZAHB lo-

cus. (b) As for (a), but only for ZAHB stars more massive than 0.85 $M_\odot$, with various metallicities (the line coding is the same as in Figure 4.25). (c) and (d) show the location of the same ZAHB loci plotted in (b), but in the $(M_I, V-I)$ and $(M_K, V-K)$ CMDs.

reduce the mass at the beginning of the central He-burning phase, making the ZAHB location only slightly fainter (see Figure 4.23).

This characteristic morphology of the core He-burning phase in the mass range we are now considering has an important consequence: if a simple stellar population has an age in the range $\sim (1-4)$ Gyr, the core He-burning phase is represented in the CMD by an *ensemble* or *clump* of red giant stars: this observational feature is commonly named the *Red Giant Clump* (hereinafter red clump or RC). As discussed in Section 7.2.4, the mean magnitude of RC stars has been employed as a distance indicator.

Figure 4.26 shows that, at metallicities lower than about $Z = 0.004$, the ZAHB location becomes progressively hotter when decreasing Z. Therefore, metal-poor ZAHB models do not produce a true red clump. In addition, for metallicities $Z > 4 \times 10^{-3}$, the central H-burning evolutionary lifetime of stars with mass lower or of the order of $0.7-0.8 M_\odot$ is too long to allow them to reach the core He-burning stage in a Hubble time; as a consequence, one can expect that only the core He-burning stage of populations with metallicity larger than this critical value produces a red clump in the CMD. In this context, the RC can be considered as the counterpart of the HB in old, metal-poorer populations.

When considering the large metallicity range from $Z = 0.004$ to 0.03, RC stars span a T_{eff} range from ~ 4000 to ~ 5500 K (Figure 4.26b). In this range, the bolometric corrections to the I-band for surface gravities typical of RC stars, are constant within ~ 0.1 mag, and do not depend significantly on the metallicity. As a consequence, the absolute I magnitude of red clump stars at their ZAHB location nicely traces the behaviour of the bolometric luminosity, that is, essentially determined by the change with metallicity of the He-core mass at the He-ignition. On the other hand, in the same effective temperature range, the bolometric corrections to the K-band change by about 3 mag due to the huge sensitivity to T_{eff}, while they are largely independent of Z [283]. Due to the combined effect of the change in luminosity and effective temperature at the ZAHB for red clump stars of different masses, the sensitivity of the K-band bolometric corrections on T_{eff} partially compensates for the dependence of the stellar luminosity on the He-core mass, reducing the magnitudes range spanned by RC stars with different progenitors.

4.6.1
Properties of the Red Clump in Simple Stellar Populations

The behaviour of the average properties of the RC as a function of the age (t) and metallicity Z of a simple stellar population can be easily determined by appropriate numerical integrations along isochrones [283–285]. Uncertainties in the RGB mass loss is not a crucial issue for proper RC stars, as discussed before. The corresponding properties of older populations are, however, dependent on the chosen RGB mass loss law.

For a generic pair (t, Z), the average (mean) RC absolute magnitude in a generic photometric band, λ, is given by

$$\langle M_\lambda(t, Z) \rangle = -2.5 \log \left[\frac{1}{N_{\text{RC}}} \int_{\text{RC}\star} \phi(m_i)\, 10^{-0.4 M_{\lambda, i}}\, \mathrm{d}m_i \right] \tag{4.1}$$

where $\phi(m_i)$ is the adopted IMF; the integral is performed over the portion of the isochrone corresponding to the core He-burning stage. N_{RC} is the number of RC stars, at the given age t, per unit mass of born stars, given by the following integral (with an IMF normalized to produce in a single-burst of star formation a stellar population of total initial mass equal to $1 M_\odot$)

$$N_{\text{RC}}(t, Z) = \int_{\text{RC}\star} \phi(m_i)\mathrm{d}m_i \tag{4.2}$$

Using Eq. (4.1) for different photometric filters, one can also derive the mean colours of RC stars. Another useful quantity is the mean initial mass of RC stars,

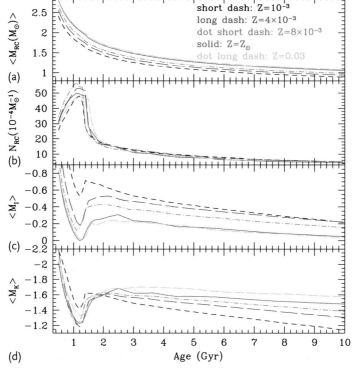

Figure 4.27 The behaviour with age of selected properties of the red clump in simple stellar populations for various metallicities (see labels) [283, 285]. From (a)–(d), the following quantities are shown: the mean mass of red clump stars, the number of red clump per unit mass of stars born at the time $t = 0$, the mean magnitudes of the RC in the I, and K photometric bands.

defined as

$$\langle m_{RC}(t, Z)\rangle = \frac{1}{N_{RC}(t, Z)} \int_{RC\star} m_i \phi(m_i) \mathrm{d}m_i \tag{4.3}$$

The trend with age and metallicity of these quantities is shown in Figure 4.27 for various metallicities. We show here the behaviour of the I and K magnitudes, often used for distance estimates (see Section 7.2.4). Regardless of the metallicity, the number of RC stars per unit mass of stars formed is maximized for an age of ~ 1.2–$1.4\,\mathrm{Gyr}$. Also, for ages below ~ 3–$4\,\mathrm{Gyr}$, the dependence on metallicity of the K magnitude is much reduced compared to the I-band. It is also interesting to note that for ages larger than about 2 Gyr, the dependence of the K-band magnitude on metallicity is reversed in comparison with the I-band, for example, the mean magnitude of the RC in K decreases for increasing metallicity because of the strong dependence of the bolometric correction on T_{eff}.

4.6.2
The Red Clump in Composite Stellar Populations

Composite stellar populations can always be considered as the combination of multiple simple stellar populations, each one characterized by its age and chemical composition, and the previous relations can be easily modified to be applied to this more general case. The RC mean magnitude for a generic photometric filter in a composite system (a "galaxy") with age t_{Gal} can be derived from

$$\langle M_\lambda(Gal)\rangle = \frac{1}{N_{RC}(Gal)} \int_{t=0}^{t_{Gal}} N_{RC}[t, Z(t)]\, \psi(t)\, \langle M_\lambda[t, Z(t)]\rangle\, dt \tag{4.4}$$

where

$$N_{RC}(Gal) = \int_{t=0}^{t_{Gal}} N_{RC}[t, Z(t)]\, \psi(t)\, dt \tag{4.5}$$

$\psi(t)$ is the adopted star formation history (SFH) that represents the number of stars formed at each time t in unit of $M_\odot$/year. When considering a composite stellar population, one has to consider the possibility that the metallicity changes with time because of Supernova ejecta (and mass loss from RGB and AGB stars) that chemically enrich the interstellar medium with time. To take this process into account, one needs to also assume a specific age–metallicity relation (AMR), that is, a functional relation $Z(t)$.

The number of RC stars of mass, m_{RC}, present in a composite stellar population, that is, $N(m_{RC})\, dm_{RC}$, is given by

$$N(m_{RC}) \propto \phi(m_{RC})\psi\left[t_{Gal} - t(m_{RC})\right] t_{He}(m_{RC})$$

where $t_{He}(m_{RC})$ is the core He-burning lifetime for the models with mass m_{RC}, $t(m_{RC})$ is their evolutionary lifetime during the core H-burning stage[22] and $\psi\left[t_{Gal} - t(m_{RC})\right]$ represents the star formation rate at the time $t_{Gal} - t(m_{RC})$ the stars with mass m_{RC} formed.

Obviously, the number of stars of a given mass located in the RC region strongly depends on the parameters chosen for the simulation such as the IMF and more importantly the SFH and AMR. Although, as previously discussed, the brightness and colour ranges spanned by RC is relatively small, different masses populate different sections of the RC, and depending on the adopted parameters, distinctive features can appear in the CMD as "blue plumes" and "secondary clumps" [284]. The metallicity and mass distribution of RC stars can also be largely affected by the adopted SFH and AMR.

Figure 4.28 displays synthetic $(I, V-I)$ CMDs of the RC, determined with realistic SFHs [286], in selected composite stellar populations, namely, the local disk,

---

22) This quantity allows one to estimate how long after formation, stars with mass m_{RC} start to populate the RC.

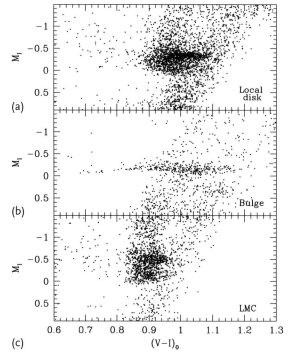

Figure 4.28 The synthetic CMD of the red clump locus in three composite stellar populations characterized by different star formation histories and age–metallicity relations: (a) the local disk, (b) the Galactic bulge in the Baade window direction, and (c) a field in the Large Magellanic Cloud bar.

the Milky Way Bulge and one field in the bar of the Large Magellanic Cloud. Differences in the RC morphology are evident, and are mainly due to the different SFHs. Figure 4.29 shows metallicity, age and mass distributions of these RC synthetic samples. It is interesting to note that, but for the case of the Galactic bulge whose age distribution is strongly peaked towards very old ages, in both the local disk and LMC bar field, a large fraction of the RC stars has a mass of the order of $\sim 2.2 M_\odot$, a value close to $M(M_{cHe}^{min})$. In general, peaks in the m_{RC} distribution are due to the combination of the SFH and the trend of t_{He} with total stellar mass. The double-peaked mass distribution in the simulation of the local disk (Figure 4.29c) coupled to the fact that the masses corresponding to the two peaks occupy well-defined regions in the CMD, separated by ~ 0.4 mag in I, cause the appearance of both a main and a secondary (in terms of star counts) clump in the CMD that agree with observations.

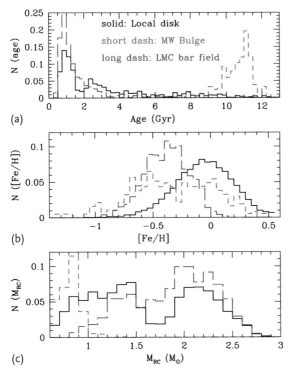

(a)

(b)

(c)

Figure 4.29 The age (a), [Fe/H] (b) and actual mass distributions (c) for RC stars populations corresponding to the composite stellar systems whose synthetic CMDs are shown in Figure 4.28. The distributions have been normalized to the number of RC stars in each simulation. In (c), the mass distribution for the Galactic bulge has been divided by a factor 3.5 to improve the clarity of the figure.

4.7
The Effects of Non-canonical Processes on HB Evolution

Although the general properties of core He-burning stars, from both a structural and evolutionary point of view, have been well-understood for many years, we still lack a sound interpretation of all the physical parameters definitively governing, together with the metallicity, the morphology of the HB in various stellar systems. In addition, in this last decade, high-precision photometry and multi-object, high-resolution spectroscopy of large samples of HB stars in Galactic GCs have revealed a number of anomalies that find no straightforward interpretation in the canonical evolutionary framework: the presence of narrow gaps, that is, regions along the HB showing a significant paucity of stars, with respect to their blue and red boundaries; the fact that HB stars hotter than a critical T_{eff} appear brighter than expected; the evidence that hot HB stars apparently show surface gravities too low; peculiar patterns in the surface heavy element abundances and in the surface rotational velocities.

Nowadays, it is clear that there exists a (strong) link between all these "anomalies" and various physical mechanisms such as atomic diffusion, radiative levitation, rotation, mass loss during the HB and RGB stages, and probably non-canonical mixings (not accounted for in standard stellar models) also cooperate to produce the peculiarities observed in HB stars and their progeny such as sdO and sdB stars. Even though significant improvements have been made in stellar model computations, we still lack a sound, physically grounded, explanation for the observational findings as a whole.

In the following, we discuss the latest observational scenario, and then we review the main theoretical results that have been obtained in this context.

4.7.1
The Observational Scenario

4.7.1.1   The "Jump" in the *u* Strömgren Photometric Passband

Accurate Strömgren photometry of a large sample of Galactic GCs has shown the presence of a jump in the *u*-band photometry of HB stars: from a morphological point-of-view, this jump is described as a systematic deviation in the *u* magnitude and/or the $(u - y)$ colour with respect to the canonical theoretical ZAHB (see Figure 4.30) [287, 288]. This observational feature shows some interesting properties:

- The jump is an ubiquitous feature, observed in all GCs with a blue-enough HB morphology. The intrinsic $(u - y)$ colour (~ 1.0 mag) at which the *u* jump occurs, and the T_{eff} value ($\sim 11\,500$ K) of this point, are independent of the GC metallicity. The size of the *u* jump is ~ 0.7–0.8 mag in all GCs;
- It is worth noting that in GCs with very extended HBs, as is the case for M13, the *u* jump occurs over a well-defined range of effective temperatures from $\sim 11\,500$ to $\sim 19\,000$–$20\,000$ K. In fact, HB stars hotter or cooler than these limits show a location in the Strömgren CMD in agreement with canonical expectations.

Additional observational analysis revealed that a jump, fully analogous to the *u* jump, is also present in $(U, U - V)$ CMDs [289, 290].

The evidence that the jump is present in all GCs with a blue HB morphology (regardless of their structural and evolutionary properties such as metallicity, stellar concentration, total mass, and so on) and that it is observed only when using specific photometric filters indicates that this observational feature is related to some specific properties of the HB stars populating the relevant T_{eff} range.

The *u* Strömgren photometric band is located just shortward of the Balmer jump, so the *u*-band flux is dominated by the opacity associated with hydrogen. If the relative contribution of metals to the opacity with respect that provided by H is increased, the modification of the stellar spectrum means that a larger fraction of the flux comes out through the *u* filter. Actually, it was suggested early on by [288] that the occurrence of radiative levitation in hot HB stars, by hugely enhancing the amount of heavy elements in the atmosphere, would strongly modify

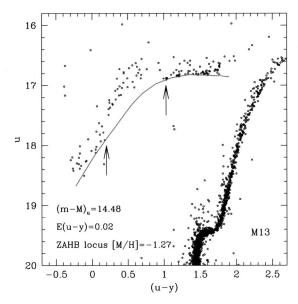

Figure 4.30 The $(u, u - y)$ CMD of the GC M13. The comparison between the observed HB and the theoretical ZAHB locus discloses the occurrence of the Grundahl's jump. The two vertical arrows mark the location of the start of the jump and its hottest point where the observed HB distribution again matches the theoretical predictions (courtesy of F. Grundahl).

the emergent radiative flux. Levitation of heavy elements decreases the far-UV flux, and by back-warming, it also increases the flux in the u passband. In this scenario, the existence of two limiting T_{eff} values for the occurrence of the jump reveals that in stars cooler than $\sim 11\,500$ K and hotter than $\sim 20\,000$ K, radiative levitation is not efficient or it is inhibited by some other physical mechanism(s).

Actually, the discovery of the u jump represents the most striking, indirect, evidence of the occurrence of diffusive processes in hot HB stars.

4.7.1.2 The Low Gravity Problem

For a long time, the comparison between theoretical and observational spectra for hot HB stars in Galactic GCs has provided surface gravity estimates that are significantly lower (by $\Delta \log g \sim 0.4-0.5$ dex) than expected on the basis of canonical model computations [291]. Interestingly enough, this problem only concerns HB stars whose effective temperature is in the range $\sim 11\,000$ to $\sim 20\,000$ K, that is, the same T_{eff} interval where the u jump is observed.

For many years, there was no clear explanation for this problem. However, it has been recently shown that the main shortcoming in the spectroscopic analysis is in the use of model atmospheres for a chemical composition consistent with the initial, metal-poor, chemical abundances of GC stars, that is, not taking into account for the large heavy element enhancement caused by radiative levitation (as is suggested also by the occurrence of the u jump) [291].

Indeed, the comparison of the observed spectra with theoretical ones with a solar chemical composition[23] largely solves the discrepancy between theory and observations, reducing the difference in $\log g$ to about 0.1–0.2 dex.

A more detailed modeling has been performed by [292], who explicitly included diffusive mechanisms in the computation of model atmospheres for blue HB stars. This analysis confirmed that the Balmer profiles computed from models accounting for chemical abundance stratification yield a decrease in $\log g$, amounting to as much as ~ 0.5 dex.[24]

This result represents additional, strong, indirect evidence supporting the presence of efficient diffusive processes in hot HB stars. Before closing this section, we note that the residual discrepancy of $\Delta \log g \approx 0.15$ dex could be possibly solved by accounting for more realistic heavy element distribution in the atmosphere of hot HB stars and/or for the effect on the Balmer line wings of mass loss during the core He-burning stage [70].

4.7.1.3 Chemical Abundance Anomalies

The appearance of chemical anomalies in the outer layers of hot HB stars was first predicted on theoretical grounds by [293]. These theoretical expectations were tentatively confirmed by [294]: an overabundance of Fe by a factor of ~ 50 and a He depletion was observed in a hot HB star with $T_{\text{eff}} \approx 16\,000$ K, whereas no chemical peculiarities were detected in a star with $T_{\text{eff}} \approx 10\,000$ K.

Recently, a large amount of direct empirical evidence for the existence of large abundance anomalies in HB stars in Galactic GCs has been collected (see [295–298] and references therein). These results can be summarized as follows:

- While in HB stars cooler than $T_{\text{eff}} \approx 11\,000$ K, the iron and other heavy element abundances are close to those corresponding to the initial GC chemical composition, the hotter HB stars show remarkable enhancements of iron and other metal species. In particular, the majority of hot HB stars show iron abundances greater than the solar value, [Fe/H] ≈ 0.0 to $+1.0$. Depending on the intrinsic (initial) cluster metallicity, these values represent enhancements of factors of ~ 30 to about 300. Titanium also shows a similar enhancement, by a factor of 100 or more in the hottest HB stars; as does calcium, but with a more modest enhancement (a factor ~ 10);

---

23) Discussing the actual effects of radiative levitation, we show that as a consequence of this diffusive process, several metals are enhanced to solar or super-solar abundance values even for metal-poor, GC stars. However, there are also elements that are less enhanced or completely unaffected by the interaction with electromagnetic radiation. As a consequence, the use of model atmospheres corresponding to a solar chemical composition (for *all* elements) is a crude approximation. Detailed model atmosphere computations based on more realistic heavy element distributions are mandatory for definitively settling this issue.

24) This investigation also confirmed that significant changes in the location of the ZAHB in the CMD (in selected photometric passbands) have to be expected, depending on the presence of diffusion-driven metal stratification in the atmosphere. When assuming that radiative levitation turns on quickly at around $\sim 11\,500$ K, these diffusive model atmospheres predict jumps because the Strömgren $(u - y)$ colour index suddenly decreases by ~ 0.2 mag, the u magnitude becomes brighter by ~ 0.4 mag, and $(U - V)$ colour decreases by about 0.1 mag.

- Other chemical elements such as phosphorus, nitrogen, chromium, vanadium, and manganese display enhancements larger than that of iron. In some GCs, there are also hints that nickel and yttrium are highly enhanced in stars in the bluest portion of the HB;
- There are other elements, such as Mg and Si that, on the contrary, show very little, if any, enhancement. These elements are apparently not affected at all by the physical mechanism(s) producing the enhancements of the other elements. The same outcome seems to also be valid for Na, Al, and Sr;
- The surface abundance of He also varies as a function of T_{eff} along the HB. Indeed, He lines are not visible in HB stars with $T_{eff} < 9000$ K; but when the stars become hot enough to excite the He I transitions at 4471, 5016 and 5876 Å, the He abundance can be measured. For the coolest HB stars for which He can be measured, a primordial He content is obtained ($Y \approx 0.24$); but as hotter and higher gravity stars are considered, the He abundance drops with a fairly monotonic trend, reaching a depletion of ~ 2.5 dex or a factor of ~ 300 for the hottest HB stars analyzed in Galactic GCs.

The empirical trends of the chemical abundances as a function of T_{eff} are shown in Figure 4.31 for the GC M13.

4.7.1.4 Rotational Rate Distribution

Since the pivotal investigation of [299], no effort has been devoted to investigating the distribution of the rotational rate of stars along the HB of Galactic GCs. This situation has changed drastically in the last decade because in combination with the study of the chemical abundance anomalies in hot HB stars, accurate measurements of their projected velocity, $v \sin i$, have been also obtained [297].

The most significant result of these rotation measurements is that hot HB stars and HB stars cooler than $T_{eff} \approx 11\,000$ K have very different distributions of $v \sin i$ value, implying a large difference in the actual rotational rate. In more detail, it has been found (see Figure 4.32) that stars cooler than $\approx 11\,000$ K can reach values of $v \sin i$ as high as $(20-40)$ km s^{-1}, depending on the selected GC;[25] while hot HB stars show significantly lower rotational rates, in almost all GCs being limited to ~ 8 km s^{-1}. This difference in the rotational rate distribution is real, and it can not be due to a chance polar orientation of the observed stellar sample. In fact, on the basis of statistical considerations, it is possible to demonstrate that the observed distribution of $v \sin i$ is not what one would expect given a single intrinsic rotation speed v and random orientation of the rotation axes since large values of $\sin i$ are more likely than small values.

25) The apparent differences observed among various GCs could be due to small sample sizes, but they could also indicate a spin-up mechanism whose effectiveness, in terms of the maximum rotational speed and the fraction of stars affected, could depend on the cluster parameters. Until now, this is an open issue waiting for a more complete observational scenario.

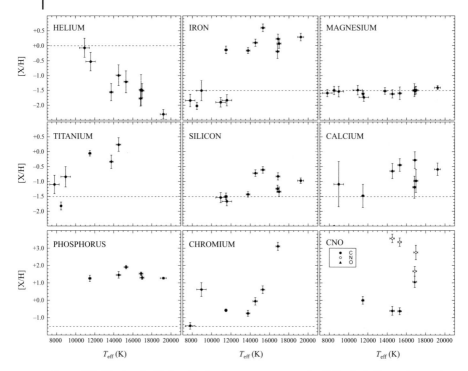

Figure 4.31 The trend with the effective temperature of the chemical abundances of selected elements in HB stars belonging to the GC M13. In each panel, the dashed line represents the expected abundance of the selected chemical element when neglecting the oc-currence of radiative levitation and assuming a scaled-solar heavy element chemical distribution; in the case of helium, the dashed line corresponds to the surface He abundance in the Sun (courtesy of B. Behr).

Therefore, it appears evident that hot and extremely hot HB stars are intrinsically slow rotators.[26]

In the case of the cooler HB stars, it appears that the rotational rate distribution changes by cluster to cluster. In addition, cool HB stars are not all rotating at the same intrinsic velocity, but a real spread in the rotational rate seems to be present for several clusters. There are also some hints for the existence of a bimodality in the rotation rate of cool stars in some GCs, but larger statistical samples are required to assess this issue. Nevertheless, the present data suggests that cool HB stars rotate, on average, faster than hot HB ones.

It is interesting to note that the same distribution of rotational rate as a function of T_{eff} has been found for field HB stars [300]; although the sample of hot HB stars

26) One has to note that there are some outliers in this general trend, for example, hot HB stars with rotational rates larger than observed in other stars with similar T_{eff}. It is worth mentioning that these peculiar objects do not show the metal-enhanced abundances commonly observed in HB stars with the same T_{eff}, suggesting the existence of a link between rotational rate and efficiency of diffusive processes.

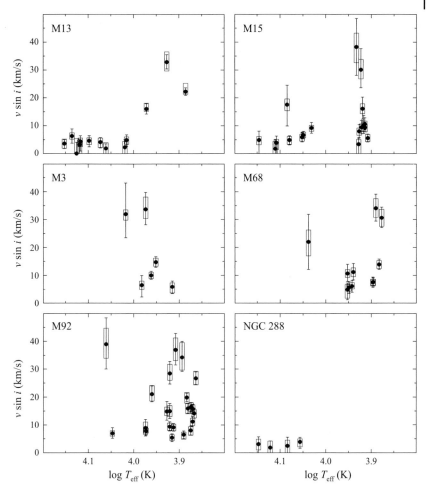

Figure 4.32 The trend of projected rotation velocity as a function of T_{eff} for selected Galactic GCs. The error bars represent the random errors, while the narrow rectangles show the sum of various systematic errors (courtesy of B. Behr).

investigated in the field is very small, it has been found that three out of five of these stars have a very low, if any, rotation rate, while the cool HB stars with $7500 \leq T_{eff}(K) \leq 11\,500$ display a bimodal distribution with a fast rotator population of $v \sin i \approx (30-35)\ \mathrm{km\,s^{-1}}$, and a slow rotators one with a $v \sin i \approx (10-15)\ \mathrm{km\,s^{-1}}$, that is, a distribution similar to that observed in GCs. In addition, the study of field HB stars has shown that the coolest HB objects, that is, those with $T_{eff} < 6000$ K, have a rotational rate lower than $10\ \mathrm{km\,s^{-1}}$.

The similarity in the T_{eff} limits for very different empirical findings, such as the rotational rate dichotomy, chemical abundance anomalies, the u jump, and the low-gravity problem, obviously points out the existence of a link between the various physical mechanisms causing these various effects. The interpretative scenario for this plethora of observational findings provided by evolutionary stellar models will be discussed in the following section.

4.7.1.5 The Gaps along the Horizontal Branch

Since the first accurate CCD observations of Galactic GCs, it has appeared that along the HB of several clusters, the stellar distribution is discontinuous, with some significantly underpopulated regions (gaps). Although some of these gaps have been known about for a while, their statistical significance is still a topic of much debate. The most compelling argument favoring significance rests on the evidence that some of these gaps appear at very similar locations in different clusters, as is shown in Figure 4.33 [301].

The most commonly accepted suggestion of the origin of the gaps is that they mark the boundaries between separate, discrete sub-populations of HB stars characterized by different origin and/or different evolutionary properties. This idea of different sub-populations is nowadays receiving further support from the evidence of the multiple population phenomenon in Galactic GCs. However, it is also possible that at least some gaps could actually mark the separation of stars with different evolutionary properties. For instance, [302] suggested that the gap at $(B-V)_0 \approx 0.0$, present in almost all GCs, is due to the occurrence of diffusive processes in the atmospheres of HB stars hotter than the location of the gap: the basic idea is that at the T_{eff} of this gap, sub-atmospheric convection, characteristic of cooler HB stars, disappears, allowing diffusive processes to significantly modify the chemical stratification and then the outgoing spectral energy distribution and, hence, the colour of these stars.

To find a physical interpretation for the gaps, it is necessary to first identify their location in the HRD. In doing this, we limit the discussion only to those gaps whose reality is well-assessed in current literature, and we adopt the commonly adopted nomenclature:

- Gap *G1*: this gap is prominent in many GCs with extended HB, such as M13, M15 and M80. It is located at $(B - V)_0 \approx 0.0$ and its effective temperature is about 11 000 K, that is, it is slightly cooler than the *u* jump and the point where the surface chemical abundance anomalies start to be evident. As a consequence, the interpretation of this gap as being due to diffusive processes in the atmosphere of HB stars appears reliable;
- Gap *G3*: is located at $T_{eff} \approx 20\,000$ K, and it is less-sharply defined than gap *G1*, particularly in optical-band CMDs. It seems to be present in all GCs with very extended blue tails, such as M13 and NGC 2808.

4.7.2
The Theoretical Framework

The best explanation for the large variations in the chemical abundance patterns observed in hot HB stars is represented by the occurrence of those diffusive processes presented in Section 2.7.

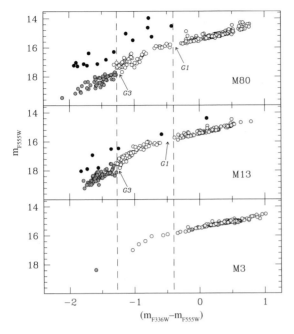

Figure 4.33 The morphology of the HB of the Galactic GCs M3, M13 and M80 from HST photometry [301]. The location of the two gaps *G1* and *G3* is indicated (courtesy of F. Ferraro).

Indeed, for a long time, it has been suggested that the under-abundance of helium in the atmosphere of field HB stars could be the consequence of He sinking under the influence of gravity from the surface to the interior layers [303]. For this process to be efficient, the outer stellar layers have to be stable against convection because the presence of any mixing (regardless of whether convective or rotational) would erase the chemical gradient produced by atomic diffusion. If the atmosphere is stable and the stellar surface is hot enough so that radiation pressure can efficiently transfer momentum to the chemical species which present sufficiently large cross sections to the outgoing radiation field, radiative levitation can be also efficient, thus producing photospheric enhancements of (some) heavy elements [293].

Therefore, it is clear that one of the most crucial properties in defining the efficiency of diffusive processes is the mass of the surface convection zone, which is thoroughly mixed. A simple look at canonical HB models shows that those cooler than $\log T_{\mathrm{eff}} \approx 3.8$ have a very deep convective envelope (see Figure 4.34). For T_{eff} larger than this limit but less than $\sim 11\,500$ K, the thin convective zones associated with the ionization of H and the first ionization of He, lie at, or slightly below, the stellar surface. For $T_{\mathrm{eff}} > 11\,500$ K, only the convective region due to He II is present and its location in mass moves outwards with increasing T_{eff}, until it approaches the stellar surface at $T_{\mathrm{eff}} \sim 23\,000\text{--}25\,000$ K.

Since convection is highly efficient at preventing diffusive processes such as atomic diffusion and levitation, the disappearance of convection at $T_{\mathrm{eff}} \approx 11\,500$ K

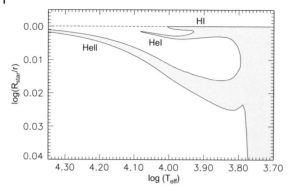

Figure 4.34 The location with respect to the stellar surface of the convective zones (shaded areas) in canonical ZAHB stars as a function of T_{eff}. The convective zones associated with the H I, He I and He II – when they are separated – ionization zones are clearly marked as well as the location of the stellar surface (dashed line). One has to note that the H I and He I convective regions are actually only separated in a very limited T_{eff} interval.

may be the "switch" that controls the appearance of the effects of diffusive processes. In this context, model predictions about the dependence of the outer convective zones on the effective temperature provide a self-consistent interpretation of why the chemical abundance anomalies, the u jump and low-gravity problem appear at almost exactly the same T_{eff} value for all studied GCs.

However, we are still faced with the question of how stellar models accounting for the diffusive processes can reproduce the empirical evidence. Due to the intrinsic difficulties in computing such stellar models, and the lack – in the past – of suitable prescriptions about radiative acceleration for the various chemical species, no large effort has been devoted to this issue since the pivotal work of [293]. However, during this last decade, the situation has been changing rapidly and many accurate investigations have been performed [234, 304, 305]. The main outcomes of these theoretical investigations can be summarized in the following way:

- The trend of radiative acceleration (g_{rad}) for various chemical elements as a function of mass coordinate are shown in Figure 4.35 for a selected HB model at different times in its HB evolution. These g_{rad} values are a factor of 10–100 (or more) larger than the local gravity over a large fraction of the star for many elements. Interestingly, the radiative acceleration on He is always lower than local gravity, so levitation is ineffective in pushing He upward.
  One also has to note that there are some elements, such as Mg and Si, for which observations do not show significant enhancements, which have a maximum value of log(g_{rad}), that is, ≈ 0.5 dex lower than that for elements that observationally appear to be strongly affected by levitation, for example, P, Ti and Fe. More importantly, we note that some elements such as P and Si have a value of radiative acceleration that is less than the local gravity in, and slightly below, the outer convective zone (see the following discussion).

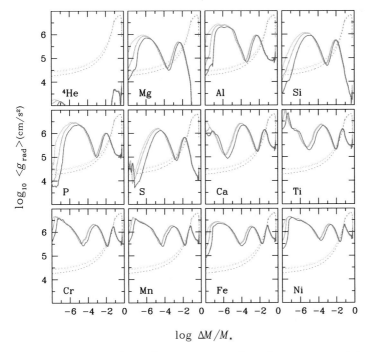

Figure 4.35 The trend of the radiative acceleration (solid lines) for various chemical species, as a function of the mass coordinate inside the outer layers of a $0.59 M_\odot$ HB star, whose ZAHB location is at $T_{eff} \approx 14\,500$ K. The dotted lines are the local gravity. The coding refers to various evolutionary stages: light-grey lines to the ZAHB, mid-grey lines to 5 Myr after the ZAHB and dark-grey lines to 23 Myr after the ZAHB (courtesy of G. Michaud and J. Richer).

The behaviour of g_{rad} with position in the stellar envelope for various elements has to be related to the dependence of the specific ionization states on the local temperature. The radiative acceleration does not depend much on the metallicity. This is because of the saturation of the metallic lines, g_{rad} drops as the metallicity is increased. Conversely, an increase of T_{eff} causes an increase of g_{rad} in the outer layers because of the dependence of the ionization states on the local temperature, thus changing the relative contribution of the various envelope layers to the global heavy element enhancement;

- Once the trend of g_{rad} with mass coordinate is known, one can estimate the expected abundance variation. However, evolutionary computations show that, when also accounting for the convective zones (which contribute to limit the efficiency of the diffusive process), the predicted heavy element enhancements are too large with respect to spectroscopic measurements. Therefore, it is unavoidable that including some ad hoc (additional) mixing process, operating in the outermost layers, is necessary with the aim of limiting the efficiency of the radiative levitation.

In fact, what is (are) the physical mechanism(s) operating in real stars which partially inhibits diffusive processes is under debate. It is commonly assumed that some kind of turbulence causes the mixing of an arbitrary fraction of the outer layers. The amount necessary is fixed by comparing various sets of model predictions with spectroscopic measurements of GC stars: it has been shown that a value for this mixed mass fraction, $\Delta M_{mix}/M_{tot}$, that will reproduce the empirical evidence, for cluster HB stars and field sdB and sdO stars, is equal to 10^{-7} of the total stellar mass.[27]

When taking this additional mixing process into account, the chemical abundance changes can be estimated (such predictions are shown in Figure 4.36). Without entering into great detail, diffusive models that include this sort of turbulence reproduce the differential abundance enhancements among the various heavy elements fairly well (obviously, the absolute values are controlled by the value adopted for the free parameter $\Delta M_{mix}/M_{tot}$). In particular, it is worth noting that the lower enhancements predicted for Mg and S with respect, for instance, to Fe, P and Ti, are in good agreement with the observations;

- The availability of diffusive models for various initial metallicities allows one to analyze the dependence of the final heavy element enhancements on the initial value of Z. For instance, when comparing, at the same T_{eff}, HB stars with an initial metal-poor composition ($Z \sim 10^{-4}$), representative of GC stars, with similar structures with solar chemical composition, the final surface abundances vary by, at most, a factor of ~ 30, despite the fact that the initial abundances (before diffusive processes start working) vary by a factor of ~ 200, and for Fe and Ni the variations are still smaller. This is a consequence of the saturation of g_{rad} with metallicity. However, on general grounds, more metal-rich HB stars end up with larger heavy element abundances and larger He under-abundances with respect to their metal-poor counterparts;

- The dependence of the metal enhancements with T_{eff} along the HB locus predicts a significant increase in heavy element enhancements when moving from $T_{eff} \approx 11\,500\,K$ to the EHB stars. When considering a relatively smaller T_{eff} interval, the changes in the abundance enhancements are smaller, being phosphorus the most sensitive to changes in the effective temperature. Because of the scatter due to observational errors, it is difficult to observationally detect these chemical abundance changes over small intervals in T_{eff}, but the global trend is confirmed by empirical data;

- All numerical experiments confirm that the efficiency of diffusive processes in modifying the surface chemical pattern is so high that in the first $\sim 5\,Myr$ after the stars reached the ZAHB, the abundance variations are already relevant (see Figure 4.36). In the subsequent evolution, the changes are significantly smaller. The time dependence of the abundance variations appears to be more linear for those elements, such as He, which become under-abundant.

---

27) Diffusive models computed under various assumptions on $\Delta M_{mix}/M_{tot}$ reveal that a decrease by an order of magnitude would produce unrealistic surface Fe abundances, while an increase by a factor of ~ 3 also seems to be excluded by abundance observations.

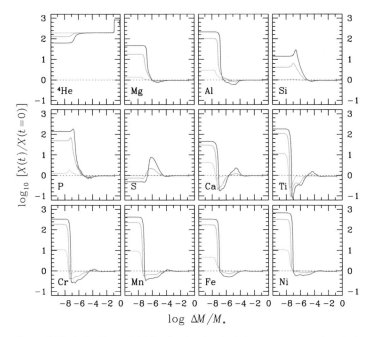

Figure 4.36 As for Figure 4.35, but for the chemical element concentration. In these HB models, the choice of mixing the outer layers up to a depth of $\Delta M/M_\star \sim 10^{-7}$ has been adopted (courtesy of G. Michaud and J. Richer).

When comparing diffusive models with suitable empirical evidence, as shown in Figure 4.37, one finds, in general, a good agreement. However, one should note that these models predict heavy element enhancements (see the case of Fe in Figure 4.37) also for stars cooler than 11 500 K, and this is at odds with the spectroscopic measurements. This clearly implies that the canonical surface convective zone predicted by canonical HB models (Figure 4.34) is not deep enough to inhibit the effect of levitation in the diffusive models (see also below). So, some other mechanism(s) has (have) to be at work in real cool HB stars reducing – or completely eliminating – the effects of diffusive processes on the surface chemical abundances.

Before closing this discussion, we note that the occurrence of these physical processes does not affect the structural properties of low-mass, core He-burning stars. However, [234] has presented compelling theoretical evidence, suggesting that the role played by convective overshoot during the phase of expansion of the convective core (see Section 4.2) could be significantly reduced or eliminated when accounting for the diffusion of C nuclei from the canonical convective core into the surrounding radiative zone. From a practical point of view, diffusion has the same effect as convective overshoot does in transporting C into the region outside the convective core, thus enhancing its radiative opacity. Interestingly, the rate at which the mass of the convective core increases in models accounting only for atomic diffusion (but not for convective overshoot) is exactly the same as obtained for stellar mod-

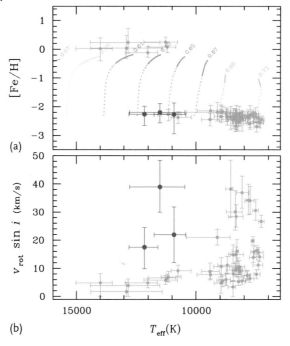

Figure 4.37 (a) The surface abundance of [Fe/H] as a function of T_{eff} for HB stars in the metal-poor GCs M15, M68 and M92. The theoretical predictions provided by HB models [305] accounting for radiative levitation and non-canonical mixing (see text for more details) for a metallicity suitable for the selected GCs are also shown. Each colour-coded segment represents the evolution of the surface Fe abundance for various evolutionary masses: the line is solid for the time interval from 10 to 30 Myr after the ZAHB, and dotted from 0 to 10 Myr. (b) The projected rotation velocity for the same GC HB stars shown in (a). The three dark-grey points identify the few stars characterized by a relatively high-rotation velocity and unenhanced iron abundance, despite being hotter than $\sim 11\,000$ K (courtesy of G. Michaud and J. Richer).

els based on the usual approach for managing the convective core expansion (see Figure 10 of [234]).

The inclusion of turbulence in diffusive HB models is then essential for decreasing the efficiency of diffusion processes, and so reproducing the empirical findings. However, one should consider that various hydrodynamic processes may compete with diffusive processes and reduce the chemical abundances anomalies they lead to. Among these various hydrodynamical processes could be: mass loss, rotationally-induced meridional circulation, turbulent diffusion. The observational evidence on the rotational properties and chemical anomalies observed in HB stars seems to suggest that rotation and the correlated mixings could be the process competing with diffusion.[28]

28) This hypothesis has already been proposed for explaining the peculiar chemical patterns observed in HgMn stars, that is, B stars with anomalously high surface abundances of mercury and manganese. Actually, HgMn stars are very slow rotators compared to normal O stars, but in the HRD, they occupy almost the same T_{eff} interval.

In this context, a possible link between peculiar rotational rates and chemical anomalies has been suggested by [306]. When neglecting non-canonical mixings, due to gravitational settling, the envelope He abundance becomes so low that the convection zone associated with He ionization disappears; when this occurs, diffusive processes can hugely alter the envelope and surface chemical stratification as they are not held back by convection. However, if the star rotates with a rate larger than some critical value, the rotationally-induced meridional circulation below the outer convective zone could prevent atomic diffusion to reduce the envelope He content below the critical value for which the He convection zone disappears.

The value for this critical velocity in HB stars is unknown. As a first order approximation, one can adopt the same prescriptions used for interpreting the chemical anomalies in rotating, main sequence O stars, with a scaling law for the critical velocity as a function of T_{eff}. In any case, this is not the only problem when investigating the impact of meridional circulation on the chemical abundances of HB stars: another issue is related to the amount of mixing occurring between the outer (canonical) thin convective layer in the outer envelope and the underlying rotationally-induced mixing zone. Due to the lack of a sound prescription, one can consider two extreme cases: (i) the meridional circulation does not enter into the convective envelope and barely reaches its lower boundary, (ii) the rotationally-induced mixing zone effectively penetrates into the outer convective zone. The two cases correspond to two extreme situations concerning the capability of meridional circulation to prevent the sedimentation of He; clearly the latter case corresponds to the higher efficiency.

Such analyses have shown that, regardless of the assumption about the capability of meridional circulation to enter into the outer convective zone, the observed rotational velocities of HB stars hotter than $\sim 11\,500$ K are always below the critical velocity; so rotationally-induced mixing cannot compete with He-settling and avoid the disappearance of the He convection zone. At the same time, stars cooler than $T_{eff} \approx 10\,500$ K show a rotational velocity larger than the expected critical value, and so in such stars, meridional circulation would prevent the diffusive processes from significantly altering the chemical stratification in the outer layers, as is indeed observed.[29] For intermediate T_{eff} values, the critical velocity for meridional circulation being efficient and the rotational rate of the stars are comparable. Therefore, the exact T_{eff} value at which chemical anomalies appear strongly depends on a more accurate evaluation of the critical velocity for efficient rotationally-induced mixing, as well as on the extent of hydrodynamical coupling with the superficial convection zone.

It is worth noting that this scenario, based on the competition between gravitational settling of He and He advection by meridional circulation, does not provide any physical explanation for the observed bimodality in the $v \sin i$ values. However, once the observed $v \sin i$ bimodality is taken into account, this working scenario

29) This link between rotation and diffusive processes, that was not accounted for in the models of [305], would provide an explanation of the reason why these diffusive HB models predict huge metal enhancements also for stars cooler than $\sim 10\,000$ K.

provides a natural explanation of the trend of the chemical abundances anomalies with effective temperature.

In passing, we note that this scenario also provides a nice explanation for the existence of the few HB stars whose T_{eff} is larger than 11 500 K, but do not show any chemical anomalies with respect to their RGB progenitors: it is a consequence of their anomalously high-rotational velocity (see the corresponding points plotted in Figure 4.37).

The empirical evidence concerning the rotational rate distribution of HB stars in GCs is more difficult to explain than the observed chemical anomalies, and more challenging for the theoretical evolutionary framework. While several scenarios have been outlined, we lack any compelling evidence to choose between them.

Actually, the main problems in interpreting the observational findings are related to the existence of a bimodality in the rotational rate distribution, with some objects showing high-rotational rates, and to the fact that the bluest stars are the slowest rotators among the HB stars. Indeed, following early suggestions [226], it has always been considered that rotating stars would delay the ignition of He at the RGB tip, these stars attaining a larger luminosity and a lower T_{eff} with respect to the non-rotating stars. All these effects will contribute to enhancing the mass loss during the RGB stage. So, one could expect that the larger the rotational rate in the RGB stage, the lower the envelope mass of the corresponding HB stars and, hence, the hotter the ZAHB location. Such a theoretical (simplified) expectation is in apparent contrast with observations showing that the hotter HB stars are in fact the slowest rotators along the HB. However, as mass loss also removes angular momentum, bluer HB stars would be expected to have a lower angular momentum than redder HB stars; nevertheless, hot HB stars have smaller radii and so they also have a smaller moment of inertia than redder HB stars. Therefore, it is not possible to predict a priori if hot HB stars should show lower rotational rates or not.

Another embarrassment arises from the mere existence of fast HB rotators: due to the large mass loss during the RGB stage, huge amounts of angular momentum should be lost during this evolutionary stage, so it is not clear what the source of the angular momentum is that is forcing the envelope of some HB stars, cooler than $\sim$ 11 500 K, to rotate as fast as observed.

Although the computation of stellar models for advanced evolutionary stages accounting for rotation and rotationally-induced mixings is still in an early stage (mainly due to the lack of suitable observational constraints for Population II core H-burning stars), some effort has been devoted to finding a solution to the conundrum of the HB star rotational state [307]. The theoretical models for rotating HB stars have been computed by assuming a solid body rotation for the MS stellar progenitor, but with various assumptions about: the initial angular momentum, the redistribution of angular momentum between the He-core and the outer envelope during the RGB phase, and the rotational law (solid body versus constant specific angular momentum) for the convective regions.

Regardless of the adopted physical assumptions, all of these models consistently predict that during the RGB stage, as a consequence of the contraction of the He-core, of the deepening of the convective envelope and the increase of the stellar radius, the core spins up and the outer envelope spins down. At the tip of the RGB, the stellar surface is rotating at a factor of ~ 100 slower than its MS progenitor. After the He flash, the structural properties of the star are reversed because the core expands, whereas the outer layers contract. This causes a flattening of the rotational profile and so the surface rotation rate is increased by a factor of ~ 10 with respect to the rate at the RGB tip. HB stars are not expected to rotate as a solid body, but the interior rotation profile is much more similar to that of a solid body than to that of the highly differentially rotating RGB stars. This general scenario is further complicated by the occurrence of mass loss during the RGB. In fact, as already mentioned, mass loss is highly efficient in also removing angular momentum: the efficiency of this process strongly depends on the assumed rotational law for the convective envelope.

The main outcomes of these numerical simulations can be summarized in this way:

- Regardless of the adopted physical assumptions on the angular momentum redistribution and rotational law for the convective zones, the adoption of an initial rotational velocity of $\sim 1\,\mathrm{km\,s^{-1}}$ for the MS progenitor, that is, a value obtained by extrapolating to the Pop. II objects the values commonly adopted for low-mass Pop. I stars, provides rotational rates on the HB too low and largely discrepant with empirical measurements for both hot and cool HB stars;
- The same outcome is also obtained when increasing the initial rotational velocity of a factor of 4, and adopting in all cases a local conservation of angular momentum while the star is leaving the RGB and settling on the ZAHB. In other words, to assume a local conservation of angular momentum means that no coupling between the core and the envelope is allowed to occur;
- On the other hand, if the larger initial angular momentum configuration is adopted, and one allows the redistribution of angular momentum from the – faster – rotating core to the outer envelope, then fast rotating HB stars with rotational rates comparable to the faster rotating HB stars in GCs can be obtained. Indeed, this scenario is also able to explain the rotation velocity increase when moving from the reddest portion of the HB to $\sim 11\,000$ K: the coolest HB stars are those which have lost less mass and thus less moment of inertia, so they have to be slow rotators; when moving toward larger T_{eff} values, the corresponding ZAHB models have lost more mass and more initial angular momentum. However, since their moment of inertia decreases more quickly than the initial angular momentum, they are expected to rotate faster as their T_{eff} increases;
- Evolutionary considerations related to the timescales of the various processes support the idea that this core-envelope coupling should occur during the core He-burning stage, and that this process would require a significant fraction of the core He-burning lifetime. As a consequence, one should expect that the HB stars (cooler than $\sim 11\,500$ K) rotating faster are those that have been on the HB longer, and have had more time to spin up. According to this scenario, one should expect that among the HB stars in GCs, the fast rotators should also be

the brighter. So far, the available observational evidence does not allow one to check this prediction;

- These rotating models are not able to provide any hint on the slow rotation rate characteristic of hot HB stars.

A viable solution for the apparent discrepancy between the core-envelope coupling scenario for fast HB rotators and the observed slow rotational rates of stars hotter than $\sim 11\,500$ K is that based on the existence of a physical process able to make the transport of angular momentum from the core to the envelope fully inefficient. Such a process has been related to the existence of a large gradient in the mean molecular weight originated by the gravitational settling of He as it actually occurs in stars hotter than the quoted T_{eff} limit. This scenario clearly suggests the existence of a strong link between diffusive processes, rotationally-induced mixings and angular momentum transport.

There is an alternative scenario for explaining the low rotational rate of hot HB stars [70]: the mass loss rates for hot HB stars are expected to be high due to the huge metallicity enhancement caused by radiative levitation; such a mass loss process could be efficient in removing the envelope mass and, hence, angular momentum, thus forcing the star to have a low rotational velocity. In passing, we note that the occurrence of mass loss during the HB stage could also be very important in modifying the chemical stratification induced by the diffusive processes: it has been shown that if the mass loss exceeds a critical value of $\approx 10^{-14}\,M_\odot$/year [308], the chemical overabundances due to levitation could be wiped out.[30]

Another scenario is based on the idea that slow rotating HB stars correspond to the normal HB population, whereas the fast rotators would be an anomaly arising from non-canonical evolutionary channels based largely on the interaction with a stellar companion in a binary system or dynamical interactions in the dense cluster environments. For the binary system case, the additional source of angular momentum required for spinning up the star is provided by the companion, either through tidal synchronization of the orbital and rotation periods or via a merger. However, this evolutionary channel is not supported by the evidence since the fraction of binary systems including a hot HB star in GCs is very small. In the "cluster dynamics" scenario, close tidal encounters with a single star or a binary system could spin up the HB progenitor or the HB star itself. Obviously, this scenario cannot explain the presence of fast rotators among field HB stars.

Alternatively, it has been pointed out that a hot HB star could be a fast rotator because its RGB progenitor was spun up when it swallowed (orbital separation ≤ 5 AU) a planetary companion: the deposit of angular momentum and energy into the stellar envelope would significantly increase the available angular momentum and enhance the mass loss [309].

---

30) Actually, the value of this critical mass loss rate is uncertain by about one or two orders of magnitude. More detailed and extensive investigations on this argument are mandatory.

5
Asymptotic Giant Branch

5.1
Overview

When the He abundance inside the convective core drops down to a low enough value as a consequence of the burning process, all stellar structures, regardless of their initial mass, move in the HRD towards lower effective temperatures and start to climb the *Asymptotic Giant Branch* (AGB). This evolutionary stage corresponds to the shell He-burning phase; however since, at a certain moment, the H-burning shell can also be efficient, it is also known as the *double shell burning* stage.

The AGB evolution of stellar structures is interesting for many reasons. It represents, together with the WD cooling stage, the final evolutionary phase for the large majority of all stars in the Universe. AGB stars contribute to the integrated starlight of many galaxies, via their winds may provide the only interstellar matter present in some galaxies and, due to their large brightness and old ages (at least for the low-mass stellar objects), they can be used as important population tracers and to probe the Galactic structure and dynamics.

The fundamental role played by AGB stellar structures in the thermally pulsing stage to the chemical evolution of matter in the Universe has been recognized for a long time. Indeed, these stars are responsible for the nucleosynthesis of the main and the strong components of the *s*-process and they contribute to the synthesis of several light elements, such as Li, C, N and F.

The evolutionary properties of stars during the following evolutionary stage, that is, the WD cooling sequence, largely depends on the physical mechanisms such as mass loss and mixing occurring during the AGB phase. For this, AGB stars are fascinating objects, where a complicated interplay between physical and chemical processes takes place; an occurrence that still makes computing reliable stellar models for this evolutionary phase a challenge.

Until a few years ago, our knowledge of the observational properties of AGB stars and their nucleosynthesis was mostly based on observations of AGB stars in the field and clusters of our Galaxy. However, the recent availability of optical and near-infrared (near-IR) photometry and spectroscopy of AGB stars belonging to nearby galaxies such as dwarf spheroidals and the Magellanic Clouds is providing a wealth of observational data for metal-intermediate and metal-poor AGB stars. In

Old Stellar Populations, First Edition. S. Cassisi and M. Salaris.
© 2013 WILEY-VCH Verlag GmbH & Co. KGaA. Published 2013 by WILEY-VCH Verlag GmbH & Co. KGaA.

addition, the growing number of investigations devoted to C-enhanced extremely metal-poor stars (the so-called CEMPs) enriched in s-elements, is opening a new window in the investigation of the nucleosynthesis that occurred in the now extinct AGB halo population.

In the following, we will discuss the main structural and evolutionary properties of AGB stars in old stellar systems: we limit our discussion to stars with initial mass lower than about $3 M_\odot$. In particular, we discuss in some detail, the mixing processes occurring in this stage and the nucleosynthesis typical of low-mass AGB stars.

5.2
Early Asymptotic Giant Branch Evolution

The evolution of the main structural and evolutionary properties of low-mass stars near the end of the central He-burning stage was outlined in Section 4.3 and shown in Figure 4.6. As soon as the He abundance in the core drops down ~ 0.10, the efficiency of H-burning in the shell increases and that of He-burning processes occurring in the convective core decreases. Quite near central He-exhaustion, from the point of view of the energy budget, the structure is supported by the H-burning process together with the energy released by gravitation via contraction. The maximum value of the central temperature is achieved at this point. From now on, the central temperature monotonically decreases; in contrast, the central density monotonically increases during the transition from the core He-burning stage to the AGB one. As a consequence of this behaviour of the central thermal conditions, the CO core, left over by the central He-burning process, experiences an increasing level of electron degeneracy, in close similarity with what it is experienced by the He-core during the RGB evolution of low-mass stars.

It has been mentioned that at the end of the HB stage, the stellar luminosity shows a quite rapid increase, followed for a while by a small drop and then a monotonic rise. This occurrence causes an accumulation of stars at the relevant luminosity interval in the HRD, known as the *AGB clump* because, as it occurs in the case of the RGB bump, the stellar structures cross this luminosity interval three times (Figure 4.6b).

As a consequence of this behaviour, theoretical models predict that the transition between the central and the shell He-burning should be marked by a clear gap where a few stars should be found in the HRD, while a well-defined clump of stars should indicate the base of the AGB. This prediction is nicely confirmed by observations of well-populated Galactic GCs as 47 Tuc [310].

The dependence of the AGB clump properties such as its average brightness and colour, on metallicity is shown in Figure 5.1. The luminosity of the AGB clump is almost independent of stellar metallicity and initial He abundance, so that this bright feature is a promising standard candle [311]. However, unfortunately, the

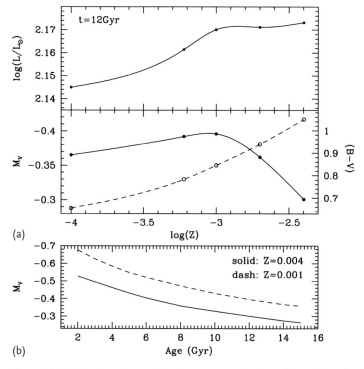

Figure 5.1 (a) The dependence of the AGB clump bolometric luminosity on metallicity for an age of 12 Gyr. (b) The same but for the V-band (solid line) and the $(B - V)$ colour. (c) The dependence of the absolute visual magnitude of the AGB clump on age for two metallicities (see labels).

identification of such a clump is not easy since the AGB phase itself is very short ($\sim 10^7$ years) and, in turn, always poorly populated.[1]

Figure 5.1 also shows the dependence of the AGB clump luminosity on the age of a given stellar population. One can note that the brightness depends significantly on the evolving stellar mass: larger stellar masses tend to generate brighter (and redder) AGB clumps. The dependence of the AGB clump luminosity on the mass of the evolving stars implies an indirect dependence on all the other parameters that could affect the mean mass and its distribution along the HB: mainly the mass loss efficiency, but also the metallicity and the cluster age. In particular, one can note that moving from metal-rich to metal-poor stellar populations, the AGB clump tends to show a hotter (i.e. bluer) location in the HRD and becomes less clumpy with decreasing stellar mass: an occurrence that makes the identification of this feature in the CMD of metal-poor stellar systems less obvious. Figure 5.2 shows a comparison between theoretical predictions of the difference between the ZAHB

---

1) In a Galactic GC with total luminosity $L_{tot} = 10^5 L_\odot$, only ~ 20 AGB stars are expected to be observed [312].

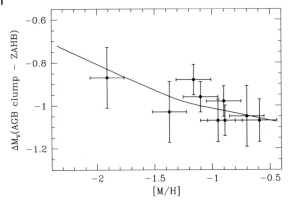

Figure 5.2 A comparison between theoretical predictions and empirical data for a sample of Galactic GCs of the V-magnitude difference between the AGB clump and the ZAHB [313].

and the AGB clump brightness, together with empirical estimates for a sample of Galactic GCs.

One also has to bear in mind that the luminosity of the AGB clump as well as the evolutionary lifetimes during the core He-burning and AGB stages depend on the treatment of the mixing processes during the central He-burning stage. As discussed in Section 4.2, the treatment of the breathing pulses phenomenon hugely affects the ratio t_{AGB}/t_{HB}, that is, the ratio between the evolutionary lifetimes during the core He-burning stage and the AGB stage:[2] roughly speaking, when accounting for the occurrence of the breathing pulses t_{HB} is increased by about 20%, whereas the AGB lifetime (t_{AGB}) is decreased by about 25% with respect a similar stellar model, but computed by inhibiting the occurrence of this convective instability. These lifetime changes are the consequence of the larger convective core and, hence, larger CO-core left over after central He-exhaustion in stellar models accounting for breathing pulses. As already mentioned, these model predictions are at odds with empirical estimates about the $R_2 = N_{AGB}/N_{HB}$ parameter. This evidence supports the need to inhibit the breathing pulses in stellar model computations. The occurrence of this convective instability also affects the AGB clump brightness: a reduction of ~ 0.06 mag is expected in stellar models accounting for breathing pulses. More importantly, it also becomes a less prominent feature, as shown in Figure 5.3. However, it is important to note that the numerical approach used for avoiding the occurrence of the breathing pulses also affect the AGB clump brightness (and the t_{AGB}/t_{HB} ratio): to adopt the alternative approach of neglecting the contribution of gravitational energy in the final phases of the core He-burning stage causes an increase of the AGB clump brightness of about 0.25 mag (see Figure 5.3).

2) For the sake of estimating the corresponding evolutionary lifetime, the AGB stage can be identified with the evolutionary phase preceding the thermally pulsing stage. This is because the thermally pulsing evolution is faster with respect to the previous evolutionary stage.

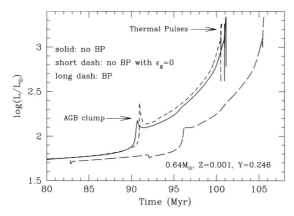

Figure 5.3 Time evolution of the surface luminosity from the late core He-burning stage to the AGB for a low-mass star computed under various assumptions for the treatment of the breathing pulses. The location of the AGB clump and the beginning of the thermal pulse stage are also indicated.

Soon after the AGB clump, the efficiency of the He-burning shell increases, taking the full control of the stellar energy budget (see Figure 4.6), while the efficiency of the H-burning shell drops. In this context, the He-burning ignition in the shell after the AGB clump marks the starting point of AGB evolution. The evolutionary stage between the onset of He-burning in a shell and the beginning of the thermally pulsing phase is named the *early*-AGB (E-AGB); while those stars that are in the thermally pulsing phase are designated as TP-AGB stars. The E-AGB stage can be easily identified in Figure 5.3.

During the early-AGB, the He-burning shell moves progressively outward and the mass of the CO-core increases. In stars with mass lower than $3M_\odot$, the exact value depends on the initial chemical composition[3]–, although the H-burning efficiency in the shell surrounding the He-rich core is not high, the burning of hydrogen is always active, thus maintaining an entropy barrier that prevents the outer convective zone from crossing the boundary between the H-rich envelope and the He-core, and increasing the size of the H-exhausted region as shown in Figure 5.4.[4] This behaviour is in contrast with what occurs in more massive stars: due to the huge energy flux coming from the He-burning shell, the H-rich envelope expands and cools, so that H-burning in the shell is completely switched off, thus allowing the outer convection zone to penetrate inward to well within the He-rich zone. This occurrence is the so-called *second dredge up*.

Figure 5.4 also shows the trend with the time of the various sources contributing to the energy budget as well as the central thermal conditions during the whole E-AGB stage of a typical low-mass AGB star.

---

3) As rule of thumb, the lower the metallicity, the lower the minimum mass for the occurrence of the second dredge up.

4) In the case of the stellar structure selected for such figures, the He-core mass increases by $\sim 10^{-2}\,M_\odot$ during the E-AGB stage, while the CO-core mass is increased by about $0.30M_\odot$.

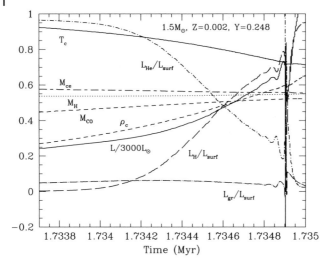

Figure 5.4 Time evolution of selected evolutionary quantities for an E-AGB, low-mass star. The mass of the CO-core (M_{CO}), of the He-core (M_H) and of the location of the lower boundary of the convective envelope (M_{ce}) are in a solar unit, while the central temperature and density are in units of 1.5×10^8 K and 2.5×10^6 g/cm^3, respectively.

As a consequence of the burning process, at a certain moment, the He-burning shell approaches the H/He discontinuity. This causes the He-burning shell to die down, while the H-burning shell comes alive again due to the heating induced by the proximity of the He-burning shell and the gravitational energy release associated with the contraction of the outer layer. The re-ignition of the H-burning process marks the end of the E-AGB stage and the beginning of the TP-AGB phase.

Before closing this section, we note that during the E-AGB stage (with the exception of the AGB clump) luminosity and T_{eff} are monotonic as a function of time. The designation "asymptotic" in the acronym AGB arises from the fact that, for masses lower than about $1\,M_\odot$, the relationship between surface luminosity and effective temperature for low-mass AGB stars is very close to, although slightly to the blue of, the equivalent relationship for low-mass RGB stars. In the case of more massive stars, due to the morphology of the evolutionary tracks in the HRD, the term "asymptotic" is largely meaningless, and it would be more appropriate to refer to this evolutionary stage as the *double shell burning* phase.

5.3
The Thermally Pulsing Stage

As a consequence of H-burning, the mass of the He-rich inter-shell between the CO-core and the H-burning shell increases with the time and, due to contraction, it is warmed up. As soon as the mass of this He-rich zone attains a critical value (ΔM_{He}), of the order of $10^{-3}\,M_\odot$ for a CO-core mass of $\sim 0.8\,M_\odot$, He-burning is ignited. The exact value of this critical mass strongly depends on the mass of the CO-core because it is the mass-size of the core that fixes the thermal conditions and, hence, the efficiency of the triple-α reaction at the base of the He-rich inter-shell region: roughly speaking, the value of ΔM_{He} increases by about an order of magnitude for a decrease in the CO-core mass of $\sim 0.2 M_\odot$.

The re-ignition of He-burning in the shell has the characteristic of a thermonu-clear runaway, somewhat reminiscent of the He flash occurring at the RGB tip in low-mass stars, as the shell reacts to the fresh energy input with a temperature increase, which causes a further increase of the nuclear reaction rate and, hence, of the energy release. The He-ignition has the characteristic of a thermal runaway despite the fact that the level of electron degeneracy at the base of the He-burning shell is low, a relevant difference with the properties of the He-core at the RGB tip in low-mass stars [314]. This occurrence must be related to two concomitant condi-tions: the small geometrical thickness of the shell and the huge temperature depen-dence of the 3α-reactions. In fact, it is generally the case that if energy is dumped into a shell, for example, as a consequence of a nuclear burning process, the shell reacts by expanding and thus lifting the stellar layers above it. As a consequence of this expansion of the outer layers, the pressure in the shell (that in the assump-tion of hydrostatic equilibrium, is determined by the weight of the layers above it) drops. If the shell is thick enough, the pressure in the shell decreases almost at the same pace as the density (assuming an homologous expansion $\Delta P \propto (\Delta \rho)^{4/3}$), and so, assuming an almost ideal gas law, the temperature in the shell also de-creases. However, if the shell is very thin, it will experience a huge expansion with a large fractional local decrease of the density, but the lifting of the outer layers will be small. Consequently, the local decrease of the pressure in the shell will be relatively small compared to the density drop. Depending on the shell thickness, it can happen that the pressure decrease is percentage-wise smaller than the density decrease, and with this the temperature will actually increase. This temperature in-crease is followed by an increase of the nuclear reaction rates, and hence a further energy gain.

Theoretical analysis of such a process allows one to predict the minimum shell thickness required for the development of a thermal runaway for the case of nuclear processes associated with both He-burning and H-burning via the *CNO cycle* [20, 314]. For H-burning occurring via the *p–p chain*, a thermal runaway can not occur regardless of the shell thickness because of the low temperature-dependence of the *p–p* nuclear process. In AGB stars, the He-burning shell is thin enough, and the He-burning process efficiency has such a large dependence on the temperature ($\epsilon_{3\alpha} \propto T^{40}$), that the thermonuclear instability always occurs, and recursively so during the double-shell burning stage.

During the early thermal pulse (TP) stage, the luminosity of the He-burning shell increases within a very short time (of the order of a few 100 years) to a value of the order of $10^6 - 10^7 L_\odot$. The expansion caused by the huge energy injection associated with He-burning takes matter to and beyond the H/He discontinuity, out to such a low temperature and density that the H-burning is shut off.

After a while, the He-burning stabilizes and the associated energy input becomes lower than the surface luminosity L, and the expanded envelope begins falling back inward. This heats the layers at the H/He discontinuity and H-burning is reignited. From now on, the He-burning efficiency continues to decrease while the H-burning rate continues to increase, until a steady state is reached during which the H-burning supplies almost (at least in low-mass AGB stars) the whole energy budget, that is, $L_H \approx L$ and $L_H \gg L_{He}$. During this steady state, whose duration depends on the value of the critical mass of the He-rich region required to reignite He-burning, the quiescent H-burning pushes the burning shell outward in mass until, after a critical mass ΔM_{He} of He has been produced and accreted to the He-core, another TP is initiated.

The time-dependence of the surface luminosity, effective temperature, H- and He-burning energy release and He-core mass are shown in Figure 5.5 for a low-mass star from the beginning of the TP stage through to the first few thermal pulses. One can see that the surface luminosity L drops immediately following the dramatic increase of L_{He}, and after reaching a relative minimum, L increases for a short period before decreasing again, at a more leisurely pace, to an absolute minimum. These surface luminosity changes with time can be immediately related to the changes in the H- and He-burning efficiencies during the development of

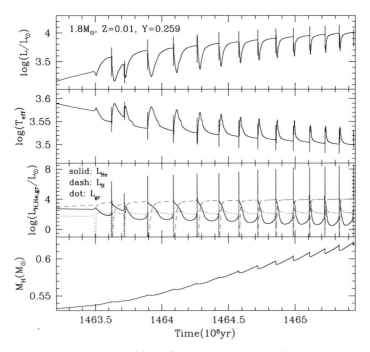

Figure 5.5 Time evolution of the surface luminosity, effective temperature, energy released by various sources, and of He-core mass during several TPs, for a selected stellar model (see labels). The sawtooth pattern of the He-core mass after the seventh TP is related to the occurrence of the TDU after each subsequent TP (see text for details).

each TP. One can also note that the pulse amplitude grows with each succeeding pulse, quite rapidly for the first 7–10 TP, then much more slowly for the next $\sim$ 10 TP, until an "asymptotic" behaviour is approached.

It is important to note that a tight correlation does exist between the surface luminosity and the mass of the He-core M_H: this is because the efficiency of the H-burning in the shell depends on the mass of the H-exhausted core. The existence of this *He-core mass–luminosity* relation for AGB stars is well-known and closely resembles the similar relation existing for RGB stars [315, 316]. The classical core mass–luminosity relation relating the He-core mass at the maximum surface luminosity, L_{max}, attained by the stellar models at the end of each inter-pulse stage is

$$L_{max} = 5.925 \times 10^4 (M_H - 0.495)$$

where all quantities are in solar units [317]. This behaviour is applicable to all stars in the mass range we are presently considering. Minor changes, such as in the time-dependence of the surface luminosity during the inter-pulse phase, are possible as a consequence of different envelope masses; but these are second order effects, and the general behaviour is not altered. Typically, a thermal pulse cycle lasts from a few 10^4 years up to some 10^5 years, depending on the mass of the CO-core.

Figure 5.6 shows the run of the most important physical quantities, such as pressure and temperature, inside a low-mass AGB star during a TP. From data shown in Figure 5.6, it is easy to note how compressed both the CO-core and the He-rich inter-shell are, while the outer envelope is quite expanded. At the same time, pressure, temperature and density drops by many orders of magnitude when moving from the H-exhausted region to the outer envelope.

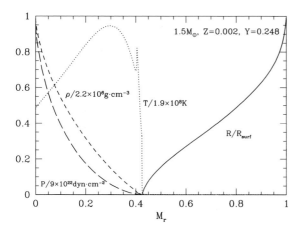

Figure 5.6 The behaviour of some selected physical quantities inside a model with initial mass equal to $1.5\,M_\odot$ after the beginning of its sixth TP. The local radius is normalized to the radius of the star.

Empirical data provide plain evidence of the fact that AGB stars suffer from continuously increasing mass loss (see the discussion in Section 2.6). Observational findings indicate mass loss rates of $\sim 10^{-7} M_\odot$/year for short period Mira variables and up to $\sim 10^{-4} M_\odot$/year (*superwind stage*) for the more luminous, longer pulsational period AGB variable stars. These winds are most likely dust-driven supported by shock waves generated by the pulsational process in the stellar envelope, and can lead, for the larger mass loss rates, to the complete obscuration of the star by a dusty circum-stellar envelope. Actually, the modulation with time of the surface luminosity and radius during the thermal pulses stage leads to a corresponding modulation of the mass loss rate.

Any star can remain on the AGB until its H-rich envelope has a mass larger than a critical value that is of the order of $\sim (10^{-3}-10^{-2}) M_\odot$. When, as a consequence of the H-burning and (mostly) of the mass loss phenomenon, the envelope mass drops below this limit, the star moves off the AGB and towards the realm of central stars of planetary nebulae and, will eventually reach the WD stage (see Section 5.8 and Chapter 6).

We note that in low-mass AGB stars, the lower layers of the outer convective zone never attain the high temperatures ($T \sim 8 \times 10^7$ K) required for the occurrence of *hot-bottom burning* (HBB) as it occurs in massive ($M > 6 M_\odot$) AGB stars.

5.3.1
Mixing Processes

In the previous discussion, for the sake of clarity, we have avoided discussing the important mixing processes occurring during the TP stage. These convective processes are actually extremely important for determining the structural and evolutionary properties of these stellar objects, and are also essential for properly predicting the nucleosynthesis occurring in these stars and to obtaining realistic predictions about the evolution of surface chemical abundances.

Let us start with a discussion on the appearance of a convective shell inside the He-rich region soon after the ignition of He-burning in the shell. During this stage, the He-rich shell becomes convectively unstable due to the huge amount of energy (and thus the large energy flux) produced by the He-burning. As a consequence, a *pulse-driven convective zone* (PDCZ) develops which comprises almost the whole inter-shell region, although without crossing the H/He discontinuity at the upper boundary of the He-rich inter-shell region (see Figure 5.7). The main consequence of this convective process is the mixing in the inter-shell zone of He-burning products, in particular, (and mostly) carbon. The abundance of carbon is the most affected with respect to other He-burning products because an incomplete He-burning process takes place at the base of this convective shell. As an order of magnitude, once the thermonuclear runaway has been quenched, the resulting mass abundance of C in the upper layers of the He inter-shell is $X_{12C} \sim 0.25$.

During the thermal pulses, the two burning shells move outward in mass, and at each TP the PDCZ appears. An important property is the degree of overlap between successive (in time) convective shells; in fact, the degree of overlap largely controls

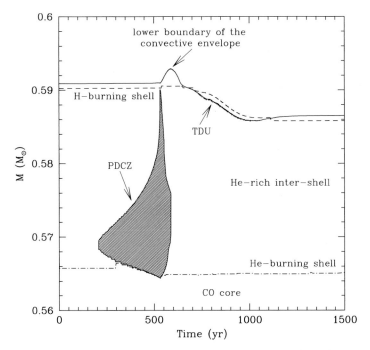

Figure 5.7 Time evolution of the convective zones during the tenth TP of a $2M_\odot$ star with $Z = 0.015$ and $Y = 0.275$. The zero point of time is arbitrary. The occurrence of the TDU after the main He flash is marked (courtesy of O. Straniero).

the distribution of the chemical elements related to the nucleosynthesis occurring in AGB stars (products of both He-burning and neutron-captures elements) inside the inter-shell region. The degree of overlap is provided by the parameter r, defined as $r = 1 - \Delta M_H / \Delta M_{PDCZ}$, where ΔM_H is the amount of mass processed by the H-burning shell during the inter-pulse phase, and ΔM_{PDCZ} is the mass of the PDCZ during the same pulse. For a given stellar mass and initial metallicity the value of r decreases with each TP until a limiting value of ~ 0.5–0.6 is reached. For a given stellar mass with lower metallicities, the limiting value of r decreases slightly, and the same occurs when fixing the metallicity and increasing the initial stellar mass [318].

There is another important convective process in AGB stars: once the H-burning has been shut off soon after the He-shell burning runaway, the outer convection can move inward in mass, a consequence of the expansion and cooling of the stellar envelope, and eventually cross the H/He discontinuity, thus entering in the H-exhausted core. The inward penetration of the convective envelope is self-sustained: an increase of the local opacity at the bottom boundary of the convection zone occurs because fresh H, characterized by an opacity larger than that of an He-rich mixture, is brought into the He-rich inter-shell by convection.

This mixing processes is known as the *third dredge up* and it is shown in Figure 5.7. The importance of the third dredge up (TDU) for the evolutionary properties of AGB stars is noteworthy due to the manyfold implications for both the evolutionary and structural characteristics of these stellar objects. In fact, if during the TDU, convection is able to penetrate deeply enough in the inter-shell zone that had been previously mixed by the PDCZ process, then huge amounts of He and products of He-burning such as C and s-elements (see the discussion in Section 5.3.2) are dredged up to the surface where they can be observed via spectroscopy. Since the TDU is actually a recursive process, its occurrence can drastically modify the chemical stratification of the outer envelope of an AGB star, hugely increasing the C/O ratio up to value larger than unity. This is the evolutionary scenario for the creation of C-stars.[5]

The significant change of the surface C/O ratio due to the TDU causes a relevant modification of the envelope opacity, as has been extensively discussed in Section 2.4. In more detail, the achievement of a C/O ratio larger than unity causes a drastic increase of the radiative opacity within the range of temperatures relevant for the outer layers of AGB stars. Detailed evolutionary computations performed by adopting radiative opacity tables that account for the appropriate C/O ratios in the heavy element mixture, predict that when the C/O ratio becomes larger than 1–1.2, the effective temperature of the AGB models decreases with time at a rate larger than for evolutionary models that do not account for the C enhancement due to the TDU[6] [46].

The C enhancement associated with the occurrence of the TDU has two important consequences:

- The enrichment of carbon in the outer stellar layers allows the formation of carbon-molecules and dust. This occurrence, together with the concomitant increase in the radiative opacity and decrease of the effective temperature leads to a stronger mass loss, which eventually results in a strong "superwind";
- The change in the thermal stratification of the outer convection zone, related to the variation of the opacity profile, induces a more efficient TDU.

Since the TDU is driven by the expansion and subsequent cooling of the envelope that occurs during the thermal pulse, it penetrates deeper when the strength of the pulse is stronger. The strength of a pulse is fixed by the thermal conditions at the bottom of the He-rich layers, which are strongly dependent on the rate at which He is accreted by the H-burning shell. As a general rule, the slower the H-burning, the higher the density at the bottom of the He-shell and, in turn, the stronger the thermal pulse. Therefore, the strength of a thermal pulse is essentially regulated by the H-burning rate: any parameter, such as initial chemical composition, envelope

5) C-stars are stellar objects whose surface C/O number ratio is larger than one. Other classes of AGB stars are: the M- and the S-stars, whose C/O ratios are ~ 0.4 and ~ 0.6, respectively.

6) Until a few years ago, the lack of suitable opacity tables for C-enhanced mixtures, required to neglect the effect of C-enhancement on the opacity, or to mimic

a C-enhancement by means of a larger total metallicity (but with the same, scaled-solar, C/O ratio). Indeed, AGB models based on this crude approximation did show a sudden decrease of T_{eff} when C/O $= 1$ that is larger than that predicted by more realistic models based on appropriate C-enhanced opacity tables.

mass and mass of the H-exhausted region, which affects the pace at which H is burnt, affects the thermal pulse strength.

The amount of material which is dredged up in each single TP initially increases because the core mass increases. After attaining a maximum value, the mass of dredged up material decreases because the envelope mass decreases due to the combined effects of H-burning in the shell and of the very efficient mass loss process. Numerical simulations reveal that the minimum envelope mass allowing the occurrence of the TDU is $\sim (0.4–0.5) M_\odot$ at solar metallicity and $\sim 0.3 M_\odot$ at the lower metallicities. This means that in stars of initial mass below a given threshold, that is, of the order of $(1.0–1.2) M_\odot$, the residual envelope mass at the beginning of the thermally pulsing stage is already too small to allow the occurrence of the TDU. This theoretical prediction provides a sound explanation for the empirical fact that the surface chemical abundances of low-mass ($< 0.9 M_\odot$) AGB stars do not appear to be affected by the occurrence of the TDU, whose signatures are C- and s-element enhancements.

For a fixed core and envelope mass, the TDU penetrates deeper into the star with decreasing metallicity because the H-burning is less efficient and so the rate of He-rich matter accretion of the inter-shell region is lower: as a rule of thumb, at $Z \sim 0.002$, the amount of mass brought to the surface during the TDU is a factor ~ 2 larger with respect to the same quantity in solar metallicity AGB stars. This dependence goes quite some way in explaining the relative difference in the fraction of C-stars between metal-rich stellar systems such as the Galactic field and metal-poor ones such as Local Group dwarf galaxies.

The efficiency of the TDU process is commonly parameterized with $\lambda = \Delta M_{TDU}/\Delta M_H$, where ΔM_{TDU} and ΔM_H represent the mass of freshly processed matter that is mixed throughout the envelope by the TDU, and the mass growth of the H-exhausted core during the preceding inter-pulse phase. Alternatively, the efficiency of the TDU can be represented as $\Delta M_{TDU}/M_{env}$. This mass ratio for various values of metallicity and initial stellar mass as predicted by updated AGB stellar models is shown in Figure 5.8 [318].

It is important to note that, from an evolutionary point of view, the occurrence of the TDU has a very important implication: when the TDU occurs, a significant amount of the H-exhausted core is "removed" and mixed to the envelope. This amount is related to the *assumed* (see below) efficiency λ of this convective process, being equal to $\Delta M_H = \lambda \times \Delta M_{TDU}$. This reduction of the core mass prevents the build-up of massive H-exhausted cores and, hence, of massive WDs; the occurrence of an efficient mass loss process during the AGB, reducing the envelope mass and, in turn, the number of possible TPs, also limits the mass increase of the He-core during this evolutionary stage.

5.3.1.1 The Numerical Problem of Modelling the Third Dredge Up

Despite the clear physical description of the sequence of events that contribute to determining the occurrence of the TDU in AGB stars, the actual appearance of this convective instability and its efficiency have been (and partially still are) a crucial

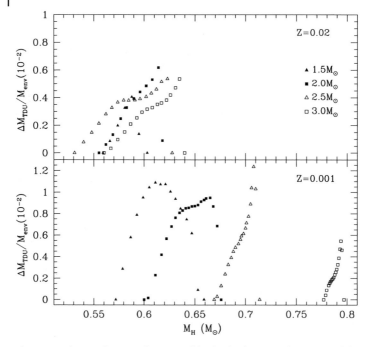

Figure 5.8 The ratio between the mass of the dredged up He-rich matter and the envelope mass as a function of the He-core mass at each TP episode for various assumptions on the initial stellar mass and metallicity (courtesy of S. Cristallo).

problem in stellar model computations. The question of how to obtain the dredge up, and, more importantly, a realistic prediction about its efficiency in dredging up fresh matter to the surface in AGB models, is tightly related to the delicate question of the determination of the convection boundaries. It was demonstrated earlier that in low-mass AGB stellar models, with appropriate He-core mass and envelope mass, the TDU can occur when adopting the classical Schwarzschild criterion, provided that a high spatial and temporal resolution is used in the evolutionary computations [319]. However, a simple physical evidence (see below) suggests that the use of this classical criterion for convection stability should not be applied in the treatment of the deepening of the convective envelope during the TDU in AGB stars. In fact, we have already noted that, due to the increase in local opacity when H-rich matter is brought down into the He-rich inter-shell zone, an increase in the local temperature gradient has to occur, with the consequence of increasing the penetration depth of the convective envelope.[7]

A different view of the same problem is the following: when evaluating the velocity of the convective elements in the framework of mixing length theory, one finds

---

7) We note that the physical mechanism causing this instability is exactly the same as that responsible for the growth of the convective core during the core He-burning stage (see Section 4.2).

that this velocity is proportional to the difference between the radiative tempera-ture gradient and the adiabatic temperature gradient. Thus, the average convective velocity usually drops to zero at the stable boundary of a convective layer, where, ac-cording to the Schwarzschild criterion, the temperature gradient coincides with the adiabatic one. However, when convection penetrates into a region of lower opacity (as in the case of the TDU in AGB stars), the difference between the actual temper-ature gradient and the adiabatic gradient is larger than zero, and hence a positive average convective velocity is found at the inner border of the convective envelope, which then becomes unstable. In principle, as soon as He-rich matter is mixed with the envelope, the opacity and, in turn, the difference between ∇_{rad} and ∇_{ad} is decreased. However, as the mass of the convective envelope is much larger than the amount of the dredged-up material, the stabilizing effect of the additional mixing is almost negligible.

From this discussion, it is clear that a simple thermodynamic criterion which is suitable to be included in stellar models computations cannot be used to determine the real extension of the convective instability. The only simple prediction that can be done is that due to the steep pressure gradient present just below the formal border of the convective envelope, the non-canonical penetration of the convective instability should be strongly limited, that is, the average convective velocity should rapidly drop to zero.

From a practical point of view in stellar evolutionary codes, this problem is solved by accounting for some amount of non-canonical extra mixing below the bottom boundary of the canonical convective envelope. Roughly speaking, three approach-es are used:

- It is assumed that in the region underlying the formal convective bound-ary, the average convective velocity follows an exponential decline, that is, $v = v_{bce} \exp\left[-z/(f \cdot H_P)\right]$, where z is the distance from the formal canon-ical convective boundary, H_P and v_{bce} are the pressure scale height and the velocity of the convective elements, obtained from the MLT, at the inner side of the Schwarzschild border, and f is a free parameter whose value is fixed on the basis of spectroscopic measurements of s-elements at the surface of AGB stars. Once the average convective velocity is estimated at each point in the convective region, the actual distribution of the chemical elements at each mesh inside this region is estimated by accounting for a damping factor in the degree of mixing that accounts for the local velocity. From a practical point of view, this approach ends with a complete mixing in the fully-convective zone (coinciding with the canonical convective envelope) and a partial mixing in the zone where an exponential decline for the convective speed has been assumed.
- An alternative approach consists of managing the extra mixing below the formal canonical boundary for convection as a diffusive process with a diffusion coef-ficient of the form: $D = v_{bce} H_P \exp\left[-z/(f \cdot H_P)\right]$, where the various quan-tities have the same meaning as above but the free parameter f can assume a completely different value. Obviously, as for the first method, this also pro-

duces a chemical profile of the various chemical species in the inter-shell region reached by this extra-mixing process.

- A third approach relies on the inclusion of some overshoot below the formal boundary of the convective envelope, so extending the position of the base of the outer convection zone downward by an arbitrary amount $f \times H_P$.

Regardless of the method adopted, as well as of the exact value used for the free parameter, to account for this non-canonical extra mixing has some notable consequences: (i) the TDU is more efficient, that is for any given envelope mass and He-core mass, the value of λ is larger than in the case of canonical models; (ii) a smooth H profile in the upper layers of the inter-shell region is left over by the TDU – as will be discussed in the following this has a crucial relevance to the production of s-elements in low-mass AGB stars; (iii) the first TDU episode occurs earlier, that is the minimum core mass for the TDU occurrence is lower.

Before closing this section, we wish to note that, regardless of the scheme adopted in modelling of the TDU, there is a free parameter that has to be calibrated in some way in order to reproduce some empirical constraints (see the discussion in the following sections). This has to be related, once again, to our poor knowledge of how to treat convection in model computations, as this physical process is highly non-linear, non-local and time-dependent.

5.3.2
Nucleosynthesis

Our knowledge of how s-process isotopes (s denotes *slow*) are produced has improved hugely in this last decade. In the seminal work of Cameron, s-elements were defined as those elements produced by neutron capture on lighter elements (the seeds) in an environment of sufficiently low neutron flux so that, in most cases, any β-unstable isotope that is formed will experience a β-decay more rapidly than it can experience a further neutron capture[8] [320]. Cameron identified two key reactions as primary neutron sources: $^{13}C(\alpha, n)^{16}O$ and $^{22}Ne(\alpha, n)^{25}Mg$. For the sake of brevity, it is common to refer to these reactions simply as the ^{13}C source and the ^{22}Ne source, respectively.[9]

---

8) For an AGB star, the typical neutron capture timescale ranges from 10 to 100 years.

9) An interesting feature of the ^{13}C source is that it is a *primary* neutron source because it is produced from the H and He originally present in the stars. Since the total time-integrated neutron flux ϕ from the ^{13}C source is proportional to the ratio $^{13}C/Z$, the s-process element production has a strong dependence on the stellar metallicity. In more detail, since ϕ increases with decreasing metallicity, heavier elements are produced at lower metallicities (see discussion in Section 5.3.3). On the other hand, the ^{22}Ne neutron source is traditionally considered to be a *secondary* neutron source since it relies on the initial CNO nuclei abundance present in the star. However, for [Fe/H] < −2, the ^{22}Ne source also has an important primary component because the N abundance in the H-burning ashes increases significantly from its initial value due to the effects of the TDU that dredges up the surface primary ^{12}C, which is then converted into N by H-burning.

However, the correct identification of the stellar site for s-process nucleosynthesis in the He-rich[10] inter-shell region of thermally pulsating AGB stars had to wait for the pivotal work of [314] and the subsequent detailed evolutionary computations of AGB models [318, 321, 322].

From an evolutionary point of view, the ^{22}Ne source is a promising neutron source for s-elements nucleosynthesis because a significant amount of ^{22}Ne can be produced by nitrogen burning. In fact, at the beginning of the TP stage, the amount of ^{14}N left by the H-burning at the top of the inter-shell zone is large, practically equal to the sum of the abundances (by number) of the CNO elements in the initial chemical composition. As discussed previously, during the early phase of the convective thermal pulse, the material within the inter-shell is mixed inside the PDCZ and the ^{14}N is totally converted into ^{22}Ne via the nuclear reaction chain ^{14}N$(\alpha, \gamma)^{18}$F$(\beta^+, \nu)^{18}$O$(\alpha, \gamma)^{22}$Ne. During the following TP, near the peak of the thermonuclear runaway, if the temperature is high enough, that is, $\sim 3.5 \times 10^8$ K, the ^{22}Ne$(\alpha, n)^{25}$Mg nuclear reaction may provide an important neutron flux. However, stellar model computations clearly show that only intermediate-mass AGB stars attain such a high temperature inside the inter-shell region, whereas low-mass AGB stars (i.e. $M < 3 M_\odot$) barely achieve a temperature of $\sim 3 \times 10^8$ K, and hence the ^{22}Ne source is marginally effective.

The ^{13}C source requires a significantly lower temperature, $\sim 9 \times 10^7$ K, and this is easily attained in the He inter-shell region of low-mass AGB stars. Indeed, between two subsequent TPs, the H-burning occurring in the shell leaves some ^{13}C in the upper region of the He inter-shell. However, the burning of this ^{13}C is almost completely ineffective in producing a flux of neutrons. This is because in the stellar matter that has experienced H-burning via *CNO cycle*, the abundance of ^{14}N is about two orders of magnitude larger that that of ^{13}C, and ^{14}N is a very efficient neutron poison via the resonant ^{14}N$(n, p)^{14}$C neutron capture reaction.

However, since the ^{22}Ne source is so weak in stellar AGB models with He-cores of small mass, and since observations reveal that the surface of low-mass AGB stars is actually hugely enhanced in s-process isotopes, some mechanism has to exist to activate the ^{13}C source in such stars. This means that an alternative source of ^{13}C is needed, in a zone where ^{14}N is depleted.

The ^{13}C isotope is produced via the nuclear reaction chain ^{12}C$(p, \gamma)^{13}$N$(\beta^+, \nu)^{13}$C. After each TP, the stellar layers within the He-rich inter-shell are quite rich in ^{12}C, while all the ^{14}N has been transformed into ^{22}Ne; however the problem is that in such a region the hydrogen has been completely exhausted. So, the problem is how to inject a few protons into the He-rich inter-shell in order to produce ^{13}C via proton capture on ^{12}C.

It was envisaged early on that the PDCZ developing at the peak of the thermonuclear runaway could extend beyond the H/He discontinuity, so dredging down some protons inside the He-rich inter-shell as requested by the previous scenario [323, 324]. However, detailed evolutionary models showed that during the thermonuclear runaway, the H-burning shell remains active and generates

---

10) He nuclei cannot capture neutrons as ^{5}He is unstable.

an entropy barrier that prevents the penetration of the convective instability into the H-rich envelope. The same AGB models reveal that only in very metal-poor stars, owing to the lack of CNO elements,[11] ingestion of protons in the PDCZ may effectively occur [325].

A potentially more promising way of activating the ^{13}C source is by means of some kind of diffusive mixing during the inter-pulse stage at the epoch of the TDU and during the extended post-flash dip, that is, the period that immediately follows the TDU. In fact, as a consequence of the TDU, a sharp discontinuity between the H-rich outer envelope and the He-rich inter-shell is in place. In addition, during this period, H-burning is extinct. The time elapsed between the maximum penetration of the convective envelope and H-burning re-ignition is about 10^4 years for a star of $2\,M_\odot$; so long enough to allow, via an unknown physical process, the downward diffusion of a few protons into the underlying radiative layer. When the H-burning shell reignites, the H nuclei present in the upper layers of the He-rich inter-shell interested by the previous H downward motion will be captured by ^{12}C to form ^{13}C. The ^{13}C-enriched region, namely, the so-called ^{13}C *pocket*, will shortly be left behind as H-burning progresses outward. This process is shown in Figure 5.9.

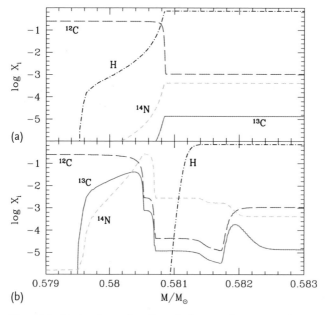

Figure 5.9 (a) The chemical profile in the layers at the transition between the He-rich inter-shell and the outer envelope just after the TDU, in a AGB model with mass equal to $2\,M_\odot$ and $Z = 0.001$. (b) As with (a), but after the formation of the ^{13}C *pocket* (courtesy of S. Cristallo).

---

11) In order to compensate for the reduction of CNO catalysts, the temperature of the H-burning shell increases, and consequently the entropy barrier associated with this burning shell becomes lower. This occurrence does not allow the H-burning shell to act as a barrier against convection.

In order to reproduce the spectroscopic data for *s*-elements at the surface of low-mass AGB stars, the *unknown* physical process must be able to diffuse about $10^{-6} M_\odot$ of protons into a region of $\sim 10^{-3} M_\odot$. It is worth noting that a very well-defined amount of H has to be injected into the inter-shell region: high enough to allow a substantial production of ^{13}C but not so large as to avoid that the production of ^{13}C is followed promptly by the production of ^{14}N via the $^{13}C(p,\gamma)^{14}N$ nuclear reaction.

The ^{13}C *pocket* formed in this *unknown* way is completely burnt by the $^{13}C(\alpha,n)$ ^{16}O reaction in *radiative* conditions during the inter-pulse stage once the temperature reaches the required value of $\sim 9 \times 10^7$ K [326]. Such an occurrence produces a large neutron exposure with a maximum neutron density of $\sim 10^7$ cm^{-3}. In passing, we note that the occurrence of ^{13}C-burning under radiative conditions is a fundamental requirement: early evolutionary models in which it was assumed rather that the *s*-element production occurred when the ^{13}C *pocket* was engulfed into the PDCZ generated by the subsequent TP, predicted a neutron density too high (up to $10^9 - 10^{10}$ cm^{-3}) and abundances for the *s*-elements in striking contrast with the spectroscopic measurements.[12)]

Despite the importance of obtaining the ingestion of some protons in the inter-shell region, we are still facing with the problem of identifying the suitable physical mechanism(s) at the basis of this process. Thus far, the most promising mechanisms are: convective overshooting, rotationally-induced mixing, and gravity waves [327–329].

Due to the lack of a firm understanding of the physical processes governing the formation of the ^{13}C *pocket* and, hence, of any possibility of predicting a priori its mass and chemical profile (fundamental knowledge for understanding the *s*-element nucleosynthesis), these are treated as free parameters. The properties of the ^{13}C *pocket* can be modified by changing the value of the free parameter, f, entering in the numerical scheme adopted for managing the problem of the location of the bottom boundary of the convective envelope during the TDU.

Regardless of the numerical approach adopted for generating the ^{13}C *pocket* in stellar model computations, we now wish to briefly comment on some properties of this process that is very important for *s*-element production in low-mass AGB stars: (i) as a consequence of proton ingestion in the inter-shell zone ^{13}C is produced via proton capture by ^{12}C, but due to some further proton capture ^{14}N is also produced; (ii) at the end of the ingestion process there are two pockets inside the He-rich inter-shell zone, that is, a ^{13}C *pocket* and a partially overlapping ^{14}N *pocket*.

Remembering that ^{14}N is a very efficient neutron poison, it can be easily understood that for the sake of estimating the *s*-element production associated with the ^{13}C source, the relevant quantity is not the *global* ^{13}C abundance in the ^{13}C *pocket* but the so-called *effective* ^{13}C abundance that is defined as $X(^{13}C_{eff}) = X(^{13}C) - (13/14)X(^{14}N)$. Actually, it is the $X(^{13}C_{eff})$ quantity (the global effective ^{13}C abun-

---

12) In more detail, the ratio Rb/Sr, being very sensitive to the neutron density, poses a severe constraint on the AGB star nucleosynthesis. In particular, the observed low Rb abundance compared to Sr implies a quite low, $< 10^8$ cm^{-3}, neutron density.

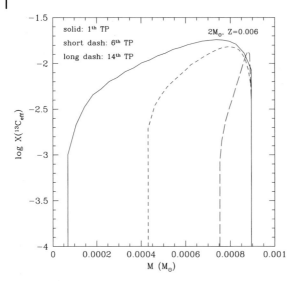

Figure 5.10 The mass-extension of the *effective* ^{13}C *pocket* during selected TP events for a star with an initial mass equal to $2M_\odot$ and $Z = 0.006$. The origin of the x-axis is arbitrary, and the location of each individual ^{13}C *pocket* has been arbitrarily shifted to make the outer edges coincide (courtesy of S. Cristallo).

dance in the ^{13}C *pocket*) that is directly related to the total number of neutrons available for the *s*-process nucleosynthesis occurring when the temperature in the ^{13}C *pocket* becomes high enough to activate α capture on ^{13}C nuclei.

The mass-size of the ^{13}C *pocket* decreases during the recursive TPs as it also occurs to the whole He-rich inter-shell region. The trend of the mass of the *effective* ^{13}C *pocket* with the He-core mass for selected assumptions on the initial stellar mass and metallicity is shown in Figure 5.10. The ^{13}C *pocket* shrinks in mass as the star climbs the AGB, and this occurrence has the important consequence that the first pockets, the largest ones, leave the major imprint on the overall *s*-element nucleosynthesis.

As previously discussed, evolutionary models predict that the bulk of the ^{13}C-burning via α-captures occurs in a radiative environment, and this is very important in achieving a neutron density low enough to correctly reproduce the spectroscopic evidence. This notwithstanding, there could be some exceptions: AGB models at relatively high metallicity ($Z > 0.003$) show that during the first TP, the first ^{13}C *pocket* is only partially burnt during the inter-pulse phase, so that the residual ^{13}C is burnt during the following TP after it has been engulfed in the PDCZ. As a consequence, the residual ^{13}C burns at a significantly higher temperature, so producing a neutron density ~ 2–3 orders of magnitude larger than that associated with radiative ^{13}C-burning. This produces specific signatures in the resulting *s*-element nucleosynthesis.

At the beginning of this section, we stressed that the ^{22}Ne source is largely ineffective in low-mass AGB stars. However, although this neutron source plays a secondary role with respect the ^{13}C source, it can still be important in explaining the abundances of some specific *s*-elements. Indeed, detailed low-mass AGB stellar models show that the ^{22}Ne$(\alpha, n)^{25}$Mg nuclear reaction is marginally activated in metal-rich stars; but with decreasing metallicity, the TPs become stronger, and so larger temperatures are achieved inside the inter-shell region during the inter-pulse stage, so that the ^{22}Ne source can also become important. For instance, for metallicity $Z \sim 0.001$ in all stellar models more massive than $\approx 2 M_\odot$, the second neutron source has relevant effects on the resulting nucleosynthesis.

The final *s*-element nucleosynthesis and hence surface compositions, predicted by AGB stellar models of distinct metallicity and/or initial total mass, is the result of the combination of the differences in the initial iron-seed content and in the physical structure of the stars belonging to different stellar populations. To summarize: at large metallicities, the nucleosynthesis of light *s*-elements (*ls*, see the discussion in the following section) such as Sr, Y, Zr is largely favoured; at decreasing metallicity, the overproduction of heavy *s*-elements (*hs*) such as Ba, La, Ce, Nd, Sm and Pb becomes progressively more and more important, with Pb becoming the main product of the *s*-element nucleosynthesis in very metal-poor regime, that is, $Z \leq 10^{-4}$.

Before closing this section, we remark that despite their importance in determining the final neutron density, and hence the resulting *s*-element production, the nuclear reactions ^{13}C$(\alpha, n)^{16}$O and ^{22}Ne$(\alpha, n)^{25}$Mg have a nuclear cross section that is still poorly known. In the case of ^{13}C$(\alpha, n)^{16}$O, the existence of a sub-threshold resonance makes difficult (and unreliable) to extrapolate this reaction rate to low energies. The case of ^{22}Ne$(\alpha, n)^{25}$Mg is even more complex due to the possible existence of not-well-defined resonances in the low-energy regime. For a recent update on this issue, we refer to [330].

5.3.3
Evolution of Surface Chemical Abundances

In the previous sections, we described the various mixing processes occurring inside the He-rich inter-shell region and the outer convective zone. As the stars climb the AGB, a significant overlap can occur between regions which have been mixed during the recursive TPs. At the same time, in the inter-shell zone the nucleosynthesis of stellar matter via H- and He-burning and also neutron captures occurs. Therefore, it is evident that the final chemical composition in the inter-shell zone during the various TPs is determined by the combination of nuclear burning efficiencies and the mixing processes. Since as a consequence of the TDU, the ashes of the nuclear burnings are dredged up to the stellar surface, chemical stratification in the outer envelope evolves with time during the AGB stage as we have already discussed for the case of the C/O ratio as soon as the TDU begins. In this context, it is also evident that mass loss plays a fundamental role:

- An obvious consequence of mass loss is that by reducing the stellar envelope mass together with the H-burning occurring in the shell, it determines when the AGB stage ends and so limits the maximum number of TPs and, hence, the maximum allowed change in chemical composition of the envelope.

- Since for a given He-core mass there exists a minimum envelope mass for the occurrence of the TDU, mass loss can force stars to not experience additional TDU; so freezing out the surface chemical composition to that of the last TDU experienced by the star.

Therefore, the evolution of the surface chemical abundance of AGB stars is actually determined by the mutual efficiency and link between nuclear burning processes, convection and mass loss.

Keeping these general considerations in mind, we can attempt to provide a schematic outline of the expected (to an order of magnitude) changes in the chemical composition of low-mass AGB stars after a number of TPs. Let us consider the resulting chemical composition of a solar metallicity, $2 M_\odot$ AGB star after 10 TP. The most remarkable changes with respect the original chemical composition are those related to the *light* elements such as C, Ne, Na and F, which show abundance changes of ~ 0.5, ~ 0.15, ~ 0.1 and ~ 0.3 dex, respectively. The huge amount of C at the surface is, as already discussed, a consequence of the dredge up of matter that has been partially processed via He-burning in the inter-shell zone. The case of fluorine production is very interesting as it is associated with the capability of ^{14}N to efficiently capture some neutrons released by ^{13}C in the pocket: indeed F is synthesized in the PDCZ via the ^{15}N$(\alpha, \gamma)^{19}$F reaction which is possible as a consequence of the accumulation of ^{15}N in the ^{13}C *pocket*, mainly via the chain ^{14}N$(n, p)^{14}$C$(\alpha, p)^{18}$O$(p, \alpha)^{15}$N. A secondary contribution to the synthesis of ^{15}N in the pocket may eventually come from ^{14}N$(n, \gamma)^{15}$N and from the ^{14}N$(p, \gamma)^{15}$O nuclear reactions.

N is enhanced at the stellar surface by about 0.3 dex, although this is due only to the occurrence of the *first dredge up* during the RGB stage. This is because all the additional N left in the inter-shell zone, as a consequence of the H-burning via *CNO cycle* occurring in the shell, is transformed into ^{22}Ne during the TPs, and so rather causing the Ne enhancement. The predicted Na enhancement is partly due to proton captures on ^{22}Ne inside the H-burning shell (the already discussed *NeNa cycle*), and partly due to neutron captures on ^{22}Ne promptly followed by ^{23}Ne decay occurring in the inter-shell zone.

All the *s*-elements from Sr to Pb are predicted to be enhanced by various degrees ranging from ~ 0.2 to ~ 1 dex. At least in the solar metallicity, $2 M_\odot$ AGB model, the abundance of *ls* elements is comparable to that of *hs* ones, but Pb appears to be enhanced with respect the initial chemical composition to a lower extent than the other *hs* (for instance Ba) elements.

Let us discuss, in a little more detail, the theoretical expectations concerning the different *s*-element groups. On the basis of nuclear physics considerations, one can expect the presence of three distinct main peaks in the *s*-element distribution. These peaks correspond to nuclei with the so-called "magic number of neutrons", that is, 50, 82 and 126; these nuclei have a very high binding energy and, hence, their nuclear cross section for neutron capture reaction is extremely low.[13] The mean abundances of the s-elements with a number of neutrons in the nucleus around 50 and 82 correspond to the so-called *ls* and *hs* indices, respectively. The third peak corresponds to the heaviest elements (with number of neutrons ~ 126) and is related mainly to the Pb abundance. The spectroscopic measurements of the abundance of *ls*, *hs* and Pb, that is, [*ls*/Fe], [*hs*/Fe] and [Pb/Fe], can hence provide important information about the specific neutron capture processes, and so the physical conditions at the stellar places where the nucleosynthesis processes occur.

It has been demonstrated early on that the final abundances of *ls*, *hs* and Pb mainly depend on the number of neutrons available per iron seed at the first neutron exposure, in the sense that the production of the heaviest s-elements is favoured when a larger number of neutrons per iron seed is in place [331]. Although in the stellar interiors, the physical conditions are extremely complex and the nucleosynthesis details are affected by the link between convection and burnings, one can derive some useful clues about the expected relative number of neutrons per iron seed. For low-mass AGB stars, the main neutron source is ^{13}C, but, see the previous discussion, the actual number of neutrons available for s-product nucleosynthesis is related to the *effective* abundance of ^{13}C. Thus, a rough estimate of the number of neutrons per iron seed for these stars is:

$$\frac{n(\text{neutrons}) - n(\text{poisons})}{n(\text{iron seeds})} \sim \frac{n(^{13}\text{C}) - n(^{14}\text{N})}{n(^{56}\text{Fe})} = \frac{n(^{13}\text{C}_{\text{eff}})}{n(^{56}\text{Fe})}$$

As previously discussed, the abundances of ^{13}C and ^{14}N depend on the primary abundance of ^{12}C produced by He-burning in the inter-shell zone, and on the abundance of H ingested from the H-rich envelope. Therefore, the numerator in the expression depends on the initial stellar mass as well as on the physical processes causing the formation of the ^{13}C *pocket*. The denominator only depends on the initial metallicity (Z). Obviously, with decreasing global metallicity the number of iron seeds decreases significantly, whereas the number of neutrons is only affected to a second order: so the number of neutrons per (primordial) iron seed increases by a large factor at lower metallicity.

Evolutionary predictions for low-mass AGB star nucleosynthesis show, as a consequence of this behaviour, that the surface abundances of the heavier s-elements, and in particular of Pb, are more significantly enhanced when the metallicity decreases,[14] whereas the *light* s-element enhancements seem to be largely independent of the initial stellar metallicity [318, 321, 322, 332, 333]. Spectroscopical measurements of s-elements in the various metal regimes, including very metal-poor stars, nicely confirm these theoretical expectations. However, there are many still unsettled issues related to the nucleosynthesis of s-elements, for instance, the ob-

13) A closed neutron shell at the neutron magic numbers is quite reluctant to accept an additional neutron.

14) As already mentioned, in low-mass AGB stars the ^{13}C source is a primary source and therefore it is not affected by the initial stellar metallicity. However, since the number of iron seeds directly scales with Z, the lower the metallicity, the larger the number of neutrons available per seed; as a consequence, neutron capture processes preferentially produce Pb.

servational evidence that stars with similar mass and metallicity do show a spread in the *hs/ls* ratio that can be explained only assuming a range of physical conditions in what are otherwise similar stars.

5.4
Calibration of Evolutionary AGB Models

The previous discussion on the physical properties of AGB stars during the thermally pulsing stage and the related nucleosynthesis has shown that is crucial to manage the stringent link that exists between convection, nuclear burning (both p-, α- and neutron captures) and mass loss (which is also intimately related to the occurrence of radial pulsations) in the numerical evolutionary computations. It is enough to remember here how critical the treatment of the lower boundary of the envelope convection is in forming a sizable ^{13}C *pocket,* mandatory for an efficient *s*-element production.

As has already been mentioned, due to the lack of a reliable physical understanding of how to manage with these physical processes in stellar model computations, their treatment in numerical codes is made by adopting some free parameters. The most important free parameters are those related to the treatment of the lower boundary of the outer convection zone during the TDU (we named this parameter f in Section 5.3.1, but due to differences in the adopted scheme this parameter can assume different meaning/values) and the mass loss efficiency.

We have already noted that, regardless of the scheme adopted (diffusive process, exponential decay of the convective element velocity, convective overshoot), to include some amount of extra mixing at the bottom of the canonical convective envelope, has a dramatic impact on the efficiency of the TDU and on the formation of the ^{13}C *pocket* and, in turn, on the *s*-product nucleosynthesis. In particular, the variation of the TDU property not only affects the amount of dredged-up matter, but also the minimum He-core mass at which the TDU occurs for the first time and, hence, the stellar luminosity at which the structure appears as a C-star and/or starts showing a significant enhancement of *s*-elements.

The occurrence of mass loss not only affects the time at which the star leaves the AGB, but also, by drastically changing the mass on the envelope, the physical properties in the interiors.

It is evident that in order to reliably use AGB stellar model predictions when analysing various spectroscopic and photometric evidence as well as when including them in population synthesis analysis, these free parameters have to be properly calibrated. A detailed description of how this calibration is performed is not within the aim of the present review, but we wish to provide a general overview of how this is done.

There are three main observational constraints that can be used to calibrate AGB models: (i) the observed *s*-element enhancements, and in particular the [*hs/ls*] and [Pb/*hs*] indices; (ii) the Luminosity Function (LF) of C-stars both in the Galactic field and in the Magellanic Clouds and, (iii) the initial–final mass relation for WD stars, that is, the relation between the initial stellar mass and the final mass of the corresponding WD when it starts cooling down along the WD cooling sequence [334].

Concerning the first constraint, the possibility of performing a meaningful comparison between spectroscopic measurements and theoretical predictions requires the knowledge of main properties of the observed stars such as their initial mass and metallicity as well as their evolutionary stage (for instance, how many TPs have been experienced). However, model computations (see also the previous discussion) show that the two indices [hs/ls] and [Pb/hs] attain their asymptotic values after just a few TPs (if the TDU is efficient in the early thermally pulsing stage); this is an important advantage because in comparing theory with observations, a precise determination of the evolutionary stage of the stars is no longer required. So one can compare the predicted neutron capture element enhancements with the observed ones, and in the case of a mismatch one can modify properties of the ^{13}C *pocket* (e.g. its mass extension and chemical profile) to increase or decrease the neutron density and so improve the match with the observations. From a practical point of view, this can be done by changing the value of the free parameter f introduced in Section 5.3.1. The ^{13}C *pocket* properties critically depend on such a parameter: if it is too small, too few protons are ingested and the resulting abundance of ^{13}C is too small, but if it is too large, the ^{13}C production is promptly followed by the neutron poison ^{14}N. Numerical tests have shown that, at least in case the exponential decay of the convective velocity is adopted, the appropriate value[15] of f is strongly peaked at ~ 0.1.

Although this observational approach for constraining models seems to be promising, one should note that from an observational point of view, the [hs/ls] and [Pb/hs] indices are affected by a significant spread at the same metallicity (the problem of the origin of such a spread, that is, if real or due to observational errors, is still unsettled), and this occurrence obviously hampers the possibility of obtaining a firm constraint on the adopted theoretical framework.

An alternative and more promising approach is the one based on the comparison of models with the luminosity function of carbon stars in various stellar environments, so allowing one to also check on the dependence of the (assumed) TDU efficiency on the initial metallicity. Indeed, the C-stars LF is a very accurate benchmark for model predictions, given that it relates the stellar luminosity with chemical properties (in particular the surface C abundance); in fact, a star contributes to this LF, in the observed and theoretical one, *only* when the C/O ratio becomes larger than unity as a consequence of the recurrent occurrence of the TDU. So via this comparison, we can simultaneously check on the reliability of theoretical predictions about internal structural properties such as the He-core mass (and hence

---

15) Stellar models based on the diffusive approach show that a value of f of ~ 0.016 should be used. However, in such cases, the situation is not very clear. In fact, at odds with the case of exponential decay of the convective velocity, the adoption of the diffusive scheme affects not only the actual boundary of the outer convective zone but also that of the internal convective shell (as the PDCZ), and the numerical simulations suggest that a non-universal value of f should be used at these different locations inside the stellar structure [46].

luminosity) and the envelope chemical stratification (which is mainly affected by the TDU efficiency and the adopted mass loss recipe). However, in order to perform this comparison, a reliable Initial Mass Function (IMF) and a Star Formation History (SFH) have to be assumed for the stellar environment under scrutiny; but numerical simulations reveal that for realistic assumptions for these physical ingredients, the theoretical C-star LFs are non-significantly affected by the adopted choices for the IMF and the SFH.

Since there is a minimum He-core mass at which the TDU begins, the faint end of the LF provides a strong constraint on the minimum initial stellar mass experiencing the TDU. When moving to brighter magnitudes, the observed C-star LF shows a maximum at $M_{bol} \sim 5.0$ mag, and then a monotonic decrease. The rise of the C-star LF until the peak is due to the "accumulation" at the same magnitude of stars with different initial mass that have already experienced a sufficient number of efficient TDUs so that they can appear as a C-star. From a theoretical point of view, the rate of increase with the magnitude of the LF, and the location of the peak, depend on the efficiency of both the mass loss and of the TDU. The drop of the LF at the brightest magnitude is due to the combination of two processes: the occurrence of mass loss forces low-mass AGB stars to leave the AGB, progressively decreasing their number while increasing the luminosity; in intermediate-mass AGB stars at a some point along the AGB *Hot Bottom Burning* begins to be effective, and as a consequence of this burning process the envelope C abundance is converted into N, decreasing the C/O ratio below unity [20]. The star is no longer a C-star and it does not contribute to the LF.

Calibrated AGB models are able to reproduce the C-star LF both in the Galactic field (at $\sim$ solar metallicity) and in the Large and Small Magellanic Clouds (whose metallicities are $Z \sim 0.008$ and ~ 0.004, respectively). This is not a trivial result because there is no a priori reason why the calibration of the TDU efficiency and mass loss rate at a given metallicity should hold also at a different metal content. Interestingly enough, it also seems that the calibration of the relevant AGB parameters obtained by using the C-star LF is also appropriate for reproducing the empirical evidence on the *s*-element enhancements.

The empirical initial–final mass $(M_i–M_f)$ relation provides an important constraint for AGB stellar models because the theoretical predictions strongly depend on two important physical assumptions[16] entering in the numerical simulations: the mass loss efficiency during both the RGB and AGB stages and the efficiency of the TDU. The impact of mass loss is obviously the most direct one because the initial mass of a star is very much reduced by mass loss. As a consequence, any change in the assumed efficiency of mass loss directly impacts the obtained $M_i–M_f$ relation. In passing, we note that in the past, the huge discrepancy between the empirical and theoretical estimates for the $M_i–M_f$ relation mainly resided in the adop-

---

16) Indeed, in the case of intermediate-mass stars, the occurrence of the HBB has also an important effect on the initial-final mass relationship. Since we do not discuss the properties of intermediate-mass AGB stars, it is suffice to note that the occurrence of HBB causes a huge increase of the surface luminosity with respect the standard *He-core mass–luminosity relation*, so strongly enhancing the mass loss efficiency in these massive AGB stars.

tion of an unsuitable mass loss prescription for the AGB stage, often the Reimers law. The efficiency of the TDU is also a crucial ingredient in predicting the M_i–M_f relation for low- and intermediate-mass AGB stars: actually, as already mentioned, at each TDU, the mass of the H-exhausted core is decreased by an amount that is proportional to the assumed efficiency of the TDU, that is, $\Delta M_H = \lambda \cdot \Delta M_{TDU}$. This occurrence clearly limits the maximum mass of the H-exhausted core and, actually, of the WD progeny at the end of the AGB stage.

The comparison between reliable empirical measurements of M_i–M_f relation and suitable model predictions can set constraints on the efficiency of the mass loss and of the TDU, as well as constrain how these efficiencies eventually depend on the initial stellar mass and chemical composition (see Chapter 6).

5.5
Synthetic AGB Stellar Models

Despite the dramatic increase in computing power, full evolutionary computations of AGB stellar models are extremely demanding because of the extreme complexity of the inner structure as well as the need for a fine temporal spacing to follow the TP outcome. Therefore, the computation of complete and extended sets of full evolutionary AGB models for population synthesis purposes is difficult, if not impossible. In addition, as emphasized in the previous sections, the evolutionary and structural properties of AGB stellar models critically depend on a number of uncertain parameters, that is, the mass loss efficiency, the treatment of convective boundaries, opacity effects induced by chemical abundances variations. To compute full evolutionary models while exploring these parameters is prohibitive as is exploring the complete mass and chemical composition ranges.

A complementary approach to full evolutionary AGB models is that provided by *synthetic AGB stellar models*. A synthetic AGB model provides a simplified, but still acceptable, description of at least the most important observables such as surface luminosity, T_{eff}, total mass, He-core mass, evolutionary lifetime and the main chemical abundances in the stellar envelope.

A synthetic AGB model is obtained by starting from recipes (usually analytic fitting formulae) provided by full evolutionary models, sometimes complemented with some information/constraints coming from observations. One should, however, always keep in mind that synthetic AGB models are a "surrogate" of the full evolutionary calculations they are based on, and in this sense, they can never be better (or more reliable) than the evolutionary models. It is also important to note that synthetic AGB models cannot provide the wealth of information about the physical structure and nucleosynthesis details, provided by full evolutionary computations. Aside from these considerations, synthetic AGB simulations have some indisputable advantages, such as:

- Computational speed. A synthetic AGB model can be computed in a few seconds compared with the weeks, or sometimes months, required for a detailed full evolutionary model;
- The effects of poorly constrained physical processes, such as the mass loss and TDU efficiency, can be quickly and easily analyzed by changing the value of the corresponding free parameters entering the synthetic simulations;
- The predictions of synthetic AGB models can be compared with empirical benchmarks as the C-star LFs, allowing the calibration of key model free parameters, such as the TDU efficiency. This allows one to shed light on the properties of the corresponding physical processes, and so providing very useful guidelines on how to modify the physical assumptions of full evolutionary models. In this sense, synthetic AGB simulations can provide a very important feedback for the complete evolutionary models.

During the last years, updated sets of synthetic AGB models have been provided and extensively used in population synthesis codes and analysis [286, 335].

As mentioned, synthetic AGB models are built by using the prescriptions provided by full evolutionary computations. These prescriptions are usually accounted for in the form of fitting analytical relations. These theoretical recipes are often supplemented with empirical data such as those corresponding to the mass loss rates in AGB stars and/or the TDU efficiency estimated on the basis of observational constraints. The operative approach adopted for computing a synthetic AGB model can be fully analytical when only analytical relationships are used, or semi-analytical when the use of fitting formulae is supplemented by computations of numerical models such as convective envelope models.

By starting with the fundamental ingredients that will be detailed next, a synthetic AGB model is able to quickly provide information about the evolution with time of: the surface luminosity (hence the magnitude in various photometric bands); effective temperature (hence colours); the growth of the H-exhausted core during the inter-pulse stage as a consequence of the H-burning, and its mass decrease as a consequence of an eventual TDU episode; the change in surface chemical stratification due to the occurrence of the TDU; the mass reduction of the outer envelope as a consequence of mass loss. Indeed, these synthetic models can also account for the occurrence of HBB in massive AGB stars as well as provide some information about the pulsational properties of an AGB star when it enters the pulsational instability strip of the *Long Period Variables* (LPV).

Here, we provide only a schematic description of the computation of synthetic AGB models. There have been several investigations devoted to providing (starting from full evolutionary calculations) the relevant ingredients of synthetic AGB models, such as the analytical relationships relating observational quantities such as surface luminosity, T_{eff} and surface chemical abundances to the He-core mass, initial metallicity and residual envelope mass [336, 337].

5.5.1

Outline of the Method

All synthetic AGB computations start at the beginning of the TP stage, that is, at the re-ignition of the H-burning shell. It is customary to consider the start of the TP phase to correspond to the achievement of the asymptotic regime in the temporal behaviour of the "average" surface luminosity and effective temperature; this means that the synthetic simulation begins after the first 4–5 (pre)-TPs.

The fundamental stellar parameters for the AGB computations at the beginning of the TP stages are: the total mass M, the surface luminosity L, the effective temperature, the He-core mass (M_H) and the envelope chemical abundance stratification. The whole set of input parameters are provided by full evolutionary computations.

Let us now discuss the evolution of the stellar luminosity. We will focus our discussion on the maximum surface luminosity achieved during the quiescent H-burning during each TP stage. It is possible, by using the same approach, to also obtain detailed predictions about the evolution of the surface luminosity at well-defined evolutionary phases within the inter-pulse cycle, but these are second-order effects when applying synthetic AGB models to more general astrophysical problems.

As previously discussed, AGB stars display a well-defined *He-core mass–luminosity* relation that is violated only in the case of massive AGB stars by the occurrence of HBB. To account for this effect, it is customary to express the maximum luminosity as a function of the luminosity that one gets from the standard *He-core mass–luminosity* (L_{CMLR}) relation and a term due to the HBB eventually occurring in the stellar envelope: $L_{max} = f(L_{CMLR}, L_{env})$. This approach has the advantage of easily accounting for the effect of HBB. The occurrence of HBB is flagged by the temperature achieved at the bottom of the convective envelope, whose value is given by an analytical relation that is a function of the He-core and envelope masses. As for L_{CMLR} and L_{env}, detailed analytical relations obtained from full evolutionary models provide their dependence on the He-core mass, actual total mass and chemical composition [337].

To evaluate how these quantities change with time during the AGB evolution, one needs to predict the temporal increase of the He-core mass. If the TDU is not occurring, the He-core mass increase (ΔM_c) in a given time step, Δt, is equal to $\Delta M_c = \Delta M_H = LQ\Delta t$, where L is the surface luminosity and Q is the effective nuclear burning efficiency that takes into account H- and He-burning as well as gravitational energy release and chemical composition effects (from full evolutionary computations, one derives $\sim 1.585 \times 10^{-11} M_\odot L_\odot^{-1} \text{year}^{-1}$).

When the TDU is occurring, one has to account for the decrease of the He-core mass associated with this mixing process. Since the TDU efficiency is defined as $\lambda = \Delta M_{TDU}/\Delta M_H$ over a whole inter-pulse period, one obtains $\Delta M_c = \Delta M_H - \Delta M_{TDU} = (1 - \lambda)\Delta M_H$. We are still faced with the problem of how to decide if the TDU is effectively occurring; in synthetic AGB simulation, this is done by providing an analytical relation to be calibrated on observations or evolutionary computations that give the minimum He-core mass (as a function of the initial mass and chemical composition, as well as of the assumed TDU efficiency) at which the TDU can begin.

One should note that to follow the occurrence of the TDU, one has to correctly choose the time step Δt that has to be smaller than the inter-pulse period, which is largely a function of the actual He-core mass and TDU efficiency [337].

The C/O ratio (and other chemical abundances) evolution with time as a consequence of the TDU (for a given adopted TDU efficiency that is value of the parameter λ), can be determined by knowing (from full evolutionary models) the chemical composition of the inter-shell region. The evolution of the C/O ratio affects the evolution of T_{eff}, and can be followed in two ways. One can use analytical relationships derived from full evolutionary models that provide the dependence of T_{eff} on the actual He-core mass, envelope mass, C/O ratio and global metallicity, or by performing stellar envelope integrations that provide the HBB luminosity (in case HBB is efficient), and T_{eff} as a function of total and He-core mass [338]. This latter approach allows a straightforward evaluation of the impact on the model T_{eff} of C-enhanced mixtures (due to the TDU), or of using different values α_{ml}.

Mass loss can also be included in synthetic AGB calculations by choosing an appropriate analytical recipe (see Section 2.6) that provides the mass loss rate as a function of properties like luminosity, effective temperature and actual total mass. In this way, it is easy to evaluate the reduction of the total mass (actually the envelope mass) at each time step. The indirect effect of mass loss on the efficiency of the TDU and, eventually, of the HBB are automatically accounted for via the dependence on the envelope mass in almost all the previously quoted analytical relationships. On the other hand, the mass loss rate is also affected by the TDU and, eventually, by the HBB, due to the T_{eff} decrease associated with the TDU, and the huge luminosity increases caused by HBB.

Once theoretical predictions for the minimum luminosity required for the activation of the various radial pulsational modes in AGB stars have been adopted, as well as an analytical relationship providing the pulsational period as a function of fundamental stellar parameters such as L, T_{eff} and actual mass (see the next section), synthetic AGB models can be very useful for predicting the pulsational properties of LPVs in a given stellar population [335].

5.6
Long Period Variables

Long period variables (LPV) form an important class of red giant stars. They show, more or less, regular photometric variability with amplitudes reaching up to $\sim$ 8 mag and periods up to $\sim$ 600 days. This class of pulsating stars traditionally comprises Miras variables, semi-regular variables (SR), and irregular (L) variables, classified on the basis of the amplitude and the regularity of their visual light curves.

LPV stars are known to be either O-rich or C-rich, and comprise thus M, S and C-stars.[17]

Within the LPV class, the Mira variables are characterized by a large visual amplitude ($A_V > 2.5$ mag) and their light curves are quite regular; SRs, which are further divided in two subgroups: SRa and SRb, have a smaller visual amplitudes and light curves showing a hint of periodicity; while L variables show little, if any, regularity in the shape of their light curves, although it is not yet fully clear if this is due to some intrinsic reason or to the lack of an appropriate observational sampling of the pulsational cycle.

Even if the Mira variable pulsations show quite large visual amplitudes (up to 6 mag or larger), the bolometric light curves reveal a quite lower modulation (of the order of ~ 1 mag, and increasing with pulsational period). This is due to two concomitant causes: (i) these stars are characterized by a low T_{eff} value, so that the largest portion of the radiative flux is emitted in the infrared bands, mainly at ~ 1 µm; and (ii) the light curve modulation is much smaller (few tenths of mag or lower) in the IR photometric bands than in the visual ones.

The large modulation of the visual light curves of Mira variables is due to both the huge variation of the radiative opacity associated with TiO,[18] and the large change in the V-band flux (due to the shape of the blackbody flux distribution, that is, peaked at IR wavelengths for these cool stars), when the T_{eff} changes during the pulsational cycle.

The various pulsational behaviours of the different families of LPVs can be caused by different masses, chemical compositions, or by being in different evolutionary stages along the AGB. For instance, the observational distinction between Mira and SR variables does not necessarily have a deep physical meaning.

These variable AGB stars are observed over the whole mass range characteristic of AGB stars: from GCs as in the case of the stellar system 47 Tuc (where the evolving mass is $\sim (0.8-0.9) M_\odot$), up to the upper mass limit of AGB stars, that is, $(6-8) M_\odot$, in the Galactic field and in the Magellanic Clouds.

On the basis of the properties of the stellar populations hosting the various kinds of LPVs, it has been possible to estimate their typical masses. For Mira stars, it is nowadays accepted that their typical mass is $\sim (1.1-1.2) M_\odot$, increasing slightly with the pulsational period P. In more detail, Miras with $P > 300$ days should have a mass larger than $\sim 1.1 M_\odot$, while those with a lower period should have a mass lower than this limit; clearly, the Mira stars with $P \approx 200$ days, observed in the more metal-rich GCs, have a mass of the order of the evolving mass, that is, $\sim 0.85 M_\odot$ or lower. The mass-range of SR pulsators appears to be as broad as that of Mira stars.

---

17) In the last decade, thanks to infrared and radio observations several OH-IR sources have been found to belong to LPV population, and they have pulsational periods up to 2000 days. They emit at the infrared and radio wavelengths, and are not associated with any detectable counterpart in optical wavelengths.

18) The TiO molecule is one of the most important opacity sources in the visual spectral bands for cool stellar objects.

As far as the intrinsic properties of the pulsational processes in LPV stars are concerned, one can safely assume that the most probable driving mechanism of the pulsations is the combined action of partial H and He I ionization zones in the envelope. However, over a large fraction of their envelopes, energy transport is via convection, and the characteristic convection timescale is of the same order as the pulsation cycle period. Hence, the pulsational driving mechanism(s) and the extension of the instability region in the HRD depend crucially on the coupling of pulsation and convection and on how convection affects the equilibrium of these stars. Unfortunately, due to the poor understanding of how to manage the complex physics of convection in stellar computations, and of the correct coupling between pulsation and convection, full-amplitude, non-linear pulsational models for AGB stars are still at in an initial stage, although significant improvements have been made in the last few years.

These knowledge gaps mean that we are still faced with the important problem of understanding what is the most probable pulsation mode for LPV variables, that is, if higher over-tone pulsation modes are more stable than the fundamental one. The most direct way to determine the pulsation mode of an LPV is based on the pulsation equation, $P = Q\rho^{-1/2} \propto QR^{3/2}M^{-1/2}$ that, to a first order approximation, relates, for a given mode, the corresponding period with the stellar mass and radius. Theoretical pulsation models predict that for a Mira, the pulsation constant Q for the fundamental mode is approximately twice as large as the value of the constant for the first overtone mode. So, for a given pulsation period P and total mass M, fundamental mode pulsations require a smaller radius than first overtone ones. Unfortunately, the comparison between theory and observations has provided contradictory results [339].

One has to note that the approach based on the pulsation equation is feasible because as a class, Mira and SR variables are old, low-mass stars and so it is possible to "safely" assume that they have a mass of the order of $\sim 1 M_\odot$. The main source of uncertainty in this approach is the stellar radius estimate. Despite huge improvements in interferometric techniques for stellar radii measurements, firm results are still lacking, and this hampers a firm establishment of the Mira pulsation mode; although, the most recent evidence seems to support the idea that at least the brightest Mira variables preferentially pulsate in the fundamental mode [340].

From the theoretical point of view, pulsation models for Mira stars show that the dominant growth rate[19] among different pulsation modes is correlated with luminosity: modes of higher order have the highest growth rate at lower luminosities [340–342]. The available theoretical framework for pulsations in AGB stars suggest that as a star climbs up the AGB, attaining larger luminosities, it first becomes unstable in the second or third overtone, then it switches to a lower pulsation mode and, eventually, at the largest luminosities it starts pulsating in the fundamental

---

19) The growth rate is a quantity provided by pulsation models that allows one to estimate the stability of a pulsation mode: the larger it is, the more unstable the pulsation mode. However, it is worth noting that the evidence of a large growth rate does not necessarily imply that a star is pulsating in such a mode.

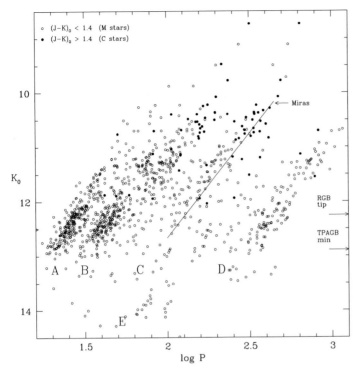

Figure 5.11 The near-IR period–luminosity relation for a sample of SR and Mira variables in a field of the Large Magellanic Cloud [343]. The letters mark the most evident sequences in the figure. The two arrows mark the brightness of the RGB tip and the start of the thermally pulsing stage for a low-mass AGB star ($1M_\odot$). Different symbols for the observational points are used to distinguish the AGB stars based on their intrinsic $(J-K)$ colour that is adopted as a criterion for separating M-type AGB stars from C-stars. The solid line shows the $\log P - K$-band luminosity relation for Mira variables provided by [344] (courtesy of P. Wood).

mode. Pulsation models suggest that there is an overlapping between the luminosity intervals where the various pulsation modes are unstable. This suggests that Mira stars could be affected by multi-mode periodicity as, indeed, it is observed.

An important property of LPVs is the fact that they obey a *Period–Luminosity* relation [345]. The extensive observational surveys related to micro-lensing experiments such as OGLE and MACHO, have clearly demonstrated that there are "several" *Period–Luminosity* relations for this class of variables. Figure 5.11 shows the *Period–Luminosity* relations for a large sample of SR and Mira variables in the Large Magellanic Cloud: there are at least four parallel period-luminosity sequences.[20] SR variables are located on all sequences from A to C, while Mira variables are confined to the brighter portion of the C sequence. The sequence D is populated by stars showing an evident multi-periodicity: they have a pulsation period that lo-

---

20) Empirical evidence based on larger observational datasets supports the idea that some of these individual sequences could actually split into additional sub-sequences.

cates them on the D sequence and another period related to sequence A or B (or sometimes C).

The most naive interpretation of this series of sequences is that they correspond to variable stars pulsating in different pulsation modes: so after identifying (on the basis of their light curve) the variables stars along the various sequences, one notes that SR variables can pulsate in many modes, whereas Mira (or at least, the majority of them) should pulsate only in one characteristic mode [339, 346]. The origin of sequence D is still not understood. Some, though thus far very speculative, interpretations of its existence are related to the occurrence of some still-unknown (radial or non-radial) pulsation mechanisms, rotation in an interacting binary system, or dust formation.

The periods (and periods ratios) predicted by pulsation models offer an alternative approach for determining the pulsation mode of an LPV. This is shown in Figure 5.12 for the same observational dataset used in Figure 5.11. The comparison between predicted *Period–Luminosity* relations for the various pulsation modes and the empirical relations supports the identification of the various observed sequences with the different pulsation modes, as well as giving evidence that (the bulk of) Mira(s) pulsate(s) in the fundamental mode [339].

We note that, as for the case of other classes of variables, several Mira stars show clear evidence for a secular period change [347, 348]. Three different classes of secular period change have been defined:

- *Continuous change:* a continuous decrease or increase of the pulsation period during a temporal interval of about 100 years or more, with no evidence for epochs of a stable period. The change in period is in all cases large, $\sim 15\%$ or more over 100 years;
- *Sudden change:* after a long phase of stability, the period suddenly begins to change (by decreasing or increasing) at a large rate. The total change in period is similar to that of the previous class, but the rate is ten times faster;
- *Meandering change:* a number of long-period Miras show evidence of meandering or fluctuating periods. The periods change by $\sim 10\%$ over several decades, followed by a return to the previous period. The rate of change is comparable to that of the first class, but the total period change is somewhat smaller than that of both classes.

The interpretation of these pulsation period changes is important in understanding AGB evolution because the period changes could, for instance, be driven by sudden changes in observational properties of these stars such as their luminosity, effective temperature and actual mass. In fact, it has been suggested that the main cause of the observed secular period change could be the occurrence of a TP. The fraction of Miras ($\sim 2\%$) with a large period change is in agreement with the expected ratio of the pulse duration to the inter-pulse time (which is of the order of 1–2%). There are other explanations for the observed period changes that do not invoke the occurrence of a TP. In any case, not all observed period changes can be understood as being due to TPs [342].

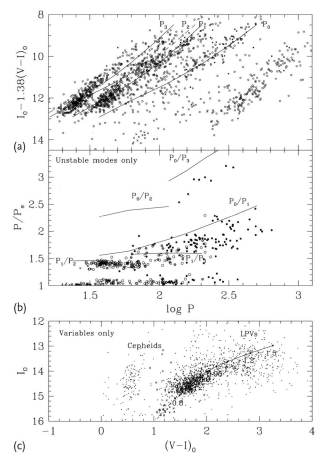

Figure 5.12 (a) The Wesenheit $W(I, V-I)$ index as a function of the pulsation period P (in days) for a sample of AGB variables in the LMC. The solid (dashed) lines denote theoretical predictions for unstable (stable) radial pulsation modes: the fundamental mode P_0, the first overtone P_1, the second overtone P_2, and the third overtone P_3. (b) As with (a), but for the Period ratios (only unstable modes are now considered). (c) The corresponding CMD of the AGB stars considered in (a) and (b). The solid line corresponds to a theoretical AGB sequence. The location of various masses is also reported (courtesy of P. Wood).

To shed more light on the possible correlation between the secular period changes and changes in the internal structures occurring during the AGB evolution, recent attempts have been made to look for correlations between the period changes and the abundances of specific chemical elements which are useful for tracing the occurrence of a TP and/or a TDU event.

One of the most important chemical elements in this context is technetium (Tc), an s-element produced during AGB evolution. Since its longest-lived isotope (^{99}Tc) has a short half-life time ($\sim 21 \times 10^4$ years) in comparison with stellar evolution timescales before the AGB stage, if it is observed at the surface of an AGB star, it provides strong evidence that the star is experiencing the TP phase. Hence, the detection of Tc would support the hypothesis of a recent TP as the cause of the observed period change. However, the absence of Tc does not exclude that TPs are ongoing in a given star, although it does place a constraint on the TP strength: only the strongest TPs on the upper AGB will be followed by a TDU event.

So far, investigations devoted to this issue have not found any clear correlation between the secular period changes and the surface abundance of Tc.

In the previous sections, we emphasized that AGB stars are affected by a very efficient mass loss phenomenon with rates typically of the order of $(10^{-6}-10^{-7})$ $M_\odot$/year, and as large as $(10^{-4}-10^{-5})\,M_\odot$/year for the brightest LPVs. However, compared to winds of other types of stars, those from AGB stars are relatively slow, with typical terminal velocities of the order of $(10-20)\,\mathrm{km\,s^{-1}}$. Empirical evidence suggests that there is a strong link between the occurrence of radial pulsations and mass loss efficiency. Indeed, the most generally accepted mass loss scenario nowadays is the so-called *pulsation-enhanced dust-driven winds* scenario. In such a scenario, the "source" of momentum for the wind is the combination of the dynamical extension of the stellar atmosphere by shock waves due to pulsations, and radiation pressure on dust grains [349].

The available sets of models predict that shocks associated with pulsations do not eject material from the star directly, but they do deeply change the structure of the atmosphere and, effectively, produce conditions which lead to mass loss. In particular, the atmosphere becomes enormously extended, with greatly increased density at large radii. The increased density in the cool outer regions greatly increases the amount of dust that is formed. This dust serves as an effective agent for absorbing momentum from the stellar radiation field and transferring it via collisions to the gas, thus driving a stellar wind.

5.7
Nucleocosmochronology

Nucleocosmochronology is the use of abundances and abundance ratios of radioactive nuclides, coupled with information on the chemical evolution of the Galaxy, to obtain information about the timescales over which elements formed in the solar system, as well as at determining a lower limit for the age of the Universe. Starting from the seminal work of Fowler and Hoyle [350], where the basic ideas of nucleocosmochronology were provided, many attempts have been made to utilize the abundances measured in meteorites in order to infer the age of the chemical elements that compose the bulk of the solar system material, and therewith set some constrains on the nucleosynthesis occurring in the Galactic disk.

Roughly speaking, the basic idea of nucleocosmochronology is to compare the expected initial abundance of a given radioactive chemical element with the observed current abundance. If its radioactive decay timescale and how its abundance could increase with the time, due to production processes, are known, then the comparison between initial and current abundance provides an estimate of the age of the formation of this radioactive nuclide. Actually, since an estimate of the initial abundances of these elements is a thorny problem, it is preferable to place a constraint on the duration of nucleosynthesis in the Galaxy from the knowledge of: (i) the known production ratio of a long-lived radioactive element (such as ^{232}Th) to a stable element (as Eu) or other long-lived isotope (such as ^{238}U), (ii) the known abundance ratio in the early solar system, and (iii) the time evolution of the chemical abundances over the Galactic history. So, the idea is the same as that of ^{14}C dating for archaeological specimens, in which one uses the known steady-state ^{14}C/^{12}C abundance ratio present in all still-living organisms, the well-known half-life time ($\tau_{1/2}$) of ^{14}C, and the measured ^{14}C/^{12}C abundance ratio to estimate how long ago that organism died.

A reliable chronometry based on radioactive nuclides requires accurate (isotopic) abundances from spectroscopic measurements, reliable nucleosynthesis yields from stellar evolution computations, and accurate predictions about the radioactive half-life times. Another important ingredient is obtaining detailed models for the chemical evolution of the Galaxy [351, 352].

The radioactive nuclides derived from *r*-processes,[21] and in particular the long-lived nuclei ^{232}Th, ^{235}U and ^{238}U, are the commonly adopted nuclear chronometers.

However, AGB stars can also produce, via the *s*-process, radioactive nuclides such as Tc (see the previous section) and ^{176}Lu. In particular, ^{176}Lu has a half-life of $\tau_{1/2} = 3.6 \times 10^{10}$ years, and its "daughter" is ^{176}Hf. In principle, this radioactive nuclide has the advantage that it is the only potential chronometer that can be attributed unambiguously to *s*-process nucleosynthesis. However, its use as a chronometer requires an accurate evaluation of the ^{176}Lu/^{176}Hf production rate which is, in practice, very difficult to obtain. The complication arises from the existence of a short-lived (~ 3.7 h) ^{176}Lu isomeric state which, depending sensitively on the temperature of the *s*-element nucleosynthesis site, can be (or not be) in equilibrium with the ground state (the relevant one for the use of ^{176}Lu as chronometer). In passing, we note that this temperature sensitivity could be useful for using the ^{176}Lu–^{176}Hf pair as an *s*-process thermometer.

Since neutron capture on ^{175}Lu leads preferentially to ^{176}Lu in the isomeric state, thus leaving only a small probability for the long-lived ground state, the partial cross section to the isomer needs to be known with high accuracy. Owing to a huge experimental effort, the uncertainty on the quoted cross section has been significantly reduced, but it is still too large to allow a reliable estimate of the ^{176}Lu/^{176}Hf production rate. One has to also take into account that the production of ^{176}Lu strongly depends on the physical properties of the He-rich inter-shell in the AGB stars.

21) The *r*-processes are those nuclear reactions involving neutron captures in the presence of a huge neutron density so that the neutron capture nucleosynthesis can proceed on a shorter timescale than the β-decay process. Correspondingly, the *r*-elements are the chemical species produced as a consequence of *rapid* neutron captures to distinguish them from the *s*-elements. The most important astrophysical site for *r*-element productions is the interiors of massive stars during both the hydrostatic and explosive evolutionary stages.

Due to the quoted difficulties, the possible chronometric virtue of ^{176}Lu has been largely dismissed, and this *s*-element continues to challenge further efforts from both the experimental and theoretical sides before it can safely be used as a nuclear clock.

5.8
The End of the AGB and Post-AGB Evolution

As discussed in Section 5.3, there are several empirical facts that suggest that strong mass loss occurs during the late stages of the AGB. The most important evidence for this is the lack of AGB stars brighter than $M_{bol} \sim -7$ mag,[22] and the observational properties of Planetary Nebulae (PNe).

Sound theoretical models support the scenario that once the envelope mass drops below a critical value, stars are forced to leave the AGB, and move towards the hottest side of the HRD. In this way, they cross the region of the HRD corresponding to the "realm" of PNe nuclei. A PN is actually the observational consequence of the presence (around a dense stellar nucleus) of material lost by an AGB star via winds, and lately excited by radiation coming from the hot post-AGB stellar remnant.

There are two conditions that must be satisfied in order to have a PN: (i) a very hot ($T_{eff} > 30\,000$ K) central stellar nucleus, and (ii) a dense circum-stellar envelope ($n \geq 1000\,\mathrm{cm}^{-3}$) which will be excited by the UV photons emitted by the central remnant.

Although there is some evidence supporting the idea that (at least) some PNe could be formed by means of stellar envelope ejection via a normal AGB wind, it is commonly assumed that the mass loss responsible for the envelope ejection, suitable for creating a PN, has to be characterized by an efficiency quite larger with respect to the standard mass loss occurring during AGB. In particular, observational measurements of the mass, radius, and expansion velocities (~ 20 km/s) of PNe allow one to infer that the mass loss rate during the process leading to nebular ejection has to be of the order of $10^{-4}\,M_\odot$/year. The term "superwind" was coined to denote this evolutionary phase [246].

On theoretical grounds, one can expect that an AGB star undergoing such dramatic mass loss should be embedded in a dusty and optically thick circum-stellar envelope. Nowadays, the observations of thin, detached CO shells around some C-stars, of IR and OH/IR sources without any optical counterpart, and measurements of the mass loss rates of the brighter LPVs support this evolutionary scenario. However, we still lack of a firm understanding of the physical reasons at the basis of the *superwind* phenomenon: the most feasible interpretation being related to the pulsation mode switching from the first overtone mode to the fundamental one (see the

---

22) The maximum observed bolometric magnitude for C-stars is of the order of ~ -6 mag; this is the brighter magnitude limit for low-mass AGB stars.

discussion in Section 5.6) and/or to atmospheric shocks produced by the pulsation itself (regardless of the pulsation mode).

As a consequence of a huge effort in the computation of accurate, pulsating and dust-driven stellar wind models, we can now provide some hints on the evolution of AGB stars near the termination of this evolutionary stage [72, 353, 354]. The available simulations reveal that:

- In order to drive the superwind process, a minimum outward directed force is required. This translates into the existence of a critical (minimum), Eddington-like,[23] luminosity for the onset of this enhanced mass loss. This critical luminosity depends on the initial stellar mass and actual effective temperature. For an AGB star with a mass of $\sim 0.8 M_\odot$ near the termination of this stage, this critical luminosity corresponds to a bolometric magnitude $M_{bol} \sim -4.3$ mag;
- For stars with initial mass larger than $1.2 M_\odot$, huge mass outflows (of the order of $0.3 M_\odot$ at the low-mass end, and increasing by a factor of ~ 2 at the massive end) are obtained in a short timescale ($\sim 3 \times 10^4$ years);
- The superwind process shows a significant modulation in its efficiency as a consequence of the huge changes in the surface luminosity and T_{eff} occurring during a TP; a temporary interruption of the superwind process is also effectively possible;
- The superwind does not stops abruptly when the star increases its T_{eff} when definitively leaving the AGB, but there is a gradual decline of the mass loss rate. This is because updated models predict a strong dependency of the superwind mass loss rate on the effective temperature, namely, $\dot{M}_{sw} \propto T_{eff}^{-7}$ [72, 355]. We note that this issue is crucial in understanding the observed properties of PNe as well as for understanding the reasons why some post-AGB stars become the nuclei of a PN and others do not develop a PN at all.

5.8.1
The Evolution from the AGB to the WD Cooling Sequence

Once the envelope mass has reached a critical value, the star departs from the AGB moving toward the hot side of the HRD. The evolutionary rate at which the star crosses the HRD is large and it depends on the residual envelope mass: the larger the envelope mass, the lower the evolutionary rate. Typical evolutionary tracks for post-AGB evolution show a shift towards large T_{eff} values (i.e. lower stellar radii), but at an almost constant luminosity. During this stage, the star is mainly supported by the H-burning occurring in the shell and by the energy released by gravitation contraction. The typical timescale for the evolution along the horizontal portion of these post-AGB evolutionary tracks, from the termination of the AGB stage to the beginning of the WD stage, is of the order of 10^4–10^5 years for a He-core mass of about $0.56 M_\odot$.

---

23) The Eddington luminosity limit is the maximum surface luminosity possible for a star with the outer layers in radiative equilibrium: $L_{Edd} \propto (M/\kappa)$ where M is the total stellar mass and κ is the mean radiative opacity in the external layers.

During the movement of the AGB remnant towards the blue side of the HRD, the surrounding material that was ejected during the superwind stage expands. If the central stellar remnant eventually achieves a $T_{\text{eff}} \approx 30\,000$ K before the surrounding circum-stellar envelope has expanded too much, ionization of the ejected material by UV photons from the remnant is efficient, and the object becomes a PN. On the basis of this qualitative description, it appears clear that a crucial parameter for determining whether a post-AGB star can effectively develop a PN is the so-called transition time, t_{tr}, which is the time required by the star to attain a $T_{\text{eff}} \approx 30\,000$ K after leaving the AGB. If t_{tr} is too long, the material ejected during the superwind phase is dispersed before the central remnant becomes hot enough to ionize it, and so a PN will not be formed.

Simple theoretical considerations show that the transition time critically depends on the residual mass of the envelope at the moment when the star leaves the AGB, and hence also on the efficiency of the superwind, as well as on the possible occurrence of mass loss (although at a lower rate than the superwind) during the post-AGB stage [317]. In fact, evolutionary models not accounting for mass loss during the post-AGB phase predict that a structure with a He-core mass of about $0.6\,M_\odot$ and a residual envelope mass of $\sim 10^{-3}\,M_\odot$ will cross the HRD from the AGB to the PNe region in $\sim 15\,000$ years. However, since the PNe can have ages as low as 1000 years, mass loss has to occur during this evolutionary stage to speed up the evolution. As a consequence, an appropriate interplay (not yet well-understood) between the AGB superwind and mass loss during the post-AGB stage is crucial in understanding the PN formation scenario.

In conclusion of this brief overview on the link between the post-AGB evolution and the occurrence of PNe, we note that the PN associated with a post-AGB stellar remnant has to disappear as soon as the circum-stellar shell disperses as a consequence of its own expansion.

Let us also discuss various channels for post-AGB evolution that are important in understanding empirical evidence related to the spectroscopic properties of various types of WD, and the existence of a few, but very fascinating, exotic stellar objects: H-deficient post-AGB stars and Sakurai's object.

The evolutionary and structural properties of post-AGB stars under various physical assumptions have been studied extensively [355, 358–362]. Three main evolutionary channels can be envisaged for post-AGB stars during the transition to the final WD evolutionary stage:

- *The non-flash scenario:* post-AGB stars belonging to this first group depart from the AGB during the extended H-burning phase occurring during the inter-pulse stage. They continue to burn H in the shell, and lie on the hot side of the HRD with very high effective temperatures, $\log T_{\text{eff}} \approx 4.9$–$5.2$, until the H-rich envelope mass becomes lower than the critical value that does not allow for an efficient H-burning. When this occurs, the star contracts and fades, thus entering the WD cooling sequence stage. When H-burning cannot be sustained any longer, the surface luminosity must drop very fast, by at least one order of magnitude, until the release of gravitational energy by contraction starts to supply a significant contribution to the energy budget. The fading time of AGB remnants down to $\sim 100 L_\odot$ is thus controlled by the gravothermal energy release and neutrino energy losses. Both processes depend on the thermo-mechanical structure of the core, and thus on the complete evolutionary history.

The typical evolutionary track for stars with this fate is shown in Figure 5.13a. Although these stars are expected to suffer from some amount of mass loss, the combination of a (not high enough) mass loss rate and a (fast) evolutionary lifetime should allow them to maintain a significant fraction of their H-rich envelope; they will evolve into normal DA (H-rich atmospheres) WDs;

- *The late He flash scenario:* this scenario envisages that a late TP (LTP) occurs while the star evolves, with roughly constant luminosity, from the AGB towards the WD domain, during which the H-burning shell is not switched-off. Evolutionary computations show that this outcome is possible if the star leaves

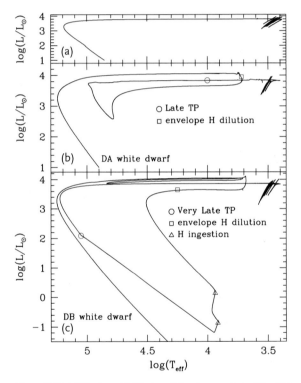

Figure 5.13 HRD of post-AGB stars following different evolutionary channels: a model that after experiencing TPs along the AGB branch cools down along the WD cooling sequence (a); a stellar model experiencing a LTP before ending its evolution as a DA WD (see Chapter 6); and (b) a model experiencing a VLTP that becomes, after proton ingestion and surface H dilution, a non-DA WD [356, 357]. The start of the LTP and VLTP as well as of the dredge up causing the surface H dilution are marked (see labels). In (c) the open triangles mark the phase of proton ingestion. All models correspond to an initial chemical composition: $Z = 0.02$, $Y = 0.275$ and various initial total masses (courtesy of M.M. Miller Bertolami).

the AGB during the last $\sim 25\%$ of the time spent between the two last pulses (this is commonly expressed in terms of the thermal pulse cycle phase ϕ, thus here $\phi \geq 0.85$, while it is supported by the stationary H-burning shell. This final TP is similar to those commonly experienced by AGB stars, but the envelope mass is smaller, and so the TDU does not occur. Therefore, the surface chemical abundance of these stars should not be affected by the occurrence of this last TP. It is worth noting that due to the occurrence of the He flash in the shell, the star performs a loop in the HRD: its surface cools and expands, so again achieving an AGB configuration. This occurrence corresponds to the *born-again* scenario.

Until few years ago, this scenario has been considered feasible for explaining the existence of H-deficient, post-AGB stars only if these stars can experience a very efficient mass loss during the *born-again* stage and the ignition of the last TP. We note that LTP stars have longer (by a factor of ~ 3) evolutionary timescales with respect to H-burning AGB stars; therefore, in the presence of an efficient mass loss, they can lose the H-rich envelope and expose the He-rich layers. However, standard evolutionary models predict that this occurs only at T_{eff} larger than $\sim 100\,000$ K.

However, there are some indications that post-AGB stellar models accounting for some amount of convective overshoot at the base of the outer convective envelope can experience an efficient dredge up of He-rich material during the *born-again* stage when the star attains its minimum effective temperature during the loop in the HRD [363]. As a consequence of this mixing, the surface abundance of H would be diluted up to $\sim 3\%$, while the envelope abundances of He, C and O would become similar to those present in the He-rich inter-shell zone during the last portion of the AGB stage. In conclusion, the LTP scenario would be able to produce H-depleted stellar objects, but only via the occurrence of huge mass loss and some amount of (non-canonical) mixing.

After the occurrence of the LTP and the subsequent evolutionary stage, the star again crosses to the PNe region and cools down along the WD sequence;

- *The very late He flash scenario*: according to this scenario, the last TP only occurs after the star has already begun moving down the WD sequence, so it is a very late TP (VLTP). As a consequence, this VLTP occurs when H-burning is already off. This has a crucial consequence: the pulse-driven convection zone can reach and penetrate the H-rich envelope because of the lack of an entropy barrier built up by H-burning. Protons are then ingested into the hot, He- and carbon-rich inter-shell region and are burnt via $^{12}C(p, \gamma)^{13}N$. We remark that this evolutionary scenario closely resembles the *He flash induced mixing* scenario described for the blue hook stars in Section 4.4, the only difference being that in this case, the flash occurs at the base of the He-shell and not in the core. Since the typical convective timescale and the nuclear burning timescale are comparable during this phase, the ingested protons are burnt *on fly* during their way towards the interior. The ingestion, and the consequent high-temperature burning of hydrogen finally causes an H flash, and the energy released by this flash leads to a splitting of the convection zone into an upper zone powered by H-burning and a lower one powered by He-burning. The upper convection zone is, however, short-lived because the available hydrogen in the envelope is quickly consumed. Finally, the star becomes H-free and exposes its inter-shell abundances (large C and He enhancement) at the surface. This scenario can feasibly predict the existence (and surface chemical abundances) of H-deficient stellar objects. The evolutionary tracks of post-AGB stellar models experiencing the LTP or the VLTP are shown in Figure 5.13. This scenario should involve stars that depart from the AGB when $0.75 < \phi < 0.85$ [360].

Actually, these theoretical evolutionary predictions are extremely useful for explaining the presence of H-deficient post-AGB stars and their extremely peculiar surface chemical patterns. About 20% of the whole population of post-AGB stars appears to be H-deficient (at various levels of depletion) while the rest shows non-peculiar chemical patterns, and $\sim$ 25% of the WD population have He-rich atmospheres (non-DA WDs). Some important classes of H-deficient stars are the so-called Wolf–Rayet (WR) central stars, the hot PG 1159, and the R CrB stars. All these stars show atmospheres hugely enhanced in He, C and O (although PG 1159 stars are characterized by a smaller O enhancement). Although the surface abundances of these stars show a very large spread, the typical abundance by mass of He, C and O are $\sim$ 0.4, $\sim$ 0.4, $\sim$ 0.1, respectively.

The comparison between the observed surface chemical abundances and those predicted by the scenarios discussed here would allow one to establish whether an observed stellar object is offspring of a LTP or of a VLTP event. For instance, the spread in the surface abundance of ^{14}N in PG 1159 objects supports the idea that these objects are indeed produced by both the VLTP scenario (which predicts a huge enhancement of this element) and the LTP one (where no significant N enhancement is expected). The surface O abundance observed in PG 1159 stars has for some time been a challenge to evolutionary models of post-AGB stars because canonical models do not predict the abundance of O that is observed in real stars. This conundrum has been explained as due to the occurrence of some convective overshoot at the bottom of the PDCZ, that brings some O present at the base of the He-burning shell into the upper layers of the inter-shell region [363]. The occurrence of convective overshoot in LTP models, as already mentioned, would be also of help in explaining the small, but still detectable, abundance of H observed in some peculiar PG 1159 stars (the so-called *hybrid*-PG 1159 objects).

From an empirical point of view, some of the best studied post-AGB stars are FG Sge and the Sakurai's object (V4334 Sgr) [364, 365]. These stars have been observed performing a loop in the HRD and changing their surface chemical abundances. The peculiarities observed in these objects are largely explained within the VLTP scenario, although the LTP one cannot definitively be ruled out.

Post-AGB stars cross the Cepheid instability strip at larger luminosities than Cepheid stars; hence pulsations are expected to occur in these objects, as is observed. The study of the pulsational properties of post-AGB variable stars, such as RV Tau and UU Her stars, provides many useful diagnostics on their internal structure.

Some PG 1159 stars show also evidence of photometric variability, with periods ranging from $\sim$ 10 to 30 min [366]. They are termed GW Vir or DOV variable stars. The high-luminosity members of this class have periods about four times longer than those of stars close to the WD cooling sequence. The longer-period PG 1159 stars are usually variable central stars of PNe. The GW Vir variability is attributed to the over-stability of low-degree, high-order gravity (g) modes. Their observed frequency spectra are quite rich, that is, many modes are excited simultaneously, and they are used with considerable success in asteroseismic analyses.

A major unresolved problem is posed by the observation that even if a PG 1159 star lies in the instability region, this does not necessarily mean that it is a pulsating star. Only about half of the population of the known PG 1159 stars inside the instability domain pulsate. The problem increases when considering that spectroscopic twins are observed, of which one is variable and the other is stable.

6
White Dwarf Cooling Sequences

Low-mass and also intermediate-mass stars with initial masses up to $6-7M_\odot$ end their life as WDs, following the AGB and PAGB phases. The structure of a WD is deceptively simple. About 99% of more of their mass is contained in an electron degenerate core made essentially of carbon and oxygen, produced during both central and shell He-burning phases; the core is surrounded by a thin (in both mass and geometric size) non-degenerate envelope, left over by the combined effect of core growth and strong mass loss during the TPs.

Due to our imperfect knowledge of mass loss processes during AGB and PAGB phases, one cannot predict accurately the mass of a WD produced by a progenitor with a given initial mass M (hence the precise chemical stratification in the electron degenerate WD core), the WD envelope composition and mass thickness. Observational constraints are therefore absolutely necessary for calculating WD models to be compared with real stellar populations.

On general grounds, very efficient mass loss during the late stages of TP evolution (mass loss rates up to $\approx 10^{-4} M_\odot$/year) cause the departure of the model from its AGB location. This occurs when the envelope mass around the electron degenerate CO core is reduced to $\approx 0.001-0.01 M_\odot$ (e.g. [367]); about $\approx 10^{-4}-10^{-5} M_\odot$ are contained in the H-shell and overlaying H-rich layers.

The exact relationship between the final WD mass and the initial MS mass of its progenitor (IFMR) depends on the competition between surface mass loss, growth of the CO core due to shell He-burning, and the mixing episodes between envelope and inter-shell region that limit the outward (in mass) movement of the He-burning shell. Semi-empirical determinations of WD masses (M_{WD}) in local binaries with a WD component and in the closest star clusters provide a range between ~ 0.55 and $\sim 1.0 M_\odot$, with M_{WD} increasing with the initial progenitor mass M_i (e.g. [334, 368]). These estimates of the IFMR in star clusters work as follows. Spectroscopy provides T_{eff} and surface gravity estimates for the WDs. Fits to theoretical models provide the WD age t_{WD} (from T_{eff}) and mass M_{WD} (from the surface gravity). At the same time, fits to the cluster CMDs with theoretical isochrones provide the cluster age t_{cl}. The difference $t_{cl} - t_{WD}$ gives the age of the WD progenitor, and hence its mass M_i. In the case of binaries, the age of the less evolved companion provides the age of the system, and the difference with the WD age again gives the age and hence the mass of its progenitor. An impor-

Old Stellar Populations, First Edition. S. Cassisi and M. Salaris.
© 2013 WILEY-VCH Verlag GmbH & Co. KGaA. Published 2013 by WILEY-VCH Verlag GmbH & Co. KGaA.

tant issue here is the knowledge of the initial chemical composition of the cluster (or the binary companion), given the strong dependence of the derived ages on the assumed metallicity of the models. Figure 6.1 displays a composite IFMR for metallicities around solar, obtained by combining the results by [334] with three objects from [368] (the three points corresponding to the lowest M_i values). Notice the lack of data for M_i below $\sim 1.5\,M_\odot$. For lower progenitor masses, there is only an estimate of $M_{WD} = (0.53 \pm 0.01)\,M_\odot$ for $M_i = (0.80 \pm 0.05)\,M_\odot$, in the globular cluster M4, with [Fe/H] ~ -1.0 [369]. Values of the order of $0.54-0.55\,M_\odot$ are also derived from theoretical calculations of the evolution through the first few TPs of progenitors with initial masses below $\sim 1.0-1.5\,M_\odot$.

As for the mass and composition of the WD envelopes, they depend on when the off-AGB evolution starts during a TP cycle (i.e. during a pulse or in the inter-pulse phase, see Chapter 5), the interplay between mass loss during the PAGB phase, H-burning during the PAGB and early WD evolution, evolutionary timescales for the transition to WD. If a final thermal pulse occurs during the PAGB transition (denoted as late thermal pulse, or LTP, see, e.g. [317, 370]), the model is pushed back towards the AGB, the convective envelope gets deeper and surface H is mixed with the underlying inter-shell region enriched in He, C and O during the He-shell convective phase. The surface abundance of H drops to values of a small percentage because of the dilution of the thin H-dominated layers with the underlying inter-shell. At the end of the pulse, the H-shell is reignited, the model moves back along its PAGB track and starts the WD evolution with a reduced H content in the envelope, down to $\approx 10^{-6}\,M_\odot$ from an initial typical value of $\approx 10^{-4}\,M_\odot$. If a final TP instead occurs at the beginning of the WD cooling sequence (very late thermal

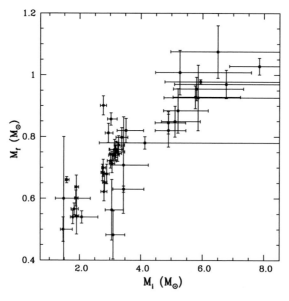

Figure 6.1 Semi-empirical IMFR (data from [334], plus three points for M_i below 2.0$M_\odot$ from [368]).

pulse, or VLTP), the H-shell is practically inactive and during the pulse the flash driven convection zone penetrates into the overlying H-rich layers, with the consequence that most of the H content is violently burned in the convective region via proton captures on ^{12}C [357, 363]. During this event the model moves back to the AGB location twice; the first time due to the onset of the violent H-burning, and the second time due to the quiescent He-burning in the shell. When the model finally reaches its WD sequence again, the total H-mass has been reduced by several orders of magnitude. Starting from an initial value $\sim 10^{-4} M_{\odot}$ for a $\sim 0.6 M_{\odot}$ remnant of a $\sim 2.5 M_{\odot}$ progenitor with initial solar metallicity, the H-mass can be reduced to $\sim 10^{-9} M_{\odot}$ through a VLTP [357]. Asteroseismology studies in a sample of local WDs have inferred a range of H-envelope thickness (M_H) between $\approx 10^{-4}$ and $\approx 10^{-10} M_{\odot}$ [371].

It is common practice to consider WD envelopes with pure H and/or pure He compositions, even though they should be retaining the metals arising from the original chemical composition of the progenitor (eventually modified by mixing episodes). Actually, the steep pressure gradients cause efficient diffusion of the heavier elements in the envelope towards the core. The end product is a chemical stratification on timescales of the order of 10^8 years or less. Figure 6.2 displays results from detailed calculations that include the effect of atomic diffusion [372]. The separation of the envelope into essentially pure H and pure He layers is evident. The transitions are, however, not sharp, and are marked by a characteristic diffusive tail that crosses the chemical discontinuity.

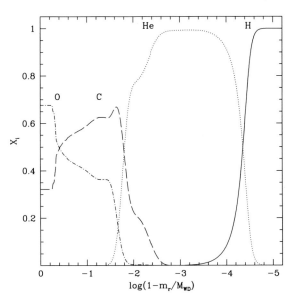

Figure 6.2 Effect of atomic diffusion on the inner chemical stratification of a WD model with mass $\sim 0.70 M_{\odot}$ before the onset of crystallization [372].

6.1
General Properties and Simple Models

Typical radii of WD stars are of the order of $\sim 0.01\,R_\odot$, with central densities of the order of $10^6\,\mathrm{g\,cm^{-3}}$. In these conditions, matter is pressure ionized and electrons are degenerate. Also, the mass thickness of the non-degenerate layers is only a small fraction of the total WD mass ($\approx 10^{-2}$ at most). To a first approximation, one can use the equation of state of a zero temperature degenerate electron gas to describe the WD mechanical structure. This leads [373] to a well-defined mass–radius relationship of the form:

$$R \propto \frac{1}{G\mu_e^{5/3}} M^{-1/3} \tag{6.1}$$

and to the existence of an upper mass limit for WDs in hydrostatic equilibrium, the Chandrasekhar mass, given by

$$M_{\mathrm{ch}} = \left(\frac{2}{\mu_e}\right)^2 1.459\,M_\odot \tag{6.2}$$

where μ_e in the previous two equations denotes the mean electron molecular weight of the electron degenerate core.

An approximate description of WD evolution comes from the virial theorem applied to an electron degenerate objects. The evolution of a bound system in hydrostatic equilibrium, where nuclear reactions are inefficient (and the pressure at the surface is negligible) has to satisfy the following relation between internal energy per unit mass E and gravitational energy Ω

$$E = -\frac{\Omega}{2} \tag{6.3}$$

The total energy E_t of the system is $E_t = E + \Omega$, and given that the hydrostatic equilibrium is perturbed by the energy radiated away into the interstellar medium from the much hotter stellar surface, the total energy E_t will decrease, thus providing the following relation

$$L \equiv -\frac{dE_t}{dt} = -\frac{1}{2}\frac{d\Omega}{dt} \tag{6.4}$$

In summary, Eqs. (6.3) and (6.4) show the well-known feature that loss of energy from the surface causes the contraction of the system; half of the gain in gravitational energy goes into internal energy, the other half is radiated away.

In the case of an electron degenerate system, it can be shown (see, e.g. [20]) that half of the gravitational energy that is not radiated away increases the Fermi energy of the electrons, whereas the thermal energy of the non-degenerate ions, proportional to T, decreases. WD evolution can therefore be interpreted as a cooling process: the energy lost by radiation is nearly equal to the rate of decrease of the thermal energy of the ions. Therefore, as a WD evolves, its radius decreases,

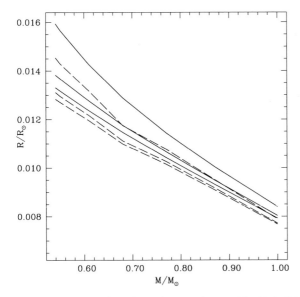

Figure 6.3 Mass–radius relationship (in solar units) for H- (solid lines) and He-atmosphere (dashed lines) WD models, at T_{eff} equal to (from top to bottom) 30 000, 10 000 and 5000 K, respectively.

the temperature decreases and E_F increases. The contraction is, however, hard to detect, and WDs can be assumed to evolve at approximately constant radius, for the equilibrium configuration of these objects is at small radii. Thus, $|\Delta\Omega| = |\Delta R|/R^2$ is comparable to L (as prescribed by the virial theorem) for small values of ΔR (see Figure 6.3).

Assuming the CO core is made of a zero-temperature degenerate electron gas plus an isothermal perfect gas of ions ($c_V = (3/2)K_B$ per ion), due to the very efficient electrons conduction that erases any temperature gradient, and that the envelope (with mass M_{env} negligible compared to the total mass M_{WD}) is well-described by a perfect gas of ions and electrons in radiative equilibrium, with Rosseland opacity of the form $k = k_0\rho T^{-3.5}$, one obtains [374]

$$L \propto M_{\text{WD}} T_c^{3.5} \tag{6.5}$$

and

$$T_c = \frac{1}{4.25} \frac{\mu G m_H}{K} \frac{M_{\text{WD}}}{R} \left(\frac{R}{r_b} - 1 \right) \tag{6.6}$$

where R is the total radius and r_b is the radius of the degenerate core. For cool WDs ($T_c \approx 10^6$ K), with a typical mass $M_{\text{WD}} = 0.6 M_\odot$ and $R = 0.01 R_\odot$, the radius of the envelope is of the order of 10 km.

Finally, the resulting cooling law (Mestel law) is of the form

$$t_{\text{cool}} \approx \frac{4.5 \times 10^7}{\mu_i} \left(\frac{L M_\odot}{M_{\text{WD}} L_\odot} \right)^{-\frac{5}{7}} \text{years} \tag{6.7}$$

where μ_i is the mean molecular weight of the ions in the CO core and it is assumed that at the actual luminosity L, the central temperature is much lower than at the beginning of the cooling. From Eq. (6.7), one derives that the cooling times of a WD of mass M_{WD} decrease with increasing molecular weight in the core, for an increase of μ_i decreases the number of ions in the model, and hence the reservoir of thermal energy. Also, for a given core chemical composition, more massive WDs have longer cooling timescales because of a larger thermal energy reservoir. Overall cooling timescales according to Eq. (6.7) are long, even in comparison with the MS phase. For a $0.6\,M_\odot$ WD with carbon core and a luminosity $\log(L/L_\odot) = -4.5$ (corresponding approximately to the luminosity of the faintest WDs observed in the solar neighbourhood), the Mestel law provides ages of the order of 10^{10} years. Given that most stars are or will become WDs, the existence of a well-defined relationship between cooling time and luminosity, and the long cooling timescales predicted by the Mestel law, WDs are very attractive candidates to unveil the history of star formation in the Galaxy. However, before using WDs as reliable probes of the early evolution of stellar populations, one needs to go beyond the simplifications of the Mestel formalism.

6.2
Detailed Models

Detailed calculations of WD evolution by means of stellar evolution codes (see, e.g. [372, 375–379]) have to go beyond the Mestel simplified analytical treatment by including the following additional processes.

6.2.1
Neutrino Emission, Envelope Contraction and Residual H-burning

The contribution of the neutrino energy losses plays an important role during the early WD evolution. At high luminosities ($\log(L/L_\odot) > -1.0$) and high central temperatures, energy is lost from the WD not only as radiation from the surface, but also through neutrino emission from the degenerate core. Plasma neutrino emission, in particular, is very efficient during the hot WD phase, and this neutrino energy losses tend to accelerate the rate of cooling compared to the predictions of the Mestel law. Also, due to the higher efficiency of neutrino losses in the densest central regions of the core, a small temperature inversion appears, the core temperature profile displaying a local maximum off-centre. When the luminosity drops below $\log(L/L_\odot) \sim -1.0$, neutrino losses become negligible, the cooling slows down, and the core temperature profiles become roughly isothermal. The WD loses its memory of the neutrino cooling phase, its evolution and structure being now determined exclusively by the properties of its ions and degenerate electrons.

Given that the electrons in the external WD layers are non-degenerate, the envelope tends to contract and release energy during the cooling according to the virial theorem. This small energy contribution is negligible, but for cores cooler than their Debye temperature for which the ionic specific heat drops fast with decreasing temperature (see below).

If the H-envelope is thicker than $\sim 10^{-4} M_\odot$ (the exact value depend on the WD mass), H-burning reactions at the base of H-layers are efficient. They contribute to the energy budget, and also decrease the mass of the H-envelope. Depending on the thickness of the envelope, this burning can also be sustained at faint luminosities $(\log(L/L_\odot) \sim -3.0$ and lower) where it could be still giving non-negligible contribution to the energy content of the WD.

6.2.2
EOS for the Ions

A major difference between a detailed WD evolution and the Mestel law (apart from corrections to the degenerate electron EOS accounting for the fact that T is not zero) stems from the treatment of the EOS for the ions. Mestel simplified treatment assumes that ions are an ideal gas with $c_V = (3/2) K_B$ per particle, but this assumption breaks down when the core cools down. Due to the steady decrease of the temperature, the ions in the core tend to move less freely because Coulomb interactions play an increasingly major role. During the WD evolution, the core cools down with the density changing only slightly, and hence Coulomb interactions get increasingly more effective. Starting from a perfect gas, the ions tend to progressively behave like a liquid, and then, with still decreasing temperatures, they form a periodic lattice structure that minimizes their total energy. This latter phenomenon is called crystallization, that is, a transition from liquid to solid phase.

Let us consider the simplest (albeit unrealistic) case of a WD core made of a single chemical element, that is, the case of a one-component plasma (OCP). A description of the most general case of multi-component plasmas can be found in [380]. We assume that the system is in equilibrium and the evolution at constant radius, hence constant volume. In these conditions, the equilibrium configuration is determined by the state of lowest Helmholtz free energy (F). The OCP free energy F of the liquid and solid phase can be described as a function of only the temperature T, the number of ions N and the Coulomb coupling parameter $\Gamma \equiv (Ze)^2/(a\, K_B\, T) = Z^{5/3}\Gamma_e$. We denote as Ze the ion charge, a is the mean ion separation, and K_B is the Boltzmann constant; $\Gamma_e \equiv e^2/(a_e\, K_B\, T)$ is the electron coupling parameter, where $a_e = [3/(4\pi n_e)]^{1/3}$ is the mean electron spacing and $n_e = Z N/V$ is the electron density (V being the volume of the system).

The ideal gas contribution to the OCP free energy (F_{ideal}) is given by

$$F_{ideal} = (N K_B T)\left[3\ln\Gamma + \frac{3}{2}\ln\left(\frac{K_B T h^2}{2\pi^2 m_i Z^4 e^4}\right) - 1 - \ln\left(\frac{4}{3\sqrt{\pi}}\right)\right] \quad (6.8)$$

where $m_i = A m_p$ is the mass of the ion.

The OCP free energy of the liquid phase F_l^{OCP} is well-approximated for $\Gamma \in [1, 200]$ by

$$F_l^{OCP} = (N K_B T)\left[-0.899\,172\Gamma + 1.8645\Gamma^{0.32301} - 0.2748\ln(\Gamma) - 1.4019\right] \quad (6.9)$$

whilst for the solid phase F_s^{OCP} is well-approximated for $\Gamma \in [160, 300]$ by

$$F_s^{OCP} = (N K_B T) \left[-0.895\,929\Gamma + 1.5\ln(\Gamma) - 1.1703 - \frac{10.84}{\Gamma} \right] \qquad (6.10)$$

When $\delta F^{OCP} \equiv (F_l - F_s)^{OCP} < 0$, the OCP is in the liquid state, whilst when $\delta F^{OCP} > 0$, it is in the solid state. In equilibrium, the system is in the state of lowest free energy. The free energy difference is therefore equal to

$$\delta F^{OCP}(\Gamma) \equiv (N K_B T)$$
$$\times \left[-0.003\,243\Gamma + 1.8645\,\Gamma^{0.32301} - 1.7748\ln(\Gamma) - 0.2316 + \frac{10.84}{\Gamma} \right]$$
$$(6.11)$$

When $\delta F^{OCP} = 0$, there is a phase transition between the liquid and solid state that occurs at

$$\Gamma_{cryst} \sim 178.6$$

A variation of c_V with Γ (hence the core temperature) is the first modification of the Mestel cooling law due to a detailed EOS for the ions. Before the transition to liquid, the specific heat per ion is close to $c_V = (3/2)\,K_B$. After this transition, a lattice progressively forms and c_V increases, reaching a maximum value $c_V \sim 3\,K_B$ at about crystallization due to the extra degrees of freedom associated with lattice vibration. At increasingly lower temperatures, fewer modes of the lattice are excited. As a consequence, c_V decreases according to the Debye law once the temperature T of the core falls below the Debye temperature ($\theta_D = 4 \times 10^3\,\rho^{1/2}$), resulting in a fast cooling phase ($c_V \propto T^3$) once $\theta_D/T > 15$.

An additional effect is the release of latent heat upon crystallization. When the WD cools down, since the density is higher in the centre than at the boundary of the core and T is almost uniform, this crystallization front will move progressively from the centre towards the core boundary. At crystallization, an amount of latent heat q equal to the difference in entropy between the two phases is released. Typically, $q \sim K_B T$ per crystallized ion. As a consequence, during the crystallization of the core, extra energy is released, first at the centre and then at progressively more external layers. This latent heat slows down the cooling with respect to the Mestel law, for it is an extra energy contribution to the energy budget. Less massive WDs start crystallizing at lower luminosities, and hence lower core temperatures because they are less dense and attain Γ_{cryst} at lower temperatures. The earlier the crystallization, the shorter the delay Δt of the cooling (with respect to the Mestel law) caused by this extra energy input. This is easy to see, for $\Delta t \sim \Delta E/L$, where ΔE is the energy injected during the crystallization and L the WD luminosity; earlier crystallization means higher L and hence shorter Δt for a given ΔE.

Let us now consider a more realistic situation whereby the WD core is made of two chemical species, carbon and oxygen, all other elements being negligible. The free energy of this two-component plasma (TCP) can be described as a function of N, T, the Coulomb coupling parameter $\Gamma_i = Z_i^{5/3} \Gamma_e$ ($i = 1, 2$, with the label 1 denoting the ion with the smallest charge) and fractional composition $x_i = N_i/N$ of either species of ion. For a TCP, the total number of ions is $N = N_1 + N_2$, and the total electron density is given by $n_e = (Z_1 N_1 + Z_2 N_2)/V$. We label the composition of the TCP by x_1 and the Coulomb coupling parameter by Γ_1, for one can express x_2 and Γ_2 as functions of these values: $x_2 = 1 - x_1$ and $\Gamma_2 = (Z_2/Z_1)^{5/3} \Gamma_1$.

The TCP free energy of the liquid phase is well-approximated by

$$\frac{F_1^{TCP}(\Gamma_1, x_1)}{N K_B T} = \sum_{i=1}^{2} x_i \left[\frac{F_1^{OCP}(\Gamma_i)}{N K_B T} + \ln\left(x_i \frac{Z_i}{\langle Z \rangle} \right) \right] \tag{6.12}$$

where $\langle Z \rangle = \sum_{i=1}^{2} x_i Z_i$ is the average ion charge. The free energy for the solid phase is well-approximated by

$$\frac{F_s^{TCP}(\Gamma_1, x_1)}{N K_B T} = \sum_{i=1}^{2} x_i \left[\frac{F_s^{OCP}(\Gamma_i)}{N K_B T} + \ln\left(x_i \frac{Z_i}{\langle Z \rangle} \right) \right] + \Delta f_s(\Gamma_1, x_1) \tag{6.13}$$

For charge ratios $R_Z = Z_2/Z_1$ in the range $R_Z \in [1 : 5]$, the term Δf_s is equal (to a very good approximation) to

$$\Delta f_s(\Gamma_1, x_1) = \Gamma_1 x_1 x_2 \Delta g(x_2, R_Z) \tag{6.14}$$

where

$$\Delta g(x, R_Z) = \frac{C(R_Z)}{1 + \frac{27(R_Z - 1)}{1 + 0.1(R_Z - 1)} \sqrt{x} \left(\sqrt{x} - 0.3 \right) \left(\sqrt{x} - 0.7 \right) \left(\sqrt{x} - 1 \right)}$$

and

$$C(R_Z) = \frac{0.05 (R_Z - 1)^2}{[1 + 0.64 (R_Z - 1)] \left[1 + 0.5 (R_Z - 1)^2 \right]} \tag{6.15}$$

In the case of a carbon-oxygen mixture with a generic value $\Gamma_1 = \Gamma_1'$ (for this specific mixture, the label 1 denotes carbon), there are typically two minima in the free energy–x_1 diagram. One is associated to the minimum for the liquid phase F_1, the other one to the minimum for the solid phase F_s, corresponding to the compositions a_1 and b_1, respectively (and also $a_2 = 1 - a_1$ and $b_2 = 1 - b_1$). Therefore, when $\Gamma_1 = \Gamma_1'$, only the compositions a_1 and b_1 are stable. One can then repeat this procedure by varying the value of Γ_1, and the resulting sequence of stable a_1 and b_1 abundances provides the so-called phase diagram of the carbon-oxygen TCP. An example of carbon–oxygen phase diagram is displayed in Figure 6.4 [381]. The horizontal axis displays the carbon abundance X_C (corresponding to x_1 in our notation of a TCP), while the vertical axis displays the temperature T in units of the crystallization temperature of a pure carbon mixture T_C. There is obviously a one-to-one correspondence between T/T_C and Coulomb coupling parameter for the pure carbon mixture.

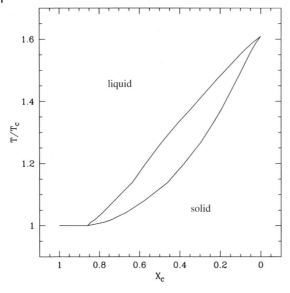

Figure 6.4 Phase diagram [381] for a CO binary mixture. T_C denotes the crystallization temperature of a pure carbon mixture, and X_C is the carbon mass fraction.

In addition to latent heat release, crystallization of a TCP displays an additional, important effect, the so-called phase separation (see, e.g. [382], [383]). In fact, the phase diagram tells us how the abundances of the two elements have to be changed during the phase transition at a given point within the WD. As an example, we can assume that $X_C = 0.50$ uniformly throughout the core. When the mixture starts to crystallize at the centre, to determine the chemical composition in the solid phase, one needs to draw a vertical line with horizontal coordinate equal to 0.50 that runs through the region belonging to the liquid phase until it intersects the upper line describing the phase diagram. The vertical coordinate of the intersection point gives the crystallization temperature of the WD centre. From this point, one has to draw a horizontal line that will intersect the lower segment of the phase diagram in correspondence of a carbon abundance $X_C \sim 0.30$. This value corresponds to the equilibrium abundance of carbon in the solid phase (the corresponding oxygen abundance will be $X_O = 1 - X_C$).

Given that X_C in the now crystallized centre is lower than the initial value, conservation of mass dictates that the carbon abundance in the liquid phase at the crystallization boundary is increased with respect to the original value. This means that right above the crystallized boundary the molecular weight is lower than in the overlying layers still in the liquid phase (where the ratio X_O/X_C is higher). An increase of molecular weight with increasing distance from the centre causes an instability to develop. The resulting mixing rehomogenizes the liquid phase on very short timescales, overall enhancing the average X_C value. This mixing region extends outwards in mass as long as the new rehomogenized average X_C is higher than the abundance in the next, unperturbed layer.

If we now suppose that the new value of X_C at the boundary of the solid core is equal to 0.55 when this layer crystallizes (at a lower temperature than the core because of lower density), the abundances in the solid phase can be derived in the same way as before, and it is equal to $X_C = 0.35$. This implies that right outside the centre, the abundance of X_2 is lower than at the centre. The instability in the liquid phase again ensues (for the same reasons explained before) and the cycle is repeated (the mixing in the liquid phase eventually stopping when the carbon abundance of the newly crystallized layer becomes equal or lower than the overlying layers still in the liquid phase) until the whole degenerate core is crystallized. The final profile of X_C and X_2 after crystallization is completed is no longer homogeneous; X_O will display central values higher than in the liquid phase, and decrease from the centre outwards, while the opposite is true for X_C. Figure 6.5 shows two examples of how the WD carbon profile in the liquid phase is modified at the end of the crystallization according to the phase diagram of Figure 6.4. This modification of the core chemical profile provides an additional contribution to the energy budget through the term $(dU/d\mu)_{T,V}(d\mu/dt)$ in the ϵ_g coefficient of the energy generation equation. An initial flat $X_C = X_O = 0.50$ abundance stratification maximizes the phase separation contribution to the energy budget because it maximizes the displacement of oxygen towards the central regions of the core (compare the upper and lower panels of Figure 6.4).

Within the carbon oxygen core, there are also small amounts of other metals, that is, so-called minor species with mass fractions of at most the order of the initial progenitor metallicity, that have been either processed through the previous

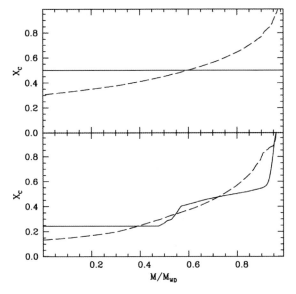

Figure 6.5 Effect of the CO mixture phase diagram [381] on the post-crystallization carbon profile (dashed lines) for two different assumptions about the chemical stratification (in mass fraction) in the liquid phase (solid lines).

burning phases (i.e. neon), or are unchanged since the formation of the progenitor (i.e. iron). Given the complexity of the calculation of multi-component phase diagrams, ternary CONe or COFe mixtures are often assumed to behave as an *effective binary* mixtures composed of neon (or iron) plus an element of average charge $\langle Z \rangle = 7$ determined from the C and O abundances. Neon is particularly important, given that ^{22}Ne is the most abundant species after carbon and oxygen; it is produced during the progenitor He-burning phase through the reactions ^{14}N$(\alpha, \gamma)^{18}$F$(\beta^+)^{18}$O$(\alpha, \gamma)^{22}$Ne. This results in a ^{22}Ne mass fraction $X_{22\,\text{Ne}} \approx X_{\text{CNO}} \sim Z$, where Z is the initial metallicity of the progenitor. A detailed phase diagram for a three component mixture CONe [384] shows that the effect of Ne on the final WD cooling times is negligible when the separation of carbon and oxygen is accounted for.

6.2.3
Neon Diffusion in the Core

The slow diffusion of minor species in the core can, in principle, also make a contribution to the white dwarf luminosity, again through the $(\partial U / \partial \mu)_{T,V} (\partial \mu / \partial t)$ term in ϵ_{g}. As discussed in [385], the diffusion of ^{22}Ne can provide a non-negligible energetic contribution in WDs originated from metal-rich progenitors (metallicity around or above solar). In fact, ^{22}Ne has two extra neutrons compared to the $A/Z = 2$ ratio of the background ions, and this results in an unbalance between gravitational and electric fields that leads to the diffusion of neon towards the centre in the liquid interior of WDs. Diffusion is assumed to be no longer efficient in the crystallized layers due to the abrupt increase of viscosity expected in the solid phase. The calculations by [386] employ a diffusion coefficient of this form (following the treatment adopted by [385] for the self-diffusion coefficient in one component plasmas)

$$D_{22\,\text{Ne}} = 7.3 \times 10^{-7} \frac{T}{\rho^{\frac{1}{2}} \langle Z \rangle \Gamma^{\frac{1}{3}}} \; \text{cm}^2/\text{s} \tag{6.16}$$

where $\langle Z \rangle$ is the average atomic number of the mixture.

A recent reevaluation of $D_{22\,\text{Ne}}$ for a CONe mixture can be found in [387], valid for $\Gamma > 5$:

$$D_{22\,\text{Ne}} \sim D_0 \, 0.53 \left(\frac{\langle Z \rangle}{Z_{22\,\text{Ne}}} \right)^{\frac{2}{3}} (1 + 0.22\Gamma) \exp\left(-0.135 \Gamma^{0.62} \right) \tag{6.17}$$

where

$$D_0 = \frac{3\omega_p a^2}{\Gamma^{\frac{4}{3}}}$$

$$\omega_p = \left(\frac{4\pi e^2 \langle Z \rangle^2 n}{\bar{M}} \right)^{\frac{1}{2}}$$

with n denoting the ion density and $\bar{M}$ the average mass of the ions.

6.2.4
Envelope Convection

During the WD cooling, owing to the decrease of temperature, the element in the superficial layers (H and/or He) recombines. This increases the local radiative opacity and as a consequence of the increased temperature gradient, convection eventually sets in, and a surface convection zone develops that gradually moves deeper into the star.

As long as convection does not reach the electron degenerate regions, the central temperature basically only depends on the radiative opacity at the edge of the degenerate region, as in the approximation that leads to the Mestel cooling law. However, at low luminosities, the surface convection reaches regions deep in the envelope that have become electron degenerate, and the subsequent evolution becomes sensitive to the details of the atmospheric layers, for the degenerate regions are roughly isothermal and convection is essentially adiabatic. This so-called "convective coupling" modifies the relationship between the luminosity of the white dwarf and its core temperature, and hence, the cooling rate. In particular, when convective coupling first sets in, the envelope becomes significantly more transparent because of the efficient convective energy transport, and there is an excess of thermal energy that the star must release. This energy release slows the cooling down for a while. However, once this energy excess is liberated, convection speeds up the cooling process compared to models with purely radiative envelopes because of the more efficient energy transport [375, 379].

6.2.5
Boundary Conditions

Although very thin in both mass and geometrical size, the envelope layers control the rate at which energy flows from the core to the surface. After convective coupling sets in, the evolution of the central temperature, and hence the cooling times, is affected by the treatment of the atmospheric layers. In particular, when computing the detailed evolution of WD models, surface boundary conditions, that is, the pressure at a fixed optical depth $\tau > 10-100$, from non-grey model atmosphere calculations are necessary [376, 388] at low luminosities (i.e. T_{eff} below ~ 6000 K), whereas the pressure obtained from the integration of a simple Eddington $T(\tau)$ relationship, together with the hydrostatic equilibrium equation with τ as independent variable, suffices for hotter models.

6.2.6
Evolutionary Results

Here, we discuss the properties of detailed cooling models [377], with masses equal to, respectively, 0.54, 0.55, 0.61, 0.68, 0.77, 0.87 and $1.0 M_{\odot}$; the only main physical effect neglected here is neon diffusion in the liquid core. Models have been calculated considering both pure H and pure He atmospheres. The H-atmosphere models have "thick" H layers, with mass fraction $q(H) \equiv M(H)/M_{WD} = 10^{-4}$ on top of an He layer with mass fraction $q(He) = 10^{-2}$. The He-atmosphere models have been calculated with an He envelope with mass fraction $q(He) = 10^{-3.5}$. With these choices for $q(H)$ and $q(He)$, the surface convective regions do not cross the H–He interface in H-atmosphere models, or the He–CO interface in He-atmosphere models. The chemical composition of the core is a pure CO mixture. Figure 6.6 displays the initial oxygen stratifications internal to the He–H discontinuity (whose mass we define to be equal to the final WD mass M_{WD}) of solar metallicity progenitor models at the first TP, calculated including convective overshooting during the MS phase. Notice how the central oxygen abundance tends to decrease with increasing WD mass, and hence increasing progenitor mass. Also, the bump in the oxygen abundance of the inner core profile produced by semi-convection during the central He-burning phase has been smoothed out following [389] because of the onset of Rayleigh–Taylor instabilities arising from the positive molecular weight gradient in the isothermal core when moving outwards from the centre of the model. As a general property of the CO profiles, the inner part of the core with a constant abundance of oxygen is determined by the maximum extension of the central He-burning convective region. Beyond this inner region, the oxygen profile is built by the thick He-burning shell moving toward the edge of the He-core. During this phase, gravitational contraction increases the temperature and density of the shell, and given that the ratio between the $^{12}C(\alpha, \gamma)^{16}O$ rate and the 3α reaction

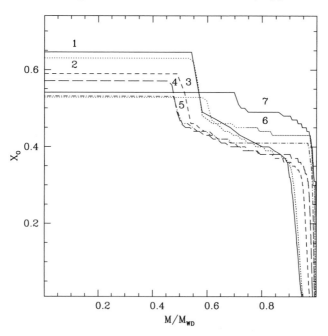

Figure 6.6 Oxygen stratification (in mass fraction) of the WD models. Labels denote, in order of increasing number, the abundance profiles for the 0.54, 0.55, 0.61, 0.68, 0.77, 0.87 and $1.0M_{\odot}$ models, respectively.

rate is lower for larger temperatures, the oxygen abundance steadily decreases in the external part of the CO core.

The initial core stratification of the models is extremely important for the rate of cooling is determined, among other factors, by the ionic specific heat which depends on the relative proportions of carbon and oxygen. The additional energy provided by the crystallization process is also affected by the CO profile, not only for the mixture-dependent contribution of phase separation, but also because, for example, a higher mean charge mixture crystallizes earlier (Γ reaches 180 at higher core temperatures, hence luminosities) and the delay induced by the latent heat release at constant WD mass is smaller.

Figure 6.7 displays the cooling times as a function of the bolometric luminosity for both H- and He-atmosphere models. The model radii for selected values of T_{eff} are shown in Figure 6.3. The fractional age difference $\Delta t/t$ between H- and He-atmosphere calculations is shown in Figure 6.8 for two selected WD masses. At luminosities above $\log(L/L_\odot) \sim -4.0$ (the exact value depending on M_{WD}), the He-atmosphere models display longer cooling times (up to 40–50%) first due to the higher opacity of their envelopes in this luminosity range and then, when their envelopes become less opaque, due to an earlier onset of crystallization and earlier latent heat release. Below $\log(L/L_\odot) \sim -4.0$, H-atmosphere WDs show progressively longer cooling times due to the much higher opacity of their envelopes,

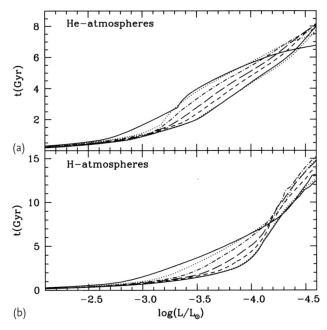

Figure 6.7 Cooling times as a function of the luminosity for sets of H- and He-atmosphere WD models. In (a), at a reference $\log(L/L_\odot) = -3.2$, from bottom to top, the different lines denote the 0.54, 0.55, 0.61, 0.68, 0.77, 0.87 and 1.0$M_\odot$ model, respectively. In (b), at a reference $\log(L/L_\odot) = -3.5$, from bottom to top, the different lines denote the 0.54, 0.55, 0.61, 0.68, 0.77, 0.87 and 1.0$M_\odot$ model, respectively.

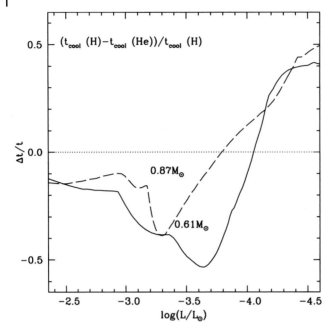

Figure 6.8 Fractional difference of the cooling times of H- and He-atmosphere WD models for two selected values of $M_{\rm WD}$.

and $\Delta t/t$ reaches values up to 50% at $\log(L/L_\odot) \sim -4.6$ to -4.7, the faintest luminosities of the He-atmosphere calculations.

The time delay $t_{\rm d}$, in the sense of increase of cooling times, caused by the inclusion of CO phase separation is displayed in Figure 6.9 for both H- and He-atmosphere models. For H-atmosphere WDs it is expected to increase with mass, have a maximum at $M_{\rm WD} = 0.77 M_\odot$, and then decrease. In the case of He-atmosphere calculations, $t_{\rm d}$ increases with mass reaching a maximum for the 0.87 and 1.0$M_\odot$ models. The values of $t_{\rm d}$ are roughly a factor of 2 larger for the H-atmosphere models because of the higher opacity of their envelopes when crystallization sets in. Overall, the delay $t_{\rm d}$ caused by phase separation is comparable to the effect of latent heat release.

Figure 6.9 also very clearly displays how the onset of crystallization is shifted to higher luminosities when $M_{\rm WD}$ increases because of the higher core densities. For a fixed $M_{\rm WD}$, He-atmosphere models crystallize earlier. It is also evident that crystallization of the core is not an instantaneous process for the advance of the crystallization front through the WD core covers roughly a 1 dex luminosity range.

More insights about the evolution of the inner structure of two selected masses (0.55 and 1.0$M_\odot$ H-atmosphere models) are provided by Figure 6.10, which displays the mass location of the H–He and He–CO interface, the crystallization front, the lower boundary of surface convection and the upper boundary of the electron degenerate region using the local Fermi temperature $T_{\rm F}$ as a diagnostic.

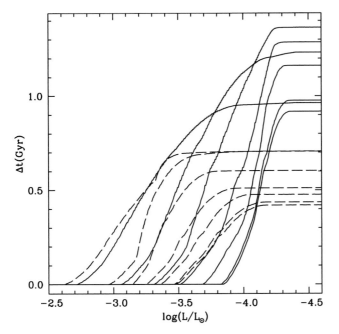

$log(L/L_\odot)$

Figure 6.9 Time delay Δt caused by phase separation, as a function of the WD luminosity. From right to left, the lines denote the 0.54, 0.55, 0.61, 0.68, 0.77, 0.87 and 1.0$M_\odot$ model, respectively. Dashed (solid) lines represent He- (H-) atmosphere WD calculations. The total time delay Δt corresponds to the final, constant value of Δt, after crystallization is completed.

In both objects, electron degeneracy advances slowly towards more external layers with decreasing luminosity. Convective coupling sets in at $\log(L/L_\odot) \sim -4.2$ for the 1.0$M_\odot$ model (during the final stages of crystallization) and $\log(L/L_\odot) \sim -3.8$ for the 0.55$M_\odot$ one (at the start of core crystallization). Notice that the lower boundary of surface convection reaches deeper layers in the less massive object. A smaller $q(H) = 10^{-6}$ for the 0.55$M_\odot$ model would have induced mixing between the H- and He-envelope for $\log(L/L_\odot) \leq -4.0$, with a change of surface composition. We will come back to this point in the next section. Figure 6.11 displays a similar plot but for a 0.55$M_\odot$ model with He atmosphere. Here, due to the higher densities, there is an earlier onset of convective coupling, at $\log(L/L_\odot) \leq -3.0$, before crystallization commences. Notice that a $q(He)$ smaller by one order of magnitude would induce mixing between the convective envelope and the external layers of the liquid CO core.

6.2.7
Theoretical Uncertainties

As discussed before, the initial chemical stratification of the models plays a crucial part in determining the predicted cooling times. Figure 6.12 displays several initial

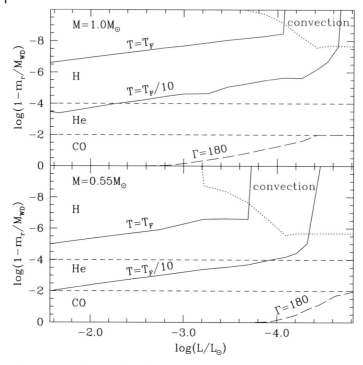

Figure 6.10 Evolution with surface luminosity of the inner mass boundary of surface convection, and the upper boundary of electron degenerate regions (denoted by lines corresponding to the local T_F and $T_F/10$) for two selected H-atmosphere models. The H–He and He–CO chemical transitions, and the upper boundary of the crystallization front are also marked.

oxygen profiles for a $0.61\,M_\odot$ WD model, determined from the progenitor evolution under several different assumptions, all taken after the rehomogenization by Rayleigh–Taylor instability. The two dashed lines correspond to progenitor models at the first thermal pulse computed with (the choice for the reference models discussed above, that is, the profile with a slightly higher central oxygen abundance) and without core overshooting on the MS. The two stratifications are roughly identical, despite the fact that in the case of MS overshooting the progenitor mass is equal to $3.0\,M_\odot$, whereas the profile without overshooting comes from a $3.5\,M_\odot$ progenitor. This can be explained by the fact that the core stratification at the start of the thermal pulse phase is determined by the value of the He-core mass at the onset of central He-burning, that is approximately the same in both progenitors, and not by the total progenitor mass. Profile number 3 displays the oxygen abundance profile obtained using a different IFMR from the reference case (models calculated with MS core overshooting, as for all other calculations discussed below). The progenitor metallicity is solar as in the reference case, but the progenitor mass is equal to $2\,M_\odot$ instead of $3\,M_\odot$, and the model has evolved through several thermal pulses until the mass internal to the He–H discontinuity reached $0.61\,M_\odot$. This

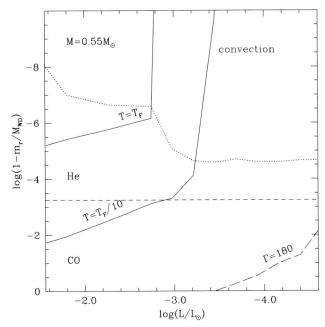

Figure 6.11 Same as in Figure 6.10, but for a He-atmosphere model.

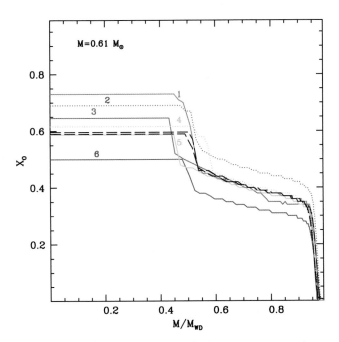

Figure 6.12 Initial oxygen stratifications for the $0.61 M_\odot$ H-atmosphere test cases discussed in the text.

choice for the progenitor is consistent with the semi-empirical IFMR displayed in Figure 6.1. This different IFMR increases the central oxygen abundance (by about 5%) compared to the reference case because of the smaller progenitor mass that produces a higher central O abundance. The extension of the inner region with the flat oxygen profile is smaller due to the reduced size of both the convective inner regions, and the whole He-core mass during the central He-burning phase.

The profiles labeled as four and five show the case of progenitors with metallicity reduced by a factor of 10 (progenitor mass reduced by $\sim 1.0 M_\odot$ compared to the reference case) and increased by a factor of 2 (mass approximately the same as for the reference case), respectively. The central values of the oxygen abundance are almost the same as the reference choice. The only major change is a different mass extension of the inner region with a flat oxygen profile.

The profiles labelled as one, two, and six display the effect of two major uncertainties. Profile number one shows the result at the first pulse for a progenitor with solar metallicity, but the breathing pulses suppressed following a different method [233] from the one employed to determine the reference profiles [238]. The main effect is to increase the central oxygen mass fraction by 15%. Profiles labelled two and six show the abundances (for solar progenitor metallicity) at the first pulse using the lower and upper limits of the reference estimate of the $^{12}C(\alpha, \gamma)^{16}O$ rate [40]. In all three cases, the progenitor mass is essentially the same as our reference case, but the final profiles are obviously affected by the different nuclear rates.

The fractional difference between the cooling ages predicted with the reference profile and values obtained with each one of the choices shown in Figure 6.12 is displayed in Figure 6.13. It is evident that selecting CO profiles from progenitor models at the first thermal pulse, and discarding metallicity effects on the progenitor evolution, does not introduce any major uncertainty in the cooling times of WD models at fixed mass. On the whole, a larger effect, within at most 7%, is caused by the treatment of convection in the late stages of the progenitor central He-burning phase, and the uncertainty on the $^{12}C(\alpha, \gamma)^{16}O$ reaction rate. The choice of the IFMR has the largest impact on the cooling timescales only during the early stages of crystallization, but the effect is within 5%.

It is also important to mention the crucial role played by the thickness of the H envelope that controls the rate at which energy is transferred from the core to the photosphere. A decrease of $q(H)$ from 10^{-4} to 10^{-5}, keeping the core chemical composition and M_{WD} unchanged, speeds up the cooling of the models, causing a maximum age difference of 7% at ages above 4 Gyr.

6.3
Colour–Magnitude Diagrams of WD Populations

Studies of the evolutionary properties of WDs in stellar populations require WD cooling models to be placed onto observational CMDs. It is therefore necessary

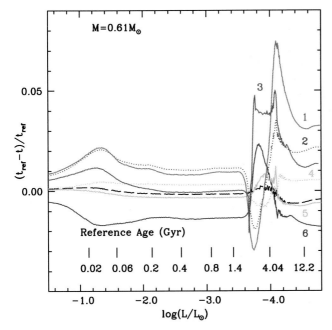

Figure 6.13 Difference in the cooling times of a $0.61 M_\odot$ H-atmosphere model, induced by the test chemical profiles displayed in Figure 6.12. The dashed line displays the effect of neglecting MS core overshooting for the progenitor evolution. Selected cooling ages for the reference stratification (see Figure 6.6) are also marked (see text for details).

to employ results of WD model atmosphere calculations that predict spectra and bolometric corrections to the appropriate photometric bands (e.g. [388, 390–392]).

Figure 6.14 shows the evolution of a $0.61 M_\odot$ WD for of a H- and a pure He-atmosphere, in different CMDs. One can notice how at faint magnitudes, the He-atmosphere models evolve steadily towards redder colours in all CMDs of Figure 6.14. The H-atmosphere models, on the other hand, display a more complex behaviour. In the M_V-$(B-V)$ CMD, the track evolves towards the red at faint magnitudes, whereas in the M_V-$(V-I)$ plane, it moves towards bluer colours at the faintest end (ages of the order of 15 Gyr, older than the age of the Universe). In the M_K-$(J-K)$ CMD, this behaviour is more prominent, and the track starts to evolve towards blue colours at an age ≈ 5–6 Gyr.

The reason for this evolution of H-atmosphere models towards blue colours, despite the fact that the model T_{eff} decreases steadily, is due to the H_2 CIA opacity discussed in Chapter 2. When T_{eff} decreases below ~ 6000 K, CIA dominates the opacity in the atmospheres and causes a depression of the flux at wavelengths above ~ 1 µm (this transition wavelength decreasing as the WD cools), and hence a shift of the model to bluer colours as discussed in Section 3.2 for VLM stars.[1]

---

1) Recently, the role played by $Ly\alpha$ quasi-molecular opacity in affecting the emergent UV and blue radiation from WD atmospheres has been addressed [393, 394].

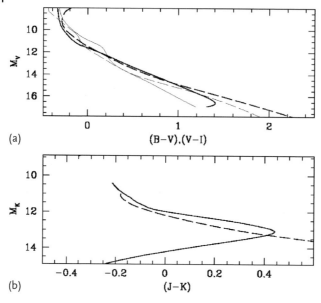

Figure 6.14 Cooling track of 0.61$M_\odot$ H- (solid lines) and He-atmosphere (dashed lines) models in three CMDs. Thin lines in (a) display the tracks in a M_V-$(B-V)$ CMD, whilst thick lines show the correspondent M_V-$(V-I)$ diagram.

It is important to notice that at bright magnitudes, in the BV and VI CMDs, the H- and He-atmosphere tracks intersect each other because of bolometric correction effects. In particular, at $M_V \sim 11$, the He-atmosphere track is bluer than the H-atmosphere counterpart in the $(B-V)$ colours, whereas it is redder in the $(V-I)$ colours (see Figure 7.21). The implication is that BVI photometric observations of WDs in this magnitude range can potentially discriminate between H- and He-atmosphere objects, without the need to resort to spectroscopy.

6.3.1
Isochrones and Luminosity Functions

Fundamental working tools for age dating stellar populations from their observed cooling sequence are WD isochrones, which are routinely employed to study the cooling sequences of WDs in star clusters. The calculation of WD isochrones requires a grid of WD models for different masses, an IFMR and evolutionary timescales of the WD progenitors plus the appropriate bolometric corrections. From a given WD isochrone, one can calculate the LF in a given passband, after assuming an initial mass function (IMF) for the WD progenitors.

Figure 6.15 displays 1, 5 and 10 Gyr isochrones (solar metallicity progenitors) in the L-T_{eff} plane, from the reference WD models discussed above, employing a linear fit to the semi-empirical IFMR by [334] (the points in Figure 6.1 with $M_i > 1.5 M_\odot$) extrapolated (when necessary) down to the smallest value of the

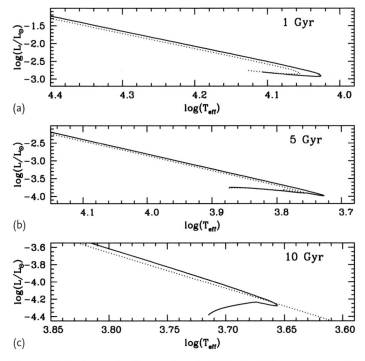

Figure 6.15 Isochrones for H-atmosphere (solid lines) and He-atmosphere (dotted lines) WD models in the HRD, for three different ages, namely, 1, 5 and 10 Gyr.

progenitor mass appropriate for the chosen isochrone age. The maximum value of the initial mass is set by the minimum stellar mass igniting carbon core burning, as derived from [148]. The progenitor evolutionary times (i.e. lifetime until the first TP) are also from [148]. The age indicator for a WD population is clearly the faint end of the isochrones. Due to the finite age of the stellar population, the more massive WDs formed from higher-mass and shorter-lived progenitors pile up at the bottom of the cooling sequence, producing the turn to the blue (i.e. a turn towards lower radii) visible at the faint end of the isochrones. Notice that at 1 and 5 Gyr the turn to the blue of the He-atmosphere isochrone has approximately the same (or slightly brighter) luminosity of the H-atmosphere counterparts. This stems from the similar (or longer) cooling times of He-atmosphere models in this luminosity range. At 10 Gyr, the He-atmosphere isochrone gets much fainter because of much faster cooling times at faint luminosities.

Another interesting feature is the general shift of the He-atmosphere isochrones towards lower radii, more pronounced when the luminosity is brighter than $\log(L/L_\odot) \sim -4.0$. There are two reasons for this. The first one is related to the fact that along a WD isochrone, the sum of the WD cooling age and the corresponding progenitor lifetime has to be equal to the isochrone age t. Above $\log(L/L_\odot) \sim -4.0$, the He-atmosphere models (at fixed M_{WD}) have longer cooling times, and hence a given luminosity along an isochrone has to be populated by a larger WD mass (smaller radius) because the lower progenitor lifetimes will compensate for the longer WD cooling times. An additional contribution to this difference stems from the fact that He-atmosphere WD models (at fixed M_{WD}) have smaller radii at any given value of T_{eff} (see Figure 6.3).

Figure 6.16 displays 5 and 10 Gyr H-atmosphere WD isochrones, and a 10 Gyr He-atmosphere isochrone, in two different CMDs. The isochrone morphology in the M_V-$(V-I)$ CMD mirrors the results in the $L-T_{eff}$ plane. The turn to the blue at the faint end of the H-atmosphere isochrones is due to the presence of increasingly massive WDs, not to the CIA. The situation is very different in the M_K-$(J-K)$ CMD. At 10 Gyr, the shape of the H-atmosphere isochrone is modulated by the onset of CIA effects (mass dependent), and the accumulation of higher mass WDs at the faint end. The turn to bluer $(J-K)$ colours that sets in at $M_K \sim 13$ is due to the CIA, for the evolving WD mass is almost constant at those magnitudes. The subsequent turn to the red and the following, almost vertical decrease of M_K, are due to a combination of the increase of WD mass and the differential efficiency of CIA. One also has to notice that at bright magnitudes, the He-atmosphere isochrone is systematically bluer in M_K-$(J-K)$, whereas H- and He-atmosphere isochrones intersect each other in M_V-$(V-I)$ because of the effect of bolometric corrections. It is important to notice that in the assumption of an IFMR that is independent of the progenitor initial metallicity, the brightness of the faint end of WD isochrones of fixed age is also almost independent of the progenitor metallicity. In fact, as

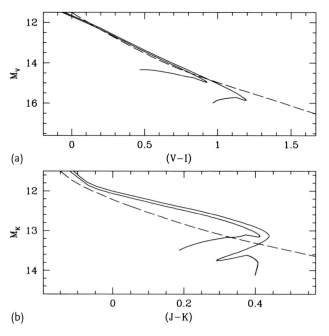

(a)

(b)

Figure 6.16 (a) M_V-$(V-I)$ and (b) M_K-$(J-K)$ CMDs for 5 and 10 Gyr H-atmospheres WD isochrones (solid lines), and a 10 Gyr He-atmosphere WD isochrone (dashed lines).

discussed before, the change of the CO stratification only induces a very minor difference in the cooling times, plus the dependence of the progenitor lifetimes on Z plays a minor role at the faint end of the isochrone, populated by increasingly massive objects, whose luminosity is determined essentially by their cooling times, not the progenitor ages.

Luminosity functions from H-atmosphere WD isochrones for a few selected ages are displayed in Figure 6.17, computed using a Salpeter mass function for the progenitors. We show LFs for the filter $F606W$ of the ACS camera on board the *HST* that has been widely used to image WDs in star clusters, and is similar to the Johnson V band. The LFs display the characteristic cut-off that corresponds to the termination of the isochrone. The sharp increase of number of objects near the cut-off magnitude is caused by the contribution of increasingly massive WDs, which pile up at the bottom end of the isochrone (along the turn to the blue). The $F606W$ magnitude of the cut-off changes with age by ~ 0.2 mag/Gyr at old ages, and ~ 0.5 mag/Gyr at intermediate ages. This, in principle, makes H-atmosphere WDs better suited for age determinations compared to the main sequence TO (the corresponding changes of $F606W$ for the TO are ~ 0.1 mag/Gyr and ~ 0.4 mag/Gyr, respectively), once the cluster distance modulus is fixed. The magnitude of the LF cut-off is weakly affected by variations of the IFMR of the order of $\Delta M_i \sim \pm 0.5 M_\odot$ for a fixed M_{WD}. The detailed shape of the LF is modified, but the brightness of the cut-off is unchanged. The reason is that the variation of progenitor lifetimes does not affect the brightness of the faint end of WD isochrones, and the WD cooling times at fixed M_{WD} are only mildly affected because of the change of the core

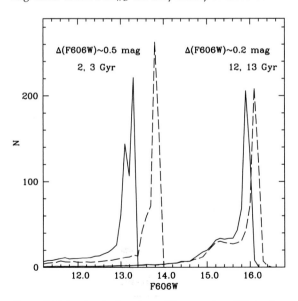

Figure 6.17 Luminosity functions of H-atmosphere WD isochrones in the $F606W$ ACS filter, for the labelled ages.

stratification, especially for ages larger than 5–6 Gyr, as discussed above (see also Figure 6.13).

6.4
Spectroscopic Classification of WDs

From the observational point of view, WDs have been classified into two broad families, according to the main chemical element identified spectroscopically in their photospheres. Spectroscopic observations reveal that the surface composition of more than 80% of the local WD population is H-rich [395]; these objects are named WDs of type DA. The other family is made of WDs with He-rich atmospheres, which are named of type non-DA. The non-DA objects are, in turn, classified as DB, DO, DZ and DQ. The precise spectroscopic definition of these WD types [396] is given in Table 6.1. We did not include the DC type, defined originally as WDs that did not show absorption lines, for UV observations have revealed the presence of carbon features, and hence these objects can be reclassified as DQ (see, e.g. [397]).

Observations also reveal that the ratio of DA to non-DA objects changes with T_{eff}, suggesting a complex synergy of several physical effects as the reason for the existence of different WD types. Figure 6.18 graphically shows the currently most accepted scenario to explain WD spectra for the local population.[2] The starting assumption of this scenario is that there are two separate channels for the formation of DA and non-DA objects.

DA objects are expected to come from progenitors that experienced their last TP on the AGB, entering the cooling sequence with a H-dominated envelope of mass thickness $\approx 10^{-4} M_\odot$ on average (see Section 5.8). Then, on very short timescales, atomic diffusion induces the formation of pure-H photospheric layers. Type DA objects with thinner layers can be produced by progenitors undergoing a LTP. When the resulting WD enters its cooling sequence, the residual H floats to the surface due to atomic diffusion producing a DA object. If the mass of the H-layer is above a threshold that depends on the WD mass, for example, $q(H) \approx 10^{-5}$ for

Table 6.1 WD spectral classification

WD type	Spectral features
DA	Only Balmer lines; no He I or metals
DB	He I lines; no hydrogen or metals
DO	He II strong; He I or hydrogen also present
DZ	Metal lines; no hydrogen or helium lines
DQ	Atomic or molecular carbon features

2) It is important to mention that the few spectroscopic observations of bright WDs in Galactic globular clusters reveal a remarkable lack of objects with He-dominated atmospheres [398].

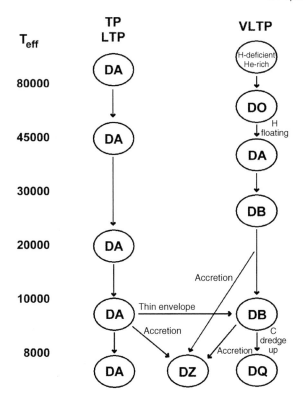

Figure 6.18 The commonly accepted scenario for the evolution of WD spectral types as a function of T_{eff} (in K, see text for details).

a $0.55 M_\odot$, and $\approx 10^{-8}$ for a $1.0 M_\odot$, surface convection never reaches the underlying He-layers, and the objects cools down as type DA, irrespective of its T_{eff}. If the envelope mass is below this threshold, cool DA objects may transform into DB due to convective mixing with the much more massive He-layers.

Non-DA objects are instead expected to be the progeny of stars experiencing a VLTP. After the VLTP, the chemical composition of the envelope of the newly born hot WD star is dominated by He, and its spectral type will appear as DO. Due to the effect of atomic diffusion, on short timescales the small amount of H left in the envelope will float to the surface, transforming the object into a DA WD. During further cooling, a convection zone mixes the external H-dominated layers with the underlying much more massive He-buffer, and hence the transformation to DB type. If the convection zone in the He-layers reaches the tail (see Figure 6.2) of the underlying carbon enriched layers (when the thickness $q(He)$ is below $\approx 10^{-3.5}-10^{-4}$), the WD will become of type DQ at low T_{eff}. For both channels, accretion of metals from the interstellar medium or circum-stellar material may transform DA or DB objects into WDs of type DZ.

The recent discovery of "hot" DQ objects, all with T_{eff} around 20 000 K, [399] may fit within this scenario if one speculates that as a product of a very violent VLTP, the surface C and O abundances are even larger than expected [399], at the expenses of He (the surface H abundance is expected to have dropped virtually to zero). Once the objects settle on the cooling sequence, diffusion creates on short timescales a DO and then a DB WD, until a recombination-driven convection zone develops in the carbon-enriched layers and dilutes from below the very thin He-layers. The end-product would be a carbon-dominated atmosphere of DQ type, for the mass in the carbon convection zone is orders of magnitude larger than the mass in the He layer, and helium would become spectroscopically invisible [400].

Grids of fully evolutionary WD calculations that take into account these changes of spectral types (i.e. envelope chemical composition) observed in local WDs are not yet available. Only very recently, a semi-analytical treatment has been developed [401] to calculate cooling curves and chemical atmosphere evolution curves as a function of M_{WD} and $q(H)$ fraction.

6.5
Low Mass CO- and He-core WDs

Let's consider stars with an initial mass above $0.50 M_\odot$ at the stage of He-ignition at the tip of the RGB. Their He-core mass stays equal to about $0.48 M_\odot$ as long as the initial mass is below $\sim 1.7 M_\odot$ (the precise values depend on the initial chemical composition). Beyond this threshold, the higher the initial mass, the weaker the electron-degeneracy of the core, and as a consequence, the He-core mass at the 3α ignition decreases (see Section 3.7.1). Eventually, when the initial mass reaches $\sim 2.3 M_\odot$, with a He-core mass of $\sim 0.3 M_\odot$, He-burning ignites almost quiescently through a weak flash. For higher masses, the He-core along the RGB is no longer degenerate, and the core mass at He-ignition increases with the initial mass. As discussed in detail by [402], an appropriate tuning of the mass loss episodes for models with initial masses around $2.0–2.3 M_\odot$ can produce He-burning objects with masses between 0.3 and $0.5 M_\odot$ and vanishing envelopes. At central He-exhaustion, all these objects will stay on the early-AGB phase every briefly before becoming WDs with essentially the same mass they had at the He-ignition.

Another class of WDs with masses between ~ 0.2 and $\sim 0.5 M_\odot$ can be produced again through strong mass loss episodes, for example in binary systems. If stars with mass below $\sim 2.3 M_\odot$ lose their envelope along the RGB before He-ignition, they will settle on the WD cooling sequence as a He-degenerate core, surrounded by a very thin (in mass) H-dominated envelope. The models by [403] predict envelope masses between $\approx 10^{-2}$ and $\approx 10^{-4} M_\odot$, decreasing with increasing core mass, as these values are of course affected by the modelling of the mass loss event. Atomic diffusion along the cooling sequence then creates a pure H envelope on timescales of order $10^7–10^8$ years, on top of a He-dominated (plus a small fraction of metals) layer surrounding the electron degenerate core. The inner tail of the hydrogen abundance profile (the transition between H- and He-layer is not sharp because of the active atomic diffusion) digs into deeper layers as cooling proceeds. During the early stage of cooling, He-core WDs experience so-called

Table 6.2 Evolutionary times of selected He-core WD cooling sequences [403]. The zero point of the ages corresponds to the first arrival onto the cooling sequence.

$M(M_\odot)$	$\log(L/L_\odot)$	t (10^6 yr)
0.1604	1.586	0.0001
	1.000	0.003
	0.000	0.108
	−1.000	1082.1
	−1.492	4497.6
	−1.864	10 000.8
	−2.096	15 003.7
0.3515	1.760	0.00
	3.440	13.51
	2.542	13.53
	2.006	13.54
	1.001	13.56
	0.000	13.88
	−1.005	78.29
	−2.008	322.6
	−2.999	1254.0
	−3.500	2408.3
	−4.019	6578.2
0.4481	2.291	4.40
	3.532	6.62
	2.771	6.63
	2.022	6.64
	0.995	6.66
	0.006	19.8
	−1.004	64.6
	−2.004	345.2
	−2.999	1387.8
	−3.500	2751.7
	−4.045	8360.9

"hydrogen flashes" due to the contraction of the H-burning shell region throughout the transition to the cooling sequence, the continuous increase of the temperature of the core that is still moderately degenerate, and the largely reduced mass size of the envelope [247, 404, 405]. The consequent thermal runaways, favoured by the inner tail of the hydrogen profile that reaches hotter regions, reduce the mass of the H-envelope (by factors typically of order 3–5), and prevent stable nuclear burning from being a sizeable energy source during the fainter cooling stages. In the majority of cases, these flashes briefly move the track back to the RGB, where mass loss episodes may eventually reduce the envelope mass even more. After a time of the order of 10^7 years, the hydrogen flashes subside. The lowest mass He-cores (below $\sim 0.17 M_\odot$) with their more massive envelopes and more electron degenerate cores do not experience these flashes, rather a quiescent H-burning [403]. At luminosities below $\log(L/L_\odot) \approx -2.0$ to -3.0 (the exact value depending on the WD mass and envelope thickness), when the contribution of nuclear reaction becomes negligible, the internal energy of the degenerate He ions in the core (that do not crystallize within an Hubble time, due to their much reduced atomic number compared to a CO mixture) controls the WD energy budget. Evolutionary times as a function of the bolometric luminosity of selected He-core WD models are reported in Table 6.2. Notice the temporary increase of luminosity during the first evolutionary stages due to the ongoing thermal runaways.

As a consequence of these two different channels for the production of low-mass WDs, both He- and CO-core degenerate objects can exist in the mass range $0.33-0.5 M_\odot$. The cooling times of these two classes of WDs are, however, very different, as can be easily guessed considering that He-cores have a larger energy reservoir in terms of ion internal energy (because of an atomic weight lower by a factor of 3–4) with respect to a CO-core with the same mass.

7
Resolved Old Stellar Populations in the Galaxy

7.1
Introduction

In the previous chapters, we have extensively discussed the structure and evolution of low-mass stars. These results can be employed to "build" a range of tools appropriate to study old stellar populations in the Universe. In this context, Galactic GCs, old field halo stars and old open clusters can be used both as test bench of low-mass stellar models and also as the first targets for stellar population analyses aimed at investigating the early evolution of the Universe and galaxies. Moreover, the application to these clusters of tools devised for studying old stellar populations will also enable us to gauge the limitations of these methods, as well as a better appraisal of their power.

Star clusters are usually considered as a perfect example of SSPs, made of stars all born with the same initial chemical composition in the same burst of star formation, and hence they all share the same age. Stars in a cluster are also all at, to a very good approximation, the same distance from us. These properties make their CMD a very close counterpart of theoretical isochrones, rigidly shifted in magnitude because of the object distance (and in colour due to the reddening). In general, GCs also host large numbers of stars, typically $\sim 10^5 - 10^6$, and as a consequence, all post-MS evolutionary phases, apart from the AGB, are reasonably well-populated, and can be used in evolutionary analyses. In the following, we describe in detail a range of methods to determine distance, age and initial abundance of some key chemical elements for old stellar populations, all fundamental parameters to investigate the early stages of the evolution of galaxies and test the standard cosmological model, with emphasis on their application to Galactic GCs. The next sections will highlight both the power and shortcomings of these techniques.

We will treat at first GCs as simple SSPs, and then we will discuss the *multipopulation phenomenon* observed in Galactic GCs – and probably common to any GC system – and its impact on the techniques presented in this chapter.

Old Stellar Populations, First Edition. S. Cassisi and M. Salaris.
© 2013 WILEY-VCH Verlag GmbH & Co. KGaA. Published 2013 by WILEY-VCH Verlag GmbH & Co. KGaA.

7.2
Old Star Cluster Distances

The determination of star cluster distances is important not only for investigating their spatial distribution (that has implications for the mechanisms of cluster and galaxy formation), but also because they can be used as calibrators of standard candles that cannot be directly calibrated from accurate parallaxes, and because distances are necessary to determine a cluster age from the absolute magnitude of its TO. Geometric parallax distances to several old clusters in the Galaxy will be available with the GAIA satellite, while eclipsing binary distances are only available for very few old clusters. In the following, we discuss a range of methods based on stellar evolution and stellar pulsation theories.

7.2.1
Main-Sequence Fitting

Even a cursory comparison of theoretical isochrones at a given metallicity (and He content) clearly shows that the brightness of the lower MS is typically unaffected by age. If one considers an upper limit to the stellar ages equal to 14 Gyr (the age of the Universe in the current standard cosmological model), magnitudes and colours of the MS at M_V larger than $\sim$ 5.0–5.5 are age independent, and are only affected by the initial chemical composition of the models. These properties are exploited by the MS-fitting method, a technique widely employed to determine star cluster distances (see, e.g. [406–409] for just a few examples). In "young" (based on the CMD morphology) clusters, brighter sections of the MS can be employed, and this allows one to apply the MS-fitting technique to reach distances out to the closest galaxies in the Local Group (typically LMC and SMC).

A representation of the method is displayed in Figure 7.1. One needs to determine observationally a deep CMD of the target population, and consider a template MS of known absolute magnitudes and dereddened colours, with the same chemical composition, usually labelled just in terms of [Fe/H], assuming a universal metal distribution and $\Delta Y/\Delta$[Fe/H] law, of the target. The difference between the absolute magnitudes of the template MS and the apparent magnitudes of the observed one (corrected for the effect of foreground reddening) provides the population distance modulus. If both template and target MS are available in several different CMDs (i.e. $V - (B-V)$, $V - (V-I)$, $V - (V-K)$, etc.) a simultaneous fit in the different magnitude-colour planes can in principle provide simultaneously $R_V = A(V)/E(B-V)$, $E(B-V)$ and $(m - M)_0$ (see, e.g. [406]).

It is obvious that to determine accurate distances, anything that may introduce systematic differences between intrinsic and observed colours and magnitudes of the target and the template MS must be accounted for. A recent analysis quantifies in detail (for the distances to a sample of three Galactic globular clusters using subdwarfs with accurate Hipparcos parallaxes) the uncertainties on this method, by taking also into account the contributions due to parallax errors and associated

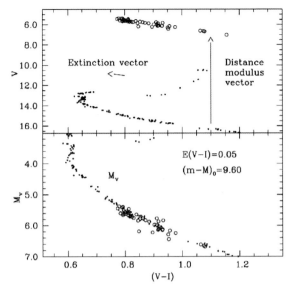

Figure 7.1 Visual representation of the MS-Fitting method applied to the Galactic open cluster M67. The template MS made of local stars with accurate parallaxes is displayed as large open circles. The cluster CMD is denoted by small filled circles.

biases [410],[1] field binary and cluster binary contamination, reddening/metallicity scale and photometric calibration. A total uncertainty on the clusters distance moduli equal to ± 0.07 mag is reported. The largest contributions arise from uncertainties in reddening/metallicity scale and photometric calibration.

In the following discussion, we will focus on the accuracy of the template MS and an additional source of potential uncertainty that has not yet been explored.

Ideally, one could use as a template the MS of theoretical isochrones calculated for the specific [Fe/H] of the target population. This choice is, however, hampered by the existing uncertainties in the T_{eff} scale of stellar models with convective en-

---

1) The issue of biases associated to parallax measurements is an example of the "classical" problem of estimating corrections to recover the true distribution of some physical quantity, from an observed distribution in the presence of observational errors. A recent analysis that contains all relevant references makes use of Monte Carlo simulations to clearly show the origin of parallax biases [411]. For any realistic choice of the stellar distribution in the solar neighbourhood, the number of stars in a given "true" parallax interval – for example distance interval – has to increase with decreasing parallax π – for example increasing the distance – for larger spatial volumes are sampled by the same parallax interval. This means that, in presence of a symmetric observational error on π, more candidate stars will be erroneously thrown into the observed ranges from larger distances ("true" parallaxes smaller than observed) than thrown out from smaller distances. It is clear that this effect (usually denoted as the Lutz–Kelker bias) is minimized when parallax errors become very small, and of course also depends on the exact spatial distribution of the sample. An additional potential bias (denoted as Malmquist bias) is introduced when stars are selected by a lower limit in the observed parallax value, for example, a cut-off is imposed at some distance limit, usually with the opposite sign of the Lutz–Kelker bias.

velopes and colour transformations. An uncertainty of only 0.02 mag in colours translates into a ~ 0.10 mag error in the derived distance modulus because in the widely employed $V - (B-V)$ and $V - (V-I)$ CMDs, the MS slope is equal to ~ 5.0–5.5 at M_V between ~ 5.0 and ~ 8.0, the typical magnitude range employed to determine distances to old populations. Theoretical isochrones that do not match observational constraints can in principle be "tuned" (if necessary) to reproduce the colours of field stars and/or star clusters with accurate metallicity estimates and parallax distances. At the moment this is feasible more rigorously only at metallicities typical of the solar neighbourhood, after the Hipparcos satellite has provided accurate parallaxes of several local field stars and the closest clusters (see, e.g. [406] for an example).

The standard method is to build an empirical MS by considering samples of field dwarfs (or subdwarfs) of known [Fe/H] with accurate parallax distances and negligible (or well-known) reddening. The main difficulty with this approach is that, given the current sample of stars with accurate parallaxes and spectroscopically determined [Fe/H], only a small number of objects will generally have the exact [Fe/H] of the target population, making it difficult to determine an appropriate template MS. To overcome this problem, one needs to shift the position of many field stars of varying [Fe/H] in reasonably narrow ranges around the target metallicity to the location they would have at the [Fe/H] of the target population. The effect of varying [Fe/H] is to change the colour and magnitude of stars of fixed mass along the unevolved MS; that is, magnitude and colours at fixed mass increase when the metal content increases. Adjustments to both magnitude and colour would therefore be necessary to match each field object to a star of equivalent mass at the metallicity of the cluster. However, the slope of the MS, at least in the M_V range between ~ 5.0 and ~ 8.0, appears to be nearly independent of [Fe/H] in both the metallicity regime of Galactic GCs and in the typical [Fe/H] range of open clusters. This implies that templates constructed using only colour shifts are a reliable choice.

Metallicity dependent colour shifts are then calculated by first determining the colour that each field dwarf (or subdwarf) would have at a fixed absolute magnitude, that is, a reference $M_V = 6$ is often used, and hence one determines $(B-V)_{M_V=6}$ or $(V-I)_{M_V=6}$, for example, using the value of the MS slope in the chosen CMD, and then calculating the variation of $(B-V)_{M_V=6}$ with [Fe/H].

The studies described in [408, 412] empirically determine

$$\Delta(B-V)_{M_V=6} = 0.154\Delta[Fe/H]$$

$$\Delta(V-I)_{M_V=6} = 0.103\Delta[Fe/H]$$

$$\Delta(V-K)_{M_V=6} = 0.190\Delta[Fe/H]$$

$$\Delta(J-K)_{M_K=4} = 0.078\Delta[Fe/H]$$

for field stars with $-0.4 < [Fe/H] < 0.3$. In the case of Galactic halo metallicities, the number of subdwarfs with accurate parallax and [Fe/H] determination, even after Hipparcos, is still small, and colour shifts need to be determined with the help of theoretical isochrones. As an example, [410] use $\Delta(colour)/\Delta[M/H]$ relationships derived from the isochrones by [413], considering $(B-V)_{M_V=6}$ and the

Strömgren colour $(b - y)_{M_V=6}$. The quantity $[M/H]$ denotes the total metallicity determined as $[M/H] = [Fe/H] + \log(0.638f + 0.362)$, where $f = 10^{[\alpha/Fe]}$ [42].

Figures 7.2 and 7.3 display theoretical relationships derived from scaled-solar (dashed lines) and α-enhanced ($[\alpha/Fe] = 0.4$, solid lines) isochrones, including the effect of α-enhancement both in stellar models and colour transformations. These colour-$[M/H]$ relations are used differentially for calculating colour shifts to apply to individual subdwarfs. As seen empirically for stars around solar metal-licities, the effect of $[Fe/H]$ (and $[M/H]$) on colours depends on the selected filter combination. The $(b-y)$ colour is better suited than $(B-V)$ because $(b-y)_{M_V=6}$ is very weakly sensitive to metallicity, and the size of the colour shifts is minimized. Also, the $(B-V)$ and $(b-y)$ colour sequences as a function of $[Fe/H]$ (or $[M/H]$) for scaled-solar and α-enhanced models are essentially parallel. Therefore, the colour shifts are unaffected by the choice of the metal mixture, at least in the displayed metallicity interval.

It is, however, crucial to notice the zero point offset between the scaled-solar and α-enhanced theoretical $(B-V)_{M_V=6}$ and $(b - y)_{M_V=6}$ vs $[M/H]$ relationships. The offset means (using again theory in a differential way) that there would be a systematic colour difference between a template MS made of α-enhanced stars, and a target scaled-solar population at the same $[M/H]$. This offset is reduced at the level of ~ 0.01 mag in both $(B-V)$ and $(b - y)$, if the comparison between target and template sequence is performed at constant $[Fe/H]$ rather than con-stant $[M/H]$. This effect should be obviously investigated in other colours. The field subdwarfs plotted in Figures 7.2 and 7.3 (employed by [410]) display various levels

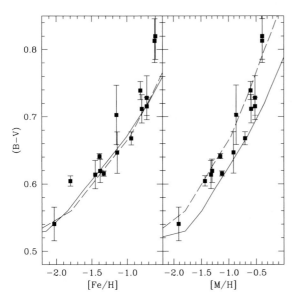

Figure 7.2 Theoretical relationship between the $(B-V)$ colour at $M_V = 6$ and $[Fe/H]$ or $[M/H]$ as derived from scaled-solar (dashed lines) and α-enhanced (solid lines) isochrones. Filled squares with errors denote the subdwarfs used in [410].

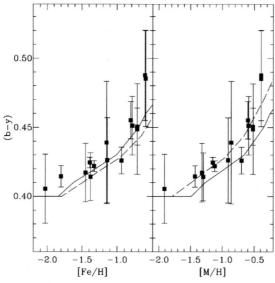

Figure 7.3 The same as for Figure 7.2, but for the $(b - y)$ colour.

of α-enhancement, with $[\alpha/Fe]$ ranging between ~ 0.10 and ~ 0.5, and are distributed between the two theoretical sequences in $(B-V)$. In $(b-y)$, the error bars are typically very large compared to the colour difference between scaled-solar and α-enhanced sequences.

An important process to consider in terms of its possible impact on MS-fitting distances is atomic diffusion. The occurrence of diffusion in the Sun has been demonstrated by helioseismic studies. It seems that neither turbulence nor other hydrodynamical mixing processes substantially reduce the efficiency of element diffusion in the Sun. Spectroscopic observations of globular cluster stars show, however, that atomic diffusion from the convective envelope is partially or totally inhibited (see, e.g. [414, 415] and references therein), but of course there are no clues about whether it is fully efficient in the star interiors.

With this caveat in mind, one can investigate the effect of fully efficient diffusion on MS-fitting distances to globular clusters [416]. The crucial point is that globular cluster metallicity scales are determined from spectroscopy on RGB stars whose surface [Fe/H] is basically unaffected by diffusion (even when fully efficient) and is roughly equal to the initial value on the ZAMS. However, the surface metallicity of cluster MS stars is affected by diffusion, and will be typically lower than RGB estimates.[2] As a consequence, when fitting the local template to the MS of a cluster with the same observed [Fe/H] (measured on the cluster RGB), one is introducing a potential source of bias in the derived distances.

---

2) Only for [Fe/H] around and below -2.0 to -2.3 radiative levitation can moderate and even reverse, at the lowest metallicities, the effect of atomic diffusion on the surface chemical abundances.

Additionally, the colour of the template subdwarfs at the observed spectroscopic [Fe/H] will depend on their (usually unknown) age because older ages cause a larger decrease of the surface [Fe/H], and hence a larger initial Fe content is required to reproduce the observed value of [Fe/H], compared to the case with no diffusion. This larger initial [Fe/H] affects the predicted subdwarf colours and, in turn, the predicted $\Delta(B-V)_{MS}/\Delta$[Fe/H] relationship. Table 7.1 reports the distances obtained in [416] for some globulars. The distances have been estimated both without (standard case) and with the inclusion of the effect of diffusion on the cluster MS and the subdwarf colours, for two realistic values of the subdwarf and cluster ages. The differences with respect to the standard case are generally within the small formal error bars associated to the fit (the error bar only takes into account the error on the fit due to the uncertainties on the subdwarf photometry). The reason is that both subdwarfs and cluster stars are all sufficiently faint (M_V between ~ 5.0 and ~ 8.0) and that the effect of atomic diffusion on the surface abundances is minimized because of their extended convective envelopes.

A final issue to be considered is the impact of sub-populations with varying degrees of CNONa anti-correlation and the possibly associated increased He abundance, as observed in individual GCs, but largely absent in the halo field subdwarf population (see Section 7.8 for an in-depth discussion). This problem has been addressed in a recent theoretical analysis [417]. The effect of CNONa anti-correlations overimposed to a standard α-enhanced mixture at fixed [Fe/H] is negligible in the typical magnitude range employed for MS-fitting distance determination for photometric filters longwards of the U-band wavelengths. An enhancement of He compared to the template MS is, however, able to shift the MS to the blue, at fixed metal content and magnitude, because of a hotter T_{eff} of the parent stars (bolometric corrections are largely unaffected), causing systematic overestimates of the derived MS-fitting distances. We can, for example, consider the case of a globular cluster with no obvious multi-modality in He content, but with a spread of initial He abundances. Depending on the size of the spread and the number of stars affected, the mean locus of the cluster lower MS will be shifted in colour compared to a template sequence built from stars with a uniform initial He abundance, and the resulting MS-fitting distance will be overestimated.

Table 7.1 MS-fitting distance moduli $(m - M)_V$ of selected clusters obtained without (standard) and including the effect of fully efficient atomic diffusion, and two different ages for both clusters and subdwarfs.

| Cluster | Standard | $t_{cl} = 8$ Gyr | | $t_{cl} = 12$ Gyr | |
		$t_{sbdw} = 8$ Gyr	$t_{sbdw} = 12$ Gyr	$t_{sbdw} = 8$ Gyr	$t_{sbdw} = 12$ Gyr
M92	14.76 ± 0.04	14.75 ± 0.04	14.74 ± 0.04	14.75 ± 0.04	14.74 ± 0.04
M5	14.59 ± 0.03	14.56 ± 0.03	14.53 ± 0.03	14.58 ± 0.03	14.54 ± 0.03
NGC288	14.93 ± 0.03	14.90 ± 0.03	14.88 ± 0.03	14.92 ± 0.03	14.89 ± 0.03
47 Tuc	13.58 ± 0.04	13.59 ± 0.04	13.56 ± 0.04	13.63 ± 0.04	13.59 ± 0.04

7.2.2
Tip of the Red Giant Branch

As we have already discussed, the TRGB marks the He-ignition in electron degenerate cores of low-mass stars, that is, stars with masses below $\sim 2 M_\odot$, corresponding to population ages above ~ 1.0–1.5 Gyr. The TRGB bolometric magnitude is weakly dependent on the initial stellar mass (and hence the isochrone age) for ages above ~ 4 Gyr, due to the fact that at a given initial chemical composition, the TRGB level is determined by the He-core mass at the He flash that is fairly constant in this age range. The core mass decreases for increasing metallicity, while the TRGB bolometric luminosity increases because of the increased efficiency of the H-burning shell, that overcompensates for the reduced core mass. When considering the TRGB magnitude in a photometric system as a function of metallicity and age, one finds that the behaviour of the bolometric correction to the *I*-band as a function of [Fe/H] and effective temperature compensates for the variation of the bolometric luminosity with metallicity.

The net effect is that M_I^{TRGB} is predicted to be basically constant at values around $M_I^{\mathrm{TRGB}} = -4$ in the Cousins filter, with both age (for ages above 4–5 Gyr) and metallicity (for [Fe/H] below ~ -0.7), as displayed in Figure 7.4. This is the reason why the *I*-band magnitude of the TRGB is one of the standard candles of choice for a stellar system hosting old populations. It is applicable for distances of the order of 10 Mpc, with present observational capabilities.

Empirically, the signature of the TRGB is a sharp discontinuity – or cut-off in low-mass systems that display a very sparse AGB population – along the luminosity function of red giant stars because of the shorter evolutionary timescales of AGB stars that can potentially reach luminosities brighter than the TRGB. Its detection has been formalized in [418], who made use of an edge-detection algorithm, that is, a kernel $[-1, -2, 0, +2, +1]$.[3] Convolution of this kernel with a well-populated luminosity function (LF) (star counts per magnitude bin as a function of the magnitude) gives a spike in the output at the TRGB level, as shown by Figure 7.5. Due to the relatively fast evolution along RGB and AGB, one needs at least ~ 100 stars within one magnitude below the TRGB, according to [419], to detect a discontinuity in the luminosity function. Smaller samples cause large oscillations between consecutive points along the LF due to Poisson statistics. In this case, the output of the edge-detection algorithm gets noisy, showing many large spikes not associated with the TRGB. A too small number of stars can even leave the magnitude level of the TRGB completely devoid of objects. More recently, alternative (parametric and non-parametric) methods have been proposed and applied to a host of stellar systems (see, i.e. [420–423]) in the quest for more stable techniques against noise effects and to improve the accuracy of the TRGB detection.

---

3) As an example, working in the *I*-band, for a given magnitude I_i the result of the convolution is simply the weighted sum output$^i = -1 N(i-2) - 2 N(i-1) + 2 N(i+1) + 1 N(i+2)$, where $N(i)$ denotes the star counts at magnitude I_i.

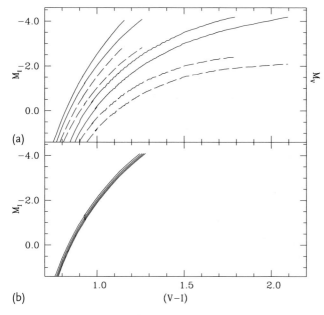

(a)

(b)

Figure 7.4 (a) RGB scaled-solar isochrones in the M_I-$(V-I)$ and M_V-$(V-I)$ CMDs (solid and dashed lines respectively) for [Fe/H] = −2.0, −1.7, −0.7 and −0.4. Notice the strong dependence of M_V^{TRGB} on [Fe/H], compared to M_I^{TRGB}. (b) M_I-$(V-I)$CMD of [Fe/H] = −1.7 RGBs with ages between 8 and 14 Gyr.

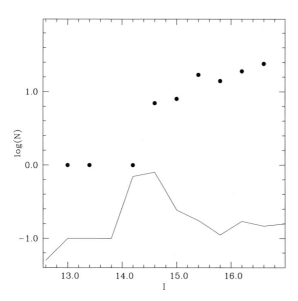

Figure 7.5 Observed RGB I-band luminosity function in a LMC field (filled dots) and the corresponding output of the edge detection algorithm (solid line) to determine the TRGB magnitude.

Comparisons of several sets of recent stellar evolution models display a $\sim$ 0.15 mag range of $M_{\text{bol}}^{\text{TRGB}}$ values, with a consequent sizable spread in the predicted M_I^{TRGB}. On the empirical side, a calibration of M_I^{TRGB} as a function of [Fe/H], considering the TRGB in the "massive" globular clusters ω Cen and 47 Tuc, is presented in [424, 425]. ω Cen is probably the only globular cluster with enough stars to ensure a clear detection of the TRGB, whereas for 47 Tuc, a small correction (based on numerical simulations) has to be applied to the "observed" TRGB level (see also Figure 3.49). In the case of ω Cen that displays a range of [Fe/H] values, the metallicity of the main component ([Fe/H] ~ -1.7) is assigned to the TRGB population. The clusters TRGB apparent magnitudes have then been translated into absolute values considering, for ω Cen, the distance modulus obtained from the analysis of one eclipsing binary system harboured by the cluster, and for 47 Tuc the average of seven HB-independent distance estimates from the literature. The final distance modulus $(m - M)_0 = 13.31 \pm 0.14$ employed for 47 Tuc agrees, within the errors, with the distance moduli later inferred from two eclipsing binary systems belonging to the cluster [426].

A different approach was followed in [427] to calibrate the TRGB as a distance indicator. Values of M_I^{TRGB} as a function of the dereddened colour $(V-I)_0^{\text{TRGB}}$ of the TRGB were determined on a sample of well-resolved nearby galaxies as follows. The apparent TRGB magnitude has been first determined in several colour slices for each galaxy to derive the slope of the m_I^{TRGB}-$(V-I)_0^{\text{TRGB}}$ relationship. The zero point of the absolute magnitude scale was fixed by employing a MS-fitting based calibration of the HB absolute magnitudes applied to the galaxy HBs [407]. The resulting relationship in the Johnson–Cousins system is given by

$$M_I^{\text{TRGB}} = -4.050 + 0.217 \left[(V-I)_0^{\text{TRGB}} - 1.600\right] \tag{7.1}$$

Analogous calibrations in the *HST* systems ACS and WFPC2 are

$$M_{F814W}^{\text{TRGB,ACS}} = -4.06 + 0.15 \left[(F555W - F814W)_0^{\text{TRGB}} - 1.74\right] \tag{7.2}$$

$$M_{F814W}^{\text{TRGB,ACS}} = -4.06 + 0.20 \left[(F606W - F814W)_0^{\text{TRGB}} - 1.23\right] \tag{7.3}$$

$$M_{F814W}^{\text{TRGB,WFPC2}} = -4.01 + 0.18 \left[(F555W - F814W)_0^{\text{TRGB}} - 1.58\right] \tag{7.4}$$

$$M_{F814W}^{\text{TRGB,WFPC2}} = -4.01 + 0.15 \left[(F606W - F814W)_0^{\text{TRGB}} - 1.12\right] \tag{7.5}$$

The calibration in Eq. (7.1) is displayed in Figure 7.6, together with the corresponding relationship derived by [428] from the M_I^{TRGB}-[Fe/H] calibration [424, 425]. Despite the different functional form and the completely different calibrating procedure, the two results are in remarkable agreement, within $\sim \pm 0.05$ mag. Figure 7.6b,c shows similar relationships in near-IR filters, obtained by [428] from the [425] calibration.

A calibration of the TRGB absolute magnitude as a function of colour has the advantage of minimizing the effect of the galaxy star formation history. As discussed in detail in [429, 430], the appearance of a well-populated RGB in the CMD of a stellar system with complex formation history does not guarantee that the RGB is only

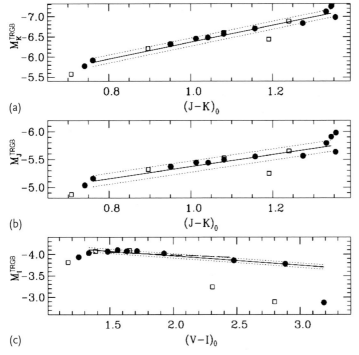

Figure 7.6 Empirical relationships between the TRGB absolute magnitude in the *I*, *J* and *K* photometric filters, as a function of the TRGB colour. Solid lines in (a) and (b) display the results from [428]. The boundaries of the ±0.10 mag interval around the empirical relationships are also shown as dotted lines. The solid line in (c) displays the [427] relationship, whilst the dashed line denotes the corresponding result from [428]. The boundaries of the ±0.05 mag interval around the [427] calibration are also marked as dotted lines. Theoretical results for ages between 4 and 14 Gyr and varying [Fe/H], metal mixture and initial He abundances are displayed (after a common zero point offset) as filled circles. Open squares denote the results from 2 Gyr models (see text for details).

populated by GC-like RGB stars, nor that the retrieved magnitude of the TRGB is determined by these very old stars. In the specific case of LMC and SMC, according to simulations similar to the ones discussed for the RC, the observed RGB is predicted to host a population dominated by ∼ 4 Gyr old stars, whose TRGB *I*-band brightness starts to differ from the case of a ∼ 10 Gyr globular cluster-like population.

An additional, more subtle effect, involves the estimates of [Fe/H] of the TRGB population. The "normal" procedure to determine TRGB distances is first to estimate the TRGB metallicity, and then use a M_I-[Fe/H] calibration as a standard ruler. The TRGB [Fe/H] is usually estimated by comparing the colours of the observed RGB with template RGBs of globular clusters with known [Fe/H]. The underlying assumption is that RGB colours are essentially unaffected by age. However, this is no longer true when there is an age difference larger than ∼ 6–7 Gyr

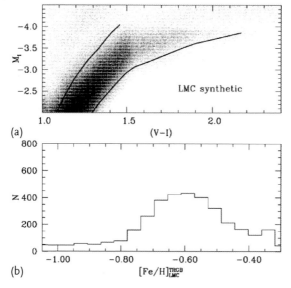

Figure 7.7 Simulation of the LMC bar population. (a) shows the synthetic M_I–$(V-I)$ CMD of the upper part of the RGB, compared to old isochrones with [Fe/H] equal to −1.5 and −0.9, respectively (thick lines). (b) shows the [Fe/H] distribution of the simulated upper RGB stars (see text for details).

between calibrating and target RGB. In this case, the galaxy metallicity would be underestimated (the younger RGBs are systematically bluer).

An error in [Fe/H] affects the TRGB distance through the dependence of M_I^{TRGB} on [Fe/H] if the real metallicity of the target population happens to be larger than [Fe/H] $=\sim$ −0.7, a threshold above which M_I^{TRGB} starts to display a strong sensitivity to [Fe/H]. The origin of this metallicity bias is perfectly illustrated by the following example. Figure 7.7a displays the bright RGB of a synthetic population with the typical star formation history of the LMC, compared to two 12 Gyr old isochrones of differing [Fe/H]. The isochrones have been calculated from the same stellar tracks used for creating the synthetic LMC population, and represent the RGBs of two Galactic GCs with the labelled [Fe/H]. Figure 7.7b displays the [Fe/H] distribution of the upper RGB stars in the simulation, for example, the real [Fe/H] distribution of the synthetic LMC sample. A determination of the mean [Fe/H] of the synthetic LMC population based on the comparison with the GC isochrones provides [Fe/H] $\sim$ −1.1, in very good agreement with the value obtained by [418] who applied the same method to I–$(V-I)$ LMC data. The "real" mean [Fe/H] is however −0.61, as obtained from the values reported in the histogram of Figure 7.7, in excellent agreement with the spectroscopic observations of bright RGB stars in the LMC bar published by [431], who found a mean [Fe/H] = −0.60. An analysis of the synthetic sample shows that the mean age of the bright RGB stars is $\sim$ 4 Gyr, implying a bluer RGB at fixed [Fe/H], when compared to typical globular cluster RGBs.

The use of TRGB magnitude-colour relationships mitigates this problem, as shown by Figure 7.6. Here, we also display, after applying a vertical shift to match the zero point of the empirical relationships, the predictions from the BaSTI scaled-solar theoretical isochrones for selected ages between 2 and 14 Gyr, [Fe/H] between −2.27 and +0.40, a few models with enhanced He ($Y = 0.40$) and models with an α-enhanced metal mixture [148, 432]. The $V-I$ linear relation of Eq. (7.1) is displayed together with the following JHK (on the 2MASS photometric system) relationships calibrated in [428]:

$$M_J^{\mathrm{TRGB}} = -4.29 - 1.08 \left[(J-K)_0^{\mathrm{TRGB}} - 1.74 \right] \tag{7.6}$$

$$M_H^{\mathrm{TRGB}} = -4.56 - 1.64 \left[(J-K)_0^{\mathrm{TRGB}} - 1.23 \right] \tag{7.7}$$

$$M_K^{\mathrm{TRGB}} = -4.29 - 2.08 \left[(J-K)_0^{\mathrm{TRGB}} - 1.58 \right] \tag{7.8}$$

Notice the much stronger dependence of the TRGB absolute magnitudes on colours in the infrared, when compared to the I-band, due to the very strong dependence of the relevant bolometric corrections on T_{eff}. All the relationships displayed have been extrapolated to cover the full $(V-I)$ and $(J-K)$ colour interval where theory predicts a well-defined narrow relationship within, respectively, ±0.05 mag in $(V-I)$ and ±0.10 mag in $(J-K)$, of the empirical results. In these colour ranges, M_I^{TRGB}, M_J^{TRGB} and M_K^{TRGB} are predicted to lie along a very tight sequence. Variations of age, metal mixture, He abundance and [Fe/H] move the TRGB magnitudes along these well-defined sequences. Beyond the red limit for the predicted validity of these relationships lie the [Fe/H] = 0.4 super-solar metallicity populations, while at the blue end one finds the youngest ages. Only the 2 Gyr models at solar and super-solar metallicity appear to be underluminous within the colour range where theory otherwise predicts tight relationships with colour.

7.2.3
Horizontal Branch Distances

The brightness of the HB in the CMD of GCs is one of the most important distance indicators for Galactic halo stars. The main reason for using the HB as *standard candle* arises from the theoretical evidence, which is extensively discussed in Section 4.3, that the HB luminosity does not depend on the cluster age. This is because the main parameter that determines the luminosity of HB models (at fixed initial chemical composition) is the He core mass at the He flash (M_{cHe}) that does not depend on the initial stellar mass (hence population age) in the low-mass star regime.

There are several reference HB levels defined in the literature that can be used as a *candle*, and several empirical, semi-empirical and theoretical calibrations of the absolute magnitudes of these reference levels, as a function of the population metallicity. They are usually defined in the V (or equivalent) photometric band, owing to the "horizontal" morphology of the HB in visual magnitudes over a large colour range that include also the RR Lyrae instability strip. The most straightforward reference luminosity derived from theory is the ZAHB level, typically taken at the average T_{eff} of the instability strip, that is, $\log(T_{eff}) = 3.83$ or 3.85. An alternative is the mean level of the HB $\langle V_{HB} \rangle$, or the mean magnitude of the RR Lyrae variables[4] in the observed population $\langle V_{RR} \rangle$. Actually, these HB luminosity levels are not coincident and are also different from the ZAHB level, the differences being a function of the GC metallicity and HB_{type}. In the following, we will clarify this issue and provide some useful analytical relations.

7.2.3.1 Which Horizontal Branch Luminosity to Use?

The ZAHB marks the start of quiescent central He-burning, and HB models evolve off ZAHB towards larger luminosity, at least for masses within the RR Lyrae instability strip. From a practical point of view, it may appear appropriate to compare a theoretical ZAHB with the lower envelope of the observed HB stellar distribution. From an observational point of view, the determination of the ZAHB level can be difficult and sometimes also ambiguous. This problem is particularly evident in case of clusters with blue HB morphology, that is, when the horizontal portion of the HB at visual wavelengths, approximately coincident with the instability strip location, that is, for $0.2 \leq (B-V) \leq 0.6$, is completely depopulated or with only very few stars. In this second case, the lower envelope of the observed HB distribution is much brighter than the ZAHB, as shown below.

We can rely on synthetic HB simulations to investigate the relation between various choices of the HB luminosity level, and how they depend on the GC metallicity and HB_{type}. Figure 7.8 shows an example of synthetic HB. The theoretical ZAHB does not coincide with the lower envelope of the HB distribution, as a consequence of a relatively fast off-ZAHB evolution in the early phases of the core He-burning. After only $\sim (8-10)$ Myr (the exact value depending on the adopted metallicity and initial He content), that is, $\sim (8-10)\%$ of the total HB lifetime, the models are already $\sim (0.05-0.1)$ mag brighter than the ZAHB. After this initial stage, they spend $\sim (70-80)\%$ of the total HB lifetime within the next ~ 0.1 mag interval.

Figure 7.9 shows the difference between the ZAHB and the mean magnitude (the "static" magnitude) of the RR Lyrae as a function of HB_{type} for various assumptions of the global metallicity as provided by synthetic HB simulations. The data shown in this figure clearly reveal that this difference is strongly dependent on HB_{type}. The bluer the HB morphology, the larger the magnitude difference. At fixed HB_{type}, the magnitude difference between ZAHB and the average RR Lyrae luminosity is only slightly dependent on the global metallicity, but for the redder HB morphology (though only at the level of $\sim 0.02-0.03$ mag). By using the relations shown in Figure 7.9, it is easy to compare these two different luminosity levels in old star clusters.

---

4) This corresponds to the mean value of the average magnitudes averaged over a pulsation period of a sample of RR Lyrae stars. The larger the sample of RR Lyrae stars, the more meaningful the comparison of this quantity with the theoretical counterpart.

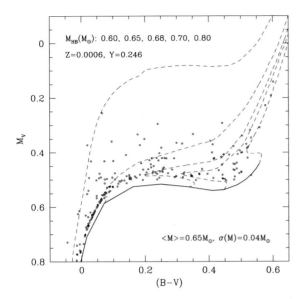

Figure 7.8 A synthetic HB for an old stellar population, with the labelled average mass, mass dispersion (Gaussian distribution) and chemical composition (the photometric error is set to zero in all bands). The thick line corresponds to the theoretical ZAHB with the same chemical composition for a RGB progenitor with initial mass equal to $0.8 M_\odot$. HB tracks of various masses (see labels) and the same RGB progenitor and chemical composition are also shown.

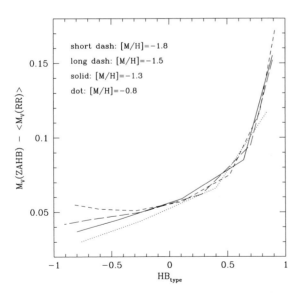

Figure 7.9 The difference between the V-band magnitude of the ZAHB at $\log(T_{\text{eff}}) = 3.83$ and the mean RR Lyrae magnitude as a function of HB_{type} for various assumptions about the total metallicities (see labels).

As for a ZAHB luminosity calibration, updated stellar models [432] with $[\alpha/\text{Fe}] = +0.4$ predict, in the range $-2.27 \leq [M/H] \leq 0.06$, the following relation

$$M_V(\text{ZAHB}) = 0.921 + 0.362 \cdot [M/H] + 0.057 \cdot [M/H]^2 \tag{7.9}$$

An alternative to using these types of relationships to determine distances from the observed HB is to use a procedure proposed by [313]: (i) compute a synthetic HB, tuning all relevant parameters of the simulation , that is, metallicity, mean HB mass and mass dispersion, see Section 4.3.3, to best reproduce the observed HB morphology; (ii) apply magnitude (to account for the GC distance) and colour (to account for the reddening) shifts to finally match the CMD of the observed HB population. The magnitude shift alone determines the population distance. A byproduct of this procedure is the determination of the apparent magnitude of the ZAHB in the observed population given by the magnitude-shifted theoretical ZAHB used in the synthetic calculations.

7.2.3.2 The RR Lyrae Brightness–Metallicity Relation

RR Lyrae stars have traditionally been one of the most widely used standard candles for old stellar systems (see, e.g. [258]) because they are easy to identify thanks to their periodic light variation, and are luminous and detectable to relatively large distances.

It is customary to assume a linear relation between the mean RR Lyrae luminosity and their metallicity

$$\langle M_V(\text{RR}) \rangle = a \cdot [\text{Fe}/\text{H}] + b$$

The calibration of this standard candle consists of determining the appropriate values for the two constants a and b. This $\langle M_V(\text{RR}) \rangle$–[Fe/H] relation is widely adopted because it only requires two observables, namely, the apparent visual magnitude (averaged over the pulsation cycle) and the metallicity. From a theoretical point of view, it is also well-defined because the RR Lyrae instability strip is located in the "horizontal" part of the HB at optical wavelengths. In spite of this, the absolute calibration of the $\langle M_V(\text{RR}) \rangle$–[Fe/H] relation is still an open problem. Current theoretical and empirical calibrations provide differences in absolute distances that range from ~ 0.1 to ~ 0.25 mag. The quoted internal errors are quite often of the order of a few hundredths of magnitude; this indicates that current methods are mainly affected by systematic errors. The main problems affecting the calibration of $\langle M_V(\text{RR}) \rangle$–[Fe/H] are the following:

- *Small sample of accurate parallaxes*: There is a lack of accurate parallax determinations for large samples of field RR Lyrae stars [433].
- *Evolutionary effects*: An accurate $\langle M_V(\text{RR}) \rangle$–[Fe/H] calibration has to rely not only on accurate distances, but also on large numbers of calibrating objects in order to properly sample the instability strip and the knowledge of the underlying HB morphology. This is due to the intrinsic width of the HB at the instability strip that becomes larger by moving from metal-poor to metal-rich GCs. As

a consequence, RR Lyrae samples at different metal contents in both clusters and the Galactic field are affected at different extents by off-ZAHB evolution as well as the HB morphology that is impossible to establish for field stars [434]. Actually, a spread in luminosity of the order of ~ 0.1 dex causes a magnitude spread of ~ 0.25 mag. To this effect, one has to add the ~ 0.1 mag change in the bolometric correction to the V-band BC_V when moving from the blue to the red edge of the instability strip.

- *Reddening*: An uncertainty of about 0.01 mag in $E(B-V)$ implies an uncertainty of the order of 0.03 mag in the visual magnitude.[5] In several cases for both field and cluster RR Lyrae stars, the uncertainty on the interstellar extinction is larger than this value.

- *Mean magnitude for a pulsating star*: The mean magnitude of RR Lyrae stars is estimated from the time average of either the magnitude or flux along the pulsation cycle. Theory and observations [435, 436] suggest that these two mean magnitudes are different from the mean "static" magnitude, for example, the magnitude of the equivalent hydrostatic equilibrium stellar model. The discrepancy for fundamental RR Lyrae stars increases from a few hundredths of a magnitude in V close to the red edge, to ~ 0.1 mag close to the blue boundary of the instability strip. This discrepancy becomes marginal in the K-band for the luminosity amplitude becomes a factor of ~ 3 smaller than in the V-band. The best approximation to the equivalent "static" value is the magnitude obtained from the time-averaged flux.

- *Metallicity*: There are still uncertainties in the metallicity scale of metal-poor stars [437]. Uncertainties of the order of ± 0.15 dex on the metallicities of RR Lyrae stars in clusters or in the field introduce an error on $\langle M_V(\mathrm{RR}) \rangle$ of ~ 0.03–0.04 mag. We still lack a detailed knowledge of α-elements abundances for RR Lyrae stars in many clusters and in the field. This parameter is crucial to estimate the global metallicity, that is, the metallicity adopted in the calculation of both stellar evolution and pulsation models.

- *Non-linearity*: Although a linear relation between the mean V-band brightness of RR Lyrae stars and their metallicity is very often employed, there are now compelling indications both from stellar evolution (see Eq. (7.9)) and pulsation models that the $\langle M_V(\mathrm{RR}) \rangle$–[Fe/H] relation is indeed a non-linear relation, and at least a quadratic dependence on the metallicity has to be taken into account.

Some of the most recent calibrations of the $\langle M_V(\mathrm{RR}) \rangle$–[Fe/H] relation are

- $\langle M_V(\mathrm{RR}) \rangle = 0.18 \cdot [\mathrm{Fe/H}] + 0.74$, from GC RR Lyrae stars and GC distances obtained from the MS-fitting method [407];

- $\langle M_V(\mathrm{RR}) \rangle = 0.214 \cdot [\mathrm{Fe/H}] + 0.88$, obtained using the slope determined from a huge sample of LMC RR Lyrae stars [438] and a LMC distance modulus ($M - m)_0 = 18.50$;

---

5) We note that extinction in the V-band is related to the interstellar reddening $E(B-V)$ according to the relation $A_V = 3.1 \cdot E(B-V)$, while the extinctions in the I- and K-band are related to the interstellar reddening according to $A_I = 1.48 \cdot E(B-V)$ and $A_K = 0.35 \cdot E(B-V)$.

- $\langle M_V(\mathrm{RR}) \rangle = 0.29 \cdot [\mathrm{Fe/H}] + 0.96$, derived by [439] with the Baade–Wesselink (BW) method.[6]

There are still significant differences among the results provided by different independent calibrations. It has become customary to compare the predicted $\langle M_V(\mathrm{RR}) \rangle$ at $[\mathrm{Fe/H}] = -1.5$; the weighted average among the several estimates is equal to 0.59 ± 0.03 [440], but they span a range of $\sim 0.35\,\mathrm{mag}$. A comparison between some selected linear $\langle M_V(\mathrm{RR}) \rangle$–$[\mathrm{Fe/H}]$ relations from the literature and empirical data is shown in Figure 7.10.

We have already mentioned that the $\langle M_V(\mathrm{RR}) \rangle$–$[\mathrm{Fe/H}]$ relation predicted by evolutionary stellar models since the early 1990s [311] is non-linear. Despite this theoretical evidence and some observational hints [266, 441]), the use of linear relationships is still widespread.

The non-linearity has been definitively confirmed by a thorough analysis that combines the results from both stellar pulsation and evolution models with

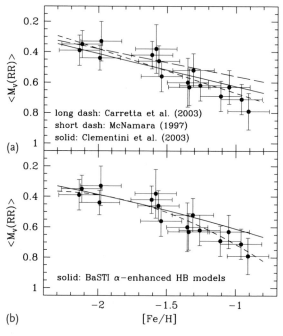

Figure 7.10 (a) The absolute mean magnitude of RR Lyrae stars in a sample of Galactic GCs [442]. Linear relations from the literature (see labels and text for more details) are also shown. (b) As with (a), but in this case, the theoretical prediction (solid line) for M_V (ZAHB) provided by α-enhanced HB models is shown, after a $+0.10$ mag correction to take into account the evolutionary effects [432]. The dashed line corresponds to a quadratic fit to the empirical data.

---

6) This method derives the distance of a pulsating star by comparing the linear radius variation estimated from the radial velocity curve, with the angular radius variation estimated from the light curve.

observed periods of first overtone RR Lyrae at the blue edge of the instability strip[7] [442]. From a practical point of view, in case of a GC with a sizeable sample of first overtone RR Lyrae stars, one can obtain a reliable estimate of its distance by considering the first overtone RR Lyrae distribution in the *mean RR Lyrae magnitude-period* diagram and matching its blue edge with the theoretical prediction. The shift applied to the theoretical result provides the GC distance modulus. A byproduct of this method is the derivation of the average absolute magnitude of the RR Lyrae in the cluster. A quadratic fit to the results of this analysis (also displayed in Figure 7.10) provides

$$\langle M_V(RR)\rangle = 1.415 + 0.893 \cdot [\text{Fe/H}] + 0.190 \cdot [\text{Fe/H}]^2 \tag{7.10}$$

7.2.3.3 Evolutionary Model Predictions and Their Uncertainty

Theoretical stellar evolution and pulsation models together with synthetic HB simulations provide estimates of the ZAHB absolute magnitudes, and the average absolute magnitude of RR Lyrae stars. The accuracy of these predictions relies on the accuracy of the predicted M_{cHe}. Table 7.2 lists the change of M_{cHe} and consequent changes in the RGB tip and HB luminosity due to variations of the input physics that most affect M_{cHe}.

Table 7.2 The variation of the He-core mass, TRGB luminosity, and ZAHB brightness for a $0.8 M_\odot$ stellar model with $Z = 0.0002$, $Y = 0.23$, due to the labelled changes in atomic diffusion efficiency, plasma neutrinos energy losses, 3α reaction rate, conductive and radiative opacities. The variations are evaluated with respect to a reference model computed with atomic diffusion efficiency from [78], conductive opacities from [51], radiative opacities from [443], nuclear cross sections from [444], neutrino emission from [445]. The data corresponding to the use of different conductive opacity sets are adapted from [446].

	$\Delta M_{cHe}(M_\odot)$	$\Delta \log(L_{Tip}/L_\odot)$	$\Delta \log(L_{ZAHB}^{3.85}/L_\odot)$	ΔM_V^{ZAHB}
No diffusion	-0.0043	-0.003	$+0.013$	-0.033
Diffusion$\times 0.5$	-0.0017	-0.002	$+0.008$	-0.021
Diffusion$\times 2$	$+0.0034$	$+0.004$	-0.016	$+0.041$
Plasma $\nu + 5\%$	$+0.0013$	$+0.006$	$+0.004$	-0.011
Plasma $\nu - 5\%$	-0.0012	-0.006	-0.003	$+0.010$
$\sigma(3\alpha) + 15\%$	-0.0010	-0.005	-0.003	$+0.008$
$\sigma(3\alpha) - 15\%$	$+0.0013$	$+0.005$	$+0.005$	-0.011
$\kappa_{rad} + 5\%$	$+0.0005$	-0.015	-0.002	-0.004
$\kappa_{rad} - 5\%$	-0.0006	$+0.016$	$+0.002$	$+0.005$
κ_{cond} [50]	-0.0050	$+0.022$	-0.017	$+0.042$
κ_{cond} [52]	-0.0066	-0.031	-0.020	-0.050

7) We note that theoretical evolutionary predictions are needed to constrain the mass range of pulsators inside the pulsational strip.

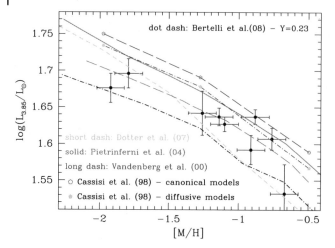

Figure 7.11 The ZAHB luminosity at $\log(T_{\rm eff}) = 3.85$ as a function of [M/H] as predicted by several sets of stellar models. Points with error bars correspond to a semi-empirical estimate of $\log(L_{3.85}/L_{\odot})$ [266].

The major sources of uncertainty are associated with the conductive opacity and the efficiency of atomic diffusion. If one decides to simply add together the change of $M_{\rm cHe}$ due to the variation of each individual ingredient, the total variation is $\sim 0.01\,M_{\odot}$, thus translating into a variation of $\sim 0.10\,{\rm mag}$ in the bolometric luminosity of the ZAHB within the instability strip. This uncertainty on the ZAHB luminosity is largely (by $\sim 40\%$) driven by the use of different tabulations of the conductive opacity, and to a slightly smaller extent ($\sim 35\%$), by the uncertainty in the efficiency of atomic diffusion.

Figure 7.11 shows a comparison of the ZAHB luminosity at $\log(T_{\rm eff}) = 3.85$ as a function of metallicity from several sets of stellar models. The theoretical predictions span a range of $\Delta \log(L_{3.85}/L_{\odot}) \approx 0.07$.

7.2.3.4 RR Lyrae Stars in the Near-IR Bands: Observational and Theoretical Aspects
It is well-known that near-IR photometry of RR Lyrae stars presents some relevant advantages compared to the optical bands, i.e.:

- A much smaller interstellar extinction, about one order of magnitude less than in the optical bands;
- A smaller dependence of the mean magnitudes on the metallicity;
- Smaller pulsational amplitudes (approximately a factor of four smaller than in the *B*-band) that allow a larger precision on time-averaged magnitudes with a relatively small number of observations. This is a relevant advantage that has to be taken into account when planning an observational survey of RR Lyrae stars in both field and star clusters.
- It was shown that cluster RR Lyrae stars do obey a well-defined *Period-K-band luminosity* (PL$_K$) relation [447].

Despite these relevant advantages, major efforts in the near-IR bands have been undertaken only in the last decade as a consequence of the strong observational limitations imposed by the previous generation of near-IR detectors.

Until recently, near-IR observations of RR Lyrae stars in Galactic GCs were sparse [447, 448], though it was sufficient to reveal that cluster RR Lyrae do follow a rather tight PL_K relation. In more detail, K-band photometry of variable stars in eight Galactic GCs have shown that the scatter around a linear PL_K relation was only due to the observational errors. Regardless of the cluster metallicity, the linear relations have a slope $dM_K/d \log P \sim -2.2$ within the observational uncertainties. The existence of such a tight relation between K-band magnitudes and pulsation periods for variables belonging to the same GC represents a clear proof that the unavoidable scatter of stellar masses and luminosities among the RR Lyrae sample only marginally affects the PL_K relation. These same analyses have provided three zero points, depending on the distances of field RR Lyrae stars as estimated with different implementations of the Baade–Wesselink method. In all cases, a mild dependence on metallicity between 0.04 and 0.08 mag dex^{-1} was derived. The most updated empirical calibrations of the PL_K relation for RR Lyrae stars are

$$\langle M_K \rangle = -0.647 - 1.72 \cdot \log P + 0.04 \cdot [Fe/H]$$

$$\langle M_K \rangle = -0.76 - 2.257 \cdot \log P + 0.08 \cdot [Fe/H]$$

$$\langle M_K \rangle = -(0.72 \pm 0.11) - (2.03 \pm 0.27) \cdot \log P + (0.06 \pm 0.04) \cdot [Fe/H]$$

from [447, 449, 450] respectively. All empirical PL_K relations agree in predicting a quite small dependence on metallicity, in contrast with the results for the V-band.

Updated pulsation models [451, 452] agree with these early empirical results about both the slope of the PL_K relation and the dependence on metallicity of the zero point.

Figure 7.12 shows the predicted visual and K-band mean magnitude of both first overtone and fundamental RR Lyrae stars as a function of the pulsation period. A glance at the data explains the existence of a well-defined *Period-Luminosity* relation. For any fixed luminosity, the period increases from the blue to the red edge of the instability strip as a consequence of the decrease of T_{eff}, as predicted by the fundamental pulsation equation (see Section 4.5.1). On the other hand, the M_V magnitude remains almost constant because the bolometric correction to the V-band does not show a significant dependence on T_{eff} in the temperature range of the RR Lyrae instability strip. At any fixed luminosity, the dependence of M_V on the period is $dM_V/d \log P \approx -0.14$. However, for M_K, the slope is much steeper, $dM_K/d \log P \approx -2.1$. This effect is due to the fact that when moving from the blue to the red edge of the instability strip, the period increases and at the same time, the pulsators become brighter in the K-band because the bolometric correction BC_K has a strong dependence on the effective temperature as shown in Figure 7.13. An increase in the luminosity for any fixed effective temperature causes an increase of the period according to $d \log P/d \log L \sim 0.83$, while the visual magnitude closely follows the expected shift in luminosity as $dM_V/d \log L \sim -2.5$. On the other

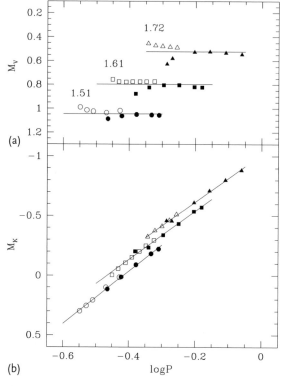

(a)

(b)

Figure 7.12 (a) Mean M_V magnitudes for fundamental (filled symbols) and first overtone (open symbols) pulsators at fixed mass ($0.65 M_\odot$) and chemical composition ($Z = 0.001$, $Y = 0.240$) and three different luminosities (see labels), as predicted by pulsation models covering the whole T_{eff} range from the blue to the red boundary of the instability strip. The period of FO pulsators has been "fundamentalized". (b) As in (a), but for the K-band. In both (a) and (b), the solid lines represent the predicted "mean" *period-luminosity* relations at fixed luminosity (courtesy of G. Bono).

hand, a change in the luminosity level causes a variation of both K magnitude and periods in such a way that the PL_K relation of bright pulsators roughly matches the relation for the fainter ones.

Pulsation models predict the following relations for F and FO RR Lyrae:

$$\langle M_K^F \rangle = 0.511 - 2.102 \log P + 0.095 \log Z - 0.734 \log \left(\frac{L}{L_\odot} \right) - 1.753 \log \left(\frac{M}{M_\odot} \right)$$

(7.11)

$$\langle M_K^{FO} \rangle = -0.029 - 2.265 \log P + 0.087 \log Z - 0.635 \log \left(\frac{L}{L_\odot} \right) - 1.633 \log \left(\frac{M}{M_\odot} \right)$$

(7.12)

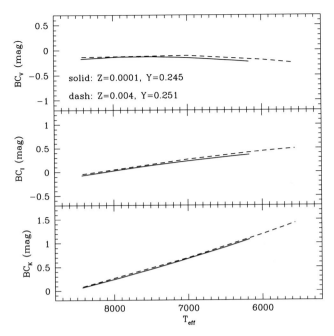

Figure 7.13 Bolometric corrections for the *V*-, *I*- and *K*-band along the ZAHB for two assumptions about the initial chemical composition as a function of the effective temperature. Only the T_{eff} interval corresponding to the RR Lyrae instability strip has been considered.

where the rms scatter is $\sigma = 0.016$ mag. The predicted slope at constant luminosity, mass and metallicity is in good agreement with GC data. These relations reveal that, for a given chemical composition, the effect on the PL_K of a dispersion in the stellar mass within the instability strip of $\pm 4\%$ is almost negligible, being of the order of ± 0.03 mag. The effect of a dispersion in luminosity is significantly smaller than in the visual relation. Even when accounting for a fairly large $\Delta \log(L/L_\odot) = \pm 0.1$, the PL_K relation provides M_K magnitudes with an accuracy better than ± 0.10 mag at a fixed period, significantly smaller than the analogous dispersion in the visual band, that is, $\Delta M_V \sim \pm 0.25$ mag. At the same time, a variation of 0.3 dex in the metallicity would imply a negligible change of $\Delta M_K \sim 0.03$ mag.

Although Eqs. (7.11) and (7.12) predict a very small dependence on both the mass and intrinsic luminosity of the stars inside the pulsation instability strip, this dependence has to be taken into account for distance determinations. To this purpose, we can use two approaches: (i) employ average values for both the stellar mass and bolometric luminosity; (ii) given that both mass and luminosity ranges of pulsators do depend on HB_{type}, one can use a combination of pulsation models and synthetic HBs [434, 453, 454].

Several *Period-Luminosity* relations in both K and J bands as a function of the HB_{type} and metallicity are listed in Table 7.3 [454]. A comparison between the latest near-IR observations of RR Lyrae stars in the GC ω Cen and these theoretical

Table 7.3 Theoretical *Period-Luminosity* relations in the K- and J-band as a function of metallicity and HB$_{type}$, for both fundamental (F) and first overtone (FO) RR Lyrae stars (adapted from [454]).

Metallicity	Pulsation mode	a	b	c
$\langle M_K \rangle = a + b \cdot \log P + c \cdot HB_{type}$				
0.0001	F	−1.32	−2.19	0.14
0.0001	FO	−1.65	−2.32	0.14
0.0003	F	−1.23	−2.34	0.04
0.0003	FO	−1.55	−2.38	0.04
0.001	F	−1.14	−2.37	0.01
0.001	FO	−1.48	−2.41	0.01
0.004	F	−1.05	−2.46	0.0
0.004	FO	−1.37	−2.43	0.0
$\langle M_J \rangle = a + b \cdot \log P + c \cdot HB_{type}$				
0.0001	F	−0.92	−1.62	0.08
0.0001	FO	−1.13	−1.64	0.07
0.0003	F	−0.83	−1.77	0.05
0.0003	FO	−1.10	−1.83	0.04
0.001	F	−0.76	−1.89	0.01
0.001	FO	−1.04	−1.88	0.01
0.004	F	−0.70	−2.13	0.0
0.004	FO	−0.94	−1.93	0.0

predictions is shown in Figure 7.14. The empirical data show a slope for the PL_K relation in very good agreement with theoretical expectations and, so that the comparison between theory and observations provides a very direct and reliable method for retrieving the distance to the cluster. We also note that the first overtone and fundamental RR Lyrae stars follow two distinct, but largely parallel, PL_K relations. The same results also hold for very metal-poor GCs as demonstrated in case of M92 [455].

The relations listed in Table 7.3 show some interesting features worthy of being mentioned: (i) the slopes of both the PL_K and PL_J relations are, at any given metallicity, independent of the HB$_{type}$, while they become slightly steeper with increasing metallicity; (ii) the zero points of both *Period-Luminosity* relations become, at a fixed metal content, brighter when moving from blue to red HB morphologies, that is, decreasing values of the HB$_{type}$ parameter. This dependence decreases significantly with increasing the metal content; (iii) the zero points of the $PL_{J,K}$ relations become fainter with increasing metallicity at fixed HB$_{type}$ (but see also [453]).

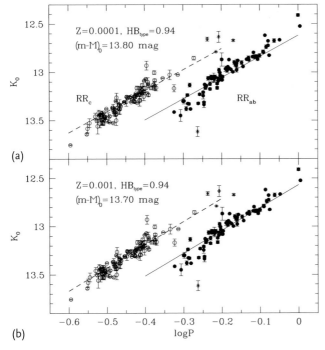

Figure 7.14 (a) Dereddened average K-band magnitudes of RR Lyrae stars in ω Cen for both fundamental (filled circles) and first overtone (open circles) RR Lyrae [454]. Solid and dashed lines represent the theoretical PL_K for F and FO pulsators, respectively, for the labelled metallicities, HB_{type} and distance modulus $(m - M)_0$. (b) As in (a), but for different assumptions about the cluster metallicity and distance. These distance modulus estimates provide the best match between predicted and observed PL_K relations (see Table 7.3). In both panels, asterisks mark the location of suspected objects affected by blending and/or variables with unreliable mode identification (courtesy of A.M. Piersimoni).

7.2.4
Red Clump

The use of red clump (RC) stars "for crude determinations of cluster distances" was first advocated over 40 years ago in [456], following evidence that the V magnitude of the RC appeared to be constant within ~ 1 mag (half of this spread may have been due to observational errors) in open clusters with ages above ~ 0.5 Gyr. In recent times, RC-based distances, that with present observational capabilities can reach extragalactic objects within the Local Group, have been the subject of several investigations after the publication of the Hipparcos parallaxes (see Section 4.6).

The RC is recognizable in the CMD of populations with ages larger than $\sim 0.5-1$ Gyr as a clump of red stars close to the RGB (see Figure 7.15). It is populated by objects experiencing central He-burning after igniting He in an electron degenerate core or around the transition to non-degeneracy. They are the counterparts of HB stars in globular clusters, and in fact, the HB of metal-rich globulars are

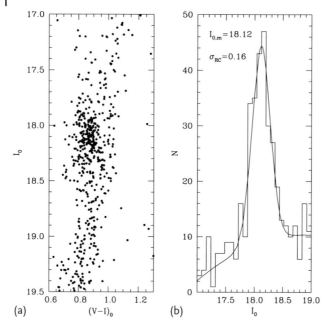

Figure 7.15 Observed CMD of the RC of a LMC field (a), and the fit to the corresponding luminosity function (b) to determine the RC level.

usually (but not always) a RC (see, for example, 47 Tuc). Figure 7.16 displays the absolute magnitude – bolometric and in the VIK filters, for which there is an accurate empirical estimate of the local disk RC mean absolute magnitude –, as a function of effective temperature at the beginning of quiescent central He-burning phase of stellar isochrones for two scaled-solar metallicities and several ages ranging from 0.5 to 10 Gyr. There is a number of features to be noticed that are relevant for the application of RC stars as distance indicators:

- At a given age, the models become cooler (redder) for increasing metallicity. The bolometric magnitude, and the magnitudes in the I and V bands become fainter with increasing metallicity when the age is fixed. The reverse is true for the K band. This different behaviour is obviously due to the effect of the bolometric corrections (see Figure 4.26).
- At fixed metallicity, RC models tend to cover a limited range of brightness, of the order of 0.2–0.3 mag. When the age decreases from 3 to 1 Gyr, the luminosity decreases due to the transition between degenerate and non-degenerate He-cores at He-ignition, followed by a sharp increase at an age of 0.5 Gyr, that signals the complete transition to non-degenerate cores.

Following the release of the Hipparcos catalogue, the mean absolute magnitude of the local RC can now be measured in several photometric bands with a precision of the order of 0.01 mag, and used as zero point of the RC distance scale. The determination of the absolute magnitude of the local RC in a given passband λ

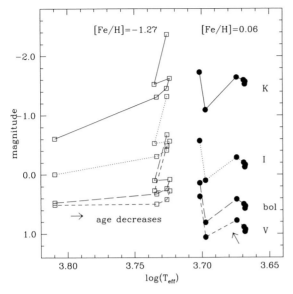

Figure 7.16 The bolometric and *VIK*-bands magnitude at the start of quiescent He-burning as a function of T_{eff}, for ages equal to 0.5, 1, 3, 6, 8 and 10 Gyr, and the labelled metallicities. Arrows display the direction of the age increase. At [Fe/H] = 0.06, the oldest age corresponds to the coolest T_{eff} (reddest colour), whereas at [Fe/H] = −1.27, the oldest age corresponds to the hottest T_{eff} (bluest colour).

$(M_{\lambda,local}^{RC})$ and the apparent magnitude m_λ^{RC} of the RC in a given stellar population are straightforward, for in both the Hipparcos database of local stars, and in CMDs covering even small fractions of a nearby galaxy, one finds hundreds of RC objects, easily identifiable from their CMD location. As proposed in [457], a non-linear least-square fit of the function

$$N(m_\lambda) = a + b m_\lambda + c m_\lambda^2 + d \exp\left[-\frac{\left(m_\lambda^{RC} - m_\lambda\right)^2}{2\sigma_{m_\lambda}^2}\right]$$

to the histogram of stars per magnitude bin in the RC region provides the value of m_λ^{RC} and its associated standard error (see Figure 7.15 for an example). Figure 7.17 displays the local RC in the *VIK* bands. It is important to notice the different shape of the RC (horizontal in *I*, sloped in opposite directions in *V* and *K*) due to the effect of the bolometric corrections. The calibration presented in [458] provides $M_{V,local}^{RC} = 0.73 \pm 0.03$, $M_{I,local}^{RC} = -0.26 \pm 0.03$ and $M_{K,local}^{RC} = -1.60 \pm 0.03$, this latter value revised recently [459] to $M_{K,local}^{RC} = -1.50 \pm 0.04$.

Once m_λ^{RC} is measured in a nearby galaxy, its distance modulus $\mu_0 = (m - M)_0$ is easily derived by means of

$$\mu_0 = m_\lambda^{RC} - M_{\lambda,local}^{RC} - A_{m_\lambda} - \Delta M_\lambda^{RC}$$

In this equation, A_{m_λ} denotes the interstellar extinction to the target population, and $\Delta M_\lambda^{RC} = M_{\lambda,local}^{RC} - M_{\lambda,target}^{RC}$ is the so-called "population effect", the difference

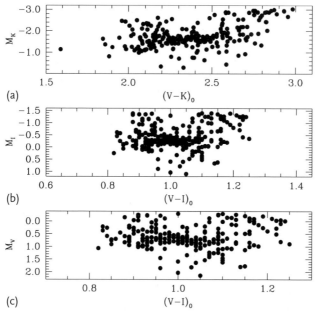

(a)

(b)

(c)

Figure 7.17 Three different CMDs of local RC stars with accurate parallaxes.

of the mean RC absolute magnitude between the local and target sample of stars due to different star formation histories. Given the complex behaviour of M_λ^{RC} as a function of age and metallicity, accurate RC distances require a proper determination of ΔM_λ^{RC} that is expected to be strongly dependent on the star formation histories of both the target and calibrating populations. A number of analyses [283–285] have addressed this issue from a theoretical point of view, making extensive use of synthetic CMDs to investigate the dependence of the RC mean magnitude in detail in several photometric filters as a function of age and metallicity in single-age, single-metallicity populations, and the case of composite stellar systems like the solar neighbourhood, LMC and SMC employing star formation history estimates from the literature. A set of "population corrections" ΔM_λ^{RC} has been then calculated from these simulations. These corrections employ theory only in a differential sense: both $M_{\lambda,local}^{RC}$ and $M_{\lambda,target}^{RC}$ entering the definition of ΔM_λ^{RC} are calculated from synthetic CMDs employing the appropriate star formation history, but it is their difference that determines ΔM_λ^{RC}.

Some general conclusions about the properties of the RC as a distance indicator can be drawn from these analyses:

- In a composite stellar population, the age (mass) distribution of stars in the RC depends in a complex way on the population SFH, the trend of the He-burning timescale with stellar mass, metallicity and IMF. As an example, for a uniform star formation rate between 0.1 and 10 Gyr ago (at constant metallicity), the RC population is dominated by ~ 1.5 Gyr old objects that will have the largest weight in determining $M_{\lambda,target}^{RC}$;

- The value of ΔM_λ^{RC} is strongly dependent on both the formation history of the target population and the chosen photometric filter because of the different behaviour of the bolometric corrections in different bands with T_{eff} (hence stellar mass). For example, in case of the RC in the LMC, the predicted population corrections amount to ~ 0.2 mag in I, and only to ~ -0.03 mag in K (see, i.e. [460] for detailed analyses of the LMC distance using the RC);

- K-band observations are best suited to determine RC distances to populations with a SFH not decreasing with decreasing look-back time. In this case, the RC will always be dominated by objects with ages around 2 Gyr, for which the population correction is close to zero and essentially metallicity independent. I-band data are best for old populations with sub-solar metallicities, since in this case ΔM_λ^{RC} is close to zero.

Figure 7.18 displays the predicted ΔM_λ^{RC} values in the VIK bands for a large range of single-age, single-metallicity populations that have been tested empirically in [461]. This investigation has made use of a sample of eight Galactic open clusters hosting RC stars plus the Galactic GC 47 Tuc, determined their distances, and hence the RC absolute magnitudes in VIK, applying a MS-fitting method based on a large sample of local field dwarfs with accurate Hipparcos parallaxes, and estimated their TO ages. The corresponding empirical values of ΔM_λ^{RC} have been then calculated by considering the empirical absolute magnitudes of the local RC. A comparison with the theoretical population corrections displayed in Figure 7.18

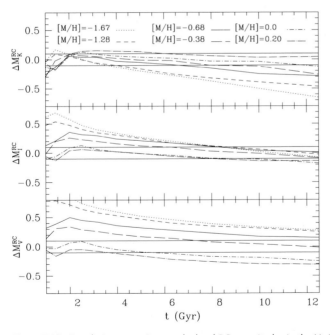

Figure 7.18 Population corrections to the local RC magnitudes in the V, I and K bands as function of the stellar population age, for the labelled metallicities (scaled-solar mixture).

for the appropriate cluster ages and metallicities discloses an agreement at the level of ~ 0.1 mag.

Empirical calibrations of M_λ^{RC} versus [Fe/H] relations determined in situ in systems harbouring composite stellar populations (see, i.e. [462]) need to be treated with caution. Depending on the method followed, they may be strongly tied to the specific SFH of the stellar system used as calibrating sample, and do not have a general validity. The best empirical approach to date has been presented in [463]. Here, RC magnitudes have been compared with the I-band magnitude of the TRGB (weakly sensitive to the galaxy star formation history) in a sample of 23 galaxies in the Local Volume. They found strong population effects in the V and I magnitude of RC stars, confirming the theoretical predictions.

7.2.5
White Dwarf Fitting

An additional empirical method to derive star cluster distances makes use of the bright CO-WD cooling sequence exactly the same way as the lower MS is used in the MS-fitting. A key assumption of this method is that the WDs of the local template are totally equivalent to the WDs of the cluster. The vertical shift applied to the local template sequence in order to fit the sequence of the cluster provides its distance modulus.

This WD-fitting method has been applied to derive the distance of, for example, the Galactic GC NGC 6752 [464]. The template sequence was made in this case of a sample of local WDs with T_{eff} ranging between $\sim 10\,000$ and $\sim 20\,000$ K, a mass $M_{WD} = 0.53 \pm 0.02 M_\odot$ and precise parallax measurements. Figure 7.19 displays two representative WD isochrones of an old population. The magnitude range corresponding to T_{eff} between $\sim 10\,000$ and $\sim 20\,000$ K is also marked. The bright sequence corresponds to the cooling track at almost constant mass, and even at two magnitudes below the lower limit of this T_{eff} range, the evolving WD mass is still constant. It is easy to understand this behaviour by recalling that at each brightness along the cooling sequence, the constraint $t_{cluster} = t_{cool} + t_{prog}$ has to be satisfied. Given that WD cooling times t_{cool} are very short at the bright end of the sequence irrespective of the WD mass, and are practically negligible compared to the cluster age t_{clust}, the progenitor age t_{prog}, and hence the progenitor mass (and the resulting WD mass, if the IFMR does not have a spread at fixed initial mass) has to be, to a very good approximation, constant and very close to the TO mass. On the contrary, towards the faint end of the cooling sequence, t_{cool} becomes a sizeable fraction of $t_{cluster}$ and the contribution of the WDs coming from higher mass progenitors (and, consequently, with smaller t_{prog}) becomes apparent in the CMD.

In the case of the template local WD sequence, it is possible to select objects on the basis of their mass and spectral type (DA or non-DA) and potentially also envelope thickness; however, this information is largely unavailable for WDs in more distant populations. This means that one has to consider the contributions arising from potential differences of the parameters affecting the location of WD

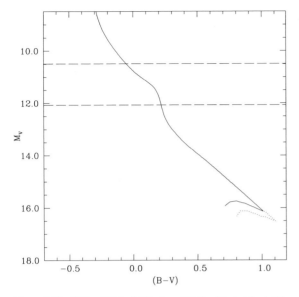

Figure 7.19 CMD of 10 (solid line) and 12 Gyr (dotted line) CO-core WD isochrones computed by assuming progenitors with solar chemical composition. The horizontal lines denote the
region with T_{eff} between 20 000 and 10 000 K.

sequences (an extensive discussion of the sources of systematic errors associated
to this method can be found in [465]) in the distance error budget:

- Mass – The derivative $\Delta M_V/\Delta M_{WD}$ at fixed colour is equal to 2.3 mag/$M_\odot$ in
  the mass range between 0.45 and 0.60 $M_\odot$. To give a visual representation of
  this effect, Figure 7.20 displays the position of cooling tracks for two masses in
  BVI CMDs;
- Envelope chemical composition – As we have seen in Chapter 6, field WDs in
  this T_{eff} range are of either DA (H-atmosphere) or DB (He-atmosphere) type. In
  the BV CMD, DA WDs are brighter than He-atmosphere WDs at a fixed colour,
  whereas the reverse is true in the VI CMD (see Figure 7.21);
- Core chemical composition – The chemical stratification of the CO core may potentially affect the CMD location of WD tracks, for the mass–radius relationship
  also depends on the electron mean molecular weight of the electron-degenerate
  core. However, as demonstrated in [465], the effect of realistic changes of the
  CO stratification, at fixed mass, is negligible;
- Envelope thickness – Theoretical models, observational constraints and scenarios to explain the field WD spectral evolution suggest that the thickness of the
  H-layers, $q(H)$, in DA objects spans a range of values between $q(H) \approx 10^{-4} M_{WD}$
  and $\approx 10^{-10} M_{WD}$. However, for DB objects, one estimates a range between
  $q(He) \approx 10^{-2}$ and $\approx 10^{-7}$. The thickness of the He-layers does not appreciably affect the location of WD tracks in the CMD, whereas the same is not true
  for H-envelopes. Theoretical calculations provide $\Delta M_V/\Delta \log(q(H)) \sim -0.035$
  for $\log(q(H))$ between -4.0 and -7.0. Figure 7.21 shows the effect of varying

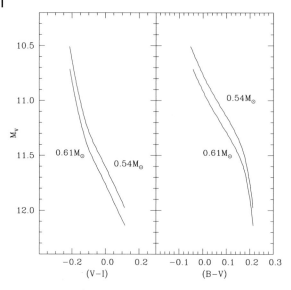

Figure 7.20 CO-core WD cooling tracks in the BV and VI CMDs for T_{eff} between 20 000 and 10 000 K. Models are for DA objects with the labelled masses.

$q(H)$ and $q(He)$ for, respectively, a $0.54\,M_\odot$ DA and DB model. In the case of a DB model, the reference value is $\log(q(He)) = -3.5$, which we have changed by ± 1.0 dex, obtaining virtually no variation in the location of the cooling track (the three tracks perfectly overlap). The representative DA models have $\log(q(H)) = -4.0$ and $\log(q(H)) = -6.0$; a sizeable reduction of the thickness of the H-envelope shifts the track, at constant colour, towards higher M_V values.

- He-core WDs – He-core WDs are typically the byproduct of strong mass loss during the evolution of RGB stars in binary systems. Their mass can range from $\sim 0.2\,M_\odot$ to values close to the He-core mass at the He flash, for example, $\sim 0.45 - 0.50\,M_\odot$. He-core WDs are shifted to larger brightness at fixed colour with respect to CO-WDs with the same envelope composition and thickness, due mainly to their lower masses compared to the CO-core counterparts.

All these effects have been included in a series of Monte Carlo simulations [465] to obtain a realistic estimate of the systematic errors involved in the application of this technique to Galactic globular clusters. The main results are:

- The unknown thickness of the H-layers in cluster DA WDs plays a non-negligible role, comparable to the role played by uncertainties on the WD masses. For reasonable assumptions (derived from observations of field WDs and constraints from globular cluster CMDs) about the unknown mass and $\log(q(H))$ values in cluster DA WDs, a realistic estimate of the maximum systematic error on the derived distance moduli is within ± 0.10 mag;
- A photometric precision better than ~ 0.05 mag is needed to distinguish DA from DB objects in the BV CMD. An even better precision is needed when

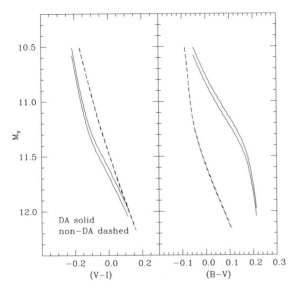

Figure 7.21 The same as Figure 7.20, but for a $0.54 M_\odot$ WD with either DA or DB atmospheres, and two different envelope thicknesses (see text for details).

using the VI CMD. For larger observational errors and no spectroscopic distinction between DA and DB cluster WDs, fitting a template DA sequence to a cluster sequence made of a mixture of DAs and DBs introduces a very small systematic error ~ -0.03 mag in the VI CMD. This error reaches $\sim +0.20$ mag in the BV CMD;

- If one accepts actual estimates of the binary frequency in GCs of the order of 10% or less [466]. Contamination by He-core WDs should not appreciably influence the WD-fitting distances;
- Due to the steep slopes of the WD cooling curves in the CMD, one obtains $\Delta(m - M)_V / \Delta E(B-V) \sim -5.5$, a dependence which is similar to the one of the MS-fitting technique in the BV CMD.

7.3
Old Star Cluster Ages

In the introduction of this chapter, we emphasized the importance of the age of old star clusters in a broader astrophysical context. The determination of absolute and relative ages of old star clusters are based on different sets of techniques that will be discussed separately.

7.3.1

Absolute Age Methods

The simplest procedure to the determine ages of SSPs would be to identify the isochrone with the cluster chemical composition that fits best (after appropriate vertical and horizontal shifts) the "shape" of the whole observed CMD, at least from the MS to the RGB. This only requires information from spectroscopy (for example, [Fe/H] and [α/Fe] along the RGB) without, in principle, any estimate of distance modulus and reddening. However, the associated uncertainties are potentially uncomfortably large, given that isochrones from different authors, even with the same initial chemical composition and age, display different shapes, especially when the metallicity approaches solar. Also, there is the problem of whether to assign the same weight to all evolutionary phases when fitting the shape of the CMD, or give more weight to the age-sensitive features. One can also try to simultaneously fit the predicted number of stars along the observed CMD, assuming an IMF. Again, a "wrong" isochrone shape can skew the result of the fit towards a wrong age. The latter approach is however unavoidable to determine ages of the old stellar component in composite systems, as described in the next chapter.

A different type of approach envisages the use of essentially just one age-sensitive feature. Section 3.9 has presented a detailed discussion of the sensitivity of the TO brightness of isochrones to age. The TO becomes fainter and redder in the CMD with increasing age. This property makes the isochrone TO the most important 'clock' provided by theoretical stellar astrophysics. In principle, one can directly compare the observed magnitude of the TO along the MS of a given old SSP with theoretical predictions and estimate the population age. In practice, things are not that simple [467, 468]. What we need to compare with the theory is the absolute magnitude of the TO, that is, assuming to work in the V-band, $M_V^{TO} = V_{TO} - (m - M)_V$, where $(m - M)_V$ is the cluster distance modulus, and V_{TO} is the *only* observable quantity. Even assuming that the uncertainty on the apparent TO brightness is very small (this is the case with the most recent photometric datasets, but one has to keep in mind that for the more metal-poor old SSPs the TO region is vertical and uncertainties up to ~ 0.1 mag on the apparent TO magnitude are possible), it is clear that the error on the distance affects M_V^{TO}. A change of $\sim 0.07 - 0.10$ mag in M_V^{TO} causes a change in the absolute age estimate (for ages larger or of the order of 6 Gyr) of about 1 Gyr.

On the theoretical side, we need a calibration of the TO brightness as a function of the age as well as of the initial chemical composition. This theoretical calibration is obtained by computing suitable sets of evolutionary stellar models and isochrones (see Figure 7.24c). One finds that isochrones predict:

$$\frac{\partial \log t_9}{\partial M_V^{TO}} \approx +0.37 \tag{7.13}$$

at fixed chemical composition,

$$\frac{\partial \log t_9}{\partial Y} \approx -0.43 \tag{7.14}$$

at fixed TO magnitude and metal content,

$$\frac{\partial \log t_9}{\partial [Fe/H]} \approx -0.13 \qquad (7.15)$$

at fixed TO magnitude and He, where t_9 represents the age in gigayears.[8] These relations allow us to estimate the impact of the uncertainty in the various ingredients entering the final absolute age estimate.

Let us start with the main contributor to the error budget. Until a decade ago, the distance to old clusters was known with an uncertainty of about (0.2–0.3) mag, and according to Eq. (7.13), this indetermination implies an error of $\sim$ 25%, that is, $\sim$ 3 Gyr for GCs. Current estimates of cluster distances have an estimated uncertainty of $\sim$ 0.10–0.15 mag that translates into an indetermination of about $\sim$ 10% on absolute age. The "classical" distance determination method employed for Galactic GCs is the MS-fitting, but also calibrations of the HB absolute magnitude (and of its RR Lyrae population, when present) are often used as distance indicators for GCs.

The initial He content is known to within $\sim$ 2% (but, see also Section 7.8) that translates into a negligible $\sim$ 2% error in the age estimate. The iron content is known with an accuracy of $\sim$ 0.15 dex, and causes an uncertainty of about 4–5% in the absolute age determination.

In Section 3.9.1, we have shown that the initial distribution of the heavy elements affects the evolutionary lifetime during the MS, and hence the calibration of the *age–TO luminosity* relation. Roughly speaking, one obtains that an enhancement of the α-elements $[\alpha/Fe]$ of the order of 0.3 dex (at fixed [Fe/H]) causes a decrease of the age (at a given TO brightness) by $\sim$ 1 Gyr. Taking into account that, on average, for old stellar clusters the exact heavy elements distribution is known with an uncertainty of $\sim$ 0.15–0.20 dex, this translates into an uncertainty of $\sim$ 4% on age. The uncertainty on the *global* metallicity therefore contributes to the total error budget at the level of $\sim$ 9–10%.

Applying a very conservative approach, one can simply add the uncertainty associated to these "ingredients" to have a rough estimate of the *maximum* indetermination on absolute age, of the order of 20–25%, that is, $\sim$ 2–3 Gyr.

It is interesting to briefly summarize the most recent estimates of the age of one of the oldest GC in the Galaxy, M92 ([Fe/H] $\approx$ −2.4). The most recent absolute age estimates range from (11 $\pm$ 1.5) Gyr [469] to (14.8 $\pm$ 2.5) Gyr [407], with a mean value of the order of $\sim$ 13.5 Gyr and a spread of $\sim$ 3.5 Gyr. Recent analyses of the absolute age of a small sample of metal-poor GCs with very accurate photometric data, metallicity estimates and MS-fitting distances have provided a mean age of $11.6^{+1.4}_{-1.1}$ Gyr [470].

---

8) Although, the theoretical calibration of the *age–TO luminosity* relation has been significantly modified in the last decade, the quantitative estimates of these differential properties are basically unchanged.

7.3.2
How Reliable Is the Stellar "Clock"?

In the following discussion, we will assume to "perfectly" know both the distance and the chemical composition of the stellar clusters, and concentrate on the accuracy of the theoretical *age–TO luminosity* calibration provided by stellar models, that is the cornerstone of age determinations in stellar systems.

The main "inputs" that affect luminosities and effective temperatures of TO models are:

- EOS → luminosity, effective temperature
- Radiative opacity → luminosity, effective temperature
- Nuclear reaction rates → luminosity
- Superadiabatic convection → effective temperature
- Treatment of the boundary conditions → effective temperature
- Atomic diffusion → luminosity, effective temperature

Some "inputs" directly affect the *age–TO luminosity* relation by modifying the bolometric luminosity of the models. Others have an indirect effect on the TO magnitude in the chosen photometric band through the change induced in the bolometric corrections by a change in T_{eff}.

EOS: the importance of an accurate EOS when computing a SSM has been largely emphasized by all helioseismic analyses. In relatively recent times, the relevance of an accurate EOS for the *age–TO luminosity* calibration has been pointed out [471]. The treatment of non-ideal effects such as Coulomb interactions significantly affects the thermal properties of low-mass stars and then their core H-burning lifetime. The OPAL EOS [472] would imply a reduction of the GC age by about 1 Gyr (about 7%) when compared with the ages derived by using models based on less accurate EOSs. More recent updates of the OPAL EOS [473] have not brought additional relevant modifications to the *age–TO luminosity* calibration. Recent calculations like the FreeEOS project [474] agree with OPAL and we think that current residual uncertainties on the EOS for low-mass stars do not significantly affect the *age–TO luminosity* calibrations.

The radiative opacity: As a general rule, increasing the radiative opacity makes stars fainter (roughly speaking, $L \propto 1/\kappa$) that then take longer to burn their central hydrogen. For a given stellar mass, the TO luminosity is decreased and the time needed to exhaust H in the centre is increased. The two effects tend to balance each other, and the *age–TO luminosity* calibration is weakly affected. However, larger opacities also favour the envelope expansion, and therefore the TO is anticipated. In these last two decades, a large effort has been devoted to a better determination of both high- and low-temperature opacities. Concerning the high-temperature opacities, the largest contribution has been provided by the OPAL group [443] whose calculations represent a major improvement over the classical Los Alamos results. Thus, the question is: How accurate are the current evaluations of radiative opacity in the high-temperature regime?

When comparing the recent results, one finds that at solar metallicity, the mean difference is of about 5%. In the metal-poor regime, the opacity difference between independent evaluations ranges from $\approx$ 1% at conditions typical of a low-mass MS star centre, to about 4% at the base of the convective envelope.

However, the existence of a good agreement between independent estimates does not guarantee that the predicted opacity is equal to the "true" one. There is a general consensus that, at least for conditions appropriate to the core of metal-poor stars, current uncertainties should not be larger than about 5%. For temperatures of the order of 10^6 K, a larger uncertainty seems to be possible: it has been recently shown [475] that, for temperatures of this order of magnitude, a difference of the order of $\sim$ 13% does actually exist between the monochromatic opacities provided by the OPAL group and those provided by the Opacity Project. This notwithstanding, when assuming a reasonable average uncertainty of about 2–3% for the high-temperature radiative opacity, stellar evolution calculations reveal that the *age–TO brightness* calibration is left almost unaffected [224].

As for the low-temperature opacities, the effect on absolute ages is negligible, of the order of 1%, because the relatively hot TO models, especially in the metallicity regime of Galactic GCs, are hardly affected by opacities in this temperature range [42].

Nuclear reaction rates: In these last years, a large effort has been devoted to increasing the accuracy of cross section measurements at energies as close as possible to the Gamow peak. The effect on the *age–TO luminosity* calibration of uncertainties on the rates of the nuclear reactions in the *p–p chain* has been extensively investigated. For a realistic estimate of the errors, the effect on the derived ages is almost negligible, lower than $\sim$ 2%.

In Galactic GC-like populations, near the end of the core H-burning stage, the energy supplied by the H-burning becomes insufficient and the star reacts, contracting its core to produce the requested energy from gravity. As a consequence, both the central temperature and density increase and, when the temperature reaches (13–15) $\times 10^6$ K, H-burning is controlled by the *CNO cycle*, whose efficiency is determined by the rate of $^{14}N(p, \gamma)^{15}O$, the slowest reaction in the cycle. The TO luminosity depends on this reaction rate. The larger the rate, the fainter the TO, for example, $\Delta \log(L_{TO}/L_\odot) \propto -0.015 \Delta \sigma_{^{14}N}$. The core H-burning lifetime is only marginally affected by the rate, and is still mainly controlled by the efficiency of the *p–p* chain. The LUNA experiment has recently provided a new estimate of this critical rate [36]. The new rate is about a factor 2 lower than previous determinations. This lower rate for the $^{14}N(p, \gamma)^{15}O$ reaction leads to a brighter and hotter TO for a fixed age. This means that for a fixed observed TO brightness, the new calibration predicts ages systematically older by $\sim$ 0.8 Gyr.

The treatment of superadiabatic convection and outer boundary conditions: A change of the mixing length *only* alters the stellar radius and, in turn, the effective temperature (mainly along the RGB due to the larger extension of the superadiabatic layers with respect to less evolved evolutionary stages), leaving the surface luminosity (as shown in Figure 7.22) unchanged. The size of this effect depends on the extension of the superadiabatic region – larger for stars in the mass range

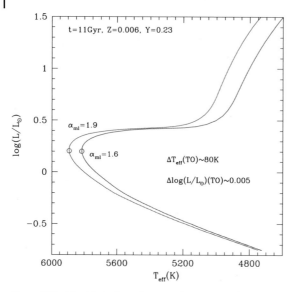

Figure 7.22 The HRD of two theoretical isochrones with the same chemical composition and age and two different values of α_{ml}: $\alpha_{ml} = 1.6$ and 1.9, respectively. The location of the MS TO is marked by the open circle. The associated changes in T_{eff} and bolometric luminosity at the MS TO are also listed.

$\sim 0.7 M_\odot - \sim 1.4 M_\odot$, because more dense less massive stars are almost completely adiabatic, while in more massive stars the convective envelope and the superadiabatic region are extremely thin – and, in turn, on the total star mass, thus altering the shape of the isochrones around the TO.

From the point of view of the *age–TO luminosity* relation, the uncertainty in the superadiabatic convection efficiency introduces an indetermination as a consequence of the induced change in the T_{eff}, and hence in the TO bolometric correction. Given that a solar calibrated value for α_{ml} seems suitable to reproduce the observed effective temperature of RGB stars in Galactic GCs over a wide range of metallicities (see discussion in Section 3.9.3), one expects that the superadiabatic convection efficiency is not a relevant source of uncertainty in the age determinations. Numerical experiments have shown that a change of 0.1 in the mixing length (with the BV58 choice of the other free parameters) causes a variation of about 1% in the absolute age of old star clusters [476]; assigning a maximum uncertainty of ~ 0.25 to α_{ml}, the age uncertainty is of $\sim 3\%$. The same numerical experiments show that the choice of the outer boundary conditions ($T(\tau)$ relations *versus* non-grey model atmosphere calculations) only has a negligible, if any, effect on the accuracy of the *age-luminosity* calibration.

Atomic diffusion and levitation: Helioseismology has brought to light the evidence that diffusion of helium and heavy elements must be at work in the Sun. It is logical to assume that this process is also efficient in more metal-poor low-mass objects like those currently evolving in Galactic GCs. On theoretical grounds, one expects atomic diffusion to be more efficient in metal-poor MS stars because of their thinner convective envelopes at fixed mass, and hence a smaller surface reservoir of He and metals. As for radiative acceleration, in the Sun it can amount to about 40% of the gravitational acceleration [477], and one expects that its value is larger in more metal-poor, MS stars due to their typically larger T_{eff}. As far as the *age–TO luminosity* calibration is concerned, however, the effect of radiative levitation is negligible compared to atomic diffusion.

For ages of the order of 10 Gyr, a change of the efficiency of diffusion by a factor of 2 corresponds to an age variation of about $-0.7/ + 0.5$ Gyr. However for larger ages the situation is even worst and, for an average age of 15 Gyr the corresponding age indetermination is equal to $-1.7/ + 1$ Gyr.

As discussed in Section 7.7, we are now facing a crucially important problem. Although helioseismology supports SSMs, including atomic diffusion, spectroscopic measurements of iron and Li in Galactic GCs (and Li in field halo stars) are in severe disagreement with the predictions of models that include diffusion and radiative levitation. These two contrasting empirical results can be at least partially reconciled if one adds to the SSM a slow mixing process below the convective envelope that could help explaining the observed Be and Li abundances [478] and improve the agreement between the predicted sound speed profile and that derived from helioseismic data [479]. This slow mixing, with the same or different efficiency, may be responsible for the apparent lack of photospheric signatures for the efficiency of diffusion and levitation in metal-poor stars in the Galaxy.

7.3.3
Relative Age Methods

We have emphasized that the calibration of absolute age for old stellar systems is still subject to observational and theoretical uncertainties at the level of at most $\sim 20\%$, although huge improvements in both the distance determinations thanks to the *GAIA* mission, and the metallicity scale are expected within the next decade. On the other hand, it is possible to employ differential methods, in the sense that magnitudes or colour differences are employed, to determine relative ages of GCs with a larger precision. These same differential methods can also be employed to estimate absolute ages, although with a precision comparable to the absolute methods discussed above [480]. These differential age indicators are independent of the population distance and reddening, and of uncertainties in the photometric zero points. When employed to determine just a relative age scale among clusters, one makes use of only differential properties of isochrones, thus minimizing the impact of uncertainties in the theoretical models.

These methods fall into two broad categories: those relying on magnitude differences, denoted as *vertical methods*, and those relying on colour differences, denoted as *horizontal methods*. Vertical methods employ, as an age indicator, the magnitude difference between TO and HB, this latter usually taken typically at a colour within the RR Lyrae instability strip, or at its red or blue edge. One makes use of V-band photometry where the HB is horizontal at the colours of the instability strip, and the age indicator is denoted as $\Delta V_{\text{TO}}^{\text{HB}}$. In *horizontal methods*, the colour difference

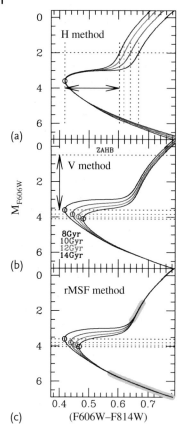

Figure 7.23 Differential age dating methods applied to a set of theoretical isochrones in the ACS photometric systems, for example the horizontal (H method) (a), the vertical (V method) (b), the relative MS-Fitting (rMSF) method. (c) method. Open dots mark the location of the TO. In the case of the rMSF method, shaded areas mark the portions of MS and RGB, employed in the "fitting" procedure.

between the TO and a selected point (at a given magnitude brighter than the TO brightness) along the RGB is adopted as age indicator; the $(B-V)$, $(B-I)$ or $(V-I)$ colours are commonly used, and these differential age diagnostics are denoted as $\Delta(B-V)_{TO}^{RGB}$, $\Delta(B-I)_{TO}^{RGB}$ or $\Delta(V-I)_{TO}^{RGB}$. Figure 7.23 displays an example of ΔV_{TO}^{HB} and $\Delta(B-V)_{TO}^{RGB}$ definitions in the equivalent *HST* filters.

In the case of ΔV_{TO}^{HB}, the HB brightness is largely unaffected by age, and the change of age at a given [Fe/H] affects this parameter through the change of the TO brightness; for increasing ages, ΔV_{TO}^{HB} increases because the TO becomes dimmer. As for a horizontal parameter like $\Delta(B-V)_{TO}^{RGB}$, one is taking advantage of the fact that the colour location of the RGB only has a minor dependence on the age of the stellar population, while the colour of the TO becomes redder with increasing age. As a consequence, the $\Delta(B-V)_{TO}^{RGB}$ decreases with increasing age.

From a theoretical point of view, the vertical method appears more robust because the luminosity of the TO predicted by theoretical models is not subject to large uncertainties, as discussed in Section 7.3.2, with the only notable exception of the efficiency of atomic diffusion. It is true that the theoretical uncertainty on the intrinsic HB luminosity can be relevant – see Section 7.2.3 – but this probably only affects absolute ages obtained with this method, not relative ages, for (almost) all sets of theoretical HB stellar models predict a similar slope of the $M_V - [M/H]$ relation. On the other hand, the horizontal method depends on the accuracy of theoretical colours that are affected by a combination of uncertainties on the mixing length calibration, choice of boundary conditions, and colour transformations. These considerations imply that the calibration of the horizontal method, even for relative age measurements, may be strongly model dependent. In passing, we notice that the vertical method is not completely immune from uncertainties affecting the T_{eff} scale of stellar models because, as discussed in Section 7.3.2, a change of the effective temperature of the TO affects its magnitude via the corresponding change of the bolometric corrections.

An additional difference between the two methods is the sensitivity of ΔV_{TO}^{HB} and $\Delta (B-V)_{TO}^{RGB}$ to age. For the vertical parameter, $d(\Delta V_{TO}^{HB})/dt \sim 0.1$ mag/Gyr for ages typical of Galactic GCs, while for the horizontal parameter, $d[\Delta(B-V)_{TO}^{RGB}]/dt \sim 0.01$ mag/Gyr. This implies that much stronger accuracy is needed in the theoretical predictions of $\Delta(B-V)_{TO}^{RGB}$ as well as in its empirical determination. Moving from $(B-V)$ to $(B-I)$, the sensitivity to age increases by a factor of ~ 2.5.

Theoretical models predict that both horizontal and vertical methods are not strongly dependent on $[M/H]$, at least for ages above ~ 10 Gyr, because the change of magnitude and colour of the TO due to a change of metallicity at fixed age is almost completely compensated by an equal change of the HB brightness and RGB colour. More precisely, Figure 7.24 shows that in case of the horizontal method, the dependence on $[M/H]$ is non-linear, especially when moving from metallicities above $[M/H] \sim -1.2$ dex to lower values. As for the sensitivity to the initial He content, the absolute values of ΔV_{TO}^{HB} as a function of age show a dramatic dependence on Y because the HB brightness strongly increases when the helium content increases.

From an observational point of view, the apparent magnitude of the TO can be hard to measure accurately, particularly in case of metal-poor systems due to the vertical morphology of their TO. This can lead to an uncertainty of ~ 0.1 mag in the measured TO apparent magnitude that translates into an uncertainty of the order of 1 Gyr in terms of relative and absolute ages. On the other hand, a vertical TO region represents an advantage in the case for the horizontal method because it enables an easier estimate of the TO colour.

An additional problem with the vertical method is related to the estimate of the HB brightness at the colours of the RR Lyrae instability strip that is problematic for old systems with blue or red HB morphology. This problem is usually overcome by comparing the observed HB with the HB of another cluster with the same metallicity and a significant population of RR Lyrae stars. A fit to the observed HB in the

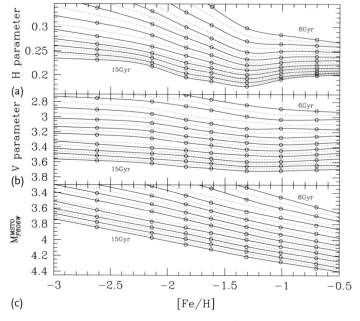

Figure 7.24 (a) The theoretical calibration of the horizontal method as a function of [Fe/H] and age in the ACS photometric system (see Figure 7.23). This calibration is based on α-enhanced stellar models from the BaSTI archive [432]. (b) As with (a), but for the vertical method. (c) The calibration of the absolute *F606W* magnitude of the TO. In each panel, solid lines denote steps of 1 Gyr; intermediate ages are displayed as dashed lines.

common colour range allows one to determine the corresponding HB level at the RR Lyrae strip.

For ages below $\sim 4\,$Gyr (the exact value depending on the chemical composition), the He core mass at the He-ignition also depends on the initial mass of the evolving stars, and hence on the age of the stellar population. Moreover, the HB is usually a red clump, far in colour from the RR Lyrae instability strip. Even if we instead consider the HB magnitude at a red colour as a reference point, there is the complication that at these ages, the ZAHB turns upward and then back towards bluer colours with increasing mass. There are therefore two possible ZAHB sequences at the same red colour, corresponding to different masses (see Section 4.5.3). Therefore, two different ΔV_{TO}^{HB} values could, in principle, correspond to the same age.

There are several detailed discussions of these differential age dating techniques [480–483]. We sketch here a practical application of these techniques, following [480]. The observed GC sample can be divided into a few metallicity intervals with $\sim 0.3\,$dex width, and in each interval, a reference cluster showing a HB well-populated at the instability strip region and/or at its red side is selected. The absolute age of these reference clusters is determined from the ΔV_{TO}^{HB}, and age differences within each metallicity bin with respect to the reference clusters are determined with the horizontal method.

More recently, a variant of these methods has been proposed that determines a ranking of cluster ages in terms of relative TO luminosity. This ranking is then translated into age differences by assuming a distance modulus for one of the clusters [245]. This new method, denoted as the *relative MS-fitting* (rMSF) method, is similar to the MS-fitting technique. In this case, one "shifts" the CMD of a given cluster until the non-evolved portion of the MS and the lower RGB (both independent of age, but dependent on metallicity) perfectly overlap those of a reference cluster with the same metallicity. This way, differences between the cluster distances and reddenings are eliminated and the different populations can be ranked in order of their relative TO magnitude, as shown in Figure 7.23. In some sense, the rMSF method represents an improved version of the traditional vertical method. The combined use of the RGB together with the unevolved portion of the MS produces unequivocal results.

Given that the slope of both the low MS and RGB depends on metallicity, the rMSF technique has to be employed among clusters with similar metallicity,[9] and hence accurate estimates of the cluster metal content need to be available. The outcome of this method is expected to be largely model independent because predictions of the *age–TO luminosity* are largely consistent among different independent sets of isochrones.

A recent implementation of this method has provided very accurate relative ages of a large sample of Galactic GCs [245]. The authors claim a formal accuracy of 2–7% on the relative ages defined in terms of ratios with respect to a reference cluster whose absolute age was determined using a distance modulus derived from MS-fitting. Their implementation can be summarized as follows: (i) the whole metallicity range of of GCs was divided into six metallicity bins and, in each bin a reference cluster was selected; (ii) in each bin, for any individual GC the TO magnitude difference with respect the reference cluster has been estimated; (iii) the TO magnitude differences between two consecutive metallicity groups have been determined by applying the rMSF method to their two reference clusters; (iv) the absolute age of one of the reference clusters has been determined assuming a MS-fitting distance modulus. This absolute age fixes the relative ages of all other objects compared to this cluster. Notice that the relative ages depend somewhat on the assumed reference absolute age given that the derivative $d\Delta V_{TO}^{HB}/dt$ depends on the age t, as clearly shown by Figure 7.24, especially when the age is below ~ 13 Gyr.

Figure 7.25 shows the derived relative ages, normalized to the reference age of $\sim$ 12.8 Gyr. We will briefly discuss this result as an example of the type of information that can be extracted from this type of analyses. Analogous investigations, along with other age dating methods, have painted a similar picture.

- Galactic GCs seem to fall in two distinct groups. The first group, containing the bulk of the GC population, consists of old, largely coeval clusters. The relative age dispersion is of the order of 5%; the average absolute age for the most metal-

---

9) The analysis in [245] has shown that the rMSF method can be safely applied in $\sim (0.3-0.4)$ dex metallicity bins.

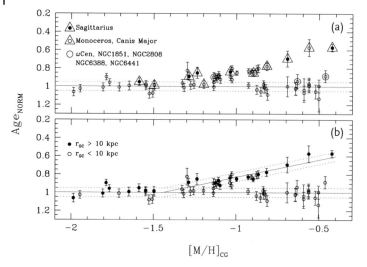

Figure 7.25 (a) Normalized relative ages for a sample of Galactic GCs as a function of [M/H] [245]. Different symbols denote the GCs associated to various tidal streams as well as those that clearly show photometric signatures of the presence of multiple stellar populations (see Section 7.8.2). (b) As with (a), but in this case, different symbols refer to GCs with a Galactocentric distance larg- er (full circles) or lower (open circles) than 10 kpc, respectively. In (a) and (b), the hor- izontal solid and dotted lines represent the mean normalized age and associated rms scatter for the old GC components. The ad- ditional sloped lines in (b) show the average age–metallicity relation and corresponding rms for the "young" GC branch (courtesy of A. Marín-Franch).

poor clusters is ~ 12.8 Gyr, and the corresponding absolute age dispersion is of the order of 0.6 Gyr. It can be reduced to an intrinsic age dispersion of ~ 0.4 Gyr when accounting for the observational errors. The second group is formed by GCs that show a clear age–metallicity relation (the so-called *young branch*), with the younger GCs being the more metal-rich ones. The intrinsic (i.e. corrected for the observational uncertainties) age dispersion around this age–metallicity relation is of the order of ~ 0.4 mag;

- Almost all Galactic GCs (with only one exception represented by NGC 5286) as- sociated to the stellar streams of Sagittarius, Monoceros and Canis Major (that probably represent the debris of dwarf galaxies accreted by the Milky Way) be- long to the *young branch*. The majority of the Galactic GCs showing the most spectacular manifestation of the "multiple populations" phenomenon (see Sec- tion 7.8) are distributed along the *young branch*. Many of these GCs are among the most massive stellar systems in the Galaxy. According to some scenarios, these massive GCs could actually represent the nuclei or the remnants of ac- creted Milky Way satellites [484]. Therefore, the majority of the GCs in the *young branch* could be tracers of process(es) of accretion during the early phase of for- mation of the galactic halo.

It is therefore plausible to associate the first stage of the formation of the Galaxy to a rapid collapse on timescales shorter than 1 Gyr [485], followed by a second phase of a larger time interval of about 6 Gyr that was able to form GCs with a range of ages and a clear age–metallicity relation. These GCs were probably formed within dwarf satellites of the Milky Way that were later accreted, although there is no clear explanation for the existence of a such a tight age–metallicity relation.

As for old clusters with ages between 1 Gyr and typical Galactic GC ages, another type of differential method can be found in the literature. The role played by these clusters in our galaxy is that they allow one to address the question of when the thin disk started to build up, relative to the thick disk and halo. A technique that can be found in the literature is described in the following (see, e.g. [486]). Two differential parameters are defined, δV and $\delta 1$; δV is defined as the magnitude difference between the cluster TO region and the He-burning RC stars, while $\delta 1$ is the difference in colour between the bluest point on the MS and the colour of the RGB one magnitude brighter than the TO luminosity (see Figure 7.26). A simple linear relationship between δV and $\delta 1$ does exist in Galactic old open clusters (OCs) that shows a RC in their CMD. This relationship can be used to estimate δV for sparse clusters without an apparent RC, but with a well-defined $\delta 1$.

The cluster δV values can be translated into absolute ages by determining a relationship δV-t-[Fe/H] based on a sub-sample of clusters with high quality CMDs, spanning the entire [Fe/H] and δV range of the full cluster sample, and for which the age can be determined with confidence. For Galactic OCs, there is a very tight relation based on a sample of eleven clusters with ages between ~ 1 and ~ 10 Gyr, as determined from their absolute TO magnitude (matched to theoretical isochrones with the appropriate metallicity) estimated from MS-fitting

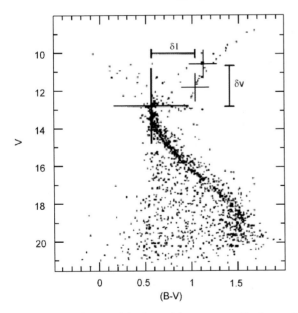

Figure 7.26 Definition of the δV and $\delta 1$ parameters for the age determinations. The CMD is of that the ~ 4 Gyr old Galactic open cluster M67 [487].

distance moduli [486]

$$\log(t) = 0.04(\delta V)^2 + 0.34\delta V + 0.07[\text{Fe/H}] + 8.76 \tag{7.16}$$

7.3.4
Ages from the WD Cooling Sequence

The termination of the WD cooling sequence observed in a thus far small sample of old Galactic open clusters (five clusters with ages above 1 Gyr) and GCs (NGC6307 and M4) provides us with an additional "clock" for stellar population age dating, complementary to the TO. As discussed in Chapter 6, the bottom end of the WD cooling sequence is sensitive to the population age once the distance modulus is fixed. The magnitude of the faint cut-off of the WD luminosity function is weakly sensitive to variation of the progenitors' initial metallicity, their mass function (for reasonable changes) and the exact form of the WD IFMR (at least for ages above a few gigayears). In particular, the IFMR and progenitor mass function tend to modify only the shape of the LF. Clearly, the most obvious source of uncertainty for this stellar clock is the CO stratification and composition and thickness of the envelope, as discussed in Chapter 6.

For typical GC ages, the relative fraction of H- and He-atmosphere objects also does not affect the observed cut-off magnitude because at these ages, He-atmosphere WDs cool down much faster than the H counterpart, and the termination of their cooling sequences are way below the detection limit of the current generation of imagers. In the observable magnitude range, the contribution of these fast evolving WDs is just an approximately uniform (with magnitude), low-number background of objects. The situation is, however, more complicated at ages of the order of 2–3 Gyr, that correspond to a luminosity range where the cooling timescales of WDs with He-envelopes are longer than for H-envelope WDs. Even more complications can arise in case of DA WDs with a sizable range of H-layer masses that undergo mixing with the underlying He-layer at luminosities where the evolutionary speed is very different for H- and He-envelopes. In this case, additional features in the stellar distribution on the CMD (and the associated luminosity function) can appear because of the changes in the evolutionary speed.

Another important parameter that can potentially alter the shape of the observed WD luminosity functions is the fraction of unresolved (non-interacting) WD+WD binaries, originated from MS binary systems. Theoretical cooling sequences that include unresolved WD+WD binaries have recently been investigated [488] by means of Monte Carlo simulations, according to the following procedure. A value of the WD progenitor mass for a generic single object along the chosen WD isochrone is extracted randomly according to a mass function (i.e. Salpeter). The mass and magnitudes of the WD are then determined from the WD isochrone. A companion to this WD is then assigned by randomly extracting a value of the MS progenitor binary mass ratio q according to an adopted statistical distribution, and then determining, if this progenitor is in the mass range that produces WDs, the corresponding WD mass and magnitudes from the same WD isochrone. The fluxes of the two components are then added, and the total magnitudes and colours of the composite system are computed. Given that the WDs at the bottom end of a cluster cooling sequence span almost the full range of progenitor masses, the most probable outcome for a WD+WD system corresponds to two WDs randomly selected from this clump. As the magnitude, at least in the visible wavelength range, is approximately constant for these objects, WD+WD unresolved binaries will populate a feature in the CMD similar to the bottom end of the cooling sequence, but ~ 0.75 mag brighter. This causes the appearance of an additional brighter peak, if the number of unresolved WD+WD systems is large enough, in the luminosity function, ~ 0.75 mag brighter than the cut off level. It is clear from this short discussion how the prediction of the exact shape of the WD LF and/or the exact "bi-dimensional" distribution of WDs in the observed CMDs are dependent on a number of parameters which are sometimes hard to predict theoretically, whereas, overall, the magnitude of the LF cut-off is a more "solid" prediction.

Figure 7.27 displays an estimate of the WD age for the GC M4 [489]. The distance and reddening have been estimated from ZAHB and MS-fitting using theoretical isochrones and ZAHB models with the appropriate chemical composition. For the assumed IFMR ($M_f = 0.54 \exp[0.095(M_i - M_{TO})]$ in $M_\odot$, where M_{TO} is the value of the stellar mass evolving at the TO point for the given cluster age), the shape of the WD luminosity function is best reproduced with a small fraction of WD+WD binaries and a 30% fraction of He-atmosphere WDs, and a power law progenitor mass function with exponent $x = 0.95$. Notice that the value of the mass function exponent depends on the IMF, but also on the dynamical evolution of the cluster. For this cluster (as for all old open clusters studied so far, with the exception of NGC6791), TO and WD ages are consistent, within the error bars.

In the case of the old super metal-rich ([Fe/H] ~ -0.3 to -0.4) open cluster NGC6791, there is instead a clear inconsistency between TO and WD ages [490, 491]. A fit of isochrones to the MS-TO-RGB-HB sequence in the CMD provides $E(B-V) = 0.17$, $(m - M)_V = 13.50$ (a distance in perfect agreement with results from the analysis of eclipsing binary systems in the cluster [492]), and a TO age of about 8 Gyr. The magnitude of the bottom end of the WD luminosity function corresponds instead to an age of ~ 6.0, when the same distance modulus is adopted. This disagreement vanishes when WD models including the effect of the so far neglected process of ^{22}Ne diffusion (expected to be efficient in the WD progeny of metal-rich populations, and discussed in Chapter 6) is included. Another peculiar feature of the NGC6791 cooling sequence is the presence of a second peak in the luminosity function, about ~ 0.75 mag brighter than the cut-off magnitude (see Figure 7.28). On the basis of the previous discussion about binaries, a natural and simple explanation for this feature is the presence of a sizable ~ 34% fraction of WD+WD unresolved binaries that corresponds to a not-too-unrealistic ~ 50% fraction of primordial MS+MS binaries [488].

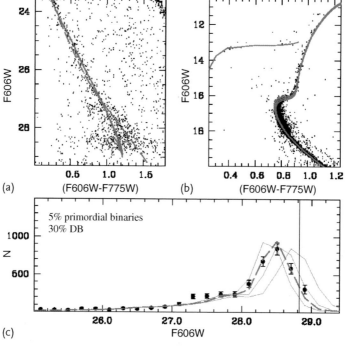

(a)

(b)

(c)

Figure 7.27 (a) Similarly for WD isochrones; solid lines denote H-atmosphere (DA) models, dashed line one He-atmosphere (DB) model isochrone. (b) The fit of theoretical isochrones to the observed CMD of M4, from the MS to the ZAHB, for ages of 11, 12 and 13 Gyr, [Fe/H] = −1.0, [α/Fe] = 0.4. (c) displays the fit of theoretical WD luminos- ity functions for 11, 11.6 (dashed line), 12 and 13 Gyr to the observed star counts. The parameters of the theoretical LFs are also displayed. The isochrone fits employ the same extinction (A_V = 1.2, with R_V = 3.8, typical of this cluster) and apparent distance modulus $(M − m)_{F606W}$ = 12.68 as the luminosity function fits.

7.4
Ages of Old Field Stars from Photometry/Spectroscopy

The determination of the age of individual field stars is less straightforward than in the case of determining the age of star clusters. Photometric observations of large samples of field stars in the galactic halo or disk will certainly contain objects distributed at different (generally unknown) distances, with different chemical compositions and possibly ages. If the distance (and reddening) is known for each star in a Galaxy field, it is theoretically possible to determine the ages (and [Fe/H]) on a star-by-star basis by fitting isochrones to the observed CMD of each object. Let's consider, for example, metal-poor stars similar to the Galactic halo population. Figure 7.29 displays scaled-solar isochrones with [Fe/H] = −1.3 and ages

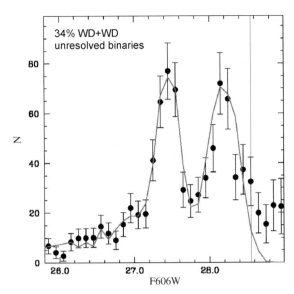

Figure 7.28 Comparison between observed (points with error bars) and the simulated, completeness-corrected WD luminosity function of NGC6791, assuming a fraction of WD+WD binary systems equal to 34% (corresponding to ∼ 50% of primordial MS+MS binaries). A vertical line shows the limit of reliability of the completeness correction.

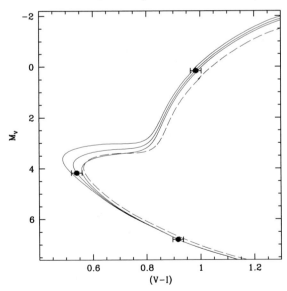

Figure 7.29 CMD of three scaled-solar isochrones with [Fe/H] = −1.3 and ages equal to 8, 10 and 12 Gyr, respectively (solid lines) plus a 10 Gyr isochrone with [Fe/H] = −1.0. The filled circles represent the distance and reddening-corrected photometry of three synthetic stars (generated from the 10 Gyr, [Fe/H] = −1.3 isochrone) with error bars equal to 0.05 mag in M_V and 0.02 mag in $(V-I)$.

of, respectively, 8, 10 and 12 Gyr, plus a 10 Gyr isochrone with $[Fe/H] = -1.0$.[10] Three synthetic stars are displayed with reasonably small error bars of 0.02 mag in $(V-I)$ and 0.05 mag in M_V. Notice how the faintest star on the low MS cannot really be assigned an age because it is very close to its ZAMS location. Its position is much more sensitive to $[Fe/H]$, and the error bars introduce an uncertainty of $\sim \pm 0.3$ dex on the derived metallicity. Also, in case of the other two synthetic stars, age determinations are not well-constrained and are degenerate with $[Fe/H]$. An increase of $[Fe/H]$ can be somewhat compensated by a decrease of the age to match the observed colours and magnitudes. In this situation, it is possible to only determine the statistical distribution of ages and initial chemical compositions rather than individual values of the observed stars by applying the techniques described in the next chapter based on fitting ensembles of isochrones/synthetic CMDs to the observed CMD of the whole population, not to the individual objects.

The additional knowledge of $[Fe/H]$ certainly improves the situation regarding the age determination of the synthetic near-TO star. Let's assume that the metallicity is known to be, to high accuracy, equal to $[Fe/H] = -1.3$. In this case, its age (with the given error bars) can be determined with an accuracy of $\sim \pm 2$ Gyr, whereas the age range for the RGB star is still extremely large.

Let's assume now that the mass of the three synthetic stars is also known, that is, they belong to eclipsing binary systems, or their masses can be derived from asteroseismic observations using the large frequency separation together with the frequency of maximum power (see, e.g. [493]). Figure 7.30 displays the run of $(V-I)$ and M_V as a function of the stellar mass for the isochrones in Figure 7.29, compared to the values for the three synthetic stars. We attached a 2% error bar to their mass estimates. An accurate knowledge of the stellar mass reduces the error bar on the near-TO star age, and also greatly improves the accuracy on the age of the RGB object. A word of caution is necessary regarding the RGB age determination. The displayed isochrones have been calculated including mass loss along the RGB with the Reimers formula, fixing the free parameter η to 0.4. The effect is a slow reduction of the evolving RGB mass starting from $M_V \sim 1.0$. The uncertainties on the prediction of RGB mass loss rates add an additional source of error to age estimates of individual RGB stars because they hamper a precise prediction of the evolving mass at a given colour/magnitude along an isochrone, especially along the upper RGB, where one expects an increasingly efficient mass loss.

If the stars belong to eclipsing binary systems, individual masses and radii can be, in principle, determined, and the mass-radius plane can be employed to determine accurate ages for the near-TO and the RGB objects (provided the chemical composition is known), as shown by Figure 7.31.

Strömgren photometry is often employed to study large samples of field stars with unknown distance. Strömgren colours like $(b - y)$, $c_1 = (u - v) - (v - b)$, $m_1 = (v-b)-(b-y)$ and $\beta = \beta_w - \beta_n$ (the two β filters, narrow and wide, measure the strength of the Balmer H_β line and its adjacent continuum) can be adopted to

---

10) In this discussion, we neglect the effect of uncertainties in stellar models and of the initial He abundance.

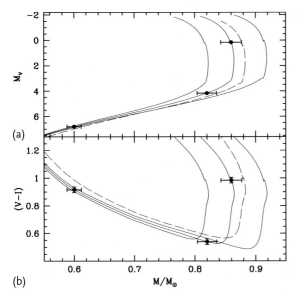

(a)

(b)

Figure 7.30 As in Figure 7.29, but for the (a) mass-M_V and (b) mass-$(V-I)$ planes. A 2% error bar on the estimated masses of the synthetic stars is also displayed.

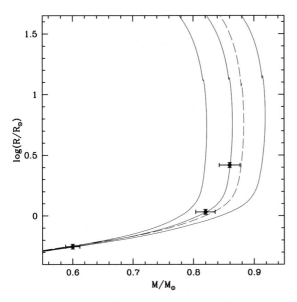

Figure 7.31 Example of age determination using the mass–radius plane. The four theoretical sequences and the synthetic observational points are the same as in Figures 7.29 and 7.30. A 5% error bar has been assigned to the estimated radii of the synthetic stars.

derive both reddening and [Fe/H] of individual stars (see, e.g. [494, 495]) based on relationships calibrated on local stars with known (generally zero) reddening and known [Fe/H]. Moreover, the c_1 index is a proxy for the stellar gravity, and a diagram like, for example, $c_1 - (b - y)$ is equivalent to a CMD, with the advantage of being insensitive to the star distances.

Figure 7.32 displays two examples of dereddened $c_1 - (b - y)$ diagrams for galactic field halo stars in two narrow [Fe/H] ranges, from the low MS to the RGB [495]. A comparison of the observed diagrams with appropriate isochrones can, in principle, provide an estimate of the age(s) of the observed population, mainly based on the colours of TO stars.

Ages of large samples of stars with available photometry, reddening and [Fe/H] determinations can be constrained by simply considering a colour-[Fe/H] diagram, as displayed in Figure 7.33 for the $(b - y)$ *Strömgren* colours of a sample of Galactic halo and thick disk stars [495, 496]. From well-populated diagrams, it is possible to estimate (i.e. by employing an edge detection algorithm) the $(b - y)$ of the blue cut-off of the stellar distribution (marked as solid line with the associated uncertainty denoted by dashed lines) that corresponds to the TO colour of the whole population for simple stellar populations, or of the youngest component in case of composite populations. A comparison of the blue cut-off as a function of [Fe/H] with isochrone TO colours provides the corresponding ages as a function of [Fe/H].

Figure 7.33 also compares the TO of the field halo stars with the TO of several Galactic GCs, showing a substantial consistency between GC ages and ages of the halo field (at least in the assumption that halo stars are coeval at a given [Fe/H]). A similar analysis of field halo stars has been recently performed [497], employing Sloan filter data from the SDSS survey. As found with Strömgren filters, also in

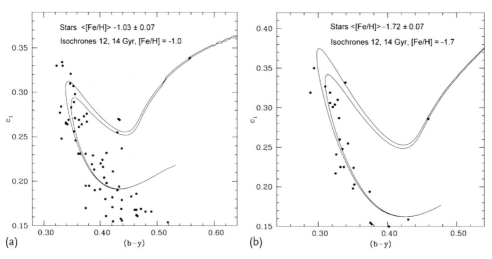

Figure 7.32 Age determination for samples of halo stars of similar metallicity with Strömgren photometry.

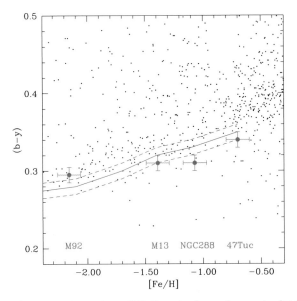

Figure 7.33 Comparison of TO $(b - y)$ colours of a sample of Galactic GCs (circles) with the estimated TO of the field halo population (solid line). The dashed lines represent the error on the TO colour estimate.

this photometric system, the TO colours of Galactic GCs are consistent with the blue cut-off of the field halo population.

A comparison of isochrones with the blue cut-off shows that when ignoring atomic diffusion in the models, ages of 14–16 Gyr are derived for the halo stars. This result is a potential argument against inhibited diffusion in old halo field objects, for it is in conflict with an age of the Universe equal to $t_U = 13.75 \pm 0.11$ Gyr, as derived from the currently accepted ΛCDM cosmological model. The age obtained, including atomic diffusion in the isochrones, is 10–12 Gyr, in agreement with the constraint imposed by t_U. This result is in agreement with an early analysis based on Strömgren colours [498]. Given the consistency of the blue cut-off for the field halo and the TO colours of GCs, cluster ages obtained from isochrones without diffusion are also in the range 14–16 Gyr, 2–4 Gyr older than ages obtained with the differential methods described before that are also less sensitive to the effect of diffusion compared to the use of the TO colour. Also, observations of [Li] and [Fe/H] in GCs, that seem to point to a very inefficient atomic diffusion, add an additional element of uncertainty to this problem.

Finally, we will briefly discuss another technique to determine the age of old populations in the field. This technique can currently be applied only to the solar neighbourhood, and makes use of the observed luminosity function of WDs the number of WDs per unit volume and luminosity/magnitude bin. Figure 7.34 displays an observational LF from [499], compared to the theoretical counterpart for two different ages. For an assumed star formation history (usually a constant star formation rate as a function of time, and single metallicity for the WD progenitors),

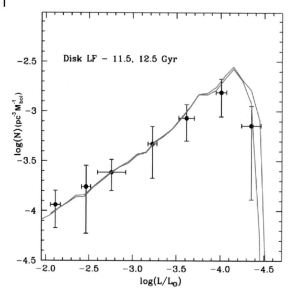

Figure 7.34 Comparison of the observed WD luminosity function in the solar neighbourhood [499] with theory. Theoretical LFs are normalized to the observed star counts at $\log(L/L_\odot) = -3.2$. A constant star formation rate with time and Salpeter IMF have been employed for the WD progenitors. The IFMR is from [334]. Only DA models have been employed [377].

IFMR and progenitor IMF, the position of the cut-off of the luminosity function is sensitive to the age of the oldest population. In this example, the position of the cut-off in the empirical LF provides an age $t = 11.5 \pm 1.0$ Gyr for the onset of star formation in the local disk. Uncertainties in this estimate (assuming the functional form of the star formation history is known) are related to the thickness of the H-layers (smaller layers increase the cooling speed and decrease the cut-off age) CO-core profile (a 50/50 mixture of C and O maximizes the effect of phase separation upon crystallization), IFMR (changes the progenitor evolutionary time at fixed cooling time), IMF (changes the shape of the theoretical luminosity function at the cutoff) and fraction of non-DA He-atmosphere objects (changes the shape of luminosity function around the cut-off, because the faster cooling times of He-envelope models tend to distribute the fainter non-DA WDs below the observed magnitude of the cut-off).

7.5
Empirical Rotation/Activity–Age Relationships

The idea of using rotation as a stellar clock is an old one, at least in the regime of solar-like stars. It is long known [500] that both stellar rotation periods increase (hence rotational velocities decrease) with age, approximately as the square root of age, due mainly to mass and angular momentum loss in a magnetized wind.

Theoretical predictions of the rotational evolution of stars are, however, difficult because the theory of angular momentum evolution in stars is very complex. One has to not only understand the origin of stellar rotation and initial distribution of angular momentum, but also its transport in stellar interiors and wind losses.

Observations of the evolution of rotational periods in nearby star clusters come to the rescue and provide important clues about how to use the evolution of rotational properties as a clock for low-mass MS stars, whose ages would be extremely difficult, if not impossible, to measure from their position in CMDs. Figure 7.35 shows measurements of Period (P) against colour ($B-V$) of samples of MS stars (mass in the range between ~ 0.6 and $\sim 1.4 M_\odot$) in the ~ 600 Myr old Hyades open cluster, and in the ~ 120 Myr old M35 open cluster. One can easily appreciate the large period spread of P values at fixed colour in M35, with two well-defined sequences. One sequence of fast rotators with period $P < 1$ day, independent of colour (denoted as sequence C in [503], or "convective" sequence, in the assumption that these objects lack large scale dynamos and are inefficient at slowing down their rotation), and a diagonal sequence of faster rotating/warmer stars and slower rotating/cooler stars (sequence I, or "interface", given the theoretical expectation that these stars are producing their magnetic flux near the convective-radiative interface). A comparison of the two clusters suggests that by the age of the Hyades, almost all stars along sequence C have moved onto sequence I. The stars populating the gap between these two sequences in the younger cluster are then interpreted as objects in transition from the C to the I sequence. The colour-P diagrams also suggest that the dependence of P on colour along the I sequence is the same in both clusters,

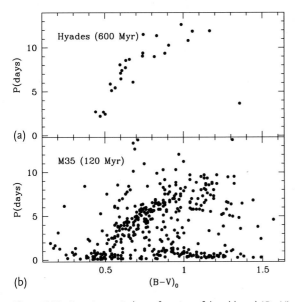

Figure 7.35 Rotation period as a function of dereddened ($B-V$) colour for a sample of MS stars in the (a) Hyades [501] and (b) M35 [502] open clusters. The corresponding mass range is between ~ 0.6 and $\sim 1.4 M_\odot$.

and hence the value of the rotation period P along this sequence can be expressed as the product of a mass (colour) dependence and other variables, amongst which age t is probably the most important one. A functional dependence of this type has been proposed [503]:

$$P = f(B-V)\,g(t) \tag{7.17}$$

with $f(B-V) = a[(B-V)_0 - c]^b$ and $g(t) = t^n$.

The most recent determinations of the coefficients a, b, c, n [502] provide $a = 0.770 \pm 0.014$, $b = 0.553 \pm 0.052$, $c = 0.472 \pm 0.027$, based on the I sequence in M35. The exponent n is determined by ensuring that the colour dependence gives the solar rotation period at the solar age, and gives $n = 0.519 \pm 0.007$. Age determinations based on this so-called "gyrochronology" rely on fitting the I sequence rotational isochrone determined by Eq. (7.17) with age as a free parameter, to individual field stars, or clusters (where one can determine the position of the I sequence even in the presence of fast rotators) in the colour-period diagram. This calibration is based on Galactic disk clusters younger than 1 Gyr, with a time-dependence covering ages up to ~ 5 Gyr based on the Sun. For single field stars, this method is obviously applicable in case of ages above the Hyades age, when the colour-P distribution is expected to converge to the tight I sequence modelled by the rotational isochrones. Fractional errors of order 10–20% are expected for ages determined from this method. Theoretical work is underway to extend the calibration to super-solar ages using an empirically constrained theoretical modelling of the angular momentum evolution in stellar models [504].

Chromospheric activity of cool stars is usually measured by the amount of emission in the cores of chromospheric lines such as the H and K lines of Ca_{II}, and the h and k lines of Mg_{II}. This emission is attributed to chromospheric temperature gradients generated by non-radiative heating generated through a magnetic dynamo mechanism induced by differential rotation, whose strength appears to scale with the observed surface rotation velocity.

Here, we describe an example of how the level of chromospheric activity can be used to date field stars, though of course the method can also be applied to clusters. Let's consider the traditional activity index, R'_{HK}, defined as the ratio of the emission from the chromosphere in the cores of the Ca_{II} H and K lines, to the total bolometric emission of the star.[11] A large set of measurements of R'_{HK} for stars in open clusters and young associations with known ages, in the range between ~ 10 Myr and ~ 6 Gyr, covering the $(B-V)_0$ colour range between ~ 0.5 and ~ 0.9, has recently been employed to determine the following relationship between age t (in years) and R'_{HK} [505]

$$\log(t) = -38.053 - 17.912 \log(R'_{HK}) - 1.6675 \log(R'_{HK})^2 \tag{7.18}$$

---

11) It is important to recall that the value of R'_{HK} is also subject to short timescale variability, for example, on timescales from days to years. Ideally, regular monitoring of active stars should provide average values of R'_{HK} to be employed in the relationships discussed in this section.

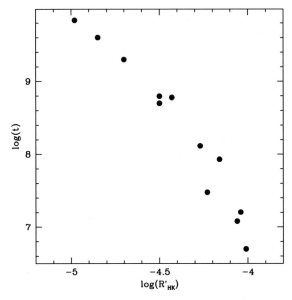

Figure 7.36 Mean $\log(R'_{HK})$ interpolated to solar $(B-V)$ versus the logarithm of the age (in years) of a sample of galactic open clusters [505].

In the absence of a simple way to parameterize age as a function of activity and colour, the R'_{HK} values used in this calibration are cluster mean values at the solar colour and are plotted in Figure 7.36 versus the cluster age. They have been obtained after reducing all observed R'_{HK} measurements in each cluster to the corresponding value at $(B-V)_0 = 0.650$, employing the $\Delta \log(R'_{HK})/\Delta(B-V)_0$ slope determined in each object. Values of $\Delta \log(R'_{HK})/\Delta(B-V)_0$ vary from cluster to cluster, ranging from ~ 3.0 to ~ -1.0, with no correlation to age. Obviously, given the dependence of R'_{HK} on colour at a given age, Eq. (7.18) is an oversimplification. Simply applying this activity–age equation to individual stars provides age estimates of $\sim 60\%$ accuracy (excluding contributions from uncertainties in the age of the calibrators).[12]

Combining activity and rotation measurements can, however, improve the situation. Rotation rate can be linked to dynamo strength via the Rossby number (Ro), which is defined as $Ro = P/\tau_c$, where τ_c is the convective turnover time

12) A metallicity dependent $R'_{HK}-t-[Fe/H]$ has been proposed [506]. It is based on samples of field disk stars with independent age determinations that provide a mean $R'_{HK}-t$ relation. Trends of the difference between stellar ages for the calibrating sample determined from this mean relation and ages from isochrone fits as a function of [Fe/H] are employed to determine the corrections for metallicity effects. This calibration is claimed to be reliable for $R'_{HK} < -4.75$ and [Fe/H] between -1.2 and $+0.4$. Relative errors associated to ages derived with this calibration are of the order of 40–60% [507]. This estimate comes from adding in quadrature the effect of a ~ 0.15 dex uncertainty in [Fe/H] (that contributes most of the final relative error) and the uncertainty due to the short term variability of activity, when only single measurements of R'_{HK} are available.

$\tau_c = H_P/v_c$. Here v_c is the convective velocity and H_P the pressure scale height near the base of the convective envelope, where dynamo activity is generated, that is, a quantity inferred from stellar models. A semi-empirical relationship [508] has been found between τ_c and $(B-V)$:

$$\log(\tau_c) = \begin{cases} 1.362 - 0.166x + 0.025x^2 - 5.323x^3 & x > 0 \\ 1.362 - 0.14x & x < 0 \end{cases} \qquad (7.19)$$

where $x = 1 - (B-V)_0$. Typical values of τ_c are of about 15 days in the colour range of interest. Empirically, one also finds a strong correlation between Ro and chromospheric activity for stars with $(B-V)_0$ in the colour range between ~ 0.5 and ~ 0.9, and $\log(R'_{HK}) < -4.35$ [505] given by

$$\log(R'_{HK}) = -4.522 - 0.337(Ro - 0.814) \qquad (7.20)$$

One can therefore use the observed mean values of R'_{HK} and stellar colour to estimate τ_c from Eq. (7.19), and then the period P after determining Ro from the inversion of Eq. (7.20). From P and $(B-V)$, one finally derived the age by applying Eq. (7.17). An application of this methodology to (presumably) coeval stars in resolved binaries and star clusters suggests that the derived ages have a precision of $\sim \pm 0.1 - 0.2$ dex (25–50% accuracy).

It is important to recognize that all these semi-empirical calibrations of rotation/activity as a function of colour/age have been obtained in the solar metallicity regime and it is not clear how reliably they can be applied to different chemical compositions.

7.6
Estimates of the Primordial Helium Abundance

The knowledge of the primordial He abundance (Y_P) is important to test the BBN predictions and determine the $\Delta Y/\Delta Z$ He enrichment ratio. The value of Y_P derived from spectroscopy of low-metallicity, extragalactic HII regions appears to be still subject to some systematic uncertainties. Current estimates range from $Y_P = 0.234 \pm 0.002$ to $Y_P = 0.244 \pm 0.002$ [509, 510]. On the other hand, according the currently accepted parameters of the ΛCDM cosmological model the baryonic matter density, coupled to BBN calculations, provide $Y_P = 0.248 \pm 0.001$ [511, 512].

The estimate of the initial He content in the oldest metal-poor stellar systems expected to be equal to Y_P, can be used to test BBN predictions, and also, in the case, for example, of Galactic globular clusters, to shed light on galactic chemical evolution. Moreover, the existence of an abundance spread at constant metallicity would strengthen the case for He to be the so-called "second parameter" that determines the HB morphology in Galactic GCs (see Section 4.3.4).

A direct measurement of the initial He abundance in HB stars is possible for stars hot enough ($T_{eff} \geq 8500$ K) to display the He feature at 5875 Å in the spectrum, but cold enough to avoid the problem of He sedimentation and radiative levitation of metals that affect HB stars hotter than $\sim 11\,000\,\text{K}$ [513, 514]. The constraint of this narrow T_{eff} and very high signal-to-noise ratio makes this analysis possible only for a few close Galactic GCs. To date, it has been applied only to NGC 6752 and NGC 6121 and the estimated He abundances are 0.245 ± 0.012 and 0.29 ± 0.01 (random) ± 0.01 (systematic), respectively. It is clear that this spectroscopic approach for measuring the initial He abundance on old star clusters is subject to strong observational limitations. As a consequence, one has to rely on indirect He indicators based on stellar evolutionary properties that are affected by the initial He content.

7.6.1
The Parameter *R*

The R parameter [515] is defined as the number ratio of HB stars to RGB stars brighter than the HB level $R = N_{\text{HB}}/N_{\text{RGB}}$. One can use different definitions for the HB level, but what is actually important is for the adopted definition to be self-consistently used for both the observational data and the theoretical calibration of the parameter. Here, we will consider the ZAHB luminosity at the instability strip. The parameter R is sensitive to the initial He abundance of the population mainly because a higher initial Y (at fixed metallicity) makes the HB brighter. This, in turn, produces a lower value of N_{RGB} because a smaller fraction of the RGB is enclosed between the HB level and the tip of the RGB, and R increases. The derivative dR/dY is equal to ~ 10.

Figure 7.37a shows the trend of R as a function of [Fe/H] for several values of Y, as predicted from theory. At a fixed age and Y, the R parameter is very slowly decreasing up to [Fe/H] ~ -1.1 – the exact value depending on the assumed He abundance – while between [Fe/H] ~ -1.1 and [Fe/H] ~ -0.9 it increases steeply, and at higher [Fe/H] values, R is again only very mildly decreasing with increasing metallicity.

The abrupt "jump" of R at [Fe/H] ~ -1.1 is due to the fact that the RGB bump, previously located at luminosities above the ZAHB, moves below the ZAHB level when increasing the metallicity, causing an abrupt decrease of the number of RGB stars brighter than the ZAHB. In fact, as discussed in Section 3.7, a RGB model spends about 20% of its RGB evolutionary time at the RGB bump location, as shown in Figure 7.37. The dependence of R on age is restricted to the interval corresponding to the "jump". This occurs because the RGB bump luminosity does also depend on the stellar age, as higher ages shift the RGB bump location towards lower luminosities, whereas the ZAHB level is basically unaffected for ages typical of Galactic GCs.

The theoretical calibration of the R parameter shown in Figure 7.37 has been obtained considering the typical HB lifetime of models with ZAHB location within the RR Lyrae instability strip, and the lifetime of RGB models brighter than the reference HB level. This simplification is possible because star counts along post-MS phases are proportional to the corresponding evolutionary timescales.

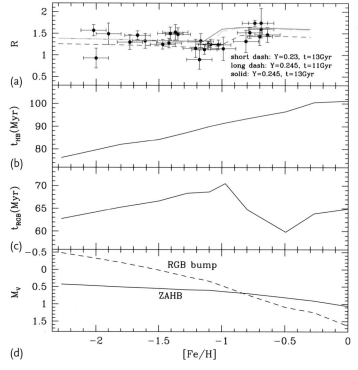

Figure 7.37 (a) Theoretical calibration of the *R* parameter as a function of [Fe/H] for various assumptions about the He abundance and the age. (b)–(d) show the trend with [Fe/H] of the core He-burning lifetime for low-mass models whose ZAHB location is inside the RR Lyrae instability strip ($Y = 0.23$), the time spent along the RGB at magnitudes brighter than the ZAHB luminosity, and the V-band magnitudes of both ZAHB and the RGB bump for $Y = 0.23$ and an age of 13 Gyr.

As discussed in Section 4.3.2 and shown in Figure 4.7, the timescale of the core He-burning phase increases (up to $\sim 20\%$ along the hotter portion of the HB) with decreasing mass. As a consequence, when applying this theoretical calibration of the *R* parameter to GCs with blue HB morphology, a correction needs to be applied. One can expect that for a cluster with $HB_{type} \sim 1$, this correction has to be of the order of 20%. The typical observational uncertainties affecting the empirical determination of *R* are of the order of 0.1–0.2, due mainly to the determination of the HB luminosity level and Poisson statistics; these observational errors cause an uncertainty of the order of $\sigma(Y) \sim 0.01$–0.02 in the estimate of *Y* for a single cluster. Recent analyses of the *R* parameter for a large sample of Galactic GCs provide values of *Y* consistent (within the current observational error bars) with a constant helium abundance, whose weighted mean is $Y = 0.250 \pm 0.005$, in agreement with Y_P and predicted by BBN calculations [474, 516].

7.6.2
The Parameter A

The pulsational properties of RR Lyrae stars provide another helium indicator for old star clusters hosting a significant population of variable stars in the instability strip. This indicator is the so-called A mass-to-luminosity ratio [517] already introduced in Section 4.5.1. The mass-to-luminosity ratio for stars inside the instability strip is defined as

$$A = \log \frac{L}{L_\odot} - 0.707 \log \frac{M}{M_\odot} \tag{7.21}$$

where L and M are the luminosity and mass of the individual RR Lyrae star. If the chemical composition is fixed, the parameter A only depends on the pulsation mode. Once the mode of pulsation is known, the parameter A is affected by He because increasing the abundance of this element increases the luminosity of the HB (thus, the luminosity of RR Lyrae stars) and also the value of the mean mass populating the instability strip at fixed metal abundances. In fact increasing the initial He content, it moves the ZAHB location of a stellar model moves towards hotter T_{eff}.

Indeed, the two effects tend to compensate each other and, as a consequence, the parameter A has a small sensitivity to Y, that is, $dA/dY \sim 1.4$, almost an order of magnitude less sensitive to He variation than the R parameter. However, in case of large numbers of RR Lyrae stars, the mean value of A can be determined with an accuracy of the order of ~ 0.01, inducing an uncertainty of only ~ 0.01 in the inferred He mass fraction. The dependence of A on [Fe/H] is practically negligible because an increase of metal abundance decreases the luminosity as well as the mean mass at the instability strip, and the two effects compensate each other.

From a practical point of view, the derivation of A for each individual variable star in a given GC relies on the fundamental pulsation equation (see Section 4.5.1) that can be rewritten as

$$\log(P) = 11.627 + 0.823A - 3.506 \log(T_{\text{eff}})$$

This relation shows that one needs to measure the pulsation period (only marginally affected by systematic uncertainties) and estimate the "equilibrium" effective temperature of the star, the T_{eff} corresponding to its non-pulsating hydrostatic equilibrium configuration. It is mainly the uncertainty in T_{eff} that hampers the use of A as a helium abundance indicator. Indeed, the T_{eff} scale for RR Lyrae stars is affected by non-negligible systematic uncertainties, partly due to shortcomings in the model atmospheres used to calibrate the relation between T_{eff} and "average" colour, and to the uncertainties regarding the choice of the photometric bands better suited to retrieve reliable mean colour estimates [518].

A possible approach to estimate the T_{eff} of RR Lyrae stars makes use of the following relation:

$$\frac{T_{\text{eff}}}{5040\,\text{K}} = (0.786 \pm 0.020) - (0.039 \pm 0.011) \cdot A_B - (0.008 \pm 0.009) \cdot [\text{Fe/H}]$$

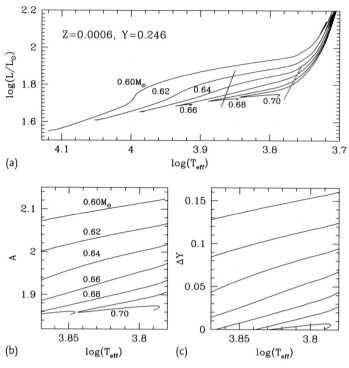

(a)

(b)

(c)

Figure 7.38 The evolutionary effects on the parameter *A*. (b) The change of *A* due to the evolution within the instability strip of selected HB stellar models shown in (a). (c) The derived spurious He increase induced by the variation of *A* due to evolutionary effects.

where A_B is the pulsational amplitude in the *B*-band. This calibration has been obtained from a sample of field RR Lyrae stars with [Fe/H] < -1.0. Due to the systematic uncertainty affecting the T_{eff} scale hence the parameter *A*, it would be safe to adopt this helium indicator for measuring relative He abundance differences rather than absolute He abundances. We note that the *A* parameter hence the corresponding He abundance determination, could be affected by the HB morphology of the selected cluster. In fact, in clusters with a very blue HB morphology (HB$_{\text{type}} > 0.7$), the majority of stars within the instability strip are expected to be very evolved, with a ZAHB location much hotter than the blue edge of the instability strip. As a consequence, these "evolved" RR Lyrae stars would have a larger luminosity and a smaller total mass with respect to RR Lyrae whose ZAHB location is inside the strip; one would therefore obtain anomalously large values of *A* that could be erroneously interpreted as due to a large initial He abundance (see Figure 7.38).

7.6.3
The Parameter Δ

Another He indicator for GCs is the so-called Δ parameter, defined as the magnitude difference between the ZAHB and the MS at a given colour, for example, at $(B-V)_0 = 0.6$ [517]; the choice of this reference colour is dictated by the requirement that it has to correspond to a point along the unevolved portion of the MS, unaffected by the age of the cluster. When the initial Y increases, the ZAHB becomes brighter and the MS at fixed colour becomes fainter because the whole MS is shifted to bluer colours when Y increases, as shown in Figure 7.39.

The sensitivity to the He abundance is reasonably large: $d\Delta/dY \sim 6.5$ mag, but its dependence on metallicity is also quite high, $d\Delta/d[\text{Fe/H}] \sim -0.5$ mag dex^{-1}. A typical uncertainty on [Fe/H] by 0.1–0.2 dex implies a 0.01–0.02 change in the derived helium abundance. This method is also affected by uncertainties in the cluster reddening because the magnitude difference between the ZAHB and the MS is measured at a specific dereddened colour along the MS.

These last considerations, together with the evidence that the Δ parameter can only be used if an accurate CMD extending ~ 2 mag below the MS TO is available, make this method less attractive.

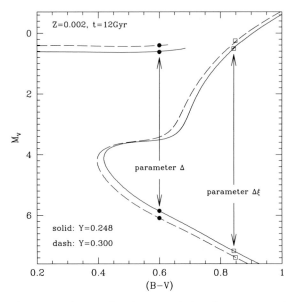

Figure 7.39 Theoretical isochrones and ZAHB for two assumptions about the initial He abundance (see labels). The arrows display the definition of the Δ and $\Delta\xi$ He indicators (see text for details).

7.6.4
The Parameter $\Delta\xi$

A new parameter has recently been defined, denoted as $\Delta\xi$ and shown in Figure 7.39. It is defined as the difference in magnitude between the RGB bump and a point along the MS at the same colour [519]. At a fixed age and metallicity, the location of the RGB bump becomes brighter when Y increases (due to the decrease of the envelope radiative opacity), while the selected point along the MS becomes fainter, and hence $\Delta\xi$ increases with increasing He content. At fixed age and metallicity, $d\Delta\xi/dY \sim 6.5$ mag, similar to the Δ parameter. This parameter is only slightly dependent on the age, that is, at fixed metallicity and Y, it decreases by ~ -0.03 mag for a 1 Gyr increase of the cluster age, and it is independent of uncertainties on photometric zero point and reddening.

Shortcomings stem from the need for accurate photometry extending from the RGB bump to the fainter portion of the MS, and sizeable samples of RGB stars to detect the RGB bump, plus the strong dependence on metallicity (RGB bump luminosity and both MS colour and slope are strongly affected by metals, due to their contribution to the low-temperature radiative opacity), that is, $d\Delta\xi/d[\text{Fe/H}] \sim -0.8$ mag/dex in the V-band. However the main concern is related to the theoretical calibration of $\Delta\xi$ in terms of Y. As discussed in Section 3.9.4, current theoretical RGB stellar models predict a location of the RGB bump ~ 0.15–0.20 mag brighter than observed, with a possible dependence on the initial metallicity. Until this discrepancy is solved, the calibration of $\Delta\xi$ remains affected by systematic errors, even though this parameter can be used to disclose He-differences between clusters with the same metallicity and known age using the relative behaviour of $\Delta\xi$ with respect to Y.

7.7
Estimates of the Primordial Lithium Abundance

Empirical estimates of the amount of lithium produced during the Big Bang Nucleosynthesis (BBN) are a powerful constraint/test of the baryon mass density Ω_b. In this discussion, we denote (as customary) the abundance of Li with [Li], defined as $[\text{Li}] = 12 + \log[N(\text{Li})/N(\text{H})]$, and only consider the main isotope ^{7}Li for the cosmological production of ^{6}Li is negligible in comparison with ^{7}Li.[13]

To date, the only way to test BBN predictions about the cosmological abundance of Li is to measure [Li] in the atmosphere of metal-poor MS stars in the Galactic halo [520]. The single accessible Li line is the Li resonance doublet at 6707.8 Å, that is only sensitive to neutral Li, even though in the stars of interest, Li is mostly singly ionized. One has to therefore introduce an ionization correction that is strongly sensitive to temperature. Typical errors on the spectroscopic determination of [Li] are $\Delta[\text{Li}]/\Delta T_{\text{eff}} \sim 0.07$ dex/100 K for MS/SGB stars, and

---

13) Spectroscopic observations of lithium abundances in stars determine the sum ^{7}Li + ^{6}Li.

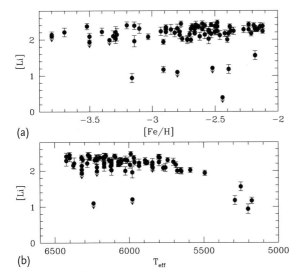

(a)

(b)

Figure 7.40 Trend of [Li] abundances with [Fe/H] (a) and T_{eff} (b) for a sample of metal-poor MS stars [521]. The flat part of the [Li] distribution in (b) is called the "Spite-plateau".

$\Delta[Li]/\Delta T_{eff} \sim 0.09\,dex/100\,K$ for RGB stars. The main result of these investigations (see Figure 7.40 that displays, as an example, data by [521]) is that metal-poor MS stars with [Fe/H] below ~ -1.5 and T_{eff} larger than approximately 5800 K show a remarkably constant [Li] value, denoted as "Spite-plateau", while there is a larger depletion at lower temperatures, increasing for decreasing temperature. Moreover, in the plateau region, a handful of stars show a much lower [Li] than the plateau counterpart. The exact T_{eff} location, the temperature extension of the plateau, as well as the existence of some weak trend of [Li] with T_{eff} and [Fe/H] are still debated. Also, the absolute average value of [Li] for plateau stars shows differences between different authors, ranging between [Li] ~ 2.1 and [Li] ~ 2.4, mainly caused by uncertainties in the T_{eff} scale for halo stars.

The simplest empirical explanation for the existence of this plateau is that it reflects the primordial Li abundance due to the low metallicity of the observed stars. A reliable interpretation of these observed abundances must however take into account the constraints posed by stellar evolution models. Lithium is a fragile element, destroyed in stellar interiors when $T \geq 2.5 \times 10^6$ K. These temperatures are already attained in the stellar cores during the contraction to the MS phase, and whenever surface convective regions extend down to these Li burning regions, the surface value of [Li] decreases rapidly. At a fixed chemical composition and for decreasing total mass, or for increasing metallicity at a given mass, PMS stellar models show that convection extends from the surface down to increasingly deeper regions, where eventually Li starts to be burned. Although the bottom of this convective region will then retreat rapidly towards the surface, in some metallicity/mass regime, there is time to burn a substantial amount of Li during the PMS phase. If the stellar mass is small enough, the bottom of the convective envelope

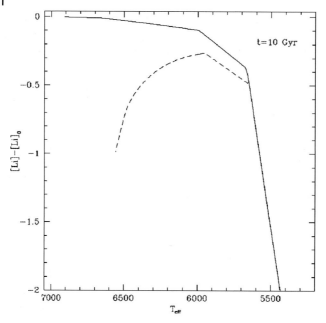

Figure 7.41 Trend of [Li] depletion (actual [Li] minus initial [Li]$_0$) with T_{eff} along the MS of a theoretical isochrone with $[\alpha/Fe] = 0.4$, $[Fe/H] = -2.0$ and $t = 10$ Gyr with (dashed line; the effect of radiative levitation on [Li] at this age and metallicity is negligible) and without diffusion for an initial lithium abundance [Li] = 2.5 (the depletion is largely insensitive to the exact value of the initial [Li]).

continues to overlap with Li burning regions also during the MS phase, and the surface Li depletion is expected to be much larger.

In the metallicity regime of "Spite-plateau" stars, objects with mass below $\sim$ 0.65 $M_\odot$ continue to deplete the surface lithium during the MS, whereas stars more massive than $\sim$ 0.70 $M_\odot$ do not appreciably deplete their surface lithium, even along the PMS. Figure 7.41 displays the predicted run of Li depletion as a function of T_{eff} along a 10 Gyr isochrone for an initial [Li] = 2.50, but a different choice of the initial value does not affect the result. The shape of this depletion curve closely mirrors the observations as the inclusion of isochrones with different metallicity spreads the theoretical points mainly in the horizontal direction, with an almost flat part and a sudden decrease of [Li] at the lowest temperatures. This sharp decrease is explained by lower mass (cooler) MS stars that undergo substantial depletion of their surface Li during both PMS and MS phases. The flatter part, that is, the theoretical counterpart of the "Spite-plateau", corresponds to those MS stars that are massive enough to avoid substantial depletion of Li during the PMS phase, and therefore reflects the initial [Li] abundance in these objects.

This picture of stellar evolution, where the only mixing mechanism is convection, can explain the appearance of the "Spite-plateau" in terms of a constant initial Li abundance in metal-poor halo stars that is approximately equal to the BBN value. The "problem" with this scenario is that [Li] values measured for the plateau are a factor 2–5 lower than the value [Li] = 2.72 ± 0.06 predicted by BBN calculations for the current accepted values of the cosmological parameters [522]. A way to address this discrepancy is to consider atomic diffusion in the calculations (plus radiative levitation, when efficient) that is necessary to improve the match of theory with helioseismic observations. The effect of diffusion is to reduce the surface abundance of Li during the MS phase, for all elements heavier than hydrogen tend to sink below the convective envelope boundary. The observed [Li] along the plateau will therefore appear reduced compared to the initial value. However, as displayed in Figure 7.41, the amount of Li depletion along the MS is strongly mass-dependent (thus, T_{eff}-): More massive MS stars, closer to the TO, display a stronger [Li] depletion because of the reduced size of the convective envelopes, and hence a smaller Li buffer in the fully mixed surface regions. Moreover, the TO depletion is also a function of the precise age and the initial metallicity (the depletion decreases both with increasing metallicity at fixed age, and with increasing age at fixed metallicity) because of the varying size of the convective envelopes. It is difficult, if not impossible, to reproduce the observed plateau and its narrow thickness when atomic diffusion is included in model calculations.[14] Notice that when diffusion is efficient, the observed [Fe/H] of MS stars is also lower than the initial value. The correspondence between observed and initial [Fe/H] will also depend on the stellar mass, age and initial metallicity, with the same qualitative behaviour of Li.

Radiative levitation plays a role only for initial [Fe/H] values below ~ -2.0, and is only able to moderate the amount of [Li] depletion compared to the case with just diffusion. The case of [Fe/H] is different. Thanks to radiative levitation, the Fe abundance at the TO of old isochrones with [Fe/H] below ~ -2.0 can reach values higher than the initial one, without erasing the drop of [Li] close to the TO of old metal-poor isochrones [524].

On the basis of these results, the focus has shifted on finding ways to moderate the efficiency of diffusion and at the same time, maintaining a flat surface [Li] profile along the plateau. The currently prevailing idea is based on the analysis of the Li abundance profile below the convective envelope of old, metal-poor TO models, shown in Figure 7.42. The solid line displays the abundances (relative Li abundance compared to the initial value) for a representative old, metal-poor model calculated without atomic diffusion. The profile is uniform both within the convective envelope and below, down to the temperature where Li is burned efficiently. The dashed line displays the Li profile for the same object, calculated this time with fully efficient diffusion. The convective envelope ends at $\log(T) \sim 6.1$, and its Li abundance is depleted by ~ 0.25 dex compared to the initial value. Below the convective boundary, the Li abundance increases, thanks to the Li diffused below the convective zone, reaching a local maximum at $\log(T) \sim 6.3$, before a steep drop due to the onset of Li burning. The idea to "solve" the "Spite-plateau" problem is

---

14) The investigation in [523] shows with extensive Monte Carlo simulations that with current uncertainties on the observed T_{eff} and [Fe/H] of plateau stars, and the current sample size of MS stars with [Li] estimates, in the assumption of no strong selection biases towards TO stars, a plateau can be also obtained from models including diffusion.

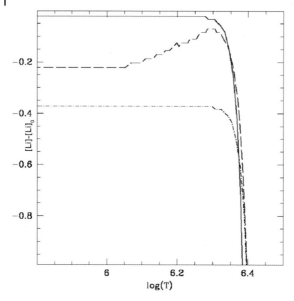

Figure 7.42 Li profile (relative to the initial value) around the bottom of the convective envelope of a MS low-mass star at the TO. The solid line displays a model without diffusion, the dashed line, the same model with efficient atomic diffusion, and the dash-dotted line denotes a model with atomic diffusion and additional turbulence effective at temperatures around $\log(T) = 6.3$ (in K).

to include some ad hoc turbulent mixing in the radiative zone below the convective boundary that limits the efficiency of Li settling. This turbulent mixing has been modelled [524] as a slow diffusive process that smooths chemical abundance gradients, with diffusion coefficient D_T given by

$$D_T = 400 \, D_{\mathrm{He}}(T_0) \left[\frac{\rho}{\rho(T_0)} \right]^{-3} \tag{7.22}$$

According to this parametrization, D_T is 400 times larger than the He atomic diffusion coefficient at a given value of T_0 (a free parameter) and scales as ρ^{-3}. If $\log(T_0)$ is set, for example, to 6.0–6.1 (labelled as T6.0 and T6.1 models), this hypothetical turbulent mixing, whose origin is not specified but may perhaps be induced by rotation, will work in the direction of moving Li back towards the convective envelope in order to erase the positive composition gradient for $\log(T) < 6.3$. The net effect is to decrease by 0.15 dex the surface depletion produced by diffusion. If $\log(T_0)$ is set around 6.3 (T6.3 models), this turbulent mixing also enables the transport of additional Li down to burning temperatures, and profiles of the type displayed with a dash-dotted line can be produced. In summary, a suitable calibration of this parameter, namely, a T6.25 model, may produce a flat Li abundance along the plateau, with [Li] depleted by about 0.4 dex with respect to the initial value, for an age of 13.5 Gyr.[15] This depletion, when applied to the range of [Li] measured by different authors, can potentially bring into agreement the Spite plateau with BBN predictions.

An alternative solution that may be viewed as a physical justification for the ad hoc turbulent mixing solution described before has been proposed, invoking the effect of a rotationally induced tachocline mixing as employed in solar models to improve the agreement between theoretical and observed sound speed [525]. The tachocline mixing is a rotationally induced instability related to the differential rotation with latitude, occurring at the transition region between the radiative core and the convective envelope of solar-like stars. It induces a slow mixing below the base of the convection zone that works in the direction of opposing the formation of chemical composition gradients, and in a first order approximation can be treated as a diffusive process, with its own diffusion coefficient. There are two free parameters to be calibrated: the width of the tachocline, that is, the width of the region where rotation goes from differential in latitude (in the convection zone) to solid body rotation (in the radiation zone) and calibrated from solar seismic measurements to 2.5% of the total radius, and the buoyancy frequency, estimated to be equal to $10\,\mu$Hz. Tachocline mixing is able to moderate microscopic diffusion at the hot edge of the plateau, and the analysis in [525] investigates the calibration of the associated free parameters and the dependence on the assumed rotation history. As a result, models including atomic diffusion and tachocline mixing appear to be able to produce a uniform [Li] along the plateau, with a depletion of ~ 0.3 dex compared to the initial abundance. Given the range of [Li] estimates, the agreement with the BBN is marginal at best.

A third, less discussed solution, is related to mass loss [526, 527]. For "small" mass loss rates (at most, about a couple of orders of magnitude larger than the solar value of $\sim 10^{-14}\,M_\odot$/year), the internal changes of the stellar model are negligible, and the only effect is to replace, inside the convective envelope, the flow of matter from the surface with matter coming from the below the convection boundary in order to satisfy mass conservation. It is therefore obvious that this process moderates the effect of diffusion from the convective region. Constant mass loss rates during the MS lifetime of metal-poor low-mass stars, with values between $\sim 10^{-13}\,M_\odot$/year and $\sim 10^{-11.5}\,M_\odot$/year for objects at the hottest end of the plateau, and rates $\sim 10^{-12}\,M_\odot$/year at the cool end are required to preserve a flat [Li] profile along the plateau, when atomic diffusion is efficient. However, the Li depletion along the plateau is found to be relatively small, of the order of ~ 0.2 dex. Observationally, we still have no information about mass loss rates in halo MS stars.

---

15) Actually, this has been demonstrated to be valid only for models with initial [Fe/H] $= -2.3$. No detailed analysis has been yet performed to assess the consistency of theoretical predictions including both diffusion, this parametrization of turbulence and varying initial [Fe/H], with measurements of both [Li] and [Fe/H] in plateau stars.

7.7.1
Li in Globular Clusters

Galactic GCs provide an alternative way to investigate the primordial Li abundance, for they host samples of stars that are approximately coeval, with uniform initial abundances of several elements (Fe among them) and whose evolutionary status is easily determined from the CMD. Also, GCs offer the additional bonus of allowing one to trace the diffusion effects along the entire evolutionary path of low-mass, metal-poor stars. In fact, the amount of diffusion can be easily measured from the offset of the surface abundance of elements such as Fe (that are not involved in nuclear burnings) in TO stars[16] (where the diffusion affects appreciably the photospheric abundances) and the RGB (where convection restores the original photospheric abundances). To clarify this point, Figure 7.43 displays the theoretical evolution of the surface Li abundance as a function of T_{eff} from the PMS to the lower RGB for a star currently at the TO of a [Fe/H] $= -1.0$, [α/Fe] $= 0.4$, 12 Gyr old GC. Solid lines represent the evolution without diffusion, and the dashed line represents the evolution with fully efficient diffusion. After the minimal PMS depletion, the surface Li stays constant until the TO is reached for the non-diffusion model. After the TO, due to the deepening of the convective envelope, the surface Li abundance decreases because convection engulfs deeper layers where, during the MS, the temperature was high enough to burn Li. After convection has reached its maximum extension (end of the first dredge up), the surface Li is predicted to stay constant because the dilution has ended. The case with diffusion is different. While approaching the TO, the surface Li decreases, reaching a minimum around the TO point. After the TO there is at first an increase of [Li] because part of the Li diffused below the envelope, that is, the local maximum in the Li abundance right above the Li burning boundary, as seen in Figure 7.42, is re-engulfed by the deepening convection, followed by a steep decrease due to dilution, to finally reach the final post dredge up constant value. The evolution of Fe is similar qualitatively to Li during the MS, but after the TO, the surface abundance of Fe can only increase up to almost the initial value because, given that Fe is not involved in nuclear burnings, a major fraction of the Fe diffused below the convective boundary is restored into the deepening convective envelope.

Spectroscopic measurements of [Li] and [Fe/H] in TO-SGB-RGB stars for a, admittedly small, sample of Galactic GCs have so far provided puzzling results. In the case of the metal-poor GC NGC6397 ([Fe/H] ~ -2.0) observations [414, 528] have disclosed a small difference between [Li] measured at the TO and on the SGB (lower abundances at the TO), a clear signature of atomic diffusion. Also for Fe, the value of [Fe/H] at the TO is lower than on the lower RGB, as expected from models with diffusion. However, from the quantitative point of view, these measurements show that diffusion cannot be fully efficient. In fact, these authors find that a turbulence as in [524], parametrized as T6.0 or T6.09, allows a good match to the spectroscopic result. With this turbulence parametrization, the estimated initial

---

16) To date, there are no data about [Li] below the TO of Galactic GCs.

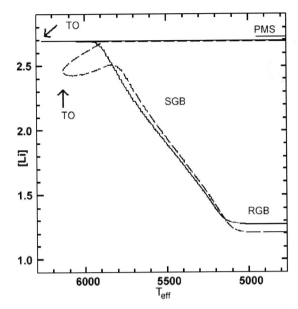

Figure 7.43 Evolution of the surface [Li] as a function of T_{eff} for 0.855 $M_\odot$, α-enhanced, [Fe/H] $= -1.0$ stellar models without (solid line) and with (dashed line) the inclusion of atomic diffusion. The PMS, TO, SGB and RGB regions are marked in the figure.

[Li] results to be about 0.2–0.3 dex lower than the BBN value. A completely different result has been found in the more metal-rich GC ([Fe/H] ~ -1.1) M4 [415]. Here, the abundance of Fe displays a uniform value from the TO to the RGB. Also, the measured [Li] is flat from the TO to the T_{eff} where surface dilution starts, without the local SGB maximum that is a clear signature of diffusion. In this case, cluster diffusion is either completely inefficient, or it has been inhibited with a turbulence model where T_0 is high enough to operate to the hot side of the local maximum seen in Figure 7.42, that is, at temperatures $\log(T) \sim 6.30$–6.35. This latter mechanism would erase the peak in the Li profile right above the edge of the Li burning region, thus erasing the local maximum in the surface Li during the SGB phase.

Figure 7.44 displays, as a summary, [Li] mean values for TO stars in the GCs studied so far, as a function of the initial [Fe/H] measured from RGB stars. Error bars denote the dispersion around the mean, that is, 47 Tuc, and NGC6752 displays a range of [Li] values that are probably a byproduct of the observed CNONa abundance anti-correlation patterns (see Section 7.8), normalized to the root mean square of the number of sampled stars [415]. The dark grey area displays the range of values obtained for field plateau stars halo stars by different recent investigations. This large range is not due to an intrinsic spread, but rather to the different values obtained by different investigations. The solid line displays the theoretical prediction for TO stars with an age of 11–12 Gyr, in case atomic diffusion is fully efficient, starting from an initial BBN value. The vertical-dashed line shows the boundary for the radiative levitation to start affecting the photospheric chemical

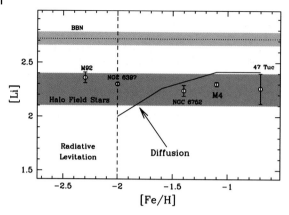

Figure 7.44 [Li] mean values for TO stars in the GCs studied to date as a function of the initial [Fe/H] measured from the RGB stars. Error bars denote the dispersion around the mean, normalized to the root mean square of the number of observed stars. The dark grey area displays the range of values obtained for "Spite-plateau" stars by different recent investigations (see text for details). The solid line displays the theoretical prediction for surface [Li] in TO stars for an age of 11–12 Gyr, in case atomic diffusion is fully efficient, starting from an initial BBN value [Li] $= 2.72$. The vertical-dashed line shows the boundary for the radiative levitation to start affecting the value of [Li]. The primordial [Li] provided by BBN calculations and the associated uncertainty are also displayed (light grey region).

abundances of Li. Notice how, apparently, consistency with BBN results can only be achieved by enhancing the TO depletion due to diffusion in the more metal rich clusters, possibly using a turbulence model that enhances the amount of Li brought into the burning region, whether the opposite is true for more metal-poor clusters.

7.7.2
Li in Lower RGB Stars

A complementary approach has been very recently developed in [529], where the use of lower RGB stars is proposed as empirical diagnostic of the cosmological Li. Both theory and observations [117] suggest that the surface Li abundance in metal-poor RGB stars after the completion of the first dredge up and before the RGB bump[17] is significantly less sensitive to the efficiency of atomic diffusion, compared with MS stars. The surface Li abundances in these objects, after dilution during the first dredge up, are predicted to be sensitive to the total Li content left in the star, that is, they are determined only by the initial Li abundance minus (apart from the small PMS depletion) the total amount eventually brought into the Li burning region during the previous MS phase. Stellar models computed under different physical assumptions (see Table 7.4 for a summary) show that the inclusion of atomic diffusion has an impact of about 0.07 dex in the determination of

---

17) After the RGB bump surface abundances of elements like C, N and Li change due to some mixing mechanism not routinely included in stellar models, probably thermohaline mixing.

Table 7.4 Li abundance depletion (Δ(Li), in dex) along the lower RGB for several [Fe/H] values and a RGB age of 12.5 Gyr. Results for models without diffusion, with the inclusion of fully efficient diffusion, and without diffusion, but accounting for overshooting below the envelope convection, are displayed. An overshooting length equal to $0.35H_P$ for the metal-poor models up to [Fe/H] $= -2.14$, and $0.10H_P$ for more metal-rich models have been considered. This calibration brings into agreement observed and theoretical RGB bump luminosities in a sample of Galactic GCs (models without overshooting predict a too bright RGB bump, due to a too shallow depth of the convective envelope at its maximum extension, see Section 3.9.4) based on [219] data.

[Fe/H]	Δ(Li) (no diff)	Δ(Li) (diff)	Δ(Li) (oversh)
-2.62	1.30	1.37	1.31
-2.14	1.33	1.40	1.34
-1.31	1.40	1.46	1.41
-1.01	1.44	1.51	1.50

the initial Li abundance – much smaller than the case of metal-poor TO stars (Figure 7.43) and it is basically unaffected by reasonable variations of other parameters (overshooting from the bottom of the convective envelope, age, initial He abundance and mixing length).

The initial Li for a sample of field halo and GC lower RGB stars has been inferred by accounting for the difference between initial and post-dredge up Li abundances in the appropriate stellar models. Different T_{eff} scales have been employed to assess the impact of this error source. The final [Li] estimate spans a relatively narrow range, between ~ 2.3 and ~ 2.5 dex. Most of this range is due to the uncertainty in the T_{eff} scale, and is ~ 0.2–0.4 dex lower than predictions from BBN calculations. This result is an independent estimate of the difference with the BBN value. Consistency with BBN rules out parametrizations of the turbulence that reduce the efficiency of diffusion. Instead, it requires turbulence to be efficient in deeper layers in order to increase the amount of Li brought into the burning region during the MS.[18]

Alternatives to explain the discrepancy with BBN have focused on either changes to the initial Li abundance at the formation of halo stars, or modifications to the predictions of BBN. Ideas involve a first generation of stars that has processed and efficiently depleted Li in a substantial fraction of the early halo baryonic matter, modifications to the BBN considering the decay of unstable particles, and modifications to the reaction cross sections for the Li production during BBN (see [530] for a recent review).

---

18) The turbulence models denoted as T6.25 or T6.28 by [524] appear to burn approximately the right amount of additional Li, at least in stellar models with [Fe/H] $= -2.31$.

7.8
Multiple Stellar Populations in Galactic Globular Clusters

For several decades, star clusters have been considered to be the best examples of SSPs. Apart from very few exceptions represented by the ω Cen and M54 that show a well-known [Fe/H] spread (~ 1.5 and ~ 0.2 dex, respectively), Galactic GCs seemed to represent ideal SSPs. The most obvious evidence supporting this identification of GCs as SSPs was the existence of well-defined MS-TO-SGB-RGB sequences – showing no splitting and/or spread apart from that associated to the photometric errors and/or to the presence of differential reddening and binary stars – in the CMD of each cluster, and a general homogeneity of the chemical composition of stars within individual clusters, as far as iron, iron-peak elements and α-elements were concerned.

However, since the beginning of this century, the paradigm of GCs as SSP prototypes has been severely challenged by a *plethora* of observational evidence collected from more accurate spectroscopic and photometric investigations that consistently show how *almost* all Galactic GCs host multiple stellar populations. This paradigm shift has been possible because of the huge improvements of the available observational facilities as the possibility to perform multi-object spectroscopy that enables one to collect with "each shot" spectra of hundreds of stars, and the exquisite photometry obtained with the most recent detectors (ACS and WFC3) on board the *HST*.

7.8.1
The Spectroscopic Evidence

The first evidence of Galactic GC chemical inhomogeneities dates back to more than 30 years ago [531, 532], and starting from these first empirical findings, many more data have been collected along the years [533, 534]. Until the advent of multi-object spectroscopy, the observed peculiar chemical patterns within individual GCs were considered just as an anomaly, involving only a small sub-sample of the cluster stellar population. In addition, given that these early empirical findings were obviously limited to the brighter RGB stars, there have been several attempts to explain the peculiar chemical abundances invoking additional physical processes that involve some amount of non-canonical extra-mixing at the bottom of the convective envelope [535]. The recent discovery that these peculiar chemical abundance patterns actually affect the majority of cluster stars in various evolutionary stages, has clarified that they must be of a more "intrinsic" nature.

The impact of these chemical inhomogeneities on the ages (and initial He abundances) determined for Galactic (and extragalactic) GCs under the "standard" SSP assumption will be discussed later in this chapter and in Chapter 9. In the following, we summarize the most important and solid results of "general validity" regarding the GC multi-population phenomenon:

- The abundance pattern of the so-called *light* elements (C, N, O, Na, Al, Mg, Si, F) observed in RGB stars belonging to Galactic GCs is a unique characteristic of cluster stars, for there is no analogue in the field. This means that the GC environment has to play a role in "shaping" the observed chemical peculiarities.

- Observations of RGB stars have revealed the existence of anti-correlations between N and C, in the sense that N increases with decreasing C. The evidence that the sum C + N increases as well with decreasing C is a proof that the *C–N anti-correlation* is not simply due to the conversion of C into N as a consequence of the *CN cycle* being at work. At the same time, for the few clusters with spectroscopic measurements available for C, N and O, it seems that the sum (C + N + O) is constant (within current uncertainties), with few possible exceptions represented by NGC 1851 (see [536, 537] for contradicting results on this cluster) and NGC 6656 [538]. This is a signature of matter processed by a fully efficient *CNO cycle*.

- Anti-correlations between O and Na, as well as between Mg an Al, the so-called *Na–O* and *Mg–Al anti-correlations*, have also been discovered [534]. The *Na–O anti-correlation* observed in a selected sample of GCs is shown in Figure 7.45. The discovery of these chemical anti-correlations represents a clear signature of matter processed by high-temperature proton capture processes like the *NeNa cycle* and the *MgAl cycle* (see Section 2.3.1). These high-*T* proton capture processes are efficient at temperatures larger than for the *CNO cycle*. Whenever these nuclear burnings are active, the *CNO cycle* also has to be efficient, and this would explain the constant C + N + O sum. A comparison of the O and Na abundances observed along the RGB of a given GC, with the corresponding abundances in field RGB stars with the same [Fe/H], defines what we denote as the cluster *primordial* stellar component (with the associated *primordial* abundance pattern), that is, the sub-sample of GC stars that share the same abundance pattern of the field counterpart.

  The existence of these anti-correlations actually challenges our understanding of how they have been produced. To retrieve the observed anti-correlations, it is necessary that the matter processed via nuclear burning(s) is mixed and diluted with a significant amount of matter with primordial chemical composition.

- These peculiar chemical patterns are not restricted to evolved stars, but are also observed in MS and SGB stars, as displayed in Figure 7.45d [539]. These results demonstrate that: (i) the chemical anomalies cannot be interpreted as a consequence of evolutionary effects because low-mass MS and SGB stars never attain in their interiors the temperature required to activate the *NeNa cycle* and *MgAl cycle*; (ii) the observed chemical patterns cannot be an effect of surface pollution because, given that the mass of the convective envelope significantly changes when the star moves from the MS to the RGB phase,[19] one should observe the effect of dilution between MS and RGB stars at odds with spectroscopic data. All these pieces of information clearly point to just one possible explanation:

---

19) In a $0.8 M_\odot$ stellar model with $Z = 0.004$, the mass of the convective envelope is $\sim 10^{-2} M_\odot$ at the MS TO and of the order of $0.54 M_\odot$ when the first dredge up occurs at the base of the RGB.

stars currently showing peculiar (with respect to field stars with the same iron content) chemical compositions were born from gas where these peculiarities were already imprinted. One can then envisage a scenario that requires two distinct sub-populations to coexist in a given GC: a "primordial" population with the same metal pattern of field stars with the same [Fe/H], and a second sub-population containing stars formed from matter processed by nuclear burnings that one is tempted to call the "second stellar generation".[20]

- The observed *Na–O anti-correlation* provides a unique tool for discriminating the primordial stars from second generation ones, and for evaluating the fraction of second generation stars hosted by each cluster. It is quite surprising to note that most GC stars do belong to the second stellar generation, while the primordial stellar component amounts to only about a third of the present total stellar population. This is crucial for understanding both how the second generation of stars actually formed, and the early cluster evolution.

- The measurements of *s*-element abundances are not yet properly understood in the framework of the multi-population scenario. Although a dispersion of *s*-elements abundances in individual clusters is present, it is not clear whether there is an (anti-)correlation between the abundances of neutron capture and light elements. Notable exceptions are represented by two objects. In NGC 1851, the *s*-element abundances seem to correlate with the light elements involved in GC anti-correlations [537]; in NGC 6656, the abundance of the *s*-elements seem to correlate with the light elements as well as with the iron and calcium abundances, as shown in Figure 7.49) [538].

This circumstantial evidence suggests that in the early stages of GC evolution, a significant amount of matter first processed by nuclear reactions inside some class of stellar objects belonging to the first stellar generation (primordial population) and then ejected, has to be present in the intra-cluster medium. From this matter, diluted with pristine gas with the same composition of the primordial population, a second generation of stars must have formed in a further burst of star formation [541]. Details of this scenario will be discussed in the following sections.

Some additional points are worth mentioning. First, there is a tight correlation between the chemical properties of cluster sub-populations and its absolute integrated magnitude, a proxy of the actual total mass. Roughly speaking, more extended anti-correlations appear in the more massive clusters, and a *Na–O anti-correlation* is present only if the actual cluster mass is above $\sim 10^4 M_\odot$ [542]. The extension of the *Na–O anti-correlation* also seems to depend on the metallicity of the cluster. In the case of the two GCs showing a large [Fe/H] spread, ω Cen and M54,

---

20) It has become customary to classify stars in a GC into primordial (P), intermediate (I) and extreme (E) objects, based on their location along the *Na–O anti-correlation* sequence, as shown in Figure 7.45 [540]. The "intermediate" sub-population is defined as those stars that show a more than 4σ overabundance of Na – where σ represents the typical star-to-star uncertainty in [Na/Fe] measurement in the given GC – above the mean abundance observed in field stars with the same iron content; the "extreme" stars are those showing a huge Na enhancement associated to a relevant O depletion (typically [O/Na] < −0.9). Extreme stars are only present in a minority of GCs, as that is NGC 2808 and M13.

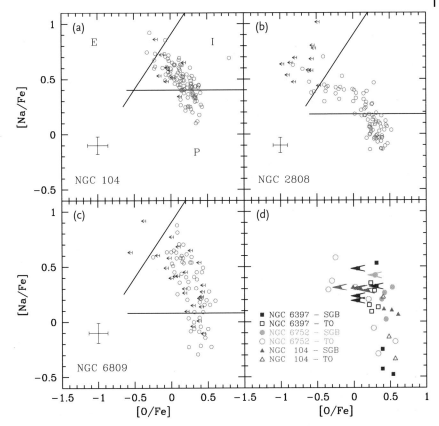

Figure 7.45 The *Na–O anti-correlation* in three Galactic GCs. The solid lines separate primordial (P), intermediate (I) stars and extreme (E) populations (see text for more details). The grey arrows correspond to stars with an accurate Na abundance measurement, but only an upper limit on the O abundance. (d) shows the same anti-correlation, but in unevolved MS and SGB stars in selected GCs (courtesy of E. Carretta).

it has been shown that the extension (and the shape) of the *Na–O anti-correlation* increases (changes) when moving from the iron-poor sub-populations to the iron-intermediate ones [543, 544].

Also, the presence of the light element anti-correlations leads to an expected He enhancement in second generation stars. The line of thought is the following: in the same layers where *CNO, NeNa* and *MgAl cycles* are expected to operate, a significant amount of He has to be present because He is the primary product of these H-burning processes. As a consequence, matter showing the signatures of these anti-correlations should be significantly enhanced in He. In the next section, we will show some compelling evidence supporting the idea that very He-enhanced stars have to be present in at least some (the more massive) GCs and that, although at a lower level, He enhancement seems to be always present together with light element anti-correlations.

7.8.2
The Photometric Evidence

The first, indisputable, photometric results showing the presence of multiple stellar populations within a GC have been collected only a few years ago thanks to *HST* observations. The scientific impact of this discovery was even more relevant than the spectroscopic evidence, for the appearance of multiple sequences in the CMD allows a direct perception of the presence of multiple populations.

In the following, we summarize the main photometric signatures of the multi-population phenomenon in Galactic GCs, one evolutionary stage at a time [545]:

- *MS:* the first photometric evidence for the presence of multiple populations in a GC was the discovery that the GC ω Cen shows a split MS, the so-called *red MS* (rMS) and the *blue MS* (bMS) [546, 547]. It was later shown that the CMD of this GC displays three distinct MS branches, as shown in Figure 7.46 [546, 547]. The most surprising result was, however, the spectroscopic evidence that the bMS is populated by stars about 0.3 dex more metal-rich than the rMS. This has been explained, invoking a huge He-enhancement in bMS stars [548]. It is currently estimated that the initial He content of the bMS should be $Y \sim 0.38$.

These results are probably not that surprising because ω Cen has always been considered to be a very atypical cluster due to its large mass (it is actually the most massive GC in the Milky Way with a mass of $\sim 10^6 M_\odot$), its large metallicity spread revealed by the huge RGB colour spread, and the possible existence of a significant age spread among the various stellar component.

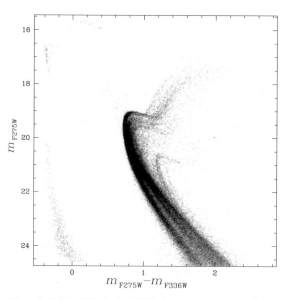

Figure 7.46 The CMD in the WFC3 photometric system of the Galactic GC ω Centauri, showing the presence of multiple MS, SGB and RGB loci (courtesy of A. Bellini).

The unexpected result was the discovery that the MS of the more normal (even if extremely massive) GC NGC 2808 also shows three distinct MS branches, well-separated in colour up the TO, as shown in Figure 7.47 [549, 550]. This evidence has been explained as due to the presence of three sub-populations characterized by three different, discrete initial He abundances: $Y \sim 0.24$ for the reddest MS, ~ 0.29 for the intermediate sequence, and ~ 0.38 for the bluest MS.

Indeed, nowadays, the splitting of the MS has also been observed in 47 Tuc and NGC 6397 [551, 552]. In these two cases, the He enhancement necessary to explain the splitting is lower, $\Delta Y \sim 0.01-0.02$.

In the case of NGC 2808, a correlation has been found between the He enhancement and the light element pattern. Accurate spectroscopic measurements of one star belonging to the red MS and one star on the bluest MS have shown that

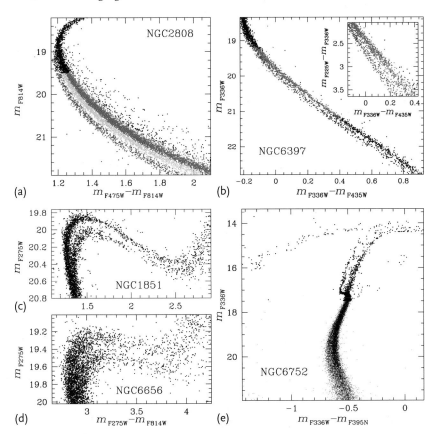

Figure 7.47 The CMDs in ACS and WFC3 photometric systems of selected GCs showing photometric signatures of multiple stellar populations along the MS, the SGB or the RGB (courtesy of A.P. Milone).

the red MS star displays a primordial chemical composition, whereas the blue MS object displays the typical chemical pattern of a second generation star [553];

- *SGB:ω* Cen has been known for a long time to show a very peculiar morphology of its SGB: this portion of its CMD is characterized by a huge spread in magnitude at fixed colour, with the clear presence of some distinct SGB sequences. These features have been explained by a combination of age and metallicity spread [554].

There are now several "normal" clusters that clearly display a split SGB, that is, NGC 1851, NGC 6656, NGC 6388 and 47 Tuc. The interpretation of this splitting is still not clear. For example, in case of NGC 1851, the problem is still unsettled. The presence of two distinct SGBs has been associated alternatively to the presence of two sub-populations with a ~ 1 Gyr age difference [554] or with the same age by two different values of the (C + N + O) total abundance [555]. In this latter scenario, the fainter SGB should be associated to a sub-population with a (C + N + O) abundance enhanced by about a factor of 2. Spectroscopical measurements of the CNO element abundance in this cluster are unfortunately still controversial.

The case of NGC 6656 is maybe more clear. In this case, combined photometric and spectroscopic observations show that the fainter SGB is populated by stars slightly more metal-rich (Δ[Fe/H] ~ 0.15 dex), s-element enhanced and also moderately enhanced (~ 0.15 dex) in CNO elements compared to stars belonging to the brighter SGB. The comparison with theoretical models accounting for these observed abundances reveals that the two sub-populations are almost coeval (see Figure 7.48);

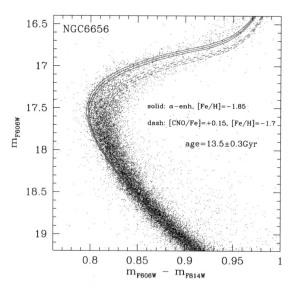

Figure 7.48 The CMD of the NGC 6656 showing the SGB splitting, compared to theoretical isochrones computed with the appropriate iron content and CNO element abundance of the two sub-populations (see labels) and for various assumptions about the age.

- *RGB:* apart from the case of ω Cen that shows multiple RGBs in all photometric bands, as well as M54, no other clusters with RGB splitting were found with optical photometry. As discussed in Section 3.9, in optical photometric bands as BVI, the mean colour (and slope) of the RGB for old stellar populations is only affected by a change of the global metallicity, but is not significantly affected by a change of the initial He content as well as by the distribution of the light elements involved in the observed (anti-)correlations typical of second generation stars (see also the discussion in Section 7.8.3). The situation changes drastically when moving to ultraviolet bands such as the Johnson U-band (or the ultraviolet bands of the WFC3 camera on board the *HST*) or the Strömgren ub filters. In this case, the RGB splitting immediately appears, as in the case of M4 shown in Figure 7.49c,d. Giant stars with the chemical pattern typical of second generation stars are distributed for the adopted combination of photometric filters along the redder RGB. This represents a clear proof that spectroscopic and the photometric evidence for multiple cluster populations are intimately linked. In general, the relative location in the CMD of the RGB and all other evolutionary sequences associated with the various sub-populations hosted by a GC strongly depends on the adopted combination of photometric filters. Sequences corresponding to different sub-populations can swap their relative location in the CMD by changing the colour baseline. This is evident when considering the data shown in Figure 7.51 for the case of 47 Tuc.
  Figure 7.49a,b show both the spectroscopy and Strömgren photometry for the GC NGC 6656. In this case, the location of the different RGBs correlates more with the difference in the iron content and s-elements abundances. Moreover, each of the two sub-populations with different s-element average abundances displays its own *Na–O anti-correlation*. This is another manifestation of the complexity of the multiple population phenomenon in Galactic GCs;
- *HB:* the HBs of several Galactic GCs show a number of peculiarities such as gaps and/or discontinuity in the stellar distribution (see Section 4.7) that can, in principle, be attributed to the presence of multiple populations. To this evidence, one can add other intriguing observational features such as: (i) the presence of extremely hot HB stars in some GCs (this evidence is intimately connected with the *second parameter* problem) as NGC 2808; (ii) the appearance of sloped HB morphologies in optical bands, that is, with the blue portion of the HB brighter (at the level of even ~ 0.5 mag in the V-band) than the redder portion of the HB, as it occurs in NGC 6388 and NGC 6441 and, to a lesser degree, also in other GCs as NGC 1851; (iii) the presence of a blue tail (in the V-band) along the HBs of some very metal-rich GCs as NGC 6388, more metal-rich by about ~ 0.1 dex than 47 Tuc, but showing a significant population of extremely hot HB stars not seen in 47 Tuc.
  On theoretical grounds, one expects that the presence of a significant fraction of He-enhanced stars can help to explain these features. At fixed total mass, MS He-rich stars burn H at a faster rate; therefore, for a given stellar population age, and in the assumption that along the RGB they lose mass at the same average rate of He-normal stars, one can predict that the average mass of their

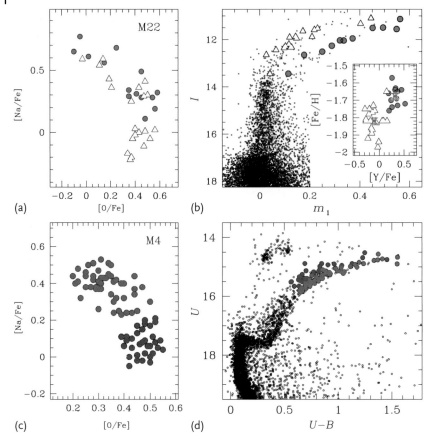

Figure 7.49 (a) and (b) display the *Na–O anti-correlation* and optical-Strömgren CMD for NGC 6656 (M22). The inset in the panel with the CMD shows the trend of the iron content with respect to the *s*-element *Y* for a sample of cluster RGB stars. The location of these stars is reported in the CMD and in the *Na–O anti-correlation* diagram. The m_1 index is defined as $m_1 = (u - b) - (b - y)$. (c) and (d) show graphs similar to (a) and (b), but for the M4. In this case, the photometric data are in the $(U, U-B)$ CMD. The various symbols light and dark grey refer to the same stars in the different plots, and highlight their location in the adopted CMD and in the *Na–O anti-correlation* diagram (and the [Fe/H]–[Y/Fe] plane for the case of M22) (courtesy of A.F. Marino).

HB progeny is lower than that of the He-normal counterpart. As a consequence (see Section 4.3), the He-enhanced stellar component is expected to be located at bluer colours compared to the He-normal one and, at least in a not-too-blue, colour range in optical bands, also at brighter magnitudes. In the case of a metal-rich GC, a large He enhancement would be required to overcome the metallicity effect that tends to make the HB morphology redder with increasing metal content (at a fixed age).

Although synthetic HB simulations accounting for both mass and initial He abundance spread (or some kind of He discretization in case of NGC 2808) seem to be able to reproduce the observed HB morphologies and pulsational properties of RR Lyrae samples, as shown in Figure 7.50 for NGC1851, these comparisons only provide an indirect proof of the existence of multiple populations along the HB [556–558].

However, in the past few years, high-resolution spectroscopy is allowing one to also chemically tag the various sub-populations along the HB. We have already mentioned the possibility of directly measuring the initial He abundance at the surface of HB stars in a narrow T_{eff} windows. In case of NGC 6121, the measured value of Y is consistent with the presence of a He-enhanced stellar component along the HB. More data are being collected on light element anti-correlations along cluster HBs, see [559, 560] and references therein. The general picture emerging from these spectroscopic analyses is that, in a given GC, red HB stars are typically O-rich and Na-poor, that is, they show the typical abundance pattern of primordial stars, while blue HB stars are typically O-poor and Na-rich, and thus they display the same abundance pattern of second generation stars observed along the RGB. Due to the already emphasized link between the presence of light element anti-correlations and He-enrichment, one is tempted to argue that moving from the red to the blue side of the HB in a given cluster, stars are characterized by progressively stronger light element anti-correlations and larger He enhancements. Indeed, it has been demonstrated the existence of a close relation between the extension of the *Na–O anti-correlation* and the maximum T_{eff} attained by HB stars in a given cluster. Figure 7.51 displays the seg-

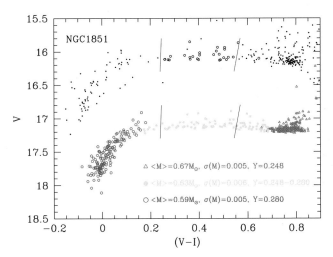

Figure 7.50 Comparison between the observed HB stellar distribution in the Galactic GC NGC 1851 (upper distribution) and a synthetic HB model computed with the labelled assumptions about average mass, (Gaussian) dispersion and initial He content. The He distribution between 0.248 and 0.280 follows a uniform probability. The boundaries of the RR Lyrae instability strip are marked. The open circles in the observed HB mark the location of the RR Lyrae stars. For the sake of clarity, the synthetic HB simulation has been shifted vertically.

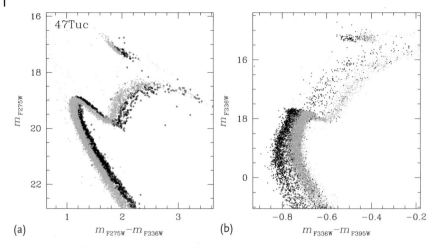

(a)   (b)

Figure 7.51 CMDs of 47 Tuc in various WFC3 photometric filters tracing the cluster sub-populations from the MS to the HB stage. The grey points denote stars belonging to the second stellar generation (courtesy of A.P. Milone).

regation along the HB of 47 Tuc of first and second generation sub-populations. It is generally difficult to trace the sub-populations identified spectroscopically on the CMD because of the horizontal nature of the branch in optical filters and the usually large colour range covered by HB stars. Suitable filter combinations therefore need to be devised to properly separate the HB sequences corresponding to the various sub-populations.

Given the existence of a tight correlation between the extension of the *Na–O anti-correlation*, and hence the expected He enhancement level and the total mass of a GC, it is possible to envisage a scenario where "the" second parameter is actually the total mass of the cluster [244]. The larger the mass, the larger the extent of nuclear processing (and He enhancement) of the intra-cluster matter from which the second (and eventually third) generation(s) formed, and the bluer the colour of the HB progeny.[21]

7.8.3
The Theoretical Evolutionary Framework

In the previous sections, we have discussed how filters in the (near) ultraviolet spectral region are the most favourable for photometrically tracing the multiple populations hosted by individual GCs, in particular along the RGB. This opportunity would allow, without the need of high-resolution spectroscopy, to identify the various sub-populations present in a cluster, follow their evolutionary sequences

21) A recent analysis reached the conclusion that the GC mass alone cannot explain the variety of the HB morphologies in Galactic GCs, but that *the* second parameter is probably a combination of the total GC mass and age [244].

in the CMD, as shown in Figure 7.51, and estimate the fraction of first to second generation stars in a GC.

To this purpose, we discuss here theoretical predictions about the photometric signatures of multiple stellar populations. This type of analysis requires the calculation not only of stellar evolution models with the appropriate initial chemical composition, but also of consistent colour-T_{eff} relations and bolometric corrections [417, 561, 562]. Let's consider the following chemical mixtures:

- A composition with $Y = 0.248$, $[Fe/H] = -1.62$, $[\alpha/Fe] \approx +0.4$, representative of the primordial population in a typical metal-intermediate Galactic GC. This will be denoted as the *reference* mixture;
- A composition with the same $[Fe/H] = -1.62$, and $Y = 0.248$ and a metal distribution where the elements C, N, O, and Na follow the (anti-)correlations observed in second generation stars. This metal mixture displays (compared to the *reference* one) enhancements of N and Na by 1.44 and 0.8 dex by mass, respectively, together with depletions of C and O by 0.6 and 0.8 dex, respectively. This mixture has (within 0.5%) the same sum (C + N + O) of the *reference* mixture. This mixture will be denoted as *CNONa* and is representative of fairly extreme anti-correlations;
- A composition similar to the *CNONa* mixture, but for the enhancement of N that in this case is equal to 1.8 dex. In this case, the (C + N + O) mass fraction is enhanced by a factor of 2 compared to the *reference* composition. In addition, since the C+N+O sum represents most of the total metal content, for the same $Y = 0.246$ and $[Fe/H] = -1.62$ of the *reference* first generation composition, the adopted global metallicity needs to be larger than the *reference* composition by a factor 1.84. We will refer to this mixture as $(CNO)_{enh} Na$;
- A chemical pattern, similar to the $(CNO)_{enh} Na$ mixture, but with a He abundance enhanced to $Y = 0.40$, $[Fe/H] = -1.62$ and Z larger by a factor ~ 1.5 than the reference mixture. It will be named $(CNO)_{enh} Na\text{-}He$.

Figure 7.52 displays the HRD of 12 Gyr isochrones with the chemical abundance patterns listed above, from the MS to the TRGB.[22]

The isochrone corresponding to the *CNONa* mixture is identical to the case of the *reference* mixture. On the other hand, the isochrone for the $(CNO)_{enh} Na$ mixture displays a fainter and redder TO and, as a consequence, a fainter SGB compared to the *reference* isochrone. This result justifies the interpretation of the SGB splitting observed in some GCs as due to an enhancement of the CNO sum. When not only the CNO sum, but also the He abundance is enhanced, as in the case of the $(CNO)_{enh} Na\text{-}He$ mixture, the MS, TO and to a lesser extent the RGB, become hotter.

As for the effect of the different chemical mixtures on the stellar spectra, helium seems to play a minor role, but the variations of C, N and O have a major effect. This effect is maximized in case of the $(CNO)_{enh} Na$ mixture, as shown in Fig-

---

22) The age is kept fixed because GC sub-populations appear in general to be largely coeval within ~ 0.5 Gyr.

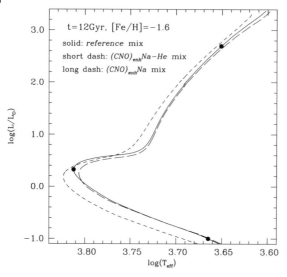

Figure 7.52 Theoretical isochrones with various chemical compositions (see labels and text for more details). The isochrone corresponding to the *CNONa* mixture overlaps perfectly with the isochrone corresponding to the *reference* mixture. Full dots mark the location of selected models whose predicted spectra are shown in Figure 7.53.

ure 7.53, which displays theoretical spectra of stars at the faint MS, TO and RGB, respectively.

For the cool RGB model, the spectrum for the $(CNO)_{enh} Na$ mixture shows much stronger NH and CN absorption bands in the spectral windows corresponding to the *U*, *B*, and *I* filters, compared to the *reference* mixture. This is essentially due to the much higher N abundance of the $(CNO)_{enh} Na$ mixture, despite the fact that the C abundance is lower. This occurrence demonstrates that the N abundance acts as a bottleneck in forming CN molecules. On the other hand, the G-band, a CH feature falling into the *B* filter, appears stronger in the *reference* mixture since the C abundance is higher. The increased opacity in the blue part of the spectrum for the $(CNO)_{enh} Na$ mixture leads to the increase of the continuum flux redwards of about 450 nm, that explains the overall higher flux observed in this spectrum between 450 and 690 nm, and further to the red in the intervals along the strong CN absorption bands.

When moving to the hotter TO, the higher temperature prevents the formation of CN and CH, and removes the most important source of the increased blue opacity in the $(CNO)_{enh} Na$ case, as well as most of the influence on the G-band. Only the very strong NH band around 340 nm is still visible. On the other hand, when considering the cool star along the fainter portion of the MS, Figure 7.53c shows features similar to the RGB model, but also a much stronger OH absorption at the blue edge of the *U* filter range in the *reference* case, due to the higher O abundance. The same band is visible, but much less prominent, also in the MS and RGB models. The Na D doublet becomes strongly wing-dominated in this spectrum and thus

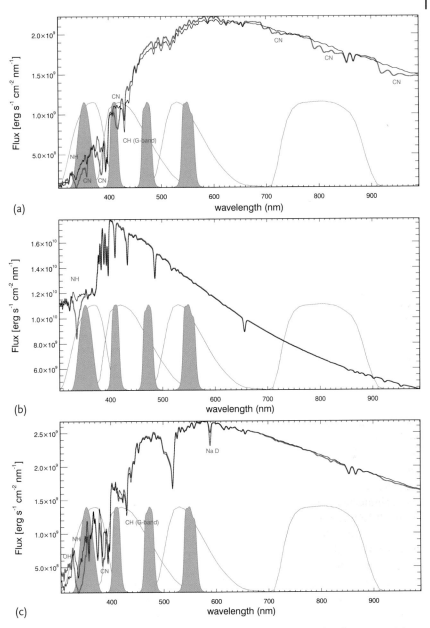

Figure 7.53 (a) The comparison between synthetic spectra for the *reference* mixture (black line) and $(CNO)_{enh}$ Na mixture (bold line) for an RGB model with $T_{eff} = 4476$ K and $\log g = 1.2$. The location of the transmission curves for the Johnson–Cousins $UBVI$ (thin lines, from left to right), and the Strömgren $uvby$ (grey-shaded regions) filters is also shown. (b) As with (a), but for a TO model with $T_{eff} = 6490$ K and $\log g = 4.22$. (c) As with (a), but for a cool MS model with $T_{eff} = 4621$ K and $\log g = 4.47$. In all panels, the molecular bands that vary significantly between the two mixtures are indicated by the name of the corresponding molecule.

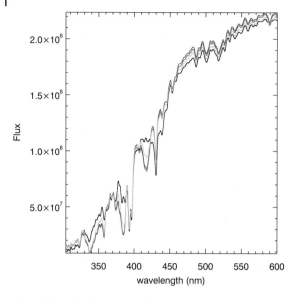

Figure 7.54 The blue portion of synthetic spectra for the same RGB model of Figure 7.53, but various chemical abundance mixtures: *reference* (black line), CNONa (light-grey line), $(CNO)_{enh}$ Na (mid-grey line), and $(CNO)_{enh}$ Na-He (dark-grey line). The flux is reported in the same units as for Figure 7.53.

appears much stronger in the spectrum for the $(CNO)_{enh} Na$ mixture, where the Na abundance has been increased to reproduce the observed *Na-O anti-correlation*. The red CN bands have a negligible effect in this cool atmosphere.

Given that the shorter wavelength region of the spectra of cool RGB stars appears massively affected by the abundance anti-correlations, Figure 7.54 shows how the blue-visible part of the spectrum changes amongst all the four mixtures, for the same RGB stellar model considered in Figure 7.53a. Two effects are evident: first, the spectrum for the *CNONa* mixture shows less prominent NH and CN bands than the $(CNO)_{enh}$ Na mixture, as well as a continuum that resembles more the one of the *reference* mixture. Second, the change in the He abundance between the $(CNO)_{enh}$ Na and $(CNO)_{enh}$ Na-He mixtures only has a minor effect on the flux distribution. This comparison among the various spectra allows one to immediately predict the photometric bands more affected by the chemical peculiarities of second generation stars, as well as to identify what chemical species (and molecules) play the most relevant role. As a rule of thumb, a major role is played by the absorption features related to CN and NH molecules in the bluer spectral windows.

The same isochrones displayed in Figure 7.52 are plotted in different CMDs and displayed in Figure 7.55. Appropriate bolometric corrections (BCs) have been obtained from the theoretical spectra discussed above, that have the same chemical pattern of the stellar isochrones. The behaviour in the optical $(M_V, B-V)$ and $(M_V, V-I)$ CMDs closely resembles the HRD. In these filters, the BCs are hardly affected by the change of the metal mixture and initial He content. As long as the

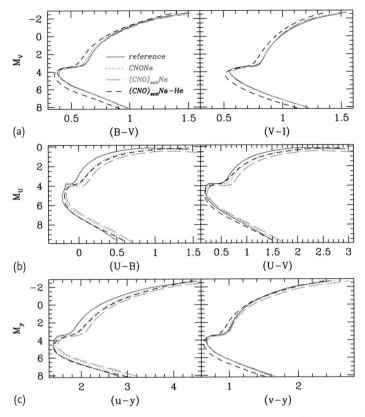

Figure 7.55 The same theoretical isochrones shown in Figure 7.52, but in various $UBVI$ and Strömgren CMDs, employing self-consistent bolometric corrections.

sum $(C + N + O)$ is constant, second generation stars are expected to overlap with first generation objects. In the case of enhanced $(C + N + O)$, only the TO and SGB regions are affected, whereas MS and RGB remain unchanged. An increase of the He abundance shifts the MS (and to a lesser degree the RGB) towards bluer colours, mainly because of hotter effective temperatures of He-enhanced stellar models.

The situation is quite different for CMDs involving the ultraviolet U-band that is most affected by the change of the metal mixture due to the emergence of strong molecular absorption in the atmosphere. The four isochrones are now well-separated along the various branches; the isochrone representative of second generation stars with enhanced $(C + N + O)$ is always the reddest, but now also the one corresponding to the *CNONa* mixture with a constant sum of CNO elements follows a distinct sequence compared to the isochrone representative of the primordial population.

The largest differences appear along the RGB, where the mixture with anti-correlations causes redder $(U-B)$ and $(U-V)$ colours, at fixed M_U. For instance, at $M_U = 2.0$, the RGBs representative of second generation stars are redder by up to ~ 0.2 mag in $(U-B)$ and ~ 0.3 mag in $(U-V)$, depending on the metal mixture considered. This theoretical expectation is in fair agreement with observations as in case of NGC 6656. It is also worth noticing that an increase of Y up to 0.40, actually a very extreme He enhancement, shifts the isochrones of second generation stars, which tend to be redder than the reference isochrone due to the CNONa-variations, bluewards and closer to the *reference* isochrone. This is an effect of the stellar interior models only, and may potentially even produce a bluer MS, depending on whether the anti-correlations are accompanied by a $(C + N + O)$ enhancement.

The comparison between the isochrones in the Strömgren filters shows a behaviour very similar to the cases previously discussed for the $UBVI$ bands. In particular, the behaviour in the $(M_y, u - y)$ CMD is very qualitatively similar to the case of the $(M_U, U-B)$ diagram, with all different sequences well-separated when the initial He is kept constant. Figure 7.56 shows the (M_V, c_y) CMD, where $c_y = c_1 - (b - y)$ while the colour index c_1 is defined as $c_1 = (u - v) - (v - b)$. This index c_1 is empirically found to be sensitive to the N abundance; on the other hand, the c_y index well represents c_1, but removes much of the temperature sensitivity of this index. As a result, (V, c_y) CMDs of Galactic GCs display an almost vertical RGB at luminosities lower than the RGB bump. Model predictions nicely agree with this empirical evidence. In addition, all isochrones corresponding to the peculiar chemical patterns of second generation stars have redder c_y colours, and the isochrone with the largest N abundance (the $(CNO)_{enh}$ Na mixture) is the reddest, in agreement with empirical results. In this CMD, an increase of the initial He abundance tends to move the RGBs of isochrones with CNONa anti-correlations further away from the *reference* isochrone, despite the fact that the $(CNO)_{enh}$ Na-He mixture has a lower N abundance than the $(CNO)_{enh}$ Na mixture due to the lower total metallicity.

Before closing this section, we comment on how the existence of the multiple stellar populations affect the GC age estimates that are obtained considering GCs as SSPs with a typical "primordial" metal mixture and He abundance. This issue is even more important when recalling that second generation stars appear to represent the bulk of the stellar component within a given GC. The impact of unrecognized differences among stellar populations in individual GCs has been, to date, the subject of only one theoretical investigation that was focused on the relative age determination [563]. The result is that age differences obtained from the horizontal and (mostly) the vertical methods are dramatically affected by unrecognized differences in the He abundance, whilst the rMSF method appears fairly insensitive to the same variations. When considering differences in the heavy elements distributions and, in particular, in the $(C + N + O)$ sum, all these methods are affected, albeit at a lower level than for the case of a He variation, with the rMSF method that once again being the least sensitive to these abundance variations. In terms of absolute ages (that indirectly also affects the exact values of the relative ages), differences in He play a major role when the $\Delta V_{\mathrm{TO}}^{\mathrm{HB}}$ or, alternatively, the absolute TO magnitude are employed as clocks. An increase of He in a dominant second generation sub-population biases the derived ages towards younger values because

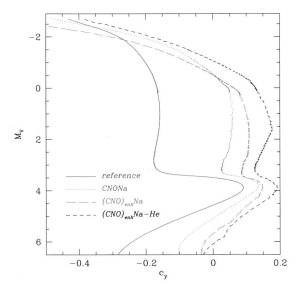

Figure 7.56 The same theoretical isochrones shown in Figure 7.52, but in the (M_V, c_y) photometric plane.

it biases the measured HB magnitude towards brighter values and the TO towards (slightly) fainter magnitudes. The size of this effect depends on the average He enhancement, the fraction of He-enhanced stars and whether they populate the horizontal part of the HB used to register the reference HB level. An enhancement of the $(C + N + O)$ acts in the same direction, mainly because it biases the TO towards fainter magnitudes.

When ages are obtained from the absolute visual magnitude of the TO, the age bias is generally expected to also be towards younger ages. In fact, distances obtained from, that is, unrecognized He-enhanced RR Lyrae stars or a He-enhanced MS component (via the MS-fitting method in visual filters) will be biased towards larger values. These too large distances, together with the effect on the empirical estimate of V_{TO} discussed above, cause an underestimate of the cluster age.

Concerning estimates of the initial He abundance, second generation stars tend to bias the estimate of "primordial" He towards higher values for all methods described in Section 7.6. In the majority of Galactic GCs, there is a strong hint that the difference in the initial He abundance among the sub-populations hosted in the same cluster is probably lower than 0.02. As a consequence, the bias on the derived initial Y should be typically smaller than these values. The situation drastically worsens when considering GCs such as NGC 2808, that hosts sub-populations with discrete and very different initial Y.

7.8.4
Formation Scenarios

So far, we haven't identified the objects able to produce the peculiar chemical patterns observed in second generation stars, nor discussed the mechanism(s) for the formation of the various sub-populations hosted by a cluster. Needless to say, to date, there are still more open questions than firm conclusions.

Let us start with a discussion on the so-called "polluters", that is, those stars able to produce, via nuclear burnings, the observed peculiar chemical patterns. The spectroscopic evidence suggests that the best candidates are those stars that experience high-temperature proton captures, without producing significant change in the iron, iron-peak, as well as α-elements. There are essentially only two candidates:[23] (i) intermediate-mass AGB and possibly super-AGB stars, with mass between ~ 4 and $\sim 10 M_{\odot}$ [566, 567]; (ii) fast-rotating massive stars (FRMSs), in the mass range $(20-120) M_{\odot}$ [568].

The following points support the AGB scenario:

- In these stars, high-temperature proton captures as those involved in the *NeNa* and *MgAl cycles* are active when hot bottom burning occurs during the thermal pulses stage. The nuclear burnings efficient in these stars do not alter the original abundances of iron and α-elements. Less massive AGB stars are not a viable choice because, as a consequence of the recurrent TDU (see Section 5.3.1), a huge amount of C is mixed in their envelope, and a large C enhancement is at odds with spectroscopic measurements in GC second generation stars;
- There is natural mechanism to dredge the nuclear processed matter from the burning region to the stellar surface; it is the extended mixing that affects the outer layers of these stellar structures during the TP-stage;
- AGB stars are characterized by very efficient mass loss, but the velocity of these winds is relatively low, $\sim 40 \, \mathrm{km \, s^{-1}}$. Mass loss during the TP-stage is therefore an adequate process for releasing the nuclear processed matter in the intracluster medium at a velocity low enough to be retained within the cluster potential well.

On the other hand, the AGB scenario has quantitative shortcomings mainly related to the current uncertainties in the treatment of convection, mass loss efficiency and the rate of some relevant nuclear reactions [322, 567]. In general, intermediate-mass AGB stellar models do not appear able to reproduce the full spectrum of chemical anomalies observed in second generation stars. This includes the very large O depletion $[O/Fe] < -0.6$ observed among some stars belonging to the "extreme" populations, as well as the very large He enhancement required for explaining the photometric properties of the blue MS in ω Cen and the bluest MS in NGC 2808 (although it seems that super-AGB stars can partially solve this problem).

---

23) Other suggested candidates are massive stars in close binary systems and Pop. III stars [564, 565]. The comparison with the spectroscopic data does not generally provide much support for these alternatives.

Concerning the FRMS scenario, the basic idea is that due to the very high rotational rate, a rotationally induced mixing is able to transport nuclear processed matter from the stellar core, to the outer envelope. From the envelope, this processed matter would be lost and injected in the intra-cluster medium via envelope ejection and/or mechanical equatorial winds, with a low enough velocity as in the case of the AGB scenario.

The FRMS scenario does not have the same shortcomings of the AGB one. FRMS models seem to indicate that these objects are probably able to produce matter also showing the most extreme chemical peculiarities observed in second generation stars. Amongst the main objections moved against this scenario, the first obvious one is that these stars will explode on short timescales as type II Supernova. These supernova explosions are expected to remove all gas from the cluster, thus preventing the formation of a second stellar generation from the ejecta of FRMS stars. Additionally, the matter lost by these stars during the hydrostatic evolution would be characterized by a variety of chemical patterns related to the spread of the initial masses as well as rotational rates [569]. The last objection that makes it difficult to imagine FRMS stars as the main polluters is related to the multiple MS observed in NGC 2808. The presence of very well-defined and "quantized" MS branches points towards a very homogeneous chemical composition in each subpopulation. Indeed, it is quite possible that, depending on the global properties of the proto-cluster at the moment of its formation, both scenarios play a role with different relative weight [570].

There is potentially some spectroscopical measurement that can help to discriminate between these two scenarios. The most important ones are lithium abundance measurements that search for (anti-)correlations with light elements. Lithium is a volatile element, promptly burnt at temperatures of $\sim 2.5 \times 10^6$ K; as a consequence, Li cannot be present in FRMS ejecta, whilst AGB stars are able to produce lithium during the TP-stage via the Cameron–Fowler mechanism [571]. Spectroscopic results are, however, not yet conclusive. Additional constraints on the properties of the polluters can come from accurate spectroscopic measurements of other elements produced with very different efficiency depending on the temperature of the nuclear burning sites; this is because the temperatures at which high-T proton captures occur, depends significantly on stellar mass.

During the last few years, various scenarios and models for GC formation have been developed, and tuned to explain the emergence of multiple populations [570, 572–576]. The basic proposed scenarios can be outlined as follows. Within the original molecular cloud that has already experienced some amount of heavy element enrichment, a first star formation burst starts, that is responsible for the formation of the "primordial" component. After a time interval of the order of few megayears in case of FRMSs, or up to hundreds of megayears in the AGB scenario, polluters of the first stellar generation inject into the intra-cluster medium a large amount of nuclear processed matter. These ejecta are expected to diffuse and accumulate in the core of the GC together with some amount of pristine gas. In the case of FRMSs, these objects are expected to migrate rapidly in the cluster core due to dynamical effects and again, their ejecta are expected to collect in the GC core.

After some delay, a second burst of star formation creates a second stellar generation from the ejecta of the primordial population. Obviously, this process can be repeated when the ejecta of AGB stars and/or FRMSa belonging to this second generation accumulate in the intra-cluster medium. However, all models seem to agree on the conclusion that a further star formation episode is disfavoured due to the huge injection of energy associated with the explosion of a significant fraction of type Ia SNe belonging to the first stellar population, although a third star formation burst may still occur in the more massive GCs, as in the case of NGC 2808.

There are two major problems with this broad picture of the GC multi-population formation. The first one is the *mass budget problem*. According to spectroscopy, the fraction of primordial stars in a given GC is about one third of the total population; if the second stellar generation formed from the ejecta of a fraction of stars belonging to the primordial population, there is obviously the problem that not enough mass from the first population is available to form the currently observed second generation. This problem exists, regardless of the nature of the polluters.

A possible solution rests on the assumption of a top-heavy IMF, that is, a higher fraction of intermediate and/or massive stars, compared to standard Salpeter or Kroupa IMFs for the primordial population. A different, more palatable solution, is based on the hypothesis that GCs in the early stage of their life were about 1–2 orders of magnitude more massive than today, and have lost a significant fraction of their mass, preferentially in the form of primordial stars. The physical mechanisms responsible for this huge mass loss could be associated with gas expulsion by SNe explosions. This gas expulsion would drastically change the potential well, making stars in the outer part of the GC unbound. Since second generation stars are expected to mainly form in the central regions of the GC, first generation stars would be preferentially lost. This scenario is also consistent with the idea that a GC (at least the more massive ones) could be formed within larger systems like dwarf galaxy satellites of the Milky Way that would be later shredded, as in the case of the ongoing accretion process involving the Sagittarius dwarf galaxy.

The other problem that affects all GC formation scenarios is the requirement of dilution of the polluting matter with primordial composition gas. This is essential for reproducing the observed anti-correlations. The required amount of dilution is lower in the FRMS scenario, but it is crucial for the AGB scenario. Without the inclusion of dilution, AGB ejecta predict a correlation between Na and O, in stark contrast with all the empirical evidence [574]. Many different hypotheses have been put forward to explain the presence of primordial gas inside the intra-cluster medium during the early stages of the GC life. For example, original gas may have survived the first star formation burst, or it is acquired via Bondi accretion from molecular clouds while the GC is orbiting around the proto-galaxy, or by injection in the proto-GC environment of the matter lost by low-mass first generation stars during the RGB stage.

Clearly, we are still at an early stage in investigations to discriminate amongst these various hypotheses and scenarios.

8
Resolved Composite Systems

The determination of the age and metallicity distributions of the old stellar component in resolved galaxies is a more difficult task compared to the case of old SSPs.
In the simplest case of galaxies with a "bursty" star formation history – the Carina
dwarf spheroidal is an example – CMDs look very much like the superposition of
several isochrones, and it is possible to get an idea of the ages and metallicities of
the various sub-populations by matching isochrones to the observed multiple sequences in the CMD, as shown in Figure 8.1. However, this type of analysis does
not provide any information about the relative strength of the various star formation episodes.

The case of more continuous star formation with varying intensity cannot be
studied with simple isochrone fittings because the CMDs do not show discrete
sequences that can be matched by individual isochrones. In this case, one has to
disentangle the effect of age and metallicity on the distribution of points along
the CMD by taking into account both morphological information about stellar
magnitudes and colours – taking advantage of the fact that different evolutionary
stages display different sensitivities to changes in age and metallicity – and star
counts along the observed evolutionary sequences – that help to disentangle age–
metallicity degeneracy and determine the intensity of the various star formation
episodes.

Let us consider in more detail the problem of determining the star formation
history (SFH) of a composite stellar population (CSP), that is, the evolution with
time of the star formation rate (SFR – mass converted into stars per unit time) and
the corresponding age–metallicity relation (AMR – initial metallicity of the stars
formed at a given time). For a given distance modulus, extinction, IMF and binary
distribution, the observed CMD can be represented by a linear combination – with
positive coefficients – of isochrones (SSPs) of different ages and initial chemical
compositions, that represent the multiple episodes of star formation experienced
by the target population. The determination of these coefficients, representing the
weights of the distinct SSPs in determining the CSP, leads straightforwardly to the
estimate of SFR and AMR.

The initial chemical composition is typically parametrized in terms of total metallicity Z or [Fe/H]. The common assumption behind this parametrization is that the
initial He mass fraction Y, that also affects the CMD location of a star, scales with

Old Stellar Populations, First Edition. S. Cassisi and M. Salaris.
© 2013 WILEY-VCH Verlag GmbH & Co. KGaA. Published 2013 by WILEY-VCH Verlag GmbH & Co. KGaA.

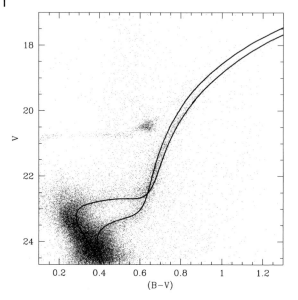

Figure 8.1 CMD of the Carina dwarf galaxy. Two isochrones from the MS to the tip of the RGB – $t = 13$ Gyr, [Fe/H] $= -2.27$ and $t = 6$ Gyr, [Fe/H] $= -1.79$ – are overimposed to the observed CMD, after applying appropriate values for the apparent distance modulus (($m - M)_V = 20.18$) and reddening ($E(B - V) = 0.06$).

the initial metallicity Z of a stellar population, and therefore a given Z fixes both Y and [Fe/H]. Typical values found in the literature for the so-called He-enrichment ratio are in the range between $\Delta Y/\Delta Z \sim 1.5$ and $\Delta Y/\Delta Z \sim 2.5$. Just considering the initial solar He determined from a theoretical calibration of the SSM together with the primordial He abundance predicted by the Big Bang nucleosynthesis provides $\Delta Y/\Delta Z \sim 1.4$.

There are two broad categories of techniques developed to determine the SFH of CSPs that will be introduced in the next two sections. It goes without saying that these same methods can in principle also be applied to a SSP; obviously, in such a case all the coefficients entering in the linear combination of SSPs will be equal to zero, but only one that is the coefficient corresponding to the target SSP.

8.1
Synthetic CMD-fitting

This widely used technique (see, i.e. [577–580] for just a few examples) is based on comparisons between synthetic and observed CMDs. The faint magnitude limit of the observed CMD sets an upper limit to the look-back time that can be reliably sampled by the data because the MS TO gives most of the information about ages. Also, if only post-MS phases are sampled in the observed CMD, this occurrence would largely limit the ability to determine a detailed SFH for the target population.

A detailed and insightful discussion of these methods can be found in [579]. Several tests based on synthetic populations have appeared in the literature, and show how this technique is able to recover the SFH of populations with both continuous and bursty star formation. Here, we sketch the main features of this technique.

The synthetic CMD computation generates a synthetic CSP populated by a large number of stars (to sample properly the underlying isochrones) with ages and metallicities following some appropriate, for instance, uniform, distribution over the full range of variation of the SFH. In more detail, samples of synthetic stars are generated with a Monte Carlo technique for a given IMF, distance modulus, extinction and binary fraction (assuming a distribution of mass ratios), starting from a library of theoretical isochrones, to create a grid of "partial models", each containing objects with an age and metallicity distribution within small intervals of t and Z. These partial models constitute a set of $n \times m$ (e.g. n values of the mean age t and m values of the mean metallicity Z for the stars in each partial model) "elementary populations" with no stars in common between any two of them.

It is paramount to include in these synthetic CMDs the effect of photometric errors, blending and incompleteness that affect the observed CMD of the target population. This is usually done by introducing, at any given magnitude, random rejection and a photometric error to the synthetic stars by accounting for both the "true" photometric error distribution and completeness level that affect the photometry of the target stellar population, as obtained from the results of artificial star tests on the observational dataset [581]. The basic idea of this technique is to inject a number of artificial stars with known brightness and colours – spanning a range of both magnitudes and colours appropriate to the whole sample of stars in the synthetic CMD – at random positions on the observed image (taking care of not altering the actual crowding properties of the image) and then recover them from the photometry by adopting the same methodological approach used for the real stars. The ratio between injected and recovered stars (in principle function of magnitude and colour) will provide a completeness function to be applied to the synthetic CMDs. The difference in magnitude and colours between the input values and the recovered ones, that is, the photometric error distribution, is a measure of the shifts to be also applied to the synthetic stars in order to complete the modelling of observational effects. The results of these experiments can be implemented by means of lookup-tables.

The next step is to define a set of boxes in the observed CMD. The most simple choice would be a uniform grid to cover equal intervals in magnitude and colour. However, a more suitable approach is based on the use of non-uniform grids to give different weights to stellar evolution phases that are modelled less accurately and/or are too sparsely populated in the observed CMD. In general, when increasing the number of boxes in a given area of the CMD, one increases the sensitivity of the solution to the finer details of the CMD; an occurrence that allows one to better recover the SFH of the target stellar population. However, the bin size of a box is limited by the number of observed stars located in that box: a suitable number of stars has to be present in order to minimize the effects of statistical fluctuations. As a consequence, in defining the grid of boxes in the CMD, one has to find a suitable compromise between the total number of boxes and their size.

The option implemented in the widely employed IAC-pop/MinnIAC code is to define several regions in the CMD, denoted as bundles [577, 582]. Each bundle is sampled by a uniform grid, but different bundles can have different grid bin size. Figure 8.2 displays an example of this procedure. Some post-MS evolutionary phases, in particular extended HBs when present, due to the current uncertainties in predicting HB morphologies from first principles, are often excluded from the derivation of the SFH. Particular attention has to be paid to the possible degeneracy between Blue Stragglers (BS) that are not included in the isochrones used to generate the elementary populations, and genuine young components (the same problem affects in principle also the second method to recover the SFH, discussed in the next section). In this case, additional considerations may help to isolate the presence of BS stars in the recovered SFH. For example, a recent analysis of the Cetus dwarf spheroidal galaxy discloses the presence of a metal-poor population 2–4 Gyr old that does not follow the general AMR determined for the galaxy, an indication that this component is not genuine, but probably a sign for the presence of BSs [583, 584].

After bundles and associated grids are defined, one can compute a matrix M_i^j that contains the number of stars in the elementary population i within the CMD box j. The same procedure applied to the observational CMD generates the vector O^j. Once M_i^j and O^j are determined, a linear combination of the M_i^j values for each CMD box j can be calculated as

$$M^j = A \sum_i a_i M_i^j$$

with i taking values from 1 to $n \times m$, subject to the constraint $a_i \geq 0$ (A is a scale factor).

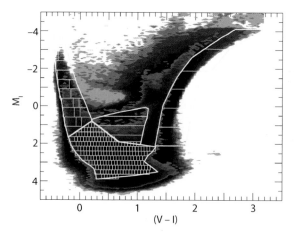

Figure 8.2 Example of bundles and grids defined on the observed CMD of a resolved galaxy (courtesy of S. Hidalgo).

Finally, the SFH that best matches the observed distribution O^j can be found by minimizing (or maximizing) a merit function. For example, the IAC-pop code employs the following statistics for parameter estimation with Poisson-distributed data [585]:

$$\chi_\gamma^2 = \sum_{j=1}^k \frac{\left[O^j + \min\left(O^j, 1\right) - M^j\right]^2}{O^j + 1} \tag{8.1}$$

where k is the number of boxes defined in the CMD, and the elements of the vector α_i are the parameters to estimate. The IAC-pop code adopts a genetic algorithm for minimizing χ_γ^2. This minimization of χ_γ^2 provides the best fitting set of α_i values as well as a test for the goodness of the fit. The solution SFH can then be written as

$$\Psi(t, Z) = A \sum_i \alpha_i \psi_i$$

A being a scale factor, and ψ_i refers to partial model i. An example of derived SFH using the IAC-pop code is displayed in Figure 8.3.

The MATCH and TALOS code minimize the so-called Poisson likelihood ratio (PLR) [578, 579]

$$-2\ln(\text{PLR}) = 2 \sum_{j=1}^k M^j - O^j + O^j \ln\left(\frac{O^j}{M^j}\right) \tag{8.2}$$

The search for the minimum of χ_γ^2 or PLR can be performed by making use of genetic – as in the IAC-pop code – annealing or downhill simplex algorithms [see

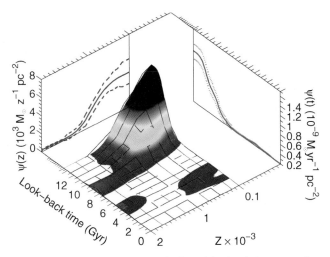

Figure 8.3 SFH of the Tucana dwarf spheroidal galaxy belonging to the Local Group, as determined using the IAC code [586] (courtesy of M. Monelli).

579, for a discussion]. One can also simultaneously consider several CMDs; in this case, the number of boxes entering Eqs. (8.1) and (8.2) will increase in order to account for the information coming from the additional CMDs, as done, for example, in [580].

It is also possible to add, as a further constraint on the SFH derivation, results from spectroscopy (see, e.g. [578]). If observations of [Fe/H] and [α/Fe] are available (current sets of isochrones are typically parametrized in terms of [Fe/H] and [α/Fe]) for stars along a portion of the CMD, one can consider additional diagrams like [Fe/H]-V and [α/Fe]-V that cover the magnitude range of the spectroscopic observations, define boxes in these diagrams, and repeat the procedure outlined above for CMD data. As in the case of the synthetic CMDs, one has to include the effect of spectroscopic errors and completeness in the Monte Carlo simulation needed to generate [Fe/H] and [α/Fe] values for the objects in each elementary population. When spectroscopic constraints are included, the functions given by Eqs. (8.1) or (8.2) have to take into account the additional boxes in these two new diagrams (increasing the range of values spanned by the index j). A different weight can be applied to the photometric and spectroscopic components in order to enhance or decrease the importance of the spectroscopic information.

Once the best matching SFH is found, it is extremely important to estimate the associated uncertainties, which can be separated into errors related to the observational dataset and parameter sampling. The errors related to the observational data are those associated to the uncertainties in the photometry such as photometric zero points, crowding conditions and Poisson statistics. Additional sources of uncertainty are those related to the determination of the distance and extinction of the target stellar population.

How these uncertainties affect the final solution has to be properly estimated. Concerning the error associated to the Poisson noise, it can be evaluated by determining several solutions by randomly changing the input data O^j, that is, the number of stars in the various boxes of the observed CMD according to a Poisson statistics. At the same time, the contribution to the final error budget coming from uncertainties on the photometric zero points, distance and extinction (in case of no differential reddening) can be estimated by applying rigid shifts in colour and magnitude to the observed CMD that account for realistic assumptions on the error affecting these "inputs", and computing a grid of solutions each one with its own "best" χ_γ^2. The analysis of the χ_γ^2 distribution allows one to identify the *best* solution for the SFH and the corresponding uncertainty associated to the errors on photometric zero points, extinction and distance to the target population.

A second source of uncertainties is parameter sampling, for example, the choice of CMD binning, and age and metallicity bins of the elementary populations, that may affect the recovery of the SFH. The size of this error source can be estimated, for example, by varying the t and Z range of the elementary populations, and the bin size of the bundles in the CMD. For all these different binnings, the SFH can be determined and the results returned to a common t, Z grid for comparison. The distribution of all the different solutions obtained by varying the various parameters will provide a realistic range of SFHs consistent with the observed CMD [577, 578].

Perhaps the largest source of error is related to the systematics in the adopted theoretical framework such as the libraries of isochrones and bolometric corrections that are difficult to incorporate rigorously in the SFH-solving algorithms. One option is to re-derive the SFH using alternatively different sets of isochrones and present the various solutions as a range of possible SFHs for the target population.

A different procedure has recently been proposed [587]. This method envisages the use of T_{eff} and luminosity shifts $(\Delta_{M_{bol}}, \Delta_{T_{eff}})$ to be applied to the isochrone set of choice – in the approximation of a single pair of values applied to the whole isochrone – and a re-derivation of the SFH with these "shifted" models. The values of the shifts are determined by comparing different sets of isochrones, and depend on the age/metallicity/evolutionary phase range sampled by the observed population. Once the shifts are determined, the synthetic CMD-fitting technique is employed by applying random shifts to the isochrones from Gaussian distributions of zero means and standard deviations specified by the pair $(\Delta_{M_{bol}}, \Delta_{T_{eff}})$. The ensemble of results provide an estimate of the uncertainty associated to isochrone systematics. Notice that this method cannot take into account the effect of differences in evolutionary timescales on the predicted star counts in a given magnitude/colour bin.

Another source of uncertainty is related to the choice of "external" inputs like the exponent of the IMF and the binary fraction. Its impact can be estimated by defining a "reasonable" range for these parameters, and determining separate SFHs for several possible combinations within the appropriate ranges.

8.2
Multiple Isochrone-fitting

An alternative approach employed in the code FIRES (that expands upon previous results [588, 589]) does not require binning CMDs and calculations of elementary populations [590]. Also, in this case, tests based on synthetic populations show that this technique is able to recover the input SFH of populations with both continuous and bursty star formation.

This technique is based on calculations of the relative probability that each star in the CMD of the target population "originates" from a particular isochrone. This probability can be defined in terms of the difference in magnitude/colour space between an isochrone of age t and metallicity Z, and the actual position of the star, accounting for the measurement errors. The question then is to find the relative number of stars that each isochrone contributes to the CMD, that is, its weight in the SFH. This can be accomplished by building a likelihood function where the free parameters are the isochrone weights.

Let us start by denoting the magnitudes of a star j in two generic passbands A and B, as A_j and B_j. The corresponding theoretical magnitudes from isochrone i are denoted as A_{im} and B_{im}, respectively. These A_{im} and B_{im} are shifted to account for the distance modulus and extinction of the observed system. One can then calculate the quantity

$$
E\left(A_{ij}\right) = \frac{1}{2\pi\sigma_A} \exp\left[-\frac{\left(A_j - A_{im}\right)^2}{2\sigma_A^2}\right]
\tag{8.3}
$$

that represents the probability that the observed star j is generated by a single point on the isochrone i, in the assumption of a Gaussian error σ_A on A_j. This error σ_A is given by $\sigma_A = \sqrt{\sigma_{A,\mathrm{phot}}^2 + \sigma_{A,\mathrm{iso}}^2}$, where $\sigma_{A,\mathrm{phot}}$ is the observational error (e.g. obtained from artificial star tests) and $\sigma_{A,\mathrm{iso}}$ accounts for the differences in magnitude between two neighbouring isochrones. We will come back to this point later on during this discussion.

Given that the mass of a star j in the observed CMD is unknown, the probability function has to be an integral over all masses along each isochrone. This probability will also need to be weighted by both the IMF $f(M)$, and a function describing the observed photometric completeness $c(A, B)$. The relative probability p_{ij} that a given star j "belongs" to an isochrone i (defined by the pair of values t, $Z^{1)}$) is therefore given by

$$
p_{ij} = \frac{1}{C_i} \int_{M_\mathrm{l}}^{M_\mathrm{u}} E\left(A_{ij}\right) E\left(B_{ij}\right) \cdot c(A, B) \cdot f(M)\, dM
\tag{8.4}
$$

where M_u and M_l are the upper and lower mass limits along the isochrone i and

$$
C_i = \int_{M_\mathrm{l}}^{M_\mathrm{u}} c(A, B) \cdot f(M)\, dM
\tag{8.5}
$$

is a normalization factor that accounts for the completeness and IMF.

This definition of the probability function makes use of two magnitudes, A and B, but it can be easily generalized to include three or more. If one considers the measurement of each magnitude to be independent, it is necessary to just include another term $E(C_{ij})$ describing the magnitude C and adjust the completeness function accordingly. Hence, the probability function for three magnitudes becomes

$$
p_{ij} = \frac{1}{C_i} \int_{M_\mathrm{l}}^{M_\mathrm{u}} E\left(A_{ij}\right) E\left(B_{ij}\right) E\left(C_{ij}\right) \cdot c(A, B, C) \cdot f(M)\, dM
\tag{8.6}
$$

The next step towards constructing the likelihood function is to determine the relative probability that a star j belongs to an ensemble of isochrones i, each of which

1) The multiple isochrone-fitting methods can account very easily for additional populations with varying Y abundances or metal mixtures, at constant Z and t. Formally, one needs only to extend the range of the values spanned by the index i. This means that one has to calculate the probability p_{ij} that star j belong to these additional potential contributors to the population SFH.

potentially contributes to the population SFH. This combined probability p_j is given by

$$p_j = \sum_{i=1}^{n_i} a_i p_{ij} \tag{8.7}$$

where a_i represents the weight of each isochrone, corresponding to a given pair of (t, Z) values, and n_i is the total number of isochrones in the adopted library. The weights represent the relative number of stars each isochrone contributes to the CMD and are subject to the obvious constraint $\sum_{i=1}^{n_i} a_i = 1$, with $0 \le a_i \le 1$.

The likelihood function can then be defined as the product of these probabilities for all stars in the CMD:

$$L = \prod_{j=1}^{n_j} p_j = \prod_{j=1}^{n_j} \left(\sum_{i=1}^{n_i} a_i p_{ij} \right) \tag{8.8}$$

hence

$$\ln(L) = \sum_{j=1}^{n_j} \ln \left(\sum_{i=1}^{n_i} a_i p_{ij} \right) \tag{8.9}$$

Figure 8.4 provides a visual representation of the calculation of the likelihood function.

A likelihood maximization technique to determine the values of the weights of each isochrone can then be employed. The code presented in [590] maximizes the log-likelihood using a genetic algorithm in order to find the combination of weights $a = (a_1, a_2, \ldots, a_{n_i})$ that most likely produced the observed CMD.

Each of these weights need then to be rescaled to account for those stars that have reached the end of their evolution and have disappeared from the CMD. For example, if a CSP has been formed by two bursts of star formation of the same intensity (same total mass formed in each one), 12 Gyr and 100 Myr ago respectively, and if the IMF is the same, there will be many more stars in the CMD from the young component. All stars with mass larger than $\sim 0.8 M_\odot$ belonging to the old component will have disappeared (they are actually along the WD cooling sequence) while most of them will still be present in the progeny of the young star formation burst. The number of "missing" stars for each isochrone, that is, the "missing" stellar mass formed in a given star formation episode, can be estimated from the IMF (once the stellar mass range at formation is fixed, for example, between 0.1 and 100 $M_\odot$) and completeness functions, and then added to its weight in the solution. The weights are then renormalized to give the correct relative number of stars formed at each age and metallicity.

It is important to note that the probability function requires an integration along the isochrone typically described by a discrete set of points rather than a continuous line. In the case of observed CMDs with small photometric errors, the sampling of the isochrones can play a significant role in determining the probabilities.

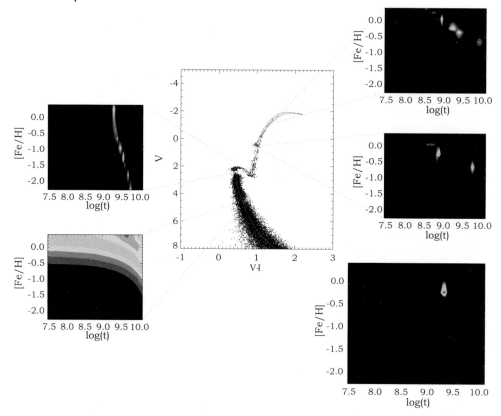

Figure 8.4 Synthetic CMD of a 2 Gyr old, $Z = 0.01$ mock population, calculated from the displayed isochrone after applying photometric errors that increase with magnitude. Four stars are marked in the diagram; the corresponding contour plots show the probability distribution for each star across the whole isochrone range used to retrieve the population SFH. The colours for the plots are scaled individually: dark grey denotes the highest probability and black denotes zero probability. The bigger contour plot shows the combined probability of the four stars (courtesy of E. Small).

Poor sampling can result in a star lying midway between two points of the correct isochrone and not registering a significant probability for the correct combination of t and Z. As a consequence, the implementation of this method requires that the magnitude spacing between adjacent points along a generic isochrone should not be greater than the photometric errors σ in order to give a reasonable approximation of the probability.

An additional technical issue is related to the fact that one must necessarily use a discrete grid, in both age and metallicity, of isochrones. Therefore, this method attempts to map a continuous distribution of possible ages and metallicities onto a discrete grid of (t, Z) pairs. This requires the introduction of the additional error term in the probability function, denoted before as $\sigma_{A,\mathrm{iso}}$. This parameter $\sigma_{A,\mathrm{iso}}$ is the minimum photometric error required for each star in the CMD to register a non-zero probability with the correct (possibly the nearest) isochrone in the grid. Its value(s) – that could vary along the isochrone – may be estimated, that is, by considering the magnitude differences at a given evolutionary stage between isochrones of neighbouring metallicities (at fixed age) or neighbouring age (at fixed metallicity) depending on the (t, Z) sampling of the adopted isochrone library. Ideally, $\sigma_{A,iso}$ could also account for the uncertainty in theoretical models and isochrones in addition to the quantization in age and metallicity.

For a given set of isochrones, distance modulus and reddening, uncertainties in the derived SFH arise from stars being associated with the wrong isochrone due to overlap of isochrones in the CMD, observational errors and the fact that one is necessarily employing a discrete set of isochrones to model an a priori continuous range of ages and metallicities.

The weights of isochrones with similar age and metallicity will be very highly correlated in the solution and cannot be treated as independent parameters. For this reason, one needs to calculate the confidence intervals of all the weights simultaneously using a Monte Carlo method. In the limit of large samples, the likelihood function approximates a n-dimensional Gaussian at the global maximum where n is the number of parameters (number of isochrones). It can be therefore assumed that a n-dimensional confidence region Q will have a χ^2 distribution. One can therefore determine Q_γ as the limit of the confidence region which has a coverage probability $(1 - \gamma) = 0.683$, and thus define a lower limit to the log-likelihood as

$$\ln\left(L_{\lim}\right) = \ln\left(L_{\max}\right) - \frac{Q_\gamma}{2} \tag{8.10}$$

The limit in log-likelihood corresponds to solutions of the SFH 1σ from the maximum. One can use a Monte Carlo approach to generate solutions with a likelihood above this limit, that depends on the the number of free parameters, and determine the range in weights for each isochrone. In general, the greater the photometric error, the greater the confidence limits since increasing the photometric error decreases the ability to distinguish between neighbouring isochrones.

The effect of uncertainties on distance, extinction and IMF can be treated as in the synthetic CMD-fitting methods.

Unlike the synthetic CMD-fitting method that can directly include a binary fraction in the computation of the elementary populations, isochrones can only model single stars given the stochastic nature of a binary population. This multiple isochrone-fitting method "sees" binaries as an additional young metal-rich component of the SFH [590]. A fully consistent way of treating unresolved binaries has yet to be developed, but it may be possible to isolate and identify the binary component in the retrieved SFH. In fact, binaries will mainly affect the MS. In certain favourable situations, a signature that isochrones entering the solution are not genuine components of the real SFH will be the absence in the observed CMD of their post-MS progeny. On the other hand, numerical tests show that the method (as for the case of the synthetic CMD-fitting technique) is robust against "reasonable" uncertainties in the IMF, and is able to reliably determine the SFH of old populations even without fitting HB stars.

9
Unresolved Old Systems

Photometric and spectroscopic observations of unresolved systems provide integrated magnitudes, colours and spectra that include the contribution of all stars belonging to the population under study. The first step to investigate the evolutionary status of unresolved populations therefore involves a theoretical modelling of the integrated properties of SSPs (and eventually of populations with a more complex star formation history). These models are usually dubbed as "stellar population synthesis" models (SPSMs – pioneered in [591]).

We have already discussed in Chapter 1 how to calculate analytically theoretical integrated fluxes F_λ and magnitudes for a SSP of a given age and initial chemical composition. This analytical computation is the standard procedure followed in the literature. However, as already mentioned in Chapter 1, this is strictly valid only when the number of stars is formally infinite. The analytical computation implies that all points along the isochrones are smoothly populated by a number of stars that can be equal to just a fraction of unity in the case of fast evolutionary phases (that correspond to extremely small mass differences between two consecutive points along the isochrone) and small values of M_{tot}.

In a real population, the number of objects at a point along the CMD is either zero or a multiple of unity, and when the number of stars is not large enough to smoothly sample all evolutionary phases, local statistical fluctuations of star counts, and hence integrated magnitudes and monochromatic fluxes, will arise. Also, the one-to-one correspondence between N_t and M_t may break down. We address this issue later on in this chapter. For the moment, we will consider the case of a smooth distribution of star counts along the isochrones.

SPSMs play a pivotal role in investigations about the formation and evolution of stellar populations, and are referred to as the "inverse" and "direct" approach, respectively. In the inverse approach, photometric and spectroscopic observations of unresolved populations provide empirical integrated spectra and/or colours/magnitudes that are compared to results from SPSMs to constrain the unknown population star formation histories. The direct approach starts from a theoretical model for the formation of the population under scrutiny that predicts a specific star formation history. SPSMs convert this theoretical star formation history into expected integrated photometric and spectroscopic properties that

Old Stellar Populations, First Edition. S. Cassisi and M. Salaris.
© 2013 WILEY-VCH Verlag GmbH & Co. KGaA. Published 2013 by WILEY-VCH Verlag GmbH & Co. KGaA.

Table 9.1 The main SPSM archives.

BaSTI [592]	Schiavon [593]
Buzzoni [594]	SED@ [595]
BC03 [8, 596]	SPEED [597]
FSPS [598, 599]	SPoT [600]
Galadriel [601]	Starburst99 [602]
GALEV [603]	Thomas [604]
GRASIL [605]	Vazdekis [606]
Maraston [11, 607]	Worthey [608]
Pegase [609]	

are then compared to their observational counterparts. Mismatches are used as guidelines to refine the theoretical formation model and the cycle is repeated.

A (non-exhaustive) list of publicly available SPSM archives is presented in Table 9.1.

9.1
Building Blocks of a SPSM

The first of the two main building blocks of SPSMs is a library of stellar isochrones that describe the predicted behaviour in the L-T_{eff} plane of SSPs, covering a range of initial chemical composition and ages. Each point along an isochrone is specified by the local value of L, T_{eff}, the initial chemical composition (parametrized by [Fe/H] and Y for a fixed metal mixture) and the mass m[1] of the star evolving at that point. From L, T_{eff} and m, one can determine the "local" stellar surface gravity g all along the isochrone, a quantity necessary to calculate the final integrated spectrum. The second main ingredient is a library of stellar spectra to transform L and T_{eff} into monochromatic fluxes, magnitudes, and colours. Individual stellar spectra are usually (in the standard plane-parallel approximation) defined in terms of T_{eff}, g and [Fe/H]. Together with these two main building blocks, the IMF has to be specified. At each point along the isochrone, interpolations among the spectral library produces the appropriate spectrum that is then rescaled according to the local number of objects predicted by the IMF. This procedure is repeated for each isochrone point, and the individual spectra are added up to produce a final integrated spectrum. At this point, if/when appropriate, one can apply corrections due to foreground dust extinction, nebular emission, and eventually K-corrections in the case of modelling populations at non-negligible redshifts. The whole procedure is schematically summarized in Figure 9.1.

---

1) Throughout this chapter, we will denote the stellar mass with m, to avoid confusion with symbols used for the integrated magnitudes.

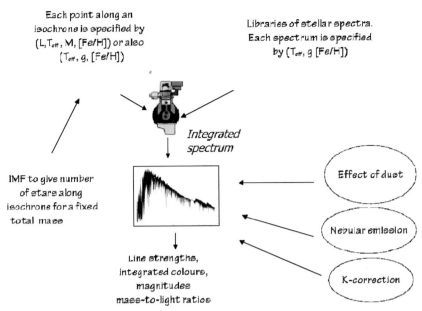

Each point along an
isochrone is specified by
(L, T_eff, M, [Fe/H]) or also
(T_eff, g, [Fe/H])

Libraries of stellar spectra.
Each spectrum is specified
by (T_eff, g [Fe/H])

*Integrated
spectrum*

IMF to give number
of stars along
isochrone for a fixed
total mass

Effect of dust

Nebular emission

K-correction

Line strengths,
integrated colours,
magnitudes
mass-to-light ratios

Figure 9.1 Schematic description of the computation of a SPSM.

A number of choices have to be made before producing SPSMs. The three main ones are the following:

1. IMF: the important characteristics are its functional form, and whether it is assumed to be universal or dependent on the stellar population properties (i.e. age, metallicity).
2. Isochrone library: the main characteristics are the coverage in terms of [Fe/H], t, metal mixture, evolutionary phase, and the choices for the mass loss history of massive stars, low-mass RGB and AGB stars. A crucial point to notice is that the HB phase of old isochrones is populated by objects all with the same mass, determined not only by the population age and initial chemical composition, but also by the chosen value of the mass loss rates during the RGB phase, that are assumed to be the same for all objects at a given point along the RGB. The colour extension of the HB phase is therefore very narrow compared to what was observed in "real" old populations like Galactic GCs. This has important consequences that will be explored later in this chapter.
3. Spectral library: empirical or theoretical. For both choices, crucial characteristics are resolution, λ-g-[Fe/H]-metal mixture-T_{eff} coverage. In the case of theoretical libraries, the adequacy of line lists and atmosphere modelling are two main issues. The cooler spectra, both at high and low gravities, are the most uncertain. Empirical spectral libraries have the obvious advantage of representing "real" stars and, for example, in the case of metal-poor RGB stars, should naturally include the observed changes of some surface chemical abundances that set in at the RGB bump, and are usually not included in the theoretical

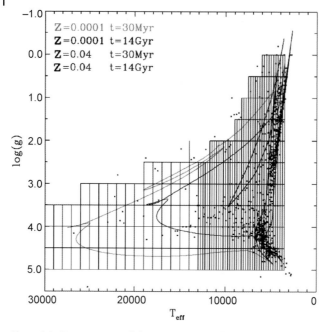

Figure 9.2 T_{eff}-g coverage of the Munari (rectangles) and MILES (dots) spectral libraries, compared to two pairs of BASTI theoretical isochrones spanning a large range of age and metallicity.

libraries adopted in SPSMs. In the case of these empirical libraries, the S/N ratio, flux calibration, and the determination of g-T_{eff}-[Fe/H] for the library stars are the main uncertainties.

Table 9.2 summarizes several publicly available libraries of empirical and theoretical stellar spectra, and isochrones suitable for computing SPSMs. A summary of the wavelength coverage of several of the spectral libraries listed in this table can be found in [8]. Depending on the wavelength, spectral resolution and age range that one aims to model (and hence the evolutionary phases that need to be covered by the underlying isochrones), SPSM calculations may need to combine two or more different spectral libraries (e.g. [592, 596]). Figure 9.2 displays, as an example, the T_{eff}-g coverage of the high-resolution MILES empirical spectral library (stars of all metallicities are displayed) and the Munari theoretical one (grid at fixed metallicity). Representative isochrones are also displayed.

Interpolations among theoretical model grids are usually simpler because of a generally more regular sampling of the [Fe/H]-T_{eff}-g parameter space. Empirical libraries, based on "local" stars, have a sparser parameter coverage, with chemical composition obviously biased towards solar or near solar metallicities, making the interpolation more challenging (see, e.g. [610] for an example of interpolation among the MILES library).

Table 9.2 Main spectral and isochrone libraries for population synthesis modelling.

Empirical spectra	Theoretical spectra	Isochrones
BaSEL (semiemp.) [611]	ATLAS9 [16]	BaSTI [148, 286, 432]
ELODIE [612]	BLUERED [613]	DSEP [614]
INDO-US [615]	Coelho-IAG [616]	Geneva [617]
IRTF [618]	COMARCS (C-stars) [619]	Padova [620, 621]
Lançon (AGB) [622]	MARCS [623]	Y2 [624]
MILES [625]	Martins [626]	Victoria [627]
NGSL [628]	Munari [629]	
Pickles [630]	PHOENIX [631]	
STELIB [632]	Rauch (post-AGB) [633]	
UVES-POP [634]	Smith (O and WR stars) [635]	
XSL [636]	TLUSTY (OB stars) [637, 638]	
	UVBLUE [639]	

9.2
Low-Resolution Diagnostics

Investigations of the evolutionary status of unresolved low surface brightness populations, like extragalactic GCs, near-field dwarf galaxies and generally, distant (high-redshift) galaxies, have to make use of diagnostics based on broad-band photometry.

Before discussing the use of integrated magnitudes and colours to estimate the age and metallicity of unresolved populations, it is worth recalling the contribution of individual evolutionary phases to the integrated flux emitted by SSPs.

Let us consider the integrated luminosity of SSPs of different ages (at fixed initial solar chemical composition) in various wavelength ranges. In the rest of this chapter, unless otherwise specified, we use results from the BaSTI population synthesis models that employ the Kroupa IMF for stellar masses between 0.1 and 100 $M_{\odot}$ [640]:

$$\Phi(m)\,dm \propto m^{-1.3}\,dm \quad m < 0.5\,M_{\odot}$$

$$\Phi(m)\,dm \propto m^{-2.3}\,dm \quad m \geq 0.5\,M_{\odot}$$

Figure 9.3 displays the case of four representative photometric bands (in the Johnson–Cousins system) from the near-UV to the infrared, namely, U, B, V, K. The contribution of the individual evolutionary phases obviously depends on the filter considered. The U and B luminosity is always dominated by MS stars. For old ages, this is mainly caused by objects close to the TO. The second most relevant phase in old populations (age above ~ 1 Gyr) is the SGB. At these ages, the HB and RGB contribute about 15–20% each to the integrated luminosity in B. The actual morphology of the HB controls this phase's contribution to the blue and UV portion of

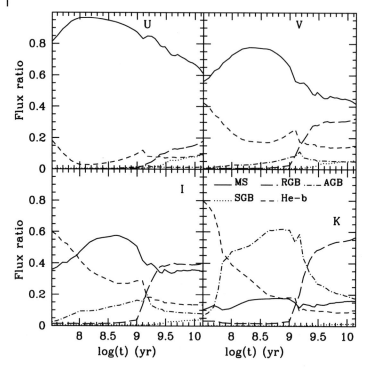

Figure 9.3 Evolution with time of the fractional contribution of different evolutionary phases to the integrated flux in the (a) *U*, (b) *V*, (c) *I*, and (d) *K* photometric bands for SSPs with solar metallicity.

the spectrum. Here, we have considered a red HB, an assumption usually true at this metallicity. Bluer HBs would provide a larger contribution to the flux in the *U* and *B* filters. Post-AGB stars and hot, bright WDs contribute to the flux in the UV below ≈ 2000 Å, shortward of the *U* and *B* bands discussed in this example.

In *V*, the situation is similar to the *B* filter, but the RGB takes over as the second major contributor at old ages. The *K* filter shows a very different picture. At ages below ∼ 200 Myr, the He-burning phase dominates the integrated luminosity because of massive stars that experience the onset of central He-burning at the red side of the CMD. Between ∼ 200 Myr and ∼ 3 Gyr, the AGB phase (mainly, the thermal pulse phase) controls the integrated flux in *K*, whilst at higher ages, the RGB takes over as the major contributor.

As a next step, we consider the predicted low-resolution Spectral Energy Distributions (SEDs) of SSPs, that is, the distribution of the integrated magnitudes in several photometric bands. Figure 9.4 displays the SED from *U* to *L* of three scaled-solar SSPs (6 and 12 Gyr old, [Fe/H] = 0.06 populations and 12 Gyr, [Fe/H] = −0.25 population) calculated assuming a total mass at birth of $1 M_\odot$.

The SEDs of the two 6 Gyr old SSPs have been shifted to have the same *I* magnitude of the 12 Gyr old one. The magnitude shifts correspond to a decrease of the

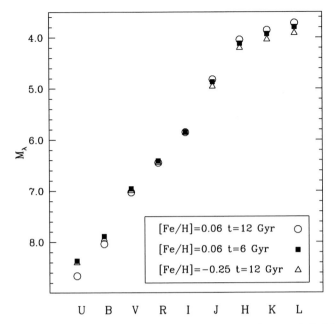

Figure 9.4 Integrated low-resolution SED from U to L, of SSPs with the labelled chemical compositions and ages. All SEDs are normalized to the same integrated I magnitude.

initial mass of the [Fe/H] = 0.06, 6 Gyr SSP by a factor 1.67, and a decrease by a factor 1.72 of the [Fe/H] = −0.25 SSP.[2]

One can first notice that, at fixed [Fe/H] = 0.06, increasing the age causes a shift of the SED towards redder colours, for the integrated magnitudes longward of I become brighter, whilst they become fainter from R to U. On the other hand, keeping the age fixed at 12 Gyr and decreasing [Fe/H] shifts the SED towards bluer colours. This behaviour gives rise to a potential age–metallicity degeneracy that is clearly illustrated by comparing by comparing the 12 Gyr, [Fe/H] = −0.25 SED with the 6 Gyr, [Fe/H] = 0.06 one. All magnitudes from U to I are essentially identical. Only when moving towards longer wavelengths do the differences between the SED of these two populations increase above 0.1 mag.

Another important degeneracy for old populations is illustrated in Figure 9.5, where we display the SED of a 12 Gyr, [Fe/H] = −1.31 population with HB colour determined using a Reimers mass loss law along the RGB with the free parameter η set to 0.4, and a 7 Gyr, [Fe/H] = −1.84 η = 0.2 SSP, both calculated with a α-enhanced metal mixture typical of galactic halo stars.[3] A higher η implies

2) Passbands are in the Johnson–Cousins system, with the exception of H, in which case the passband definition from [641] was adopted. However, the behaviour is not expected to be different in the case of equivalent filters in the HST/WFPC2, HST/ACS, 2MASS or SDSS systems.

3) Of course, as discussed for the case of Galactic GCs, bluer HB morphologies can be produced by coeval He-enhanced sub-populations hosted by the target putative SSP. This He enhancement (if moderate and if Fe and the CNO sum are unchanged) to first order has the same effect of a RGB mass loss increase at fixed chemical composition.

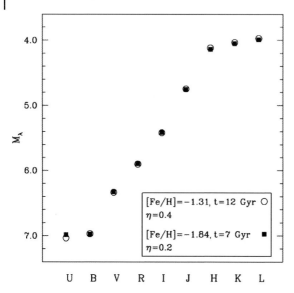

Figure 9.5 The same as in Figure 9.4, but for the effect of the HB colour. All SEDs are normalized to the same integrated *B* magnitude.

a bluer HB of the older population because of a more efficient RGB mass loss coupled to a lower TO mass. The older, more metal-rich population has a HB centred at a T_{eff} around 7800 K, while the second, younger SSP has a redder HB, centred at T_{eff} around 5500 K. The 7 Gyr SED has been shifted to have the same *B* magnitude of the older SSP (shift corresponding to a decrease of the initial M_t by a factor 1.61). After this shift, all magnitudes from *B* to *J* are practically identical (differences by less than 0.01 mag), and the differences in *H*, *K* and *L* are equal to just ∼ 0.02 mag. Only when one considers the *U*-band, does the difference increase to 0.05 mag. This age–metallicity-HB colour degeneracy is one of the most serious problems in photometric (and spectroscopic) studies of unresolved old stellar populations.

The effect of varying the SSP metal mixture is shown in Figure 9.6. Here, we display the SED of three 12 Gyr old SSPs. Two SSPs have a α-enhanced metal distribution ($[\alpha/Fe] = 0.4$) with $[Fe/H] = 0.05$ and $[M/H] = 0.06$ (corresponding to $[Fe/H] = -0.25$), respectively. The third SSP has a scaled-solar $[Fe/H] = [M/H] = 0.06$ initial chemical composition. The variation of the metal mixture is accounted for in both isochrones and stellar spectra, and the two α-enhanced SEDs have been normalized to the same *R* magnitude of the scaled-solar SED. It is very interesting to notice that the integrated magnitudes from *R* to *L* of the scaled-solar SED are reproduced within at most 0.05 mag (in the case of the *K*) by the α-enhanced SED with the same $[M/H]$, whereas for magnitudes from *U* to *R*, scaled-solar and α-enhanced SEDs with the same $[Fe/H]$ give values consistent within at most 0.03 mag. The different behaviour of these two groups of filters is clearly an effect of the variation of the with stellar spectra changing the metal mixture. At the

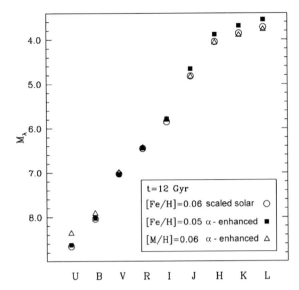

Figure 9.6 The same as in Figure 9.4, but for the effect of the metal distribution. All SEDs are normalized to the same integrated *R* magnitude.

longer wavelengths, bolometric corrections are roughly independent of the metal content, and hence differences of the integrated magnitudes (dominated by the RGB) are caused solely by differences in the initial mass distribution along the underlying isochrones, and their location in the L-T_{eff} diagram. The behaviour of the integrated magnitudes in the filters from I to L is therefore explained by the fact that old α-enhanced isochrones – especially at low metallicities – are mimicked by scaled-solar ones with the same total metallicity [42]. At shorter wavelengths (dominated by the MS), the bolometric corrections for α-enhanced metal mixtures are instead roughly mimicked by their scaled-solar counterpart with the same [Fe/H] [207].

Variations of individual element abundances have been partially investigated in [642, 643]. These investigations present isochrones between 1 and 12 Gyr from the lower MS to the TRGB, and the corresponding $(U-B)$, $(B-V)$, $(V-R)$, $(V-I)$ integrated colours obtained by varying (typically enhanced by 0.3 dex, but for C, that was enhanced by 0.2 dex) the fractions of the most abundant heavy elements for a fixed solar Z and a scaled-solar metal distribution one at a time. For most of the metals, a variation at constant Z does not substantially alter [Fe/H], and the resulting colour variation can be interpreted also as a change due to the enhancement of a single element at constant [Fe/H] (see Figure 1 in [642]). It is important to notice that the change of metal distribution was accounted for in both stellar models and spectra.

In general, the older populations show a larger sensitivity to element-by-element abundance changes, for cooler stars are more sensitive to spectral effects, in addition to the effect of abundance changes on the evolutionary tracks. Enhancing the abundances of the α elements Ne, Mg, Si, S, Ca, Ti individually by a factor of 2 with respect to a reference solar metallicity [Fe/H] stays basically unchanged, within a few 0.01 dex, because the relatively small contribution of each of these elements to the total metal content. The colour variations are typically below 0.01 mag, with a few exceptions like the variation of Ti that makes $(V-I)$ redder by 0.04–0.06 mag at ages above 2 Gyr, and the variation of Mg that makes $(V-I)$ redder by ~ 0.04 at 12 Gyr. These effects are mainly due to changes in the stellar spectra rather than the isochrones (but in the case of Mg, the RGB T_{eff} is also affected). Varying C or N (also in this case [Fe/H] stays practically constant) has again a small impact, the largest effect being again in $(V-I)$, which becomes bluer when C is enhanced, by at most 0.04 mag. One has to notice that in this case, the colour variation is due to changes in both isochrones and spectra. Variations of O and Fe have a much larger impact, but in this case, [Fe/H] is not constant. On average, an enhancement of O at constant Z (with the resulting decrease of [Fe/H]) tends to make the integrated colours bluer, whereas an enhancement of Fe tends – not unexpectedly – to make the integrated colours redder.

This brief analysis provides the background for the two main methods employed for studying stellar populations by means of broad-band photometry.

9.2.1
Integrated Colour Fitting

"Classical" distance independent (also independent of the total mass of the SSP) techniques make use of colour–colour diagrams that minimize the effect of the age–metallicity degeneracy, that is, the two colours must show different sensitivities to variations of metallicity and age. As made clear by Figure 9.4, colours restricted to the $BVRI$ bands are severely affected by the degeneracy, which can cause uncertainties by a factor of 2 on both age and [Fe/H] estimates. Extending the spectral range covered by the observations to the near infrared and/or the U band helps in minimizing the effect of the age–metallicity degeneracy, as was firstly recognized about 30 years ago [644].

Examples of colour–colour diagrams are provided in Figure 9.7. Figure 9.7a displays a theoretical $(B-V)$–$(V-I)$ diagram of α-enhanced SSPs. Dashed lines denote sequences of constant age at varying [Fe/H], while solid lines represent sequences of constant [Fe/H] and varying ages. Ages are between 2 and 15 Gyr, [Fe/H] between -2.6 and 0.05. The two sets of lines overlap almost completely, implying that this diagram is severely affected by the age–metallicity degeneracy. The "positive" side of this complete degeneracy is that on this diagram, one expects old stellar populations to lie essentially along a single, tight sequence. Comparing the position of this "observed" sequence with theoretical predictions can in principle be used as a test of the adopted SPSM, and/or of the reddening corrections applied to the data because the "degenerate" model sequences have to overlap with the observed dereddened colour sequence. Figure 9.7b displays the $(V-K)$–$(V-I)$ diagram for the same SSPs, a tool often used to study extragalactic GCs (see, e.g. [645, 646]). In this case, the two sets of lines are more orthogonal, that is, the degen-

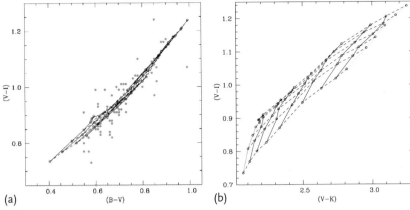

(a)

(b)

Figure 9.7 (a) Integrated $(V-I)-(B-V)$ diagram for α-enhanced SSPs spanning a large range of age and metallicity. Lines of constant [Fe/H] (solid) and constant age (dashed) are completely degenerate. Filled circles denote the colours of a sample of Galactic GCs. (b) $(V-I)-(V-K)$ diagram for the same metal mixture, ages and compositions. Moving towards increasing $(V-I)$ colours, the reference ages are equal to 2, 4, 6, 9, 12 and 15 Gyr. Moving towards increasing $(V-K)$ colours, the [Fe/H] values are equal to $-2.62, -2.14, -1.62, -1.31, -1.01, -0.70, -0.29, -0.09,$ and 0.05.

eracy is, to some degree, broken. The "ideal" colour–colour diagram would display perfectly orthogonal age–metallicity sequences. Figure 9.8 shows two other examples of theoretical diagrams (in this case for a scaled solar metal mixture) used to study old stellar populations [647–649], using UBV and BJK filters, respectively. A recent study of favourable colour combinations to minimize the age–metallicity degeneracy can be found in [650].

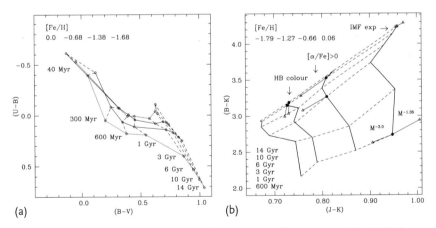

(a)

(b)

Figure 9.8 (a) $(U-B)-(B-V)$ diagram for scaled solar SSPs (from [621]) with the labelled parameters. At $(B-V) = 0.5$, [Fe/H] decreases towards decreasing $(U-B)$. (b) $(B-V)-(J-K)$ diagram from [648], showing also the effect of changing metal mixture, HB colour and IMF.

Figure 9.8 displays, also on the scaled-solar BJK diagram [648], the effects of changing the colour of the HB, moving from a scaled-solar to an α-enhanced metal mixture ($[\alpha/\text{Fe}] = 0.4$) and of changing the exponent of the IMF respectively. For ages of 10 and 14 Gyr, and $[\text{Fe/H}] = -1.27$, when doubling the value of η in the Reimers formula (from 0.2 to 0.4), the HB becomes much bluer and the populations appear to be ~ 6 Gyr old when the age is evaluated using the reference calibration with a redder HB. The metallicity estimate is only slightly affected because of the small contribution of the HB to the J and K magnitudes. The displayed α-enhanced colours are for 10 and 3 Gyr old $[\alpha/\text{Fe}] = 0.4$ populations with $[M/H] = -0.66$ (that corresponds to $[\text{Fe/H}] = -1.01$). They are shifted to the blue with respect to the reference scaled-solar counterparts with $[M/H] = [\text{Fe/H}] = -0.66$. The inferred ages – employing the reference scaled-solar SSP grid – are slightly older for the 3 Gyr population, and essentially unchanged for the 10 Gyr SSP. The inferred metallicity would correspond to $[\text{Fe/H}]$ close to the value of the α-enhanced mixture. In the case of the IMF, the reference colour grid assumes a standard Salpeter power law with exponent $x = -2.35$. In this test, the exponent has been changed to -3.0 (MS-dominated population) and -1.35 (giant-dominated population) respectively. A MS-dominated population shifts the integrated colours towards bluer values, whereas the opposite is true in the case of a giant-dominated SSP. The reason for this behaviour can be understood by recalling that low-mass MS stars do not contribute appreciably to the total integrated flux in these photometric bands, even if one decreases the IMF exponent significantly. A MS-dominated SSP will have a larger contribution of MS stars to the integrated flux, but MS objects in the relevant mass range (e.g. close to the TO) are typically bluer than bright RGB and AGB objects (that still provide the largest contribution to the J and K fluxes), and thus make both integrated colours slightly bluer compared to the case of $x = -2.35$. The opposite is true in a giant-dominated SSP. The magnitude of this effect increases with decreasing age because the extension of the MS reaches brighter magnitudes and bluer colours when the age decreases.

It is perhaps appropriate to notice that all methods employing filters from I to longer wavelengths are affected by the treatment of the RGB (and to a lesser degree, at least for ages above 1 Gyr, the AGB) phase. Considering the $(V-K)$–$(V-I)$ as an example, an increase of the RGB T_{eff} by 100 K in the target system, compared to the reference grid, leaves the age practically unchanged, but decreases the estimated $[\text{Fe/H}]$ by 0.15–0.20 dex. A decrease of the number of RGB stars by a factor of 2 (corresponding essentially to halving the RGB evolutionary times) in the target decreases the estimated $[\text{Fe/H}]$ by ~ 0.10 dex and the age by $\sim 40\%$ at 14 Gyr, and $\sim 25\%$ at 4 Gyr.

A common feature of these diagrams is the generally poorer age resolution with increasing age[4] that hampers any accurate age determination of individual old SSPs, even with photometric errors of a few hundredths of a magnitude. The best approach in this case is to statistically study the age distribution of large samples of

---

4) For the VIK diagram, this is especially true in the metal-poor regime. In [645], a $(V-K)$–$(U-I)$ diagram is used to improve the age resolution, especially at high metallicities.

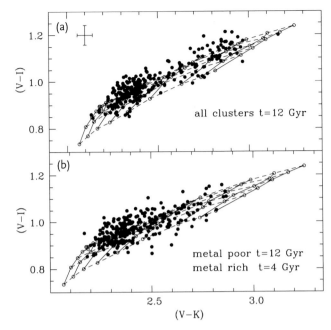

Figure 9.9 Integrated colours of clusters belonging to one single realization of the synthetic template populations discussed in the text. The $(V-I)-(V-K)$ grid is the same as in Figure 9.7.

SSPs, for example, their cumulative age distributions (CADs, see, e.g. [645, 651]). Figure 9.9 displays two Monte Carlo generated synthetic samples of SSPs overlaid onto the $(V-K)-(V-I)$ diagram of Figure 9.7. The synthetic SSPs have been calculated for a bimodal [Fe/H] distribution made of two Gaussians centred around [Fe/H] $= -1.55$ (1σ dispersion equal to 0.30 dex) and [Fe/H] $= -0.55$ (1σ dispersion equal to 0.20 dex) respectively (240 metal poor and 80 metal-rich objects) similar to the case of Milky Way GCs. The 1σ photometric error is small, equal to only 0.03 mag in all filters, comparable to the average error in the integrated colours of globulars in the Galaxy. We display the case of both single age (12 Gyr) SSPs, and a bimodal age distribution (12 Gyr for the metal-poor component, 4 Gyr for the metal rich one). It is obvious that, even in the case of coeval SSPs, the retrieved ages display a large spread – with values as low as 2 Gyr – and several points lie outside the boundaries of the grid just because of the effect of the photometric errors. We display in Figure 9.10 the input and retrieved CADs of 30 Monte Carlo realizations of these two synthetic samples, that is, the number fraction of clusters with age larger than t_i as a function of t_i. For each single realization, we associate to each synthetic SSP an age greater than t_i when it lies above sequences of constant t_i corresponding to ages equal to 0 (e.g. all SSPs younger than 2 Gyr), 2, 4, 6, 9, 12 and 15 Gyr. Filled circles represent the mean values obtained from the multiple realizations, and error bars denote the 1σ dispersion around these mean values.

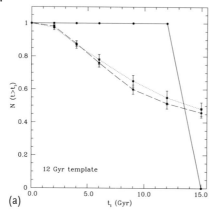
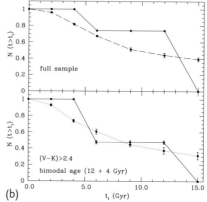

Figure 9.10 (a) CADs for the 12 Gyr template sample. The solid line displays the real age distribution, dotted and dashed lines the CAD retrieved from the colour–colour diagram for the full sample (dashed line) and for only the clusters with $(V-K) > 2.4$ (dotted line). (b) CAD for the template population with bimodal age distribution. (a) shows the CADs for the whole sample (real age distribution – solid line; retrieved age distribution – dashed line), whereas (b) shows the age distribution only for clusters with $(V-K) > 2.4$ (real age distribution – solid line; retrieved age distribution – dotted line).

Figure 9.10 very clearly shows the change from the input to the retrieved age distributions. The retrieved CADs display a smooth decrease with t_i because the synthetic SSPs tend to be more evenly distributed over the colour grid. If the analysis is restricted to the redder SSPs ($(V-K) > 2.4$), where the age-resolving power is larger, the retrieved CAD is not greatly changed.

Figure 9.11 illustrates the idea behind the use of CADs to constrain the age distribution of SSPs. One can notice how the retrieved bimodal and coeval age CADs are different. The extent of the difference depends on both the age of the young sub-population and the number ratio between the two components. Increasing the ratio of young to old SSPs – keeping the age distribution fixed – shifts the CAD to increasingly lower values for increasing t. The same happens when the age of the young sub-population is decreased, keeping the number ratio between the two components unchanged. It is very interesting to mention that a comparison of the input and retrieved cumulative metallicity distributions (CMeDs – defined analogously to the CAD) show instead only marginal differences.

Obviously, the size of the photometric error has a large impact on this type of analyses. A 1σ photometric errors equal to 0.1 mag in the VIK passbands would prevent the detection of bimodal SSP populations with age differences up to $\sim$ 8 Gyr when the $(V-K)-(V-I)$ is employed.

Figure 9.12 shows how the CADs behave in the case of changing the HB morphology and metal mixture, two potential sources of systematic errors. The synthetic samples and reference $(V-K)-(V-I)$ grid presented in Figure 9.9 have been determined from models calculated with a Reimers mass loss formula for the RGB evolution, with $\eta = 0.2$. The CAD derived from our reference grid for a synthetic 12 Gyr old coeval SSP sample, but calculated with $\eta = 0.4$ (more efficient RGB

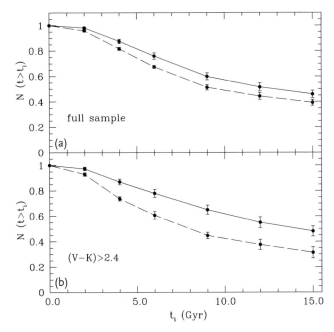

Figure 9.11 Comparison between the retrieved CADs for the unimodal (solid line) and the bimodal-age template populations of Figure 9.10. (a) displays the comparison for the full samples, while (b) only shows the comparison for objects with $(V-K) > 2.4$.

mass loss) is very close to the CAD for the bimodal (12 and 4 Gyr) age, $\eta = 0.2$ SSP. If the comparison is restricted to the redder SSPs with $(V-K) > 2.4$, differences with the coeval $\eta = 0.2$ grid disappear, for the change of η modifies the mean colour of the HB only for [Fe/H] below ~ -1.1, and therefore one does not expect any effect at larger metallicities. As a further test, we have considered a coeval SSP population, half the objects with [Fe/H] < -1.1 simulated using $\eta = 0.2$, and half using $\eta = 0.4$. This mimics the existence of SSPs with different HB colours at fixed [Fe/H] and age (like GCs in the Galaxy). The resulting CAD is still very similar to the bimodal case. The resulting CMeDs are not substantially altered compared to the input values, with only a small shift towards lower [Fe/H] (because of bluer integrated colours) at [Fe/H] below ~ -1.

As for the change of the metal mixture, we have considered 12 Gyr coeval SSPs, the metal-poor component α-enhanced as the reference grid, and the metal-rich component calculated with a scaled-solar metal distribution. The CADs retrieved with the reference α-enhanced grid are also displayed in Figure 9.12. The "wrong" metal mixture assigned to the metal-rich component simulates a bimodal-age population, albeit with a small age difference compared to our case of 12 plus 4 Gyr. The effect is obviously enhanced when the sample is restricted to the more metal-rich SSPs with $(V-K) > 2.4$. The reason is that α-enhanced lines of constant age slowly shift towards redder $(V-I)$ – for a fixed $(V-K)$ – when [Fe/H] increases, compared to the scaled-solar ones. The size of this effect increases with decreasing

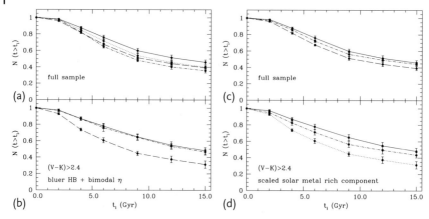

Figure 9.12 (a,b) Comparison between the retrieved CADs for the 12 Gyr template (solid line) the $\eta = 0.4$ population (dashed-dotted lines) and the bimodal η population (dotted line). (a) displays the comparison for the full synthetic samples, while (b) displays the comparison for objects only with $(V-K) > 2.4$ (the results for the bimodal η population are not displayed because they are identical to the case of the template population). The corresponding CADs for the template bimodal-age population are also displayed as dashed lines.

(c,d) Comparison between the retrieved CADs for the 12 Gyr template (solid line) and a population (dashed-dotted lines) whose metal-rich sub-population is assumed to be born out of a scaled-solar metal distribution. (c) displays the comparison for the full synthetic samples and (d) only shows the comparison for objects with $(V-K) > 2.4$. The corresponding CADs for the template bimodal-age population are also displayed as dashed (c) and dotted (d) lines.

age. This means that a metal-rich scaled-solar SSP will display a lower age than the input value when retrieved using an α-enhanced grid. Again, the retrieved CMeDs are essentially unchanged compared to the input [Fe/H] distribution.

Finally, one can also check the effect of core overshooting that is potentially relevant at ages below $\sim 4\,$Gyr. Overshooting makes both colours bluer at fixed ages and [Fe/H], and if a $(V-K)$–$(V-I)$ grid that does not include overshooting is used to retrieve ages of SSPs where overshooting is efficient, both retrieved ages and [Fe/H] will be lower than the input values.

More recently, the use of near- and far-UV filters has been advocated to minimize the age–metallicity degeneracy [652, 653]. At ages above 1 Gyr, for example, the $m(1500)$–$(B-V)$ diagram[5] has been shown to be a promising tool for age and metallicity estimates of SSPs [649]. A crucial issue related to the use of UV filters is their extreme sensitivity to the morphology of the HB, and more specifically, to the presence of hot HB stars. Calibrating the HB morphology-age–metallicity trend in the models, to match the *average* behaviour in Galactic GCs, may in principle not produce the correct colours in extragalactic globulars.

Another thorny and largely unexplored problem that affects short wavelength filters (starting from U) when hot HB stars are present is related to the efficiency of atomic diffusion and radiative levitation. Spectroscopic observations of hot HB

---

5) $m(1500) - V = -2.5 \log(\langle f_{1500}\rangle / \langle f_V \rangle)$ where $\langle f_V \rangle$ is the average flux between 5055 and 5945 Å, and $\langle f_{1500}\rangle$ is the average flux between 1250 and 1850 Å.

stars in Galactic GCs (see, e.g. [297]) reveal an increase of the surface abundance of several metal species (Fe among them) and a decrease of surface He when T_{eff} increases above $\sim$ 11 500 K. This appears to be caused by the combined effect of atomic diffusion and radiative levitation that, however, has to be moderated by additional mixing processes (see Section 4.7) because theoretical HB models including these processes predict the onset of abundance anomalies at a lower T_{eff} [306]. At present, no extended and complete – calculated until their transition to the final cooling sequence – grids of hot HB models that reproduce these observations, nor extended grids of appropriate – both theoretical or empirical – model atmospheres (see, e.g. [654] for exploratory studies) are available for population synthesis studies. Fit of theoretical models to the CMDs of HB stars in NGC2808 in the near- and far-UV shows that possibly the use of HB tracks with a "standard" metal mixture (α-enhanced in this case) and bolometric corrections calculated for a scaled-solar metal mixture and an enhanced total metallicity – so as to reproduce the observed photospheric iron abundances, disregarding the effect of the decreased He abundance – is a reasonable, albeit crude approximation [558]. Of course, this treatment cannot take into account a possible stratification of metal – and He – abundances in the atmosphere and below the photosphere, and the detailed metal pattern.

A simultaneous fit to several colours is the natural extension of techniques based on colour–colour diagrams, as discussed, for example, in [652]. The authors employed the colours $(FUV-V)$, $(NUV-V)$, $(FUV-NUV)$, $(U-B)$, $(B-V)$, $(V-R)$, $(V-I)$ – where FUV is the flux in the wavelength range 1344–1786 Å and NUV the flux in the range 1771–2831 Å as measured by the *GALEX* satellite – for a sample of 42 M31 GCs. With the HB morphology-metallicity–age trends adopted in the chosen SPSM, these authors find that ages and metallicities are, respectively, 90 and 60% better constrained than when only using combinations of $UBVRI$ filters. An improved application of multi-colour fits that include the UV part of the spectrum should make use of theoretical SSP calibrations with different HB morphologies at fixed metallicity and age, to find the age–metallicity-HB morphology combination that maximize the adopted merit function. In this respect, it is important to realize that analytical calculations of SSP integrated magnitudes including the effect of extended HB morphologies are not as straightforward as the case of a fixed value of the mass loss efficiency η. The HB part of the underlying SSP isochrone must provide a probabilistic description of where the progeny of the objects at the RGB tip is located, given the prescribed range of post-RGB masses. One possibility is to make use of Monte Carlo techniques to calculate a synthetic CMD of the HB – from ZAHB to the TP phase – for a selected post-RGB mass distribution, and calculate the integrated flux produced by all HB stars in the CMD. This flux can then be added to the integrated flux predicted from the isochrone up to the tip of the RGB. If one follows this procedure, a large number of stars must be drawn during the MC simulation in order to avoid fluctuations of the integrated flux produced by HB stars, due to poor sampling of these fast evolutionary phases. The final integrated flux can be later rescaled according to the chosen mass normalization of the SSP.

9.2.1.1 Colour-Bimodality of Extragalactic Globular Clusters

One of the most significant recent developments in the field of extragalactic GCs has been the discovery that colour distributions (i.e. $(B-I)$, $(V-I)$, $(g-z)$) of GC systems are typically bimodal. Almost every massive galaxy studied with sufficiently accurate photometry has been shown to have a bimodal GC colour distribution [655]. In the assumption of approximately coeval "old" ages typical of Galactic GCs – that seems confirmed by spectroscopic age determinations, see Section 9.3 – such a colour bimodality is generally assumed to correspond to a metallicity bimodality.

An alternative point of view that is a good example of how uncertain the interpretation of integrated colours of stellar populations can be has been put forward in [656, 657]. This scenario goes as follows. If there is a linear relationship (at constant age) between cluster [Fe/H] and, that is, the integrated $(g-z)$ colour, a unimodal [Fe/H] spread will correspond to a unimodal colour distribution, whereas a multi-modal [Fe/H] distribution will generate a multi-modal $(g-z)$ distribution. The compilation of [Fe/H] and $(g-z)$ colours of 95 GCs belonging to the Milky Way, and the galaxies M49 and M87 (the [Fe/H] of these latter two samples of "old" GCs being determined from integrated spectra, as discussed in Section 9.3) seem to display a non-linear behaviour [658], and can be modelled with the following equation [656]:

$$[Fe/H] = -33.74 + 79.81(g-z) - 66.69(g-z)^2 + 20.66(g-z)^3 - 1.08(g-z)^4$$

$$(9.1)$$

It is the non-linearity of the [Fe/H]–$(g-z)$ relationship that is able to produce a bimodal colour distribution starting from a unimodal [Fe/H] spread. Figure 9.13 shows this very clearly, illustrating the results of the following test. Figure 9.13a, b displays [Fe/H] values for three synthetic samples of 600 GCs, with Gaussian distributions centred at [Fe/H] $= -0.60$, -0.75 and -1.00 respectively, each with a 1σ spread equal to 0.35 dex, together with the [Fe/H]-colour relationship from Eq. (9.1). Figure 9.13c shows the colour distributions derived by inverting Eq. (9.1) and using the input [Fe/H] values of the synthetic GC samples. A Gaussian 1σ error of 0.03 mag has also been applied to each individual colour. It is clear from this example how a unimodal [Fe/H] distribution can in principle generate a bimodal $(g-z)$ spread. This stems from the fact that around the inflection point of the [Fe/H]–$(g-z)$ relationship given by Eq. (9.1), ([Fe/H] ≈ -0.7) equidistant metallicity intervals are projected onto larger colour intervals; it is this projection effect that creates a bimodal colour distribution. The numerical experiment of Figure 9.13 shows that the effect is maximized when the peak of the [Fe/H] distribution is located around this inflection point. Notice also that when the [Fe/H] spread is centred at [Fe/H] $= -1.00$, that is, away from the inflection point, the resulting colour distribution is unimodal.

This non-linearity of Eq. (9.1), according to [657], is due to the mild non-linearity of the integrated colours produced by stars from the MS to the tip of the RGB, plus the strong non-linearity of the HB integrated colours with [Fe/H], at least for HB

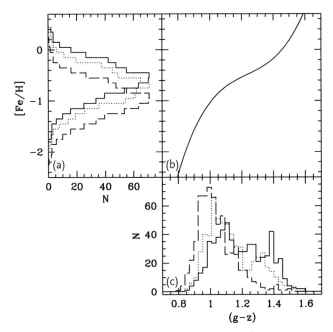

Figure 9.13 (a) and (b) display the Gaussian [Fe/H] distributions (1σ spread equal to 0.35 dex) centred at [Fe/H] = −0.60 (solid lines), −0.70 (dotted lines) and −1.00 (dashed lines) of three synthetic samples of GCs, and the $(g-z)$–[Fe/H] relationship employed to determine their integrated colours. The resulting $(g-z)$ values are displayed in (c), perturbed by a 1σ photometric error equal to 0.03 mag.

morphologies that reproduce some "typical" CMDs of Galactic globular clusters of varying [Fe/H] (NGC6624, 47 Tuc, NGC6638, NGC5904).

Another development has been the discovery that GCs in the blue peak of the colour distribution tend to be redder at brighter magnitudes, albeit not in all extragalactic GC systems that have been observed. There is also some evidence for a tilt in the red, metal-rich sub-population of GCs [659]. According to the "projection model" described before, this could be explained by the interplay between a mass-metallicity relationship for GCs – the [Fe/H] unimodal distribution is centred at increasing [Fe/H] values for brighter, more massive GCs – and the location of the peak [Fe/H] distributions relative to the inflection point of the underlying [Fe/H]–$(g-z)$ relation – which, in principle, can be different in different GC systems if the mean trend of the HB morphology with [Fe/H] is not universal.

The origin of these observed colour bimodalities is still a matter of debate. In the case of the GC system around NGC5128, a nearby giant elliptical galaxy, the metallicity of a sizable sample of objects has been determined from spectroscopic line indices (see Section 9.3), and one derives a bimodal distribution without the need to employ a [Fe/H]-colour relationship [660, 661]. From the point of view of the "projection model", the shape of derived colour distributions for a given GC system is expected to vary depending significantly on the selected passbands. This is because the shape of the [Fe/H]-colour relationship depends on the chosen colour – that is, the relevant photometric bands may be more or less sensitive to the detailed HB morphology – and multi-colour studies of GC systems may play a crucial role in validating or disproving this model [662].

Irrespective of what the real explanation is, the "projection model" is an extremely good example of how the HB morphology plays a crucial role in interpreting integrated photometry of unresolved, old stellar populations.

9.2.2
Integrated Magnitude Fitting

An alternative – conceptually equivalent – to colour fitting methods is obviously the use of the integrated magnitudes themselves. In this case, provided that an independent distance estimate is available, one can also derive stellar masses from the fitting procedure. The work by [663, 664] introduces fits to the observed broadband SED from U to H, to determine simultaneously age, metallicity, internal extinction (if any) of SSPs, once distance modulus and foreground extinction are specified. One can successfully disentangle these three free parameters based on the shape of a low-resolution SED, and determine the stellar mass of the population by scaling the model magnitudes to the observed level, with a minimum of four passbands, assuming a specific functional form of the internal extinction law and an IMF. In the tests performed by [663], the internal extinction is treated following [665] for a starburst galaxy-like behaviour, and the observed integrated flux $F(\lambda)$ and the intrinsic flux $F_0(\lambda)$ are related as

$$F(\lambda) = F_0(\lambda)10^{0.4\,E(B-V)_{\mathrm{int}}\,k'}$$

with

$$k' = 2.659\left(-1.857 + \frac{1.040}{\lambda}\right) + 4.05$$

$$\text{for} \quad 0.63\,\mu\text{m} \le \lambda \le 2.20\,\mu\text{m}$$

$$k' = 2.659\left(-2.156 + \frac{1.509}{\lambda}\right)\frac{-0.198}{\lambda^2} + \frac{0.011}{\lambda^3} + 4.05$$

$$\text{for} \quad 0.09\,\mu\text{m} \le \lambda \le 0.63\,\mu\text{m}$$

where $E(B-V)_{\mathrm{int}}$ is a free parameter. Internal starburst-like extinction is unimportant in the case of old populations. The effect of internal extinction due to circumstellar dust around cool AGB stars that reprocesses a substantial fraction of the photons emitted by these objects is, in principle, more important, but it is not accounted for in [663]. The effect – as in the more general case of the presence of dust in an astrophysical source – is to shift part of the intrinsic emitted power from the optical-UV to the mid- and far-IR region, and in principle can be accounted for by employing dust reddened spectra for the AGB section of the isochrone [666, 667]. An analysis of the models by [668] that include different treatments of AGB circumstellar dust, reveal that for ages between $\sim$ 1.5 and $\sim$ 5.0 Gyr (the precise value

Table 9.3 Preferred magnitude combinations (from U to H) for SSP SED fitting, when only four passbands are available [663]

Age (Gyr)	Preferable combinations	Combinations to be avoided
1	$UBIH, UBVH$	$BVIH, RIJH$
10	$BVIH, UBVI$	$UVIH, UBIH$

depending on the initial metallicity), the effect on filters from U to K is small, of the order of ~ 0.02 mag at most, and is essentially zero at higher ages.

For the tests performed in [663], a likelihood estimator of the form $p \sim e^{-\chi^2}$ is used to find the best fitting parameters, where

$$\chi^2\left(t, [\text{Fe/H}], E(B-V)_{\text{int}}, M_t\right) = \sum_{\text{models}} \frac{(m_{\text{obs}} - m_{\text{model}})^2}{\sigma_{\text{obs}}^2} \tag{9.2}$$

and the combination of parameters that provide the highest probability p is chosen as the "best-fitting model". In general, as shown also by Figure 9.4, one must include the U and B bands, and use the maximum available wavelength range, preferably with at least one near-IR band. A minimum of four passbands is obviously required, the most suitable combinations being shown in Table 9.3 when passbands from Johnson U to H (or equivalent) are available. If near-IR data cannot be acquired, $UBVI$ is the most suitable combination (see [663] for details). One can see, again from Figure 9.4, that this combination minimizes the impact of the age–metallicity degeneracy, provided that both the HB morphology and metal mixture in the observed population are matched by the prescriptions adopted in the models.

According to the simulations in [663], for a synthetic solar metallicity, 1 Gyr old, $M_t = 1.6 \times 10^9 M_\odot$ (employing a Salpeter IMF) SSP with $E(B-V)_{\text{int}} = 0.1$ and all seven $UBVRIJH$ magnitudes available (with an associated Gaussian 1σ error equal to 0.1 mag), this method recovers the input values of age and M_t, with 1σ error bars of ~ 0.1 dex in age, and even less in mass. The input value of $E(B-V)_{\text{int}}$ is also recovered, with error bars of $+0.1$ and -0.05 mag, and the input [Fe/H] is retrieved, with errors of the order of -0.7 and $+0.4$ dex.[6] In the case of a 10 Gyr old population with the same parameters, there is a bias towards retrieving younger (by ~ 0.2 dex) ages, lower mass (by 0.1 dex) and higher $E(B-V)_{\text{int}}$ (by 0.1 mag). The input value of [Fe/H] is retrieved, but the constraint on the metallicity is weak, the $\pm 1\sigma$ interval spanning the entire range available, between -1.7 and $+0.4$ dex. Error bars on age, mass and $E(B-V)_{\text{int}}$ are also larger compared to the 1 Gyr case. This is a consequence of the poorer age resolution at increasing ages already noticed for the colour–colour diagrams. Considering the $UBVI$ case, the recovered properties

---

6) The precise results of these tests depend on the age and [Fe/H] resolution of the underlying isochrones. Whilst there is a fine grid of ages, the [Fe/H] values available in these tests where only five, namely, [Fe/H] $= -1.7, -0.7, -0.4, 0.0, +0.4$.

of both 1 and 10 Gyr SSP compare well with the case of $UBVRIJH$, with similar error bars, but [Fe/H] is underestimated by 0.4 dex and age is essentially equal to the input value.

In the case of all $UBVRIJH$ magnitudes available, an increase of the observational error tends to bias the retrieved age, mass and [Fe/H] of the 10 Gyr population towards increasingly lower values (by up to $\sim$ 1 dex in age, $\sim$ 0.5 dex in mass and 0.7 dex in [Fe/H] when the photometric errors increase to 0.3 mag), while the retrieved $E(B-V)_{int}$ is biased towards increasingly higher values (up to 0.3 mag for a 0.3 mag photometric error). The effect on the 1 Gyr old population is much smaller.

If one searches for solutions with internal extinction fixed to the input value (a more appropriate case for old populations, where it is expected to be negligible), the results regarding the retrieved age, mass and [Fe/H] are hardly changed for both 1 and 10 Gyr populations (and all seven passbands available), whereas error bars and biases on age and mass are brought down to zero if in addition to internal extinction, also [Fe/H] is fixed to the input value. This shows how – not surprisingly – it is the residual age–metallicity degeneracy that affects the retrieved values and associated error bars.

The unknown HB morphology of the target population is also clearly a problem for this technique. One can speculate that the inclusion of a filter at shorter wavelengths and a χ^2 minimization that considers theoretical predictions for different HB morphologies – at fixed age and initial chemical compositions – can help reducing the age-HB morphology-metallicity degeneracy.

Fits to the broad-band SED of several early type galaxies have been used by [669] to compare ages and metallicities provided by seven different SPSMs. Even though the target populations may not be perfect SSPs, fits to their SED with different SPSMs do still provide a quantitative estimate of the uncertainties associated to this type of analysis. The observed sample comprises fourteen early type galaxies: Eight bright galaxies in the Coma cluster, and six intermediate luminosity objects in the Virgo cluster. Integrated magnitudes in the 2MASS J, H, K_s and Sloan $ugriz$ magnitudes have been considered, and a χ^2 minimization of the type shown in Eq. (9.2) – considering Galactic extinction corrections – provides the results displayed in Table 9.4. The observed magnitudes were not K-corrected, as the adopted fitting procedure integrated the red-shifted model (low-resolution) flux over the rest-frame wavelength filter passbands.

Let's consider for the Maraston models, the case of a red HB, and for BaSTI models the case with the Reimers parameter η set to 0.2. This implies that all seven SPSMs display a red HB at the best fit metallicities. The first thing to notice is, for each given galaxy, the large range of reduced χ^2 values. Differences can reach up to a factor of 10–20. The same range of differences can also be found in the case of the best fitting [Fe/H] values. The best fit ages – larger than $\sim$ 2 Gyr – display variations up to a factor $\sim$ 2. When the SPSM is fixed, this analysis shows strongly anti-correlated uncertainties of Δ[Fe/H] $\sim$ 0.18 and $\Delta(\log(t))$ $\sim$ 0.25, that give a measure of the residual age–metallicity degeneracy.

Table 9.4 Comparison of SPSM fitting to a sample of elliptical galaxies. For each galaxy, age, [Fe/H] and χ^2 of the best-fit model for several SPSMs are listed (see text for details).

Model:	BC03			PEGASE			Starburst99			GALEV			SPEED		
Galaxy	χ_ν^2	Age	[Fe/H]	χ_ν^2	Age	[Fe/H]	χ_ν^2	Age	[Fe/H]	χ_ν^2	Age	[Fe/H]	χ_ν^2	Age	[Fe/H]
IC 3501	0.55	2.50	0.093	2.20	3.0	−0.204	1.46	2.72	−0.40	0.93	3.16	−0.40	3.21	5.5	−0.70
NGC 4318	1.40	3.00	0.093	2.33	5.0	−0.204	2.86	2.72	0.00	1.80	6.31	−0.40	3.71	9.0	−0.70
NGC 4515	2.06	2.75	0.093	1.93	4.0	−0.204	2.60	3.37	−0.40	2.72	5.01	−0.40	6.96	5.5	−0.70
NGC 4551	1.56	7.50	0.093	2.67	3.0	0.176	2.40	2.92	0.40	4.00	6.31	0.00	12.67	14.0	−0.70
NGC 4564	1.89	8.50	0.093	2.88	3.5	0.176	2.84	3.14	0.40	2.62	6.31	0.00	23.00	14.0	−0.70
NGC 4867	2.12	8.00	0.093	2.67	3.5	0.176	2.40	6.86	0.00	1.79	3.98	0.00	12.40	14.0	−0.70
NGC 4872	1.23	8.00	0.093	2.25	6.0	0.000	2.22	6.39	0.00	1.99	6.31	0.00	10.54	14.0	−0.70
NGC 4871	0.94	8.50	0.093	2.41	8.0	0.000	2.18	6.86	0.00	2.74	6.31	0.00	17.20	14.0	−0.70
NGC 4873	0.83	6.50	0.093	1.86	4.5	0.000	1.87	3.89	0.00	4.59	13.18	−0.40	4.24	14.0	−0.70
NGC 4473	1.53	8.50	0.093	3.40	3.5	0.176	2.57	3.14	0.40	3.30	6.31	0.00	25.11	14.0	−0.70
NGC 4881	1.13	8.50	0.093	2.22	3.5	0.176	1.90	6.86	0.00	2.04	6.31	0.00	15.59	14.0	−0.70
NGC 4839	1.76	10.00	0.093	2.25	4.0	0.176	2.40	8.49	0.00	2.49	7.94	0.00	31.70	4.5	0.00
NGC 4874	1.20	9.50	0.093	2.54	4.5	0.176	2.96	8.49	0.00	2.77	7.94	0.00	33.95	2.5	0.00
NGC 4889	2.25	12.00	0.093	3.24	6.0	0.176	2.99	9.12	0.00	3.74	10.00	0.00	29.21	8.0	0.00

Table 9.4 Continued

Model:	Maraston RHB			Maraston BHB			BaSTI $\eta = 0.2$			BaSTI $\eta = 0.40$		
Galaxy	χ_ν^2	Age	[Fe/H]	χ_ν^2	Age	[Fe/H]	χ_ν^2	Age	[Fe/H]	χ_ν^2	Age	[Fe/H]
IC 3501	1.40	1.5	0.00	4.06	10.0	0.00	1.61	4.50	−0.25	1.47	4.50	−0.25
NGC 4318	4.65	1.5	0.67	6.81	10.0	0.00	1.65	8.00	−0.25	1.58	8.00	−0.25
NGC 4515	2.89	2.0	0.00	3.21	10.0	0.00	2.66	6.00	−0.25	2.56	6.00	−0.25
NGC 4551	3.44	2.0	0.67	4.26	10.0	0.35	2.62	2.75	0.40	2.50	2.75	0.40
NGC 4564	3.59	2.0	0.67	5.92	15.0	0.35	2.78	3.00	0.40	2.71	3.00	0.40
NGC 4867	3.30	2.0	0.67	7.21	15.0	0.35	2.91	3.00	0.40	2.82	3.00	0.40
NGC 4872	3.10	2.0	0.67	7.50	10.0	0.35	2.43	3.00	0.40	2.37	3.00	0.40
NGC 4871	4.12	2.0	0.67	5.95	15.0	0.35	2.09	3.00	0.40	2.07	3.00	0.40
NGC 4873	5.62	2.0	0.67	4.99	10.0	0.35	2.05	3.00	0.26	2.05	3.00	0.26
NGC 4473	3.42	2.0	0.67	5.48	15.0	0.35	2.37	3.00	0.40	2.33	3.00	0.40
NGC 4881	3.55	2.0	0.67	6.34	15.0	0.35	2.60	3.00	0.40	2.54	3.00	0.40
NGC 4839	7.83	5.0	0.35	4.81	15.0	0.35	3.90	6.00	0.40	2.70	5.00	0.40
NGC 4874	7.53	5.0	0.35	5.49	15.0	0.35	3.38	4.50	0.40	2.20	5.00	0.40
NGC 4889	9.68	6.0	0.35	9.58	15.0	0.35	4.46	6.50	0.40	4.37	6.50	0.40

If we now compare the BaSTI results for $\eta = 0.2$ and $\eta = 0.4$, the inferred ages and [Fe/H] are basically unchanged because at the relevant ages and metallicities, both values of η produce essentially the same HB morphology. In the case of Maraston models, the blue HB option includes a blue morphology also at high metallicities, and the best-fit ages increase by a factor up to 5–7, while the best fit [Fe/H] decreases typically by a factor ~ 2.

Another interesting result relates to the effect of an α-element enhancement in the initial chemical composition of the target populations. A comparison of results with BaSTI $[\alpha/Fe] = 0.4$ models and their scaled-solar counterpart reveals an increase of the best-fit ages by $\Delta(\log(t)) = 0.34 \pm 0.12$, and a decrease of the best-fit [Fe/H] by ~ 0.3 dex.

9.2.3
Effect of Dynamical Evolution

A pivotal function entering the calculation of SSP integrated magnitudes and spectra (Eqs. (1.17), (1.18)) is $\Phi(M)$, which provides the number of objects populating a given isochrone in the mass interval between m and $m + dm$. We have used for $\Phi(m)$ the term Initial Mass Function, implying that the SSP mass function is time-independent and fixed at its initial value, a common assumption in current SPSMs. This assumption is however not justified in case of populations that undergo severe dynamical evolution during their lifetimes, like star clusters. Clusters are subject to disruption – that we denote as all mass loss and destruction events, that is, due to disruptive encounters with giant molecular clouds – and dissolution – denoted as all gradual destruction processes due to tidal dissolution in a smooth external field, or multiple weak encounters with giant molecular clouds [670]. These processes alter the stellar mass function, and hence the population integrated spectra and colours.

The recent analysis in [670] estimates the difference between SSP integrated colours (calculated with the SPSM GALEV) obtained with a non-evolving Kroupa IMF, and with the same IMF, but with exponents (denoted here as α) evolving with time following the results of N-body simulations that study the dynamical evolution of clusters dissolving in a tidal field

$$m \leq 0.3 M_{\odot}$$

$$\alpha(t) = \alpha(0) - 0.1345x + 1.7986x^2 - 1.8121x^3 + 1.2181x^4 + 0.000\,003\,215xt$$

$$m > 0.3 M_{\odot}$$

$$\alpha(t) = \alpha(0) + 0.08389x + 1.9324x^2 - 0.4435x^3 + 0.734x^4 + 0.000\,0141\,43xt$$

where t is the cluster age in Myr and $x = t/t_{95\%}$ where the cluster disruption time $t_{95\%}$ is defined as the time when the cluster has lost 95% of its initial mass [671]. The main effect on the stellar mass function is a preferential loss of the lower mass objects populating the cluster, for, due to energy equipartition, low-mass stars are driven towards the outskirts of the cluster, where they are more easily removed by

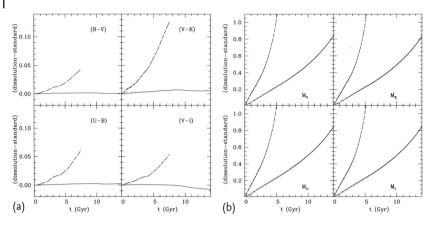

Figure 9.14 (a) Integrated colour difference between SPSMs ([Fe/H] = −0.4) at constant IMF and models with disruption times equal to 6 Gyr (dashed lines) and 20 Gyr (solid lines) respectively. (b) As in (a), but for selected integrated magnitudes.

the external tidal field. This gradual depletion of the lower mass objects leaves clear signatures in the cluster integrated spectrum.

For each metallicity, the results can be parametrized in terms of cluster age and disruption timescale $t_{95\%}$. The most obvious effect is to make integrated magnitudes in all filters progressively fainter when approaching $t_{95\%}$. This is shown by Figure 9.14. Given that we are dealing with old stellar populations, we show the case of clusters with $t_{95\%}$ = 6 and 20 Gyr (larger than the age of the Universe) respectively. As a general rule, colours for SSPs with evolving stellar mass function tend to become progressively redder when the cluster age approaches its $t_{95\%}$. The reason is that the preferential loss of the lowest mass stars affects progressively brighter and bluer parts of the MS, leaving the redder RGB population (that comprises the more massive objects in the cluster, all essentially with the same mass) comparatively unchanged because the depletion of the faintest, reddest MS stars doesn't affect appreciably the integrated colours. The net effect is therefore a general progressive reddening of the integrated colours. The magnitude of this effect decreases for increasing $t_{95\%}$. For colours like $(V-I)$ and $(V-R)$, and $t_{95\%}$ larger or equal to ∼ 10 Gyr, at ages larger than ∼ 6 Gyr, the shift is instead slightly towards the blue (by a few 0.01 mag) compared to the case of unevolving mass function. This is due to the fact that the lower MS redder than the RGB provides a small contribution to these two integrated colours. More infrared colours and magnitudes are largely unaffected by the loss of the lower mass stars.

The impact of the dynamical evolution of $\Phi(m)$ on age and metallicity determinations from colour–colour diagrams leads to overestimate both age and metallicity of the population if SPSMs with a time-independent IMF are employed. The precise values strongly depend on the passbands used, the age and $t_{95\%}$. The discrepancy is minimized when passbands from the mid-UV to K are used, but they can reach up to 30% for ages and $t_{95\%}$ above ∼ 1 Gyr and below ∼ 10 Gyr. For ages and $t_{95\%}$ above ∼ 10 Gyr, the retrieved age can be underestimated by at most

$\sim$ 10%. These discrepancies increase when mid-UV magnitudes are lacking, and the overestimates can reach up to a factor of 2.

9.2.4
Statistical Fluctuations

As discussed before, the use of Eqs. (1.17) and (1.18) to calculate integrated magnitudes and monochromatic fluxes is in principle appropriate only when the number of stars in the SSP is very large. In the case of typical star clusters, the number of post-MS objects is too low to adequately sample the isochrone. This means that an ensemble of SSPs, all with the same M_t, t and Z will display a range of integrated magnitudes (and colours) due to stochastic variations of the number of objects populating a given point along the faster evolutionary phases of an isochrone – covered by a narrower mass range compared to longer lived stages – mainly the bright RGB and the AGB for old populations. The size of these fluctuations is larger at the wavelengths most affected by the flux emitted by RGB and AGB stars, typically the near-IR and longer wavelengths. This is the origin of the colour and magnitude statistical fluctuations [594, 651, 672–675].

Let's consider a sample of N_t stars, each with luminosity L_i at wavelength λ. If $N_t = 1$, the associated Poisson error on the star counts is $\sqrt{N_t} = 1$ and the corresponding error on the total luminosity $L_{tot} = L_i$ is equal to $\sigma(L_{tot}) = \sqrt{N_t} L_i = L_i$. When $N_t > 1$, summing up over all stars (supposedly with the same luminosity L) one gets, at wavelength λ,

$$\frac{\sigma(L_{tot})}{L_{tot}} = \frac{\sqrt{N_t} L}{N_t L} = \frac{1}{\sqrt{N_t}}$$

The smaller N_t, the larger the expected fluctuations of the integrated luminosity at a given λ. For a more general case of each star having a different luminosity L_i, the previous equation becomes

$$\frac{\sigma(L_{tot})}{L_{tot}} = \frac{\left(\sum_i L_i^2\right)^{1/2}}{\sum_i (L_i)} \equiv \frac{1}{\sqrt{N_{eff}}} \tag{9.3}$$

where N_{eff} is termed as the effective number of contributors to the luminosity at wavelength λ. By definition $N_{eff} \le N_t$, and its value (and dependence on λ) can be determined analytically for a given SSP from Eq. (9.3). Once the function $N_{eff}(\lambda)$ is determined for a given total reference luminosity, it can be immediately derived for any other value, as it obviously scales linearly with the total luminosity. Just to give an example, according to [594], for a 12.5 Gyr old metal-poor population with [Fe/H] $= -1.27$, an extended HB and a Salpeter IMF, N_{eff} has a maximum around $\sim$ 3000 Å, drops by a factor $\sim$ 30 000 below $\sim$ 2000 Å (a wavelength region dominated by post-AGB stars), whilst it slowly decreases above $\sim$ 3000 Å. At $\lambda = 6000$ Å N_{eff} is lower by slightly more than a factor of 10, while at $\lambda = 10\,000$ Å N_{eff} is about 30 times smaller than at 3000 Å. Another consequence is that the relationship between N_t and M_t is also subject to fluctuations, with the relative error on M_t inversely proportional to the square root of N_t.

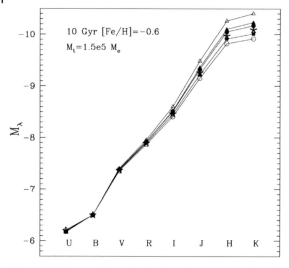

Figure 9.15 Integrated magnitudes from four Monte Carlo realizations of a GC with the labelled parameters. Starred symbols display the values for an analytical integration along the isochrone. The different SEDs are normalized to a value of B obtained from the analytical integration.

To now discuss a situation more tightly linked to observations, we consider SSP integrated magnitudes and colours obtained with a Monte Carlo technique that is particularly suited to address this problem. We have considered a 10 Gyr old α-enhanced isochrone with $[Fe/H] = -0.60$, and extracted randomly stellar masses – in the appropriate mass range – according to the Kroupa IMF, until a specified total number N_t of objects (corresponding to an observed target SSP) has been attained. We have chosen a value of N_t that corresponds – according to the analytical integration of the IMF – to a total mass $M_t \sim 1.5 \times 10^5 M_\odot$, the typical estimated mass of Galactic GCs. The position of the individual stars along the isochrone is determined by interpolation between the neighbouring tabulated mass values, the fluxes of all synthetic stars are then added, and integrated magnitudes computed. The broad-band $UBVRIJHK$ SEDs resulting from five different realizations of this synthetic SSP are displayed in Figure 9.15, together with the result obtained from the integration of Eq. (1.17) for the same value of N_t. The different SEDs are all normalized to the integrated B magnitude obtained from Eq. (1.17). Statistical fluctuations arise because the distribution of stars in the post-MS phase undergo sizable variations in number and CMD position. These fluctuations heavily affect the shape of the SED starting from the R passband, increasing in amplitude at the longer wavelengths, dominated by the RGB contribution. The obvious implication is that an ensemble of several identical clusters with these properties will be interpreted as objects with a range of age/metallicities, when these SED/colours are compared to a grid of well-sampled SPSMs. This has to be taken into account when integrated colours of local star clusters – whose properties are established from their CMDs – are used to test the reliability of SPSMs.

In general, Monte Carlo simulations show that, when N_{eff} is small, distributions of magnitudes and colours – that are non-linear functions of the luminosity – are very asymmetric (the spread obviously increasing with decreasing N_t) sometimes even bimodal, and the average value of these observables will not necessarily be the same as that predicted by a fully sampled IMF (an approximate analytical analysis of this problem in presented in [672]). The bimodality can be explained considering that in this situation different realizations of the same old SSP may or may not contain post-MS stars, and in case of, that is, the I-band, dominated by the RGB contribution, the integrated magnitude will "jump" to brighter values when RGB stars are present, compared to the case without any post-MS contribution.

The effect of these statistical fluctuations on ages and metallicities (and masses) estimated from colour–colour or SED fitting is seldom accounted for. Below, we show an example of how relevant this can be, using again the $(V-K)$–$(V-I)$ diagram. We consider the GC-like SSP sample discussed in Section 9.2.1, with the same [Fe/H] distribution. This time we have computed the integrated magnitudes with a Monte Carlo technique (employing the same isochrones used to calculated analytically the $(V-K)$–$(V-I)$ grid in Section 9.2.1) and had to specify a distribution of M_t or N_t – remembering that there isn't a one-to-one correspondence between these two quantities for sparsely populated SSPs. We have fixed M_t for each object, considering a mass function of the form $N(M_t) \propto M_t^{-\gamma}$, with $\gamma = 0.15$ when the cluster mass is below or equal to $1.5 \times 10^5 M_\odot$, and $\gamma = 2.0$ for higher masses. The resulting distribution of integrated V-band magnitudes approximates the observed luminosity function of Milky Way globulars [676]. A mass range $\log(M_t/M_\odot)$ between 3.8 and 6.3 has been considered.

We have determined with this Monte Carlo procedure one synthetic population with a single age of 12 Gyr and one bimodal-age (12 + 4 Gyr) population – in both cases the same 1σ photometric errors of 0.03 mag has been employed – as in Section 9.2.1. Ages and [Fe/H] of these SSPs have been then retrieved from the same reference grid of $V I K$ analytical integrated colours employed in Section 9.2.1. Figures 9.16 and 9.17 display the CADs for these two populations compared with the corresponding CADs obtained from analytical colours. The general trends are clear. When only objects with $(V-K) > 2.4$ are considered, the CADs are essentially unchanged compared to the case of analytical colours, an indication that at high metallicities the colour fluctuations do not significantly affect the individual age estimates in this diagram. The inclusion of the more metal-poor (bluer) clusters causes a significant change of the CAD, compared to the case of analytical colours. Appreciable differences appear starting between the 6 and the 9 Gyr age bin. The CAD that accounts for the statistical fluctuations is above the CAD for the analytical case, with a larger percentage of objects at high ages. This behaviour is very similar for both the single-age and bimodal-age populations. The estimated [Fe/H] distributions are affected to different degrees, whether the whole sample or just red clusters are considered. In the former case, [Fe/H] values for metal-poor clusters tend to be underestimated, whereas in the latter case, there is the tendency to overestimate [Fe/H].

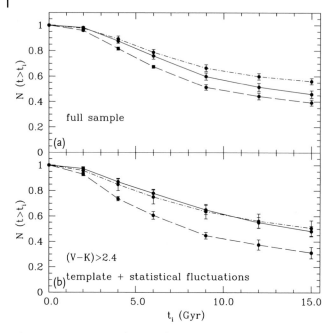

Figure 9.16 Comparison between the retrieved CADs for the 12 Gyr template (solid line) and the same population computed accounting for statistical fluctuations of the star number counts (dashed-dotted lines). (a) displays the comparison for the full synthetic samples, while (b) only offers the comparison for objects with $(V-K) > 2.4$. The corresponding CADs for the template bimodal-age population with analytical integrated colours are also displayed as dashed lines.

When M_t is above $\approx 3 \times 10^5 M_\odot$ (integrated $M_V > -8.5$), the effect of colour fluctuations on the retrieved age (and [Fe/H]) distribution of the synthetic GC samples is almost negligible, at least in this diagram. When the 1σ photometric error is equal to ~ 0.1 mag or higher, the effect of statistical colour fluctuations is completely washed out by the photometric uncertainty.

9.2.5
Binaries

Observations show that a large fraction of stars ($\sim 50\%$) populating the Galactic disk are in binary (or multiple) systems. If the components are close enough to exchange/lose substantial amounts of mass, their evolutionary path is altered compared to single-star evolution. Interactions in close binary systems are, in principle, able to produce extreme (hot) HB stars and Blue Stragglers, that may have an important effect on the integrated spectra of stellar populations. A number of recent investigations have studied the impact of the progeny of binary interactions on the integrated spectra of SSPs [677–680]. All of these studies have employed the single star evolution package of analytical formulae – that reproduce the time evolution of several stellar properties obtained from proper evolutionary calculations [681] –

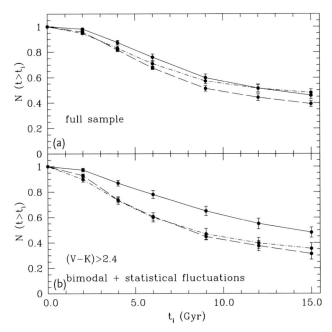

Figure 9.17 As in Figure 9.16, but this time the dashed-dotted lines represent the template bimodal-age population, with integrated colours computed accounting for statistical fluctuations of the star number counts.

coupled to a binary population synthesis code and spectral libraries, to produce Monte Carlo simulations of synthetic populations with and without binary interactions. The main physical processes that these binary evolution algorithms take into account are mass transfer, mass accretion, common envelope evolution, tidal evolution, angular momentum loss mechanisms, and supernova kicks (in young systems).

The problem with predicting the effect of binaries on the integrated spectra of SSPs is the large number of free parameters involved, for which there are very few constraints. The main parameters to fix after initial chemical composition and age are chosen, are the IMF of the primary components, the relationship between the mass of the secondary and primary components, the distribution of orbital separations, the distribution of orbital eccentricities, the efficiency of common envelope ejection (the fraction of orbital energy that is transferred to the envelope and acts against its binding energy), the mass loss efficiency for the isolated evolution of the individual component (usually the Reimers formula), and the efficiency of the tidally enhanced mass loss (compared to the Reimers value) in binary systems.

Monte Carlo simulations are usually employed because they provide an easy-to-implement way to deal with the stochastic nature of the distributions of the primary-to-secondary mass ratios and orbital separations. All investigations mentioned before start from a population made of 100% binary systems, 50% of which have orbital periods below 100 years. Stars in systems with orbital periods above this threshold evolve like single stars. The resulting integrated properties are then obtained by adding up the spectra of all components of the synthetic binary population. Samples of at most 10^6 binaries have been considered, a number of objects that surely introduces some stochastic fluctuations in the predicted integrated fluxes, especially when the infrared wavelengths are considered, but the trends should be reliable. A correspondent single star SSP is then calculated by disabling the interactions among the binary components. Comparisons of these synthetic populations for ages above 1 Gyr show that binary interactions – through the appearance of Blue Stragglers and extreme HB stars – affect the shape of the spectral energy distribution by boosting the flux below a certain threshold wavelength, that generally decreases with increasing age (at fixed metallicity). Typically, for ages of the order of 10 Gyr, binary interactions modify the shape of the integrated spectrum only below $\lambda \sim 3000$ Å, whilst for ages of the order of 1 Gyr this effect is noticeable already at $\lambda \sim 8000$ Å. The net result is to shift towards the blue broad-band colours that involve these wavelength ranges, as expected.

Figure 9.18 displays, as an example, the effect of binary interactions on $(U-B)$, $(B-V)$ and $(V-I)$ colours at varying ages for a solar metallicity SSP [679]. To assess the impact on age and metallicity estimates from colour–colour diagrams, we consider as an example, the $(J-K)$–$(B-K)$ diagram of Figure 9.8. The results by [677] for the appropriate photometric bands (and the same Salpeter IMF) provide a ~ 0.1 mag decrease of $(B-K)$ around 10 Gyr at solar metallicity when binary interactions are efficient, whereas the integrated $(J-K)$ should stay roughly unchanged. This implies an underestimate of the SSP age by ~ 4 Gyr when considering reference colours produced by single star SSPs, and a slight overestimate

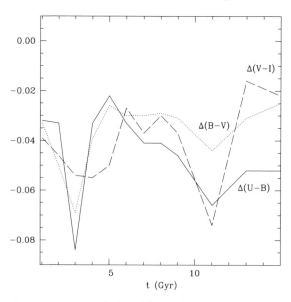

Figure 9.18 Integrated colour difference between solar metallicity SSPs with and without binary interactions.

of [Fe/H] (because of the shape of the constant [Fe/H] lines in the diagram). The size and direction of the age and metallicity biases are however dependent on the specific colour–colour diagram considered.

As another example, the $(R-K)–(u - R)$ diagram is discussed by [677], who find that ages and metallicities of binary populations obtained from single star SSPs (t_s, Z_s) correlate with the "real" age (t_b, Z_b) as $t_b = 0.24 + 0.93t_s$ and $Z_b = Z_s + 0.0037$ for t_s between 2 and 14 Gyr and Z_s between 0.001 and 0.03 for a Salpeter IMF. In the case of using the IMF by [682] (that, in spite of the different functional form, gives essentially the same results of the Kroupa IMF), the previous relationships become $t_b = 0.46 + 0.89t_s$ and $Z_b = Z_s + 0.0031$. In both cases, the inferred ages are underestimated, but less severely than for the $(B-K)–(J-K)$ diagram, and the metallicities are also systematically underestimated.

These results need to be treated very carefully. When all free parameters entering the binary calculations are fixed, the fraction of binaries undergoing interactions play a crucial role – or, in other words, the resulting fraction of Blue Stragglers and extreme HB stars – and this is a quantity virtually unknown in distant unresolved systems. On the other hand, when the fraction of interacting binaries is fixed at, that is, 50%, uncertainties on the poorly constrained free parameters entering the binary simulation can have a large impact on the final spectrum and colours. An exploration of this issue has been performed in [680], who assessed the impact on the integrated $(U-B)$ and $(B-V)$ colours at solar metallicity, of changing (one parameter at a time) the distribution of orbital eccentricities, mass ratios, common envelope ejection efficiency, and mass loss efficiency in absence of interactions (by varying the Reimers mass loss free parameter) compared to their reference case, for ages between 1 and 15 Gyr. A variation of these parameters can affect the colours of the binary population significantly. The difference with single star colours is never erased, but in extreme cases, can be reduced – on average over the full age range – or enhanced by 50%.

9.2.6
Mass-to-Light Ratios

An important parameter that helps to shed light on the dynamics and the evolutionary properties of unresolved stellar systems is the "stellar" mass-to-light (M/L) ratio. Just to give a couple of examples, the structure of dark matter haloes can be determined, for example, from spiral galaxy rotation curves only if the appropriate M/L ratios are known. Also, the evolution with redshift of the build-up of stellar mass in galaxies (a diagnostic of the processes of galaxy evolution) can be estimated from observed galaxy luminosity functions only if M/L ratios are known [683, 684]. Theoretical M/L ratios can be straightforwardly derived from SPSMs, for the actual total stellar mass (and obviously the integrated luminosity in any photometric band) of a stellar population of fixed age and initial chemical composition is immediately obtained by integrating the IMF along the appropriate isochrone. There are slightly different definitions of M/L ratios in the literature. Here we denote with M the mass locked in remnants like WDs, neutron stars and black holes, plus

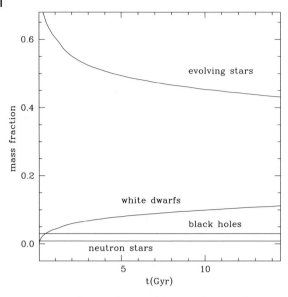

Figure 9.19 Evolution with time of the mass fraction of remnants and evolving stars, for SSPs with solar metallicity.

the mass of the stars still "evolving". By "evolving" stars we mean all objects along evolutionary stages until the end of the AGB phase in the case of old stellar populations. For younger populations, the evolving component would include all massive stars until the end of the AGB, super-AGB, or the pre-supernova stage. It is customary to provide both stellar mass and luminosity (in the chosen photometric band) in solar units.

Figure 9.19 displays the evolution with time of the various contributors to the stellar mass of a SSP with [Fe/H] = 0.06, from an age of 30 Myr to 14.5 Gyr, normalized to a stellar mass at birth equal to $1 M_\odot$. After 30 Myr all core-collapse supernovae have exploded, and the contribution of black holes and neutron star remnants remains unchanged, whereas the mass locked in WDs increases steadily, and the mass in evolving stars decreases with increasing age. At an age of 14 Gyr, the stellar mass is only $\sim 60\%$ of the total mass at birth. The precise value of the mass M at a given metallicity and age obviously depends on the treatment of mass loss during the RGB and AGB stages, and the WD, neutron star and black hole IFMRs [334, 685].

The evolution with time – from 1 to 14.5 Gyr – of M/L ratios for different initial chemical compositions in four selected photometric bands is displayed in Figure 9.20. At a fixed age, in BRI, the M/L ratio increases with [Fe/H], whereas the opposite is true for K. At a given metallicity and age, the M/L values decrease, moving to longer wavelengths because the integrated luminosities are dominated by progressively cooler and brighter stars. The same figure also displays the effect of increasing the RGB mass loss efficiency (η increasing from 0.2 to 0.4) for models with [Fe/H] = −1.27. The effect is reasonably small: for ages above 10 Gyr (the

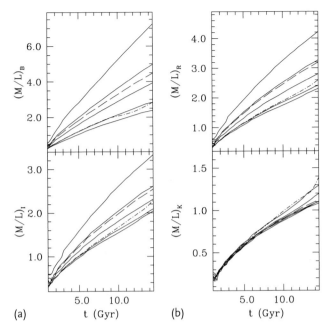

Figure 9.20 Evolution with time of (M/L) of SSPs in different photometric bands. Solid lines denote SSPs with [Fe/H] = 0.06, −0.35, −0.66, −1.27, −1.79, respectively. At a fixed age, (M/L) increases with decreasing [Fe/H] in K, whereas the opposite is true in BIR. Dot-dashed lines denote results for [Fe/H] = −1.27 and enhanced RGB mass loss (η = 0.4). Dashed lines display results for [Fe/H] = −0.70 and [α/Fe] = 0.40.

effect of RGB mass loss obviously increases with age), M/L increases on average by $\sim 10\%$ in K, $\sim 7\%$ in R and $\sim 5\%$ in I, the passbands where HB magnitudes get fainter when populated by hotter (less massive) HB stars. Switching from a scaled solar to an α-enhanced ([α/Fe] = 0.4) metal mixture at approximately constant [Fe/H] increases the M/L ratios in BRI (over the whole 1–10 Gyr age range) by on average 7–8%. The average increase in K is by only $\sim 3\%$.

With a set of SPSMs at hand, the determination of age and metallicity of an observed SSP from either the colour–colour diagrams or SED fitting also provides the desired M/L ratios. A more common approach is to make use of colour-(M/L) relationships [683, 684] where the effect of the unknown age and chemical composition of the target population being minimized, the selected colour is a proxy for the evolutionary status of the population. Figures 9.21 and 9.22 display two examples of colour-(M/L) relationships involving the $BRIK$ filters discussed before, for ages between 1 and 14 Gyr. The $(B-R)$–$(M/L)_B$ appears to be the best choice [683], for at fixed colour, the variation of (M/L) is always less than a factor of 2. All other diagrams display larger variations of (M/L) at a given colour. The analysis in [684] concludes that another suitable choice is the $(g-i)$–$(M/L)_i$ diagram in the SDSS photometric system: the g filter covers wavelengths close to B, and the i filter spans a wavelength range between R and I.

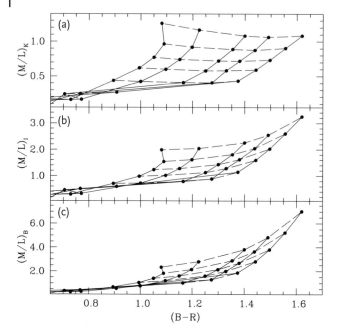

Figure 9.21 (M/L)–$(B-R)$ in several photometric bands for SSPs of different ages and metallicities. Dashed lines represent SSPs of constant age and [Fe/H] equal to – increasing from left to right – $-1.79, -1.27, -0.66, -0.35$ and 0.06, respectively. Solid lines denote SSPs of constant [Fe/H] and age equal to – increasing from bottom to top – 1, 3, 5, 7, 10 and 14 Gyr.

As for the case of integrated colours, statistical fluctuations affect the interpretation of (M/L) ratios derived from observations, even though the effect is somewhat mitigated when using BR filters. Rough estimates in [670] provide a 15, 5 and 1.5% uncertainty on $(M/L)_V$ for 12 Gyr old systems with the total mass equal to 10^4, 10^5 and 10^6 $M_\odot$, respectively.

The dynamical evolution of the target population can also have a major effect on the derived (M/L) ratios. Figure 9.23 displays the run with time of the ratio between $(M/L)_V$ calculated with and without including the dynamical evolution of $\Phi(M)$, for $t_{95\%} = 6$ and 20 Gyr respectively. The low-mass stars preferentially removed from the cluster have higher M/L ratios compared to the average cluster star. On the other hand, the fraction of massive stellar remnants – that do not contribute appreciably to the integrated luminosity L – is enhanced compared to the models with time-independent IMF. These effects cancel each other only partially, and lead to the steady decrease of the M/L for the majority of the cluster lifetime. During the final stages of the cluster dissolution, the M/L ratio is enhanced compared to the case of the time-independent IMF because of the increasing fraction of stellar remnants inside the cluster compared to the number of stars that contribute to L.

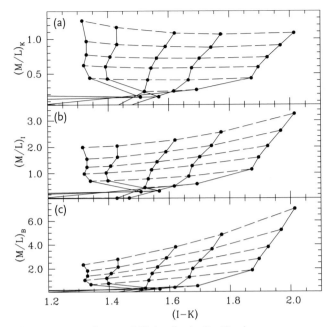

Figure 9.22 As with Figure 9.21, but for the $(I-K)$ colour.

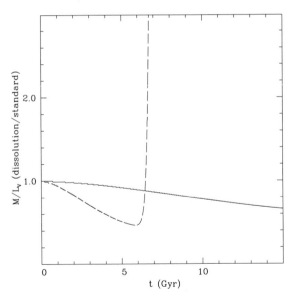

Figure 9.23 Evolution of the ratio between $(M/L)_V$ with time-independent IMF and the corresponding quantity calculated with an evolving IMF and two different disruption times (6 and 20 Gyr).

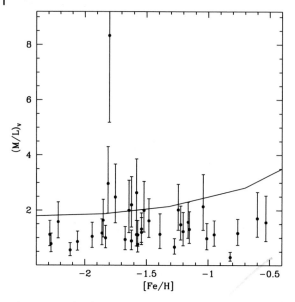

Figure 9.24 *V*-band mass-to-light ratio for a sample of galactic globular clusters [686] compared to results from 12 Gyr theoretical α-enhanced SSPs.

This dynamical evolution of $\Phi(M)$ must be taken into account when comparing theoretical M/L ratios with empirical M/L estimates for star clusters, that is, for GCs of the Milky Way [686]. Figure 9.24 displays these empirical estimates together with theoretical α-enhanced models for an age of 12 Gyr. As expected on the basis of Figure 9.23, models with a time-independent IMF overestimate the M/L ratio of these GCs.

9.2.7
Surface Brightness Fluctuations

Let's consider an image of a generic stellar population. Each resolution element (a pixel in a CCD detector) will register the flux emitted by a given number N of unresolved stars, and because N is subject to Poisson fluctuations, the measured flux will vary from pixel to pixel (assuming there is no variation of the evolutionary properties of the stars registered by different pixels, or that these differences have been taken into account). Following the same reasoning of Section 9.2.4, we can write, in the case of a population of stars all with the same luminosity $L_\lambda = L$ at wavelength λ,

$$\frac{\sigma(f_{tot})^2}{f_{tot}} = \frac{NL^2}{NL}\frac{4\pi D^2}{(4\pi D^2)^2} = \frac{L}{4\pi D^2}$$

where $f_{tot} = NL/(4\pi D^2)$ is the integrated flux at a given wavelength λ registered by a pixel, D being the distance to the population.

For a generic population of stars, each with luminosity L_i, this can be generalized as

$$\frac{\sigma(f_{tot})^2}{f_{tot}} = \frac{\sum_i L_i^2}{\sum_i L_i} \frac{4\pi D^2}{(4\pi D^2)^2} \equiv \frac{L_{eff}}{4\pi D^2} \tag{9.4}$$

From Eq. (9.3), one gets $\sum_i (L_i^2) = (\sum_i L_i)^2/N_{eff}$. Substituting this into Eq. (9.4) gives $(\sum_i L_i)^2/(N_{eff}\sum_i L_i) = L_{eff}$, and hence the relationship $L_{tot} = N_{eff}L_{eff}$ that is a function of the wavelength λ. In general, $L_{eff} \geq L_{tot}/N$, for $N_{eff} \leq N$. For stars all with the same luminosity, one recovers $L_{tot} = NL$. These considerations lead directly to a distance indicator and/or a complementary diagnostic of the evolutionary status of unresolved stellar populations.

From SPSMs, one can determine $L_{eff} = \sum_i L_i^2/\sum_i L_i$ (provided the evolutionary properties of the population are known), and when $\sigma(f_{tot})^2/f_{tot}$ is determined from the observations, Eq. (9.4) provides an estimate of the distance. In qualitative terms, the images of closer galaxies appear less smooth (higher $\sigma(f_{tot})^2/f_{tot}$) than more distant objects because a smaller number of stars contributes to the flux registered by a single pixel. This is the theoretical underpinning of the so-called Surface Brightness Fluctuation (SBF) technique developed in [687], that with present capabilities is applicable out to distances of the order of 100 Mpc.

Given that observations are performed using photometric filters, it is the value of L_{eff} in the chosen photometric band (transformed to a magnitude $\overline{M_\lambda}$ – denoted as SBF magnitude – with the same zero points of standard stellar magnitudes) that has to be provided by the theoretical models. Observationally (see, e.g. [687, 688] for detailed discussions), the SBF magnitude is estimated from the amplitude of the image power spectrum on the scale of the point spread function since adjacent pixels in a CCD are correlated through convolution with the point spread function. Care has to be taken to remove the mean brightness of the population as well as all point sources in order to measure the fluctuation signal.

Several recent theoretical calibrations for SSPs in various filters can be found in [689–693]. In general, SBF magnitudes depend on the age and metallicity distribution of the stars hosted by the target population and are heavily weighted towards high luminosities, and thus advanced evolutionary phases in old stellar populations. The treatment of RGB, AGB (and post-AGB at wavelengths shorter than U) phases, and the colour of the HB play an important part in determining the behaviour of L_{eff} as a function of λ (see, e.g. [690]), and this can at least partly explain the sizable differences among several recent theoretical calibrations (see [692, 694] for comparisons).

SBF magnitudes are representative of whole stellar systems, and cannot be associated with specific groups of stars along the CMD of a given population, but it is somewhat interesting to trace the evolutionary phases with stellar magnitudes equal to the SBF values in, for example, the U, V, I and K filters. This is shown in Figure 9.25 for a 10 Gyr, metal-rich and scaled-solar SSP. As expected, because of the quadratic dependence on the luminosity in the numerator of Eq. (9.4), the SBF magnitudes in optical/IR filters are very sensitive to cool giant stars. The full spectrum of $UBVRIJHK$ SBF magnitudes is shown in Figure 9.26 for SSPs with

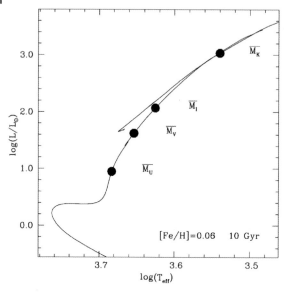

Figure 9.25 A 10 Gyr scaled-solar [Fe/H] = 0.06 isochrone. Filled circles denote the evolutionary stages whose magnitude (in different filters) is equal to the fluctuation magnitude of the corresponding SSP.

two different metallicities. As a general rule, for old ages (after the onset of the RGB at ages of the order of 1 Gyr) SBF magnitudes display a monotonic increase with age. At fixed (old) ages, a decrease of [Fe/H] causes a general decrease of the SBF magnitudes from the U to the J band (this brightening gets progressively smaller when moving to larger λ). For the longer wavelength H and K filters, this behaviour reverses, and more metal-poor SSPs display typically fainter SBF magnitudes. Overall, the dependence on [Fe/H] is very strong for $UBVRI$ filters, and much milder for JHK. The effect of age on the $UBVI$ SBF magnitudes decreases with decreasing [Fe/H] in the example shown in Figure 9.26, at least when $t > 2$–3 Gyr, whereas the opposite is true for the JHK bands.

Figure 9.27 shows, respectively, the effect of changing the metal mixture (in both stellar models and stellar spectra) from scaled-solar to α-enhanced ([α/Fe] = 0.4 – keeping [Fe/H] constant), and a bluer HB morphology for ages above 8 Gyr (Figure 9.27b) obtained by doubling the Reimers η parameter compared to the reference value used in all figures of this section ($\eta = 0.2$). SBF magnitudes from U to I increase for an α-enhanced metal distribution, whilst the opposite is true for the H and K filters. The J SBF magnitudes appear to be insensitive to the change of metal mixture. A bluer HB has an almost negligible effect on U and B SBF magnitudes, and it is most appreciable in the JHK filters that display fainter magnitudes. The quantitative behaviour depends of course on the specific morphology assumed for the HB. An increasingly bluer HB would eventually also affect the U and B bands [690].

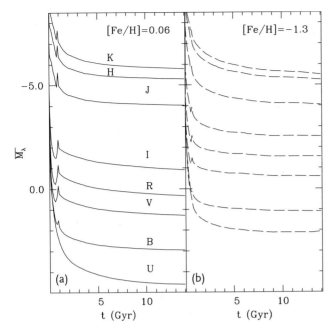

Figure 9.26 Evolution with time of SBF magnitudes in several filters, for SSPs with the two labelled values of [Fe/H].

Core overshooting has a largely negligible effect on SBF magnitudes for ages above 1–2 Gyr (i.e. after the onset of the RGB). The value of the IMF exponent also plays a minor role (for variation of a few ±0.1 around the assumed reference value).

Given the non-negligible dependence of the SBF magnitudes on metallicity and age, their use as distance indicators requires well-defined (i.e. with a small scatter) relationships between a magnitude $\overline{M_\lambda}$ and one or more colours that account for the effect of the population evolutionary status. Two empirical relationships – using Cepheid based distances [695] – between I and K SBF magnitudes, respectively, and the $(V-I)$ integrated colours have been recently determined. The I-band result reported below has been obtained from a sample of 300 E, S0 and early-type spiral galaxies [696], while the K-band result has been derived employing 19 early-type galaxies in the Fornax Cluster [697]:

$$\overline{M}_I = -1.74 \pm 0.08 + (4.5 \pm 0.25)\left[(V-I)_0 - 1.15\right]$$

$$0.95 < (V-I)_0 < 1.30$$

$$\overline{M}_{Ks} = -5.80 \pm 0.04 + (3.6 \pm 0.8)\left[(V-I)_0 - 1.15\right]$$

$$1.05 < (V-I)_0 < 1.25$$

Figure 9.28 displays the theoretical counterparts of these empirical relationships obtained from SSPs with ages between 1 and 14 Gyr, and [Fe/H] between −2.3

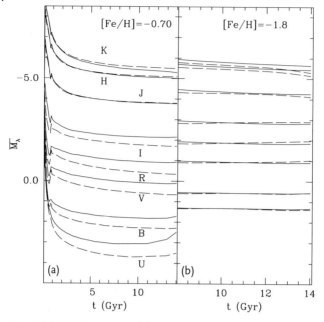

Figure 9.27 (a) Comparison of the time evolution of SBF magnitudes in several filters, between α-enhanced (dashed lines) and scaled-solar (solid lines) SSPs with [Fe/H] = −0.70.

(b) Comparison of SBF magnitudes for [Fe/H] = −1.8 scaled-solar SSPs, calculated with $\eta = 0.2$ (solid lines) and $\eta = 0.4$ (dashed lines) respectively.

and 0.4 (scaled-solar mixture). One can immediately notice the, on the whole, reasonably tight and generally linear relationship (that is, barely affected by switching from a scaled-solar to an α-enhanced metal mixture) between $\overline{M}_I$ and $(V−I)$ over the full range of ages and [Fe/H], that mirrors the empirical result (apart from a small offset). In case of $\overline{M}_K$, the relationship is approximately linear as observed only at constant metallicity. The observed relationship (after applying a +0.04 mag correction to the original zero point to account for the difference between the K_s filter used in the calibration and the K-band definition of the BaSTI models) is also displayed, and appears (apart from a possible vertical offset) to match the theoretical relationship for [Fe/H] around solar. This may be the signature of an homogeneous metallicity and age difference among the galaxies used for the calibration, but one has to be aware that – contrary to the $\overline{M}_I−(V−I)$ diagram – different SPSMs predict very different shapes of the theoretical sequences in the $\overline{M}_K−(V−I)$ plane [692]. Uncertainties in the treatment of the AGB phase – that provides a major contribution in determining SBF magnitudes at long wavelengths – probably play the major role in causing differences amongst SPSM predictions.

The dependence of SBF magnitudes on initial metallicity and age provides, in principle, a complementary tool to determine ages and metallicities of stellar populations [690, 692, 698]. Figure 9.29 displays, as an example, a theoretical SSP $\overline{M}_K−(\overline{M}_J−\overline{M}_K)$ diagram that appears to be able to largely break the age–metallicity degeneracy in the low metallicity regime (up to [Fe/H] ∼ −0.7) although, as shown

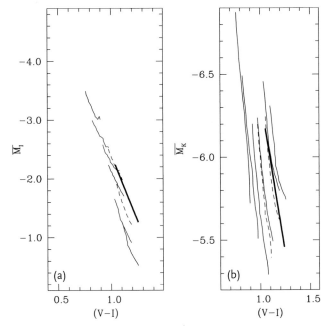

Figure 9.28 (a) $\overline{M_I}$–$(V-I)$ diagram. (b) $\overline{M_K}$–$(V-I)$ diagram. In both (a) and (b), solid lines denote scaled-solar SSPs with ages between 1 and 14 Gyr, and [Fe/H] = –2.3, –1.3, –0.7, –0.35, 0.06 and 0.4 respectively. Dashed lines display α-enhanced SSPs with [Fe/H] = –0.3 and –0.7. The thick solid lines display empirical relationships.

by [692], theoretical predictions are again not consistent, and some SPSMs show that instead ages and metallicities are almost completely degenerate in this diagram.

As the SBF signal is dominated by the brightest stars of the stellar population if the number of stars belonging to the SSP is low, stochastic effects from the rare brightest stars become important, and cause stochastic variations of the SBF magnitudes even in presence of a uniform stellar population, similar to the case of integrated magnitudes and colours discussed in Section 9.2.4. This effect can be investigated as in [690, 699]. For a SSP made of N_t stars with a given age and initial chemical composition, one can determine with Monte Carlo techniques a set of N_{sim} independent synthetic CMDs and calculate for the j-th CMD the following sum, at a given wavelength λ

$$f_{tot}^j = \sum_i L_i$$

where L_i is the flux at wavelength λ of the i-th star in the simulation.

Summing up over all N_{sim} realizations of the CMD, one can calculate

$$\langle f_{tot} \rangle = \frac{\sum_j f_{tot}^j}{N_{sim}}$$

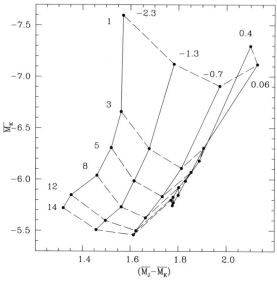

Figure 9.29 Theoretical $\overline{M_K}$–$(\overline{M_J}$–$\overline{M_K})$ diagram.

and

$$\overline{M_\lambda} = -2.5 \log \left(\overline{f_{\text{tot}}} \right) = -2.5 \log \left[\frac{\sum_j \left(f_{\text{tot}}^j - \langle f_{\text{tot}} \rangle \right)^2}{N_{\text{sim}} - 1} \frac{1}{\langle f_{\text{tot}} \rangle} \right] \tag{9.5}$$

There is a close correspondence between this procedure and the way of measuring SBFs for unresolved stellar populations. The quantity f_{tot}^j is equivalent (neglecting seeing effects) to the flux arising from N_t stars registered by a single pixel of a stellar system CCD image, while N_{sim} corresponds to the number of pixels covered by the system. According to the simulations in [690], a too small value of N_{sim} produces a large scatter in the retrieved values of the SBF magnitudes. Values of the order of 10^3 independent realizations are necessary to minimize this effect. In addition, also the value of N_t (number of stars registered by an individual pixel) has an important effect. For a very old population of solar metallicity (and $N_{\text{sim}} = 5000$) N_t of the order of 10^8, is necessary to recover the values of SBF magnitudes obtained from the analytical calculation of $(\sum_i L_i)^2/(N_{\text{eff}} \sum_i L_i) = L_{\text{eff}}$ along the SSP isochrone. Smaller values of N_t tend to produce brighter SBF magnitudes. In the I passband, $N_t > 10^5$ keep this bias within 0.1 mag, while the effect appears smaller in the K band, even for values as low as $N_t = 10^4$.

9.3
High-Resolution Diagnostics

Integrated spectra of unresolved populations, at a $\approx$ 1–10 Å resolution display a host of absorption features that can be used to constrain stellar ages and chemical compositions without the need to fit the full high-resolution SED. This is because the presence and strength of individual absorption features depend on the abundance of specific chemical elements in the stellar atmospheres and the properties and number fraction of stars in well defined evolutionary phase. The absorption feature strengths are usually measured by a system of indices – largely unaffected by dust attenuation – that are essentially a measure of their equivalent widths, and then compared to the corresponding theoretical predictions. Lists of "popular" line indices in the UV, blue-visible and IR part of the spectrum are reported in Tables 9.5–9.7, respectively.

Table 9.5 A list of UV spectral indices [700–702]. The chemical elements contributing to the feature bandpass are given in column five.

Name	Blue bandpass (Å)	Central bandpass (Å)	Red bandpass (Å)	Comments
BL1302	1270.0–1290.0	1292.0–1312.0	1345.0–1365.0	Si III, Si II, O I
Si IV	1345.0–1365.0	1387.0–1407.0	1475.0–1495.0	Si IV 1393.8; 1402.8
BL1425	1345.0–1365.0	1415.0–1435.0	1475.0–1495.0	C II 1429, Si III 1417, Fe IV, Fe V
Fe 1453	1345.0–1365.0	1440.0–1466.0	1475.0–1495.0	Fe V +20 additional Fe lines
C_{IV}^A	1500.0–1520.0	1530.0–1550.0	1577.0–1597.0	C IV 1548, in absorption
C IV	1500.0–1520.0	1540.0–1560.0	1577.0–1597.0	C IV 1548, central band
C_{IV}^E	1500.0–1520.0	1550.0–1570.0	1577.0–1597.0	C IV 1548, in emission
BL1617	1577.0–1597.0	1604.0–1630.0	1685.0–1705.0	Fe IV
BL1664	1577.0–1597.0	1651.0–1677.0	1685.0–1705.0	C I 1656.9, Al II 1670.8
BL1719	1685.0–1705.0	1709.0–1729.0	1803.0–1823.0	N IV 1718.6, Si IV 1722.5; 1727.4, Al II
BL1853	1803.0–1823.0	1838.0–1868.0	1885.0–1915.0	Al II, Al III, Fe II, Fe III
Fe II (2332 Å)	2285.0–2325.0	2333.0–2359.0	2432.0–2458.0	Fe II, Fe I, Co I, Ni I
Fe II (2402 Å)	2285.0–2325.0	2382.0–2422.0	2432.0–2458.0	Fe II, Fe I, Co I
BL2538	2432.0–2458.0	2520.0–2556.0	2562.0–2588.0	Fe I, Fe II, Mg I, Cr I, Ni I
BL2720	2647.0–2673.0	2713.0–2733.0	2762.0–2782.0	Fe I, Fe II, Cr I
BL2740	2647.0–2673.0	2736.0–2762.0	2762.0–2782.0	Fe I, Fe II, Cr I, Cr II
Fe II (2609 Å)	2562.0–2588.0	2596.0–2622.0	2647.0–2673.0	Fe II, Fe I, Mn II
Mg II	2762.0–2782.0	2784.0–2814.0	2818.0–2838.0	Mg II, Fe I, Mn I
Mg I	2818.0–2838.0	2839.0–2865.0	2906.0–2936.0	Mg I, Fe I, Cr II, Fe II
Mg_{wide}	2470.0–2670.0	2670.0–2870.0	2930.0–3130.0	Mg I, Mg II, Fe I, Fe II, Cr I, Cr II
Fe I	2906.0–2936.0	2965.0–3025.0	3031.0–3051.0	Fe I, Cr I, Fe II, Ni I
BL3096	3031.0–3051.0	3086.0–3106.0	3115.0–3155.0	Al I 3092, Fe I 3091.6

Table 9.6 Spectral indices in the blue-visible wavelength range (Lick indices) used in age and metallicity determinations of unresolved stellar populations [703]. The dominant chem-ical species contributing to the strength of each index are given in column six [704]. Elements in brackets increase the line strength when their abundance decreases.

Name	Index band	Blue continuum	Red continuum	Units	Measures
$H\delta_A$	4083.500–4122.250	4041.600–4079.750	4128.500–4161.000	Å	
$H\delta_F$	4091.000–4112.250	4057.250–4088.500	4114.750–4137.250	Å	
CN_1	4142.125–4177.125	4080.125–4117.625	4244.125–4284.125	mag	C,N,(O)
CN_2	4142.125–4177.125	4083.875–4096.375	4244.125–4284.125	mag	C,N,(O)
Ca4227	4222.250–4234.750	4211.000–4219.750	4241.000–4251.000	Å	Ca,(C)
G4300	4281.375–4316.375	4266.375–4282.625	4318.875–4335.125	Å	C,(O)
$H\gamma_A$	4319.750–4363.500	4283.500–4319.750	4367.250–4419.750	Å	
$H\gamma_F$	4331.250–4352.250	4283.500–4319.750	4354.750–4384.750	Å	
Fe4383	4369.125–4420.375	4359.125–4370.375	4442.875–4455.375	Å	Fe,C,(Mg)
Ca4455	4452.125–4474.625	4445.875–4454.625	4477.125–4492.125	Å	(Fe),(C),Cr
Fe4531	4514.250–4559.250	4504.250–4514.250	4560.500–4579.250	Å	Ti, (Si)
C_24668	4634.000–4720.250	4611.500–4630.250	4742.750–4756.500	Å	C,(O),(Si)
H_β	4847.875–4876.625	4827.875–4847.875	4876.625–4891.625	Å	
Fe5015	4977.750–5054.000	4946.500–4977.750	5054.000–5065.250	Å	(Mg),Ti,Fe
Mg_1	5069.125–5134.125	4895.125–4957.625	5301.125–5366.125	mag	C,Mg,(O),(Fe)
Mg_2	5154.125–5196.625	4895.125–4957.625	5301.125–5366.125	mag	Mg,C,(Fe),(O)
Mg_b	5160.125–5192.625	5142.625–5161.375	5191.375–5206.375	Å	Mg,(C),(Cr)
Fe5270	5245.650–5285.650	5233.150–5248.150	5285.650–5318.150	Å	Fe,C,(Mg)
Fe5335	5312.125–5352.125	5304.625–5315.875	5353.375–5363.375	Å	Fe,(C),(Mg),Cr
Fe5406	5387.500–5415.000	5376.250–5387.500	5415.000–5425.000	Å	Fe
Fe5709	5696.625–5720.375	5672.875–5696.625	5722.875–5736.625	Å	(C),Fe
Fe5782	5776.625–5796.625	5765.375–5775.375	5797.875–5811.625	Å	Cr
Na_D	5876.875–5909.375	5860.625–5875.625	5922.125–5948.125	Å	Na,C,(Mg)
TiO_1	5936.625–5994.125	5816.625–5849.125	6038.625–6103.625	mag	C
TiO_2	6189.625–6272.125	6066.625–6141.625	6372.625–6415.125	mag	C,V,Sc

The typical index definition makes use of a central feature passband, flanked to the blue and red by pseudo-continuum passbands (sidebands). The choice of the two sidebands needs to satisfy the following constraints: proximity to the absorption feature, less absorption in the (pseudo)continuum regions than in the central bandpass, and maximum insensitivity to velocity dispersion broadening. This last constraint sets a minimum width for the pseudo-continuum passbands.

Once the mean value of the flux in each of the two sidebands is determined, a straight line that connects these two flux levels is drawn through the midpoint

Table 9.7 Infrared spectral indices. Two different definitions of the NaI doublet index are provided (Na$_{FF}$ and Na$_{8190}$).

Index	Blue continuum	Index band	Red continuum	Source
Na i	21 910–21 966	22 040–22 107	22 125–22 170	[375]
Fe i A	22 133–22 176	22 250–22 299	22 437–22 497	[705]
Fe i B	22 133–22 176	22 368–22 414	22 437–22 497	[705]
Ca i	22 450–22 560	22 577–22 692	22 700–22 720	[375]
Mg i	22 700–22 720	22 795–22 845	22 850–22 874	[705]
TiO$_{6600}$	6512.1–6538.1	6617.2–6992.5	7036.9–7048.0	[706]
Na$_{FF}$	8169.0–8171.0	8172.0–8209.0	8209.0–8211.0	[706]
Na$_{8190}$	8171.5–8172	8172–8197	8233.5–8234.2	[706]
Ca$_{8662}$	8637.2–8646.2	8653.2–8668.4	8847.6–8854.0	[706]
WFB	9891.8–9895.1	9895.1–9958.6	9958.6–9962.2	[706]

Index	Continuum bands		Index band	Source
12CO(2, 0)	22 300–22 370,	22 680–22 790,	22 910–23 020	[375]
	22 420–22 580	22 840–22 910		

Ca II triplet [707]

Feature	Index band	Feature	Index band	Continuum
Ca1	8484.0–8513.0	Pa1	8461.0–8474.0	8474.0–8484.0
Ca2	8522.0–8562.0	Pa2	8577.0–8619.0	8563.0–8577.0
Ca3	8642.0–8682.0	Pa3	8730.0–8772.0	8619.0–8642.0
				8700.0–8725.0
				8776.0–8792.0

of each sideband. The flux difference between the line and the observed spectrum within the feature bandpass determines the index value.[7]

For narrow features, the indices are usually expressed in Angströms; for broad molecular bands, in magnitudes. Formally, if $F_{C,\lambda}$ represents the flux level of the straight line connecting the midpoints of the flanking pseudo-continuum levels, and $F_{I,\lambda}$ represents the observed flux per unit wavelength in the central wavelength range ($\lambda_1 - \lambda_2$), the numerical value for a narrow absorption-feature ($I_{\text{Å}}$) is defined

---

7) A recent attempt to improve on the classical sideband definition has introduced a "boosted median" method [708]. This procedure automatically chooses the largest fluxes in the sideband, and hence it defines the pseudo-continuum from those points that are least affected by absorption features that are not part of the central band. The pseudo-continuum level is therefore more robust if the spectral resolution is high enough to avoid blending of all features in the sidebands.

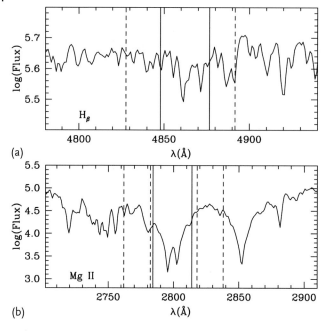

Figure 9.30 Theoretical SSP integrated spectrum (flux in arbitrary units) at 1 Å resolution, with the bandpasses of two selected Lick-type indices superimposed. Notice the presence of absorption features also within the sidebands.

as

$$I_{\text{Å}} = \int_{\lambda_1}^{\lambda_2} \left(1 - \frac{F_{I,\lambda}}{F_{C,\lambda}}\right) d\lambda \qquad (9.6)$$

The value of an index measured in magnitudes (I_{mag}) is given by

$$I_{\text{mag}} = -2.5 \log \left[\left(\frac{1}{\lambda_2 - \lambda_1}\right) \int_{\lambda_1}^{\lambda_2} \frac{F_{I,\lambda}}{F_{C,\lambda}} d\lambda\right] \qquad (9.7)$$

Two visual examples of central feature and sideband definitions are displayed in Figure 9.30, for the H_β and Mg II indices, respectively, overimposed onto a theoretical integrated SSP spectrum at a 1 Å resolution. Notice the presence of several absorption features also in the wavelength range covered by the pseudo-continuum sidebands.

An alternative system of indices that does not make use of pseudo-continua has been developed in [709], and will be denoted here as the Rose system (see Table 9.8). The indices are defined by taking the ratio of the counts in the bottom of two neighbouring lines without reference to the (pseudo)continuum levels. Other indices are formed by taking the ratio of two pseudo-continuum peaks. This definition minimizes the impact of changes in spectral resolution. Figure 9.31 displays the location

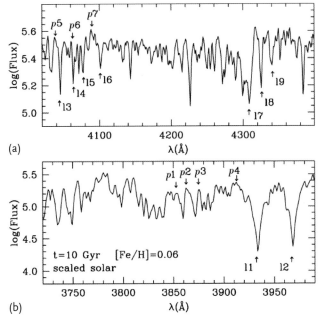

(a)

(b)

Figure 9.31 Theoretical SSP integrated spectrum (flux in arbitrary units) at 1 Å resolution. The spectral features considered by the Rose system of absorption line indices are marked. The line bottoms from $l1$ to $l9$ correspond, respectively, to the $Ca_{II}K$, $Ca_{II}H + H\epsilon$, Fe4065, Fe4063, $Sr_{II}K$, H_δ lines, the deepest absorption feature in the G-band, the line at $\lambda4325$ and the H_γ line.

Table 9.8 Definition of the Rose system of spectral indices in the blue wavelength range [709]. See Figure 9.31 and text for details.

Name	Definition
$H\delta/Fe_I$	$l6/[(l3 + l4)/2]$
Sr_{II}/Fe_I	$l5/[(l3 + l4)/2]$
$H\gamma/4325$	$l9/l8$
$H\gamma/G$-band	$l9/l7$
$Ca_{II}H + H\epsilon/Ca_{II}K$	$l2/l1$
$p(3912)/p(CN)$	$p4/[(p1 + p2 + p3)/3]$
$p(4045 + H\delta)/p(4063)$	$(p5 + p7)/(2p6)$

of the line bottoms (l) and pseudo-continuum peaks (p) used in the definition of the indices reported in Table 9.8.

To date, most of the evolutionary information gathered from the analysis of integrated spectra of old stellar populations has been obtained using the Lick indices in the blue-visual part of the spectrum. The discussion presented in this section will be mainly centred on these indices. It is worth clarifying an important issue related to this index system. As discussed in [710], the Lick system was established by observations of a large (about 500 objects) library of stars with the Image Dissector Scanner (IDS) and Cassegrain spectrograph on the 3 m Shane Telescope at Lick Observatory. The stellar spectra roughly covered the wavelength range 4000–6400 Å with a wavelength dependent resolution between 8 to 10 Å. From the non flux-calibrated spectra, the various indices were determined as described by Eqs. (9.6) and (9.7), and the derived values parametrized by simple polynomial "fitting functions" (see, i.e. [593, 710]) that provide the index strengths as functions of stellar T_{eff}, g, and [Fe/H]. When calculating indices for a SSP, the values of Lick indices at each point along the corresponding isochrone (as given by the appropriate fitting function) are weighted by the IMF and the local continuum flux in the index bandpass, and added up.

In modern SPSMs, the Lick indices are measured following Eqs. (9.6) and (9.7) on flux-calibrated integrated spectra of a given resolution (typically higher than 8 Å), and therefore are not exactly on the Lick system. When calculated in this way, they should be denoted as Lick-type indices. To transpose their values onto the Lick system – when necessary – one may follow a procedure similar to what is described in [596]. In brief, one should consider samples of stars in the Lick spectral library, broaden the theoretical counterpart of these spectra to the Lick resolution, and calculate the index values and the median offsets with respect to the index measurements on the Lick spectra. These median offsets have to be then subtracted from the index values calculated from the integrated theoretical spectra predicted by the SPSM (also degraded to the Lick resolution).

In the following, we won't make any distinction between Lick-type and proper Lick indices, unless this difference is relevant to the discussion.

The Lick system of indices provides powerful index-index diagrams that break, to a large degree, the age–metallicity relationship, and can provide, in principle, solid estimates of age, [Fe/H] and [α/Fe] of a given SSP. Examples of these diagrams are reported in Figure 9.32. Figure 9.32b shows H_β–Fe5406 index grids derived from BaSTI theoretical SSP-integrated spectra for a range of metallicities and ages, as detailed in the figure caption, and both scaled-solar and α-enhanced ([α/Fe] = 0.4) metal mixtures. It is worth noticing that the two metal mixtures are accounted for consistently in both isochrones and stellar spectra.

Lines of constant age and lines of constant Z (or [Fe/H]) are roughly orthogonal, that is, the age–metallicity degeneracy is strongly softened in this diagram. The Fe5406 index is predominantly a tracer of iron abundance – hence [Fe/H] – for the two grids closely correspond along lines of very similar [Fe/H], and is insensitive to the degree of α-enhancement. This is not the case for other Fe line indices that are also affected by other elements [643]. According to [711], if the IMF gives more weight to low-mass MS stars compared to a standard Kroupa or Salpeter one (dwarf dominated IMF), the value of the index increases. To give a quantitative example, for a 11 Gyr old SSP, changing the IMF power law from $M^{-2.35}$ to $M^{-3.35}$ increases Fe5406 by, on average, 0.3 Å for [Fe/H] between 0.0 and −2.5. The behaviour of the indices Fe5270 and Fe5335 with respect to changes of the IMF is qualitatively similar to the case of Fe5406.

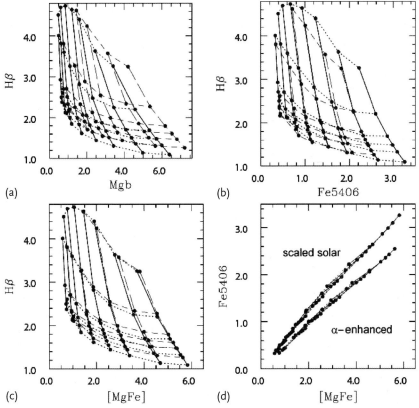

Figure 9.32 Several index-index diagrams for scaled-solar SSPs (solid and short-dashed lines) with [Fe/H] = −1.79, −1.49, −1.27, −0.96, −0.66, −0.35, +0.06, +0.40 and α-enhanced SSPs (long-dashed and dotted lines) with [Fe/H] = −1.84, −1.31, −1.01, −0.70, −0.29 and + 0.05. Metallicity increases from left to right. The ages (increasing from top to bottom) are equal to 1.25, 3, 6, 8, 10 and 14 Gyr.

The H_β index is a strong age indicator and is mildly affected by the α-enhancement in the sense that at roughly constant [Fe/H], the index shows a systematic, mild decrease at the lower [Fe/H]. With increasing [Fe/H], the behaviour is more complex; the decrease of the index becomes systematically smaller and is eventually reversed at [Fe/H] around solar, and ages above ~ 3 Gyr. The sensitivity of the H_β index on the SSP age stems from the fact that for a fixed HB morphology, the corresponding absorption feature is strongly sensitive to the T_{eff} of TO stars – the hottest MS objects – and hence the SSP age. On the other hand, at fixed age, the T_{eff} of the TO changes with metallicity, and one would perhaps expect a strong dependence of H_β also on metallicity. However, a change of the metal content also affects the flux level within the feature sidebands, which results in an overall much smaller sensitivity of H_β to metals, compared to the effect of age. Another important point to notice is that, although the feature is mainly produced by the hot MS (or HB) stars, the value of the H_β index is also affected by the much cooler (and brighter) RGB component because a significant fraction of the continuum flux of old stellar populations even in this wavelength range are provided by giant stars [712, 713]. The interplay between these contributions produces a complicated behaviour of H_β when the IMF is changed. The calculations in [711] show that when [Fe/H] is above ~ -1.0, H_β gets weaker by ~ 0.5 Å when the IMF changes from $m^{-2.35}$ to $m^{-3.35}$ (giant dominated), but stronger by up to ~ 2 Å at [Fe/H] below ~ -1.0. This latter effect is due to the progressively hotter HB, the progeny of red giants, when the metallicity decreases. This behaviour is reversed in the case of a dwarf dominated IMF.

In principle, higher order Balmer lines ($H\delta$, $H\gamma$) are also age indicators, but they appear to be more affected than H_β by the chemical composition. The calculations by [711] described before show that the alternative definitions of the $H\gamma_F$ and $H\delta_F$ indices (see Table 9.6) are less prone to abundance changes because of their narrower index definition.[8]

Figure 9.32a displays a H_β, the Mg_b index grid, that demonstrates the sensitivity of the Mg_b index to the degree of α-enhancement because of its dependence on the abundance of the α-element Mg. From the calculations in [712] at solar metallicity and 12 Gyr, the influence on the Mg_b index value of the different evolutionary phases is similar to the case of H_β. Neglecting post-MS stages produces a value higher than the case of the full isochrone. Including the RGB and post RGB phase progressively moves the value of Mg_b towards the result for the full isochrone. As a consequence, if the IMF gives more weight to the post-MS stars compared to the standard Kroupa or Salpeter one, the value of the Mg_b index decreases. Theoretical calculations also show that a giant dominated IMF decreases Mg_b by, on average, ~ 0.4 Å at [Fe/H] between solar and -2.5. The models presented in [712] show, however, that at a younger age of 4 Gyr, the contributions to the value of Mg_b from different evolutionary phases are different, and the value of the index is essentially only controlled by the MS phase.

From the indices Fe5406 and Mg_b, one can calculate, following [714], the new index $[MgFe] = \sqrt{\langle Fe \rangle \times Mgb}$, with $\langle Fe \rangle = 1/2(Fe5270+Fe5335)$, and Figure 9.32c shows the H_β–[MgFe] diagram. The [MgFe] index appears to be sensitive to the total metallicity Z, without being affected by the degree of α-enhancement. The value of H_β increases with the α-enhancement at fixed [MgFe] – hence fixed Z – for ages above 1.25 Gyr. At ages around 10–14 Gyr and Z around solar, the effect of a $[\alpha/Fe] = 0.4$ "undetected" α-enhancement in the target population would cause an underestimate of the SSP age by ~ 3 Gyr when only considering this H_β–[MgFe] diagnostic diagram.

The use of the Fe5406–[MgFe] plane, illustrated in Figure 9.32d, allows the estimate of the degree of α-enhancement. In this plane, the lines of constant age and constant metallicity are almost completely degenerate, however the scaled-solar (upper) and α-enhanced (lower) sequences are clearly separated. Therefore, using a combination of three grids – H_β–Fe5406, H_β–[MgFe], and Fe5406–[MgFe] – it is, in principle, possible to disentangle age, total metallicity Z, [Fe/H], and the degree of α-enhancement of a SSP.

---

8) The qualitative response of these indices to IMF changes is the same as for H_β.

Infrared indices that measure the strength of the NaI doublet and the Wing–Ford band (WFB – see Table 9.7) have been recently used to constrain the IMF of unresolved, old populations [715]. According to the SPSMs in [706] based on theoretical model atmosphere libraries, these two indices are affected by the population IMF – they increase for increasingly dwarf dominated IMFs – for they are sensitive to the contribution of very low-mass MS stars to the integrated spectrum. However, the WFB appears more sensitive to metallicity (increases with increasing metallicity at fixed IMF) than IMF, and it is essentially independent of age. On the other hand, the NaI doublet is much more sensitive to the IMF, but is also affected by the metal content (increases with increasing metallicity, at fixed age and IMF) and age (increases for increasing age, at fixed IMF and metallicity). In a 13 Gyr old population, an uncertainty in metallicity of 0.25 dex translates into an uncertainty of 15% in the IMF exponent $-x$, for $x > 1$, and 50% for $x < 1$.

9.3.1
HB Morphology

As for the case of integrated magnitudes and colours, a mismatch between the HB morphology in the target SSP and in the adopted SPSMs can cause major errors in the estimate of the population evolutionary properties. Especially when the observed population has a blue HB component – that enhances the strength of the Balmer lines – not represented in the SPSM counterpart, the observed value of the Balmer line indices will be matched with an age that is too young.

This issue has been exhaustively explored in [716], and we will summarize here the main points of their analysis. Two 14 Gyr old SSPs with [Fe/H] $= -1.31$ and -0.70, respectively, and α-enhanced metal mixture, have been considered. For each SSP, the integrated spectrum up to the TRGB was calculated analytically according to a Kroupa IMF. The HB was treated separately with a synthetic HB simulation, and populated with a large number of individual mass points (more than 10 000, in order to avoid the problem of statistical fluctuations) selected according to a chosen mean mass and a Gaussian mass spread. Spectra were assigned to each point in the HB simulation and then summed together. The summed HB spectrum was then scaled appropriately before being added to the spectrum of the underlying population to create the final integrated spectrum. For each of the two populations, four separate cases were created to match the extremes of HBs seen in Galactic GCs, from a red clump morphology (HB model 1) to an extremely blue HB with extended blue tail (HB model 4), plus two intermediate cases (HB models 2 and 3). Spectra from the extreme red and extreme blue cases were also combined to create the occurrence of a bimodal HB. It is important to notice that, as verified in [716], whether a blue HB morphology arises from a large mass loss along the RGB in stars with normal He abundance, or from a more moderate mass loss from stars with higher He, the resulting integrated spectrum and the diagnostic indices discussed below are largely unaffected. The only parameter that matters is the T_{eff} range covered by the HB stars, irrespective of how they have been produced. There is, however, an "indirect" effect arising from the previous evolutionary phases. If bluer HB morphologies are indeed caused by an increase of He, the TO and MS location will be hotter than in the models adopted to determine the reference index grid; this would cause a small additional increase of the Balmer lines, leaving metal lines essentially unchanged. This occurrence would not alter any of the main conclusions of this analysis.

Figure 9.33 displays the V–$(B-V)$ CMDs for the four cases, with [Fe/H] = -0.70, to illustrate the typical range of HB morphologies explored. The isochrone used to create the underlying population up to the TRGB is also plotted, extended to the HB phase and beyond for the cases with $\eta = 0.2$ and 0.4. Notice how at this intermediate-high metallicity, there is only a small difference in the HB location between $\eta = 0.2$ and 0.4 models. The $\eta = 0.4$ population has a systematically hotter HB than the $\eta = 0.2$ one, but for [Fe/H] ~ -0.70 and higher, the HB is

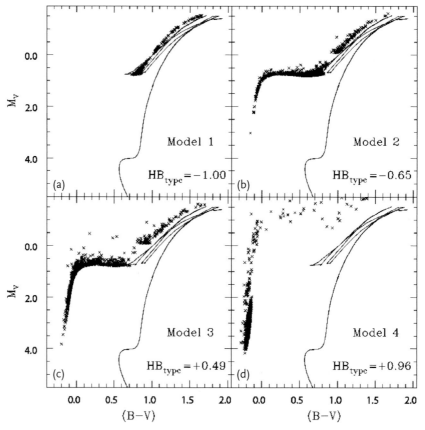

Figure 9.33 Four models of the α-enhanced, [Fe/H] = -0.70 set. For the sake of clarity, only 500 points are plotted for each HB ($\sim$ 10 000 were used in each simulation). The 14 Gyr isochrone used for the underlying population up to the TRGB is also plotted. The post-RGB later evolutionary stages of the isochrone are also shown for the two reference cases with $\eta = 0.2$ and 0.4. The HB morphology is parametrized in terms of the HB type = $(B-R)/(B+V+R)$ (see text for details).

still red and cooler than the TO, even at these old ages. It is at lower [Fe/H] values that the $\eta = 0.4$ population produces HBs much bluer than the $\eta = 0.2$ case, and bluer than the TO colour.

Figure 9.34 displays the impact of these different HB morphologies (plus a bimodal HB with 50% stars from model 1 and 50% stars from model 4) on the H_β–Fe5406 diagram, for both the [Fe/H] $= -1.31$ and the -0.70 populations. The reference grid is the same as in Figure 9.32 obtained from stellar models calculated with $\eta = 0.2$. We also display the 14 Gyr $\eta = 0.4$ case, for all metallicities covered by the reference grid.

Let's first notice the impact of changing η from 0.2 to 0.4 at a fixed age. The derived integrated spectra for [Fe/H] above ~ -0.7 produce similar values of the H_β index for the two cases, indicating that their HBs have comparable average temperatures, as both appear as red clumps in a CMD. Between [Fe/H] $= -0.7$ and -1.0, the index value suddenly becomes much stronger for the $\eta = 0.4$ case, an indication that the HB stars are now much hotter than for the $\eta = 0.2$ case, and the HB is bluer than the TO. For $\eta = 0.4$, the H_β index peaks at [Fe/H]~ -1.0 and then decreases again as the metallicity continues to decrease (and the HB gets even hotter).

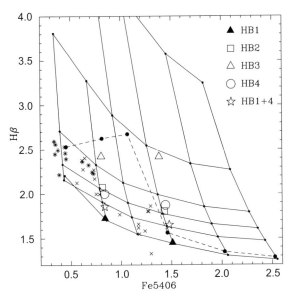

Figure 9.34 The α-enhanced H_β–Fe5406 grid shown in Figure 9.32, obtained from theoretical isochrones calculated with the Reimers mass loss parameter η set to 0.2. The dashed line denotes the theoretical calibration obtained with $\eta = 0.4$ (bluer HB morphology) and an age of 14 Gyr. Crosses and asterisks are observational data for Galactic GCs from [717], the asterisks denote clusters with HB type > 0.8 (i.e. predominantly blue). The large symbols refer to the index values predicted by HB1 refers to the red-clump-only HB model, HB4 refers to the extended blue tail case, and HB2 and HB3 are the intermediate cases. The $\eta = 0.4$ models at 14 Gyr only are also overlaid (dashed lines).

As for the sets of models calculated from synthetic HB simulations, an extended-red-only HB (HB1) only negligibly affects H_β compared to the reference grid. The index reaches a peak for one of the intermediate morphology cases (HB3), and implies ages between ~ 3 and ~ 5 Gyr when determined from the $\eta = 0.2$ grid – the real age is 14 Gyr. A bluer morphology (HB4) actually causes a decrease of the index value. This is due to the fact that H_β is maximized when a particular HB morphology is centred at $T_{eff} \sim 9000$–9500 K, and decreases when T_{eff} gets higher (or lower).

The values of the Lick metal line indices are only marginally affected by the HB extension; inferred Z and [Fe/H] from the index-index diagrams discussed in this section can however still change if H_β increases significantly, for none of the diagnostic grids is completely orthogonal. In practice, however, these inferred changes in metallicity are likely to be within the observational errors in most cases.

The obvious question to address, in light of these results, is whether there are indices that could potentially discriminate between a very old SSP with an extended blue HB and a SSP with an intermediate-old age – that is, between 3 and 6–8 Gyr; looking at the integrated spectra one immediately realizes that the range to explore is at wavelengths below ~ 4000–4500 Å. There are at least three different diagnostics proposed in the recent literature [716, 718]

The first one is the ratio between the value of the $H\delta_F$ and H_β indices, that appears to be far more sensitive to HB morphology than to age. In practice, whenever the age obtained from, for example, the $H\delta_F$–Fe5406 diagram, is not the same as the age obtained from the H_β–Fe5406 grid, this could be an indication for the presence of HB stars bluer than the SPSMs adopted in the theoretical index calculations.

The second potential diagnostic is the Mg II feature around 2800 Å (see Table 9.5). The value of the associated index displays a strong trend towards lower values as the HBs become more extended towards the blue, at fixed chemical composition. Given that Mg_b (another Mg-sensitive index) is negligibly affected by the HB morphology, any discrepancy between the metallicities derived from the Mg_b and Mg II indices for a given stellar population could be due to the presence of an extended blue HB component not included in the adopted SPSM calibration.

A third potential diagnostic makes use of the $(Ca_{II}H + H\epsilon)/Ca_{II}K$ (Table 9.8) index, that for brevity we denote here as CaII. This index was already identified in [709] as being sensitive to the presence of hot stars in a composite spectrum. Figure 9.35 displays a H_β–CaII grid, for the same ages, chemical compositions and HB morphologies of Figure 9.34. Only the case with [Fe/H] $= -1.3$ is displayed for the indices derived from the synthetic HB simulations. All the extended HB models lie outside the $\eta = 0.2$ grid, except for HB model 1, the red clump case. This means that the H_β–CaII grid could potentially discriminate between an old population with extended blue HB and an intermediate-old age – provided that the "correct" Ca abundance is accounted for in the theoretical integrated spectra. It is also interesting to consider the sample of Galactic GCs displayed both in this diagram and in Figure 9.34. The clusters with the bluest HBs fall towards the left-hand side of the H_β–CaII diagram, with lower CaII values. Clusters with HB type >

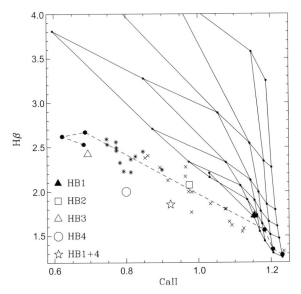

Figure 9.35 Index–index diagnostic of the HB morphology. Lines and symbols have the same meaning as in Figure 9.34.

0.8 (predominantly blue HB) are denoted with asterisks and they all fall well outside the reference grid. Looking at Figure 9.34, these same clusters would all appear to have ages around 6–7 Gyr in the H_β–Fe5406 diagram. If their large values of H_β were caused by younger ages rather than blue HBs, these clusters would need to have a higher CaII index for the same H_β strength and the points should be located farther towards the right-hand side of Figure 9.35.

An important caveat to note regarding the Ca II index is that it is strongly affected by velocity dispersion and also requires data with very high signal-to-noise ratio.

One issue still to be comprehensively explored is the effect of radiative levitation, effective in HB stars with T_{eff} above $\sim 12\,000$ K. Some approximate conclusions can however be drawn by assuming that levitation just affects the stellar spectra and can be mimicked by "standard" spectra with varied individual abundances. Observations of Galactic GCs show that Ca and Mg are basically unaffected by levitation, whereas Fe is always enhanced to solar or super-solar values. It is therefore natural to speculate whether an enhanced Fe content for hot HB stars could affect the behaviour of the models in Figure 9.34. In absence of detailed calculations, one can just notice that the Fe5406 index seems weakly sensitive to blue HB stars, especially when T_{eff} increases above $\sim 10\,000$ K. This leads to the tentative conclusion that most probably iron-sensitive indices are only very mildly affected by levitation.

9.3.1.1 UV-upturn in Elliptical Galaxies

The presence of hot HB stars in passively evolving, old unresolved stellar populations is the cause of the so-called "UV upturn" (or UVX) phenomenon found in elliptical galaxies [719]. Space-based observations show that the spectral energy

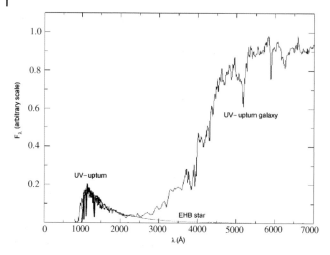

Figure 9.36 Spectrum of the elliptical galaxy NGC4552 that displays the UV upturn. Overimposed is a low-resolution theoretical spectrum for a representative solar metallicity EHB star with $T_{eff} = 22\,000\,K$ and $\log(g) = 4.5$

distribution of genuinely passively evolving ellipticals increases to shorter wavelengths over the range between 2000 to 1200 Å, as displayed in Figure 9.36 for NGC4552. The UV upturn feature can be clearly seen as an abrupt increase of the flux at $\lambda \sim 2000$ Å, followed by a maximum and a fast drop when λ decreases below ~ 1000 Å. The strength of the upturn is often quantified by means of a colour index based on the difference between the magnitude in a UV filter and in V, and there are contradicting indications [720–722] about its dependence on the galaxy metallicity – usually quantified by the strength of Mg lines. From the data obtained with the GALEX UV satellite, about 10% of local ($z < 0.2$) giant ellipticals displays the UV phenomenon [723].

The location of this feature looks remarkably consistent in all galaxies where the UV upturn is found, and can be nicely reproduced by the spectrum (Figure 9.36) of hot, extreme HB stars (EHB – $T_{eff} \sim 20\,000$–$25\,000$ K) that evolve off the HB directly to the WD cooling sequence. Neither post-AGB stars nor post-early AGB stars can contribute significantly to the UV upturn, because their spectra are respectively hotter and cooler than observations. The variations in the UV emission relative to the optical must be, therefore, the result of variations in the fraction of EHB stars, and not a variation of the type of stars producing the upturn.

Irrespective of the assumptions about the metallicity of the progenitors of these EHB stars – their surface abundance pattern, hence the signatures of metals in this portion of the spectrum are affected by radiative levitation – the cosmological limit of ~ 14 Gyr on their age requires very large values for the total RGB mass loss. Alternatively, as already discussed for the case of Galactic GCs, an increase of He at fixed RGB mass loss and age produces increasingly hot HB stars because of the steady decrease of the RGB progenitor mass. High values of Y at either low or high metallicities can potentially produce EHB stars, without extreme values for the RGB mass loss. This scenario has been explored in [724], who considered a fraction of He-enhanced (Y up to 0.38) mainly metal-poor population, with Y calibrated as a function of [Fe/H] by reproducing the trend of a UV-V colour with metallicity in the GCs belonging to M87, an elliptical galaxy that displays the UV upturn.

An alternative – or complementary – scenario involves interacting binary stars [725], and envisages three types of formation channels for EHB stars. The main feature of the binary model is that it provides well-defined mechanisms for the ejection of the envelope of RGB stars, and the production of EHB objects. The first mechanism (applicable to long orbital period systems) involves a mass donor filling its Roche lobe near the tip of the RGB, and experiencing a stable – slow – mass transfer that causes the loss of the envelope. The – almost – bare He-core, because of the proximity to the critical value of $\sim 0.5\,M_\odot$, is then able to ignite He and become a EHB star. The second mechanism (valid for short orbital periods) involves a a mass donor that fills its Roche lobe near the tip of the first giant branch and experiences a dynamically unstable mass transfer, leading to the formation of a common envelope. The ejection of the common envelope leaves a bare He-core (with thin hydrogen envelope) that ignites and produce a EHB star. In this case, the donor star needs to fill its Roche lobe closer to the tip of the RGB compared to the stable mass transfer scenario because the timescale of the common enve-lope evolution is much shorter, and the core does not grow appreciably during this phase. A third mechanism envisages a close He-core WD pair (produced by previous common envelope and/or stable mass transfer episodes) that coalesces due to angular momentum loss via gravitational wave radiation, and – if the total mass is at least $\sim 0.5\,M_\odot$ – the merged product ignites helium to become a EHB star.

Perhaps the key difference between single- and binary-star scenarios is the ex-pected evolution with redshift of the UV upturn. The interacting binary channels predict no strong evolution with age (hence redshift). Also, there is, in principle, no dependence on the metallicity, for it does not play a significant role in the envelope ejection process, even though it may indirectly affect the properties of the binary population. According to the single-star scenario one expects instead some evolu-tion with redshift, because with decreasing age (increasing redshift) and all other parameters (mass loss, Y, Z) being unchanged, the mass of stars at the tip of the RGB and of their HB progeny increases, moving objects to lower T_{eff}, hence de-creasing the strength of the UV upturn. Observationally, it is not yet clear whether the UV upturn evolves appreciably with look-back time [726].

9.3.2
Estimating Individual Element Abundances

Thus far, we have mainly discussed just a few indices that allow the estimate of age, [Fe/H] and total metallicity Z of a SSP. There are, however, several other indices – we restrict our discussion mainly to the most widely used and investigated Lick sys-tem – that are sensitive to the abundances of many individual metal species other than Fe. This opens up the theoretical possibility to determine detailed individual element abundances of unresolved SSPs, in the same way one does for single stars, without the need to assume a priori a specific metal mixture.

One widely explored avenue to accomplish this is to work in the Lick system, and calculate indices using the fitting functions obtained on local stars, "corrected" by an appropriate "response function" that accounts for the index response to individual element changes. Here, we sketch this method [704, 727].

As a starting point, a few representative stars along an isochrone of given [Fe/H] and age – corresponding typically to one point along the MS, the TO, and one point along the RGB – are selected. The abundance effects are therefore isolated at a given T_{eff} and surface gravity. For each of these points, one first calculates theoretical model atmospheres, spectra and index values I_0 for a baseline mixture (typically either scaled-solar or α-enhanced). Then atmospheres, spectra and index values I are calculated with the abundances of C, N, O, Mg, Fe, Ca, Na, Si, Cr and Ti each doubled in turn. The variation $\Delta I = (I - I_0)$ is employed to calculate the incremental ratios $\Delta I / \Delta [X_i/H]$ from which the response function $R_{0.3,i}$ for any index corresponding to a variation of the abundance of element i by $+0.3$ dex can be derived as

$$R_{0.3,i} = 0.3 \frac{1}{I_0} \frac{\Delta I}{\Delta[X_i/H]}$$

Following the derivation by [704, 728], one assumes that I can be written as

$$I \propto \exp([X_i/H])$$

After a Taylor expansion of $\ln(I)$ around $[X_i/H]$, one gets

$$\ln(I) = \ln(I_0) + \sum_{i=1}^{n} \frac{\partial \ln I}{\partial [X_i/H]} \Delta[X_i/H] + \dots$$

where the sum is over all elements i in the mixture. Neglecting higher order terms, we obtain

$$\ln(I) = \ln(I_0) + \sum_{i=1}^{n} \frac{1}{I_0} \frac{\partial I}{\partial [X_i/H]} 0.3 \frac{\Delta[X_i/H]}{0.3}$$

Using the definition of response function and taking the exponential one finally derives [9]

$$I = I_0 \prod_{i=1}^{n} \exp[R_{0.3,i}]^{(\Delta[X_i/H]/0.3)} \tag{9.8}$$

Starting from the response functions and the values I_0 for the baseline mixture, this equation provides the index value I for a generic change of the individual abundances X_i, at the chosen representative points along the isochrone. [10]

9) A slightly different equation is derived and employed in [729].
10) The basic assumption behind the derivation of Eq. (9.8) is that a generic index approaches asymptotically zero for very low element abundances. This condition is not generally satisfied by the Lick indices, and a procedure to deal with negative index values is described in [728].

Finally, these "new" values of the indices at the three reference points along the isochrone need to be appropriately weighed to provide the "global" index value for the whole SSP (see, for example, [729] and [728] for different ways to calculate "global" index values).

With these tools available, one can then determine several individual metal abundances from the integrated spectra of SSPs [730, 731]. The general idea is to use a diagram of the type H_β vs. iron lines like Fe5270, Fe5335 or Fe5406 for determining a fiducial age and [Fe/H] as they are mostly insensitive to the abundances of other elements. The key is then to proceed with the abundance estimates in such a way as to only adjust one abundance at a time. Therefore, once fiducial age and [Fe/H] of the SSP have been determined, the following step is to fit one or more Lick indices that introduce only one additional abundance, like for example Mg_b, that is dominated by Mg. In this way, step after step individual abundances of elements like Fe, Mg, C, N, Ca can be determined for a generic old SSP. Elements like O, Na and Si do not dominate any of the Lick indices and their abundances have to be fixed a priori. For example, the abundance of oxygen is largely degenerate with C and N, and affects the indices C_24668, C_1 and C_2 that are used to estimate the carbon abundance. Also, one has to be cautious with interpreting results from the Ca4227 index in terms of Ca abundances. A chemical composition with simultaneous CNONa anti-correlations overimposed onto a standard α-enhanced metal mixture (as seen in Galactic GCs) changes the value of this index – whose blue pseudo-continuum is affected by the CN molecule – compared to the value for the baseline α-enhanced mixture, even though the abundance of Ca is unchanged [732].

It is clear that this whole procedure accounts for the effect of individual abundance variations only on stellar spectra, not on the isochrones. While this approximation is justified in most of the cases [643], it doesn't have a general validity [208, 643]. For example, at fixed [Fe/H] and age, a variation of even just one of the CNO elements that leads to an increase of the sum $C + N + O$ (keeping all other abundances unchanged) leaves the lower MS and the RGB of the population largely unaffected, but shifts the TO to lower luminosities and cooler T_{eff}, to mimic an older age (see Chapter 3). Also, a variation of Mg affects the RGB T_{eff} for old ages, leaving MS and TO unchanged.

The recent work by [643] determines the effect of abundance variations of individual elements (C, N, O, Na, Mg, Si, S, Ca, Ti, Fe) on the Lick indices by calculating both isochrones and spectra with the varied abundances. The 35 T_{eff} and gravity combinations considered in the spectral calculations ensure a much better coverage of all evolutionary phases. This, together with the self-consistent isochrone calculations, provides, in principle, more reliable predictions for the response of integrated spectra and indices to single element abundance variations. However, these calculations include evolutionary phases only up to the tip of the RGB, and can be considered as only a first step towards constructing fully self-consistent SPSMs with flexible chemistry.

9.3.3
Statistical Fluctuations

Due to the finite number of stars in a given stellar system, fluctuations also in the predicted values of line indices are in principle possible for star cluster-sized populations. Figure 9.37 displays multiple results of a series of Monte Carlo simulations [716]. We display the following cases (α-enhanced): a $2 \times 10^5 \, M_\odot$, 14 Gyr, [Fe/H] $= -1.31$ SSP with both HB model 1 (see Section 9.3.1 – open squares) and HB model 3 (crosses) discussed before. These two SSPs are representative of typical Galactic GCs with different HB morphology. We also show the results for a 3.5 Gyr, [Fe/H] $= -0.7$, $2 \times 10^4 \, M_\odot$ SSP with HB model 1 (open circles) to represent a Magellanic Cloud-type of intermediate-old cluster. For the sake of comparison, points on the lower left-hand side of each diagram represent several realization of a more massive 14 Gyr, [Fe/H] $= -1.31$ SSP with HB model 3 and a total mass of $2 \times 10^6 \, M_\odot$, shifted arbitrarily for clarity – they are centred on the same location as the $2 \times 10^5 \, M_\odot$ data.

For the $2 \times 10^5 \, M_\odot$ simulations, the scatter in H_β is significant and is larger for the extended blue HB case. The associated age uncertainty depends, however, on the chosen index-index diagram. In the H_β–Fe5406 diagram, the associated fluctuation of the metal index together with the slope of the equal-age lines conspire to minimize the impact of the fluctuations on the derived ages (the resulting uncertainty on [Fe/H] is of the order of ± 0.1 dex). On the other hand, the Mg_b index looks largely unaffected by statistical fluctuations, even for the lower mass of the 3.4 Gyr SSP (as discussed before, at these ages Mg_b is dominated by the well-sampled MS),

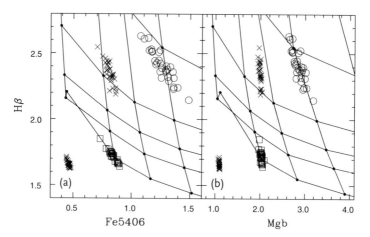

Figure 9.37 Index–index grids with results from multiple realizations of the following models (lines of constant age and [Fe/H] are as in Figure 9.32): $2 \times 10^5 \, M_\odot$, 14 Gyr, [Fe/H] $= -1.31$ SSP with both HB model 1 (open squares) and HB model 3 (crosses); 3.5 Gyr, [Fe/H] $= -0.7$ and $2 \times 10^4 \, M_\odot$ with HB model 1 (open circles). Data points on the lower left-hand side of each diagram are realizations of the 14 Gyr, [Fe/H] $= -1.31$ SSP with HB model 3, but with a total mass of $2 \times 10^6 \, M_\odot$, shifted arbitrarily for clarity – as measured, they are centred on the same location as the $2 \times 10^5 \, M_\odot$ data.

and hence the effect of the H_β fluctuations on the derived age is maximized. If we consider the $2 \times 10^5 \, M_\odot$, 14 Gyr, [Fe/H] $= -1.31$ SSP with HB model 3, the indices that display 1σ fluctuations larger than 10% are G4300, Fe5015, $H\delta_F$ and $H\gamma_F$ (as a comparison, H_β displays a 1σ fluctuation of $\sim 8\%$).

9.3.4
The Effect of Binaries

We have already discussed in Section 9.2.5 and Figure 9.18 the effect of binary interactions on integrated magnitudes and colours of SSPs. Figure 9.38 displays the corresponding effect on a few selected Lick indices. As expected, because of the presence of binary-produced Blue Stragglers and hot HB stars, the Balmer line indices increase, mimicking spuriously younger ages when binary interactions are not accounted for. Also, the effect on the Fe and [MgFe] indices is to underestimate [Fe/H] and Z when binary interactions are effective in the observed population, but neglected in SPSMs.

9.4
Unresolved Composite Systems

An unresolved stellar population that appears to be old (i.e. older than ~ 1 Gyr) when treated as SSP, can, in principle, be "hiding" one or more young(er) stellar

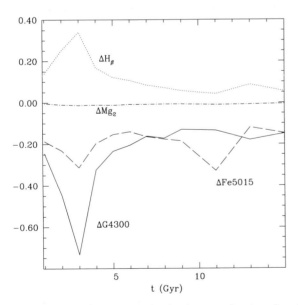

Figure 9.38 Variation of selected Lick indices as a function of age (solar metallicity) when interacting binaries are considered.

populations. Let's assume for simplicity that the target is made of an old component together with a small fraction of young stars; as discussed in [733] and [734],
SSP-equivalent ages derived from SSP grids of either integrated colours or Lick
indices are heavily biased towards the age of the young component because of the
strong sensitivity of the age indicators (i.e. Balmer lines or colours like $(B-K)$) to
the presence of hot, bright stars. As an example, in the case of two star formation
bursts 40 Myr and 10 Gyr old (keeping the chemical composition fixed), the $(B-K)$
colour is completely dominated by the youngest population, even when its contribution to the total mass of the system is only $\sim$ 3–5%. This two-bursts population
would therefore appear young, even though it is almost exclusively made of old
stars. There is generally a degeneracy between the age of the young component
and its mass fraction [734]. For a given fraction F_2 and age t_2 of the young component, the same SSP-equivalent age can be obtained with a younger t_2 as long as F_2
is reduced.

For $t_2 < 2.5$ Gyr and F_2 between 1 and 10%, different Balmer line indices display different sensitivity to the young component and a discrepancy between SSP
ages obtained with two different Balmer line indices may flag the presence of
a young component. However, this issue is complicated by the fact that different
Balmer indices have different sensitivities also to blue HB stars, and hence any
SSP-equivalent age disagreement may also be due to the presence of a blue HB
unaccounted for in the SPSMs.

The SSP-equivalent metallicities are usually dominated by the old component,
for both Lick indices and integrated colours, even though in the latter case this conclusion may depend on the chosen metallicity-sensitive colours. In fact, the $(J-K)$
colour (employed in the $(B-K)$–$(J-K)$ diagram) is strongly affected by intermediate age populations (300–600 Myr old) with a very bright AGB phase.

A more sophisticated way to infer the presence of young sub-populations in an
"old-looking" stellar system – but obviously applicable also to study pure SSPs – is
analogous to what is done to determine the SFH of composite resolved populations.
As for its CMD counterpart, the stellar spectrum of a composite population can be
represented by a sum over the spectra of individual SSPs with appropriate weights
that reflect the population SFH. The general approach to this problem involves the
definition and minimization of a merit function of this type:

$$\chi^2 = \sum_{j=1}^{n} \left[\frac{F_j - \sum_{i=1}^{l} a_i S_i^j (t_i, Z_i, A_V)}{\sigma_j} \right]^2$$

where F_j is the observed flux in n wavelength bins j, σ_j its standard deviation, a_i
are the weights of each one of the l SSPs (typically, $l = N_t \times N_Z$ combinations
of t and Z) of age t_i and metallicity Z_i that emit a flux S_i^j at each wavelength j;
the minimization process determines the values of the weights a_i and of the extinction A_V, provided that an extinction law is specified.[11] These full spectrum

---

11) The extinction law published in [735] is commonly used when dealing with old, passively evolving
    populations. The free parameter R_V is typically set to a value between 3.1 and 3.3.

fitting methods are well suited when the spectral resolution is comparable with the physical broadening. If the resolution is lower, spectral features are diluted and the method would be sensitive mainly to strong spectral features, and would practically become equivalent to fitting absorption feature indices.

An attractive feature of these methods is that no parametrization of the SFH is necessary. Codes like STARLIGHT [736], ULySS [737], STECKMAP [738], among others, (see [739] for a more comprehensive list and references) employ this kind of technique and have been widely applied to observations of different types of stellar systems. The code MOPED [740] uses a slightly different approach, starting from the evidence that it is not necessary to include all the F_j measurements in the model fitting. In fact, some of the data may carry no information because either they are not sensitive to the parameters we want to estimate, or they are very noisy. The code constructs linear combinations of the data with weights chosen carefully to avoid losing information concerning the population SFH. The observed flux distribution is reduced to a compressed data set comprising one datum per parameter, provided certain conditions are met (see [740], for more details).

9.5
Validation of Stellar Population Synthesis Techniques

In the previous sections, we have discussed SPSMs and the following three techniques to determine the evolutionary status of an unresolved stellar population:

- Low-resolution SED fitting – sensitive to the general shape of the continuum and usually performed employing broad-band filters;
- Line index fitting – sensitive to the strength of specific spectral features;
- Full spectrum fitting – makes use of all information by fitting the full flux distribution at all wavelengths.

The accuracy of the evolutionary parameters derived for unresolved stellar populations relies on the accuracy of these techniques and the SPSMs adopted. Differences in the mass distribution T_{eff} and luminosities of the adopted theoretical isochrones – keeping fixed the spectral library and IMF – will have an impact on the derived ages and chemical compositions. Changing the spectral library at fixed isochrones can also affect the predicted colours/spectra. The detailed analysis presented in [741] considers pairs (T_{eff}, g) that are representative of a 10 Gyr solar metallicity isochrone, and compares the corresponding colours obtained from several sets of theoretical and semi-empirical low-resolution spectral libraries. Differences are larger for the cooler stars, and the bluer photometric bands. In $(U-B)$, differences are of the order of a few 0.1 mag, decreasing to a few 0.01 mag when considering redder colours like $(V-I)$, $(V-R)$ or $(J-K)$. The same authors also compare 35 absorption line indices (including all Lick indices) at wavelengths between ~ 3500 and ~ 8700 Å, calculated on individual spectra from scaled-solar theoretical and empirical libraries at [Fe/H] around solar, for a large range of $(T_{eff}$,

g) pairs. The largest differences between results from theoretical and empirical spectra appear again for the very cool stars and indices in the blue region of the spectra.

It is not trivial to interpret these differences because uncertainties in the atmospheric parameters of the observed stars, and possible differences between the metal abundance pattern employed in the theoretical calculations and the "real" abundance pattern of stars in the solar neighbourhood may mislead one toward the wrong conclusions about the accuracy of theoretical spectra.

Figure 9.39 displays two scaled-solar H_β–Fe5406 grids calculated with BaSTI isochrones and, respectively, the MILES and Munari spectral libraries, for [Fe/H] = −0.35, 0.06 and 0.26 and ages between 3 and 14 Gyr. Notice how just changing the spectral library induces a massive change in the diagram. The MILES library gives systematically larger values of H_β and smaller values of Fe5406 for a given age and [Fe/H]. The differences in H_β are larger for the older ages, with the result of having a much narrower range of H_β index values when the MILES library is used.

A subtle issue, when empirical spectral libraries are employed, is the accuracy of the T_{eff}, surface gravity and [Fe/H] parameters of the library stars.

Let's consider the integrated spectra for two scaled-solar [Fe/H] = 0.06 BaSTI SSP models aged 4 and 14 Gyr, with offsets of 100 K in T_{eff}, 0.25 dex in log(g), and 0.15 dex in [Fe/H] applied to the adopted spectral library, taken as typical zero point uncertainties in these key stellar atmosphere parameters. Using simple H_β–Fe5406 diagrams to make a preliminary assessment of the impact on stellar population parameters inferred from the models, it is found that absolute ages derived

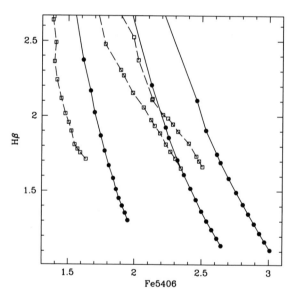

Figure 9.39 Scaled-solar H_β–Fe5406 grids calculated with BaSTI isochrones and, respectively, the MILES (dashed lines, open squares) and Munari (solid lines, filled circles) spectral libraries, for [Fe/H] = −0.35, 0.06 and 0.26 and ages between 3 and 14 Gyr.

from the H_β index can easily be affected by $\sim 20\%$ and relative ages can also be affected, albeit at a lower level [742].

Inferred ages appear to be of more problematic interpretation when also $H\delta$ and $H\gamma$ indices are employed because their behaviour in response to a systematic shift in [Fe/H] goes in the opposite sense to that of H_β. The systematic variations of the inferred [Fe/H] are minimized in the H_β–Fe5406 diagram, for it is weakly affected by the errors in T_{eff}, and log(g), and is mainly sensitive only to uncertainties in [Fe/H] for the stars in the spectral library. When multiple metallicity indicators are employed, results are more difficult to interpret because of the opposite behaviour of some metal indicators in response to systematic offsets in T_{eff} and log(g).

This behaviour has implications for methods which fit simultaneously to several absorption line indices or perform a full spectrum fit to derive ages and metallicities of stellar populations. A failure to fit several indices simultaneously could, spuriously, be interpreted as an indication of non-solar abundance ratios – or evidence for the presence of stellar components not fully accounted for in the models, rather than being ascribed to uncertainties in the fundamental parameters of the spectral library. It is also important to notice that the Fe5406–[MgFe] diagram remains a good diagnostic of α-enhancement. Altering any of the atmospheric parameters moves the scaled-solar SSP points along the scaled-solar line in this diagram (see Figure 9.32) changing the inferred [Fe/H], but not altering the inferred abundance ratios for other elements to non-solar ratios.

As for the predicted integrated colours, errors in log(g) have a negligible effect, whilst a 0.15 dex uncertainty in [Fe/H] affects colours by at most ~ 0.02 mag in optical and near-IR bands. As expected, the effect of increasing T_{eff} of stars in the spectral library by 100 K is to make all broad-band colours bluer. At old ages, the effect is of only ~ 0.01 mag in $(U-B)$ and $(B-V)$, and ~ 0.02 mag in $(R-I)$, increasing at longer wavelengths, that is, the $(J-K)$ colour gets bluer by ~ 0.04 mag. Overall, this variation of T_{eff} can significantly decrease the inferred SSP metallicity, rather than making a population look younger, if a colour–colour diagram such as $(V-I)/(V-K)$ or $(B-K)/(J-K)$ is employed.

In principle, the ideal test beds to validate the combination of methods and standard available SPSMs employed in studies of unresolved systems are nearby resolved star clusters. Galactic globular clusters, old open clusters (typically M67) and old Magellanic Cloud clusters have been traditionally employed to compare ages and chemical compositions derived from stellar spectroscopy and CMD fitting, with inferences from the analysis of their integrated light (see [592, 606, 711, 731, 743–747], for recent examples, together with the results of the "stellar population challenge" at the IAU Symposium 241: Stellar populations as building blocks of galaxies). This approach inevitably suffers from a number of intrinsic uncertainties:

- Integrated colour and/or index fluctuations are not accounted for when comparing SPSM predictions (obtained from analytical integration along the isochrone) with cluster data;

- The effect of the dynamical cluster evolution is not accounted for in standard SPSMs;
- The HB morphology in standard SPSMs is usually not an accurate representation of the extended morphologies observed in most of the Galactic GCs;
- When the effect of interacting binaries is accounted for, this is done for a fixed choice of the free parameters that regulate the effect of the interaction on the stellar population. This set of parameters is not tuned to reproduce the target clusters;
- Galactic GCs are not SSPs (see Section 7.8). CNONa anti-correlations directly affect integrated colours because they modify the stellar spectra of cold stars at wavelengths shorter than approximately the B filter [417]. The indirect effect is – very likely – through the increase of He abundance in stars affected by the anti-correlation, that may explain the presence of blue HB stars.

  As for the comparison of integrated spectra, a very recent analysis has presented detailed calculations for stellar populations made of two old (either 12 or 14 Gyr) coeval generations of stars, the first one born with the standard α-enhanced metal mixture observed in field halo objects, the second one characterized by an anti-correlated CNONa abundance pattern overimposed on the first generation, standard α-enhanced metal pattern [732].

  The effect of these two compositions (that share the same [Fe/H]) has been calculated consistently in both isochrones and spectra, for [Fe/H] $= -0.7$, typical of a metal-rich Galactic GC like 47 Tuc. The presence of a second generation of GC stars with unchanged He content – the HB morphology is the same in both populations – appreciably affects the Ca4227, G4300, CN1, CN2, and Na$_D$ metal indices. The variation of Ca4227 goes in the direction to mimic a lower Ca abundance, if this index is used as a measure of the Ca content.[12] Age, Fe-abundance, and total metallicity inferred from H$_\beta$–Fe5406 and H$_\beta$–[MgFe] diagrams are confirmed to be insensitive to the detailed chemical abundance pattern of second generation stars. If the second generation has an increased He abundance (i.e. $Y = 0.300$ compared to the reference value $Y = 0.256$ of the first generation component), the metal indices are unaffected, but the Balmer line indices increase due to the hotter TO and MS, and the slightly hotter HB – when the same value for the Reimers η parameter is used in models for the first and second generation of stars.

Together with these caveats, one has to also consider that, for example, in the case of Galactic GCs, the compilations of CMD ages often used in most SPSM validations ([245, 480, 481], that provide very similar absolute and relative ages), have employed isochrones generally different from the ones of SPSMs. As for the cluster [Fe/H], again, in the case of Galactic GCs values from the updated Harris' compilation [748], are often used, but variations of the order of 0.10–0.20 dex for individual clusters are found in the literature.

---

12) This may explain the lower Ca abundances compared to other α-elements derived recently for a sample of Galactic GCs treated as SSPs [731].

Below, the general trends for the results of these validation tests are listed, without delving into the intricate details of each individual investigation:

- Whenever codes like STARLIGHT that allow for the presence of multiple populations (all with the same metal mixture in current implementations) are used for spectrum fitting of Galactic GCs – in the range ~ 4000 to ~ 6000 Å with currently available data – the solutions for each cluster usually include several different ages and metallicities. This is clearly a sign of inadequacies of the theoretical SPSMs, and are often interpreted as a sign of inadequate modelling of the HB and/or of the presence of Blue Stragglers. When one searches for the best fit SSP solution, the derived [Fe/H] values are typically within ~ 0.2 dex of the spectroscopic ones. Ages are less reliable and the correlation with CMD-based estimates is typically weak. Differences by up to a factor ~ 1.5 are found in some studies, and ages of individual clusters are in several cases estimated to be larger than 15 Gyr (in this case, the solution is sometimes "forced" to be younger than this limit). Differences are very pronounced, in the sense of generally underestimating ages up to a factor of ~ 2 for clusters with a blue HB, but often also for these clusters, ages above 15 Gyr are found;
- Results from fitting Lick indices weakly sensitive to the initial metal mixture provide a similar picture as for full spectrum fitting. There is a tendency for ages estimated from H_β (and for some SPSMs also when the other Balmer lines are used) to be systematically well above 15 Gyr – this is sometimes dubbed the "H_β anomaly" – for some (but not all) of the SPSMs, especially when empirical spectral libraries are employed;
- Colour – of low-resolution SED – fitting is less employed. The general trends or the results are similar to the previous two methods.

It is clear that validation of SPSMs is a difficult subject, and we close this section by briefly discussing a specific work that very clearly highlights the present uncertainties [743]. These authors consider integrated spectra in the wavelength range between ~ 4800 and ~ 6000 Å for their sample of 14 SMC clusters, 9 of which are reportedly old (ages above 1 Gyr). The estimates of [Fe/H] and age for these objects are rather uncertain, but perhaps the most relevant aspect of the analysis is the inter-comparison of ages and metallicities obtained from two spectrum fitting codes (STARLIGHT and ULySS), each using three different SPSMs (Pegase, BC03 and Vazdekis).

The UlySS code as employed in this analysis returns the age and metallicity of the SSP that best matches the data. Comparisons of the results obtained with the three different SPSMs display, in some cases, large differences. Clusters that, according to the BC03 models, have ages in the range between ~ 1 and ~ 2 Gyr display huge age and metallicity variations when the other SPSMs are considered, up to factors 3–4. As an example, the cluster L113 results to be 1.4 Gyr old with BC03, 7.1 Gyr old with Pegase and 2.6 Gyr old with Vazdekis models. The metallicity range spanned by these three results is equal to 0.9 dex. STARLIGHT returns multi-component fits and the results provided in [743] refer to mean age and metallicities weighted by the normalized light-fraction of the various SSPs contained in the solution. Also, for these mean ages and metallicities, large differences appear when results from different SPSMs are compared. Considering again the cluster L113, mean ages vary between 3.4 and 8.3 Gyr (this time, the older age is obtained with the Vazdekis models) and the mean metallicities span a range of 0.7 dex.

Comparing results from the two different codes at fixed SPSM is of more difficult interpretation, for the mean values provided by STARLIGHT cannot be equivalent to best fit SSP ages and metallicities. There are large differences, for a given SPSM, between the two sets of results for individual clusters. The same cluster L113 that is 1.4 Gyr old according to UlySS implemented with BC03 models has a mean age of 8.3 Gyr with STARLIGHT implemented with BC03 SPSMs. Another example is HW1, that is, 9.0 Gyr old with ULySS, and has a mean age of 3.2 Gyr with STARLIGHT.

References

1 Komatsu, E. *et al.* (2010) Seven-year Wilkinson Microwave Anisotropy Probe (WMAP) observations: Cosmological interpretation. *Astrophysical Journal Supplement*, **192**, 18.

2 Bianchi, E., Rovelli, C., Kolb, R. (2010) Cosmology forum: Is dark energy really a mystery? No it isn't, *Nature*, **466**, 321.

3 Cyburt, R.H., Ellis, J., Fields, B.D., Luo, F., Olive, K.A. and Spanos, V.C. (2010) Nuclear reaction uncertainties, massive gravitino decays and the cosmological lithium problem. *Journal of Cosmology and Astroparticle Physics*, **10**, 032.

4 Steigman, G. (2010) Primordial nucleosynthesis: A cosmological probe, in *Proceedings of IAU Symposium 268* (eds C. Charbonnel, M. Tosi, F. Primas and C. Chiappini), p. 19.

5 Jungman, G., Kamionkowski, M., Kosowsky, A. and Spergel, D.N. (1996) Cosmological-parameter determination with microwave background maps. *Physical Review D*, **54**, 1332.

6 Baugh, C.M. (2006) A primer on hierarchical galaxy formation: The semi-analytical approach. *Reports on Progress in Physics*, **69**, 3101.

7 Benson, A.J. (2010) Galaxy formation theory. *Physics Reports*, **495**, 33.

8 Bruzual, G. (2010) *Population Synthesis: Challenges for the Next Decade*, Proceedings of IAU Symposium 262, (eds G. Bruzual and S. Charlot), Cambridge University Press, p. 55.

9 Bessell, M.S. (2005) Standard Photometric Systems. *Annual Review of Astronomy & Astrophysics*, **43**, 293.

10 Stetson, P.B. (2000) Homogeneous photometry for star clusters and resolved galaxies. II. Photometric standard stars. *Publications of the Astronomical Society of the Pacific*, **112**, 925.

11 Michard, R. (2005) Peculiarities and populations in elliptical galaxies. II. Visual-near IR colours as population indices. *Astronomy & Astrophysics*, **429**, 819.

12 Montes, D., Martin, E.L., Fernandez-Figueroa, M.J., Cornide, M. and de Castro, E. (1997) Library of high and mid-resolution spectra in the CA_{II} H & K, H_α, H_β, NA_I D_1, D_2 and He_I D_3 line regions of F, G, K and M field stars. *Astronomy & Astrophysics Supplement*, **123**, 473.

13 Rose, J.A., Arimoto, N., Caldwell, N., Schiavon, R.P., Vazdekis, A. and Yamada, Y. (2005) Radial age and metal abundance gradients in the stellar content of M32. *Astronomical Journal*, **129**, 712.

14 Bessell, M.S., Castelli, F. and Plez, B. (1998) Model atmospheres broad-band colors, bolometric corrections and temperature calibrations for O–M stars. *Astronomy & Astrophysics*, **333**, 231.

15 Girardi, L., Bertelli, G., Bressan, A., Chiosi, C., Groenewegen, M.A.T., Marigo, P., Salasnich, B. and Weiss, A. (2002) Theoretical isochrones in several photometric systems. I. Johnson–Cousins–Glass, HST/WFPC2, HST/NICMOS, Washington and ESO imaging survey filter sets. *Astronomy & Astrophysics*, **391**, 195.

16 Castelli, F. and Kurucz, R.L. (2004) *New Grids of ATLAS9 Model Atmospheres*, (eds

Old Stellar Populations, First Edition. S. Cassisi and M. Salaris.
© 2013 WILEY-VCH Verlag GmbH & Co. KGaA. Published 2013 by WILEY-VCH Verlag GmbH & Co. KGaA.

N. Piskunov, W.W. Weiss and D.F. Gray), Proceedings of IAU Symposium 210, Astronomical Society of the Pacific, poster A20.

17 Salpeter, E.E. (1955) The luminosity function and stellar evolution. *Astrophysical Journal*, **121**, 161.

18 Cox, J.P. and Giuli, R.T. (2004) Principles of Stellar Structure, 2nd edn (expanded), (eds H. Weiss and T. Ritter), Cambridge Scientific Publishers.

19 Kippenhahn, R. and Weigert, A. (1990) *Stellar Structure and Evolution*, Springer.

20 Salaris, M. and Cassisi, S. (2005) *Evolution of Stars and Stellar Populations*, John Wiley & Sons, Ltd, Chichester.

21 Böhm-Vitense, E. (1958) Über die Wasserstoffkonvektionszone in Sternen verschiedener Effektivtemperaturen und Leuchtkräfte. *Zeitschrift für Astrophysik*, **46**, 108.

22 Böhm-Vitense, E. (1981) The effective temperature scale. *Annual Review of Astronomy and Astrophysics*, **19**, 295.

23 Grossman, S.A. and Taam, R.E. (1996) Double-diffusive mixing-length theory, semiconvection and massive star evolution. *Monthly Notices of the Royal Astronomical Society*, **283**, 1165.

24 Kato, S. (1966) Overstable convection in a medium stratified in mean molecular weight. *Publication of the Astronomical Society of Japan*, **18**, 374.

25 McDougall, J. and Stoner, E.C. (1938) The computation of Fermi–Dirac functions. *Philosophical Transactions of the Royal Society*, **237**, 67.

26 Saumon, D., Chabrier, G. and Van Horn, H.M. (1995) An equation of state for low-mass stars and giant planets. *Astrophysical Journal Supplement Series*, **99**, 713.

27 Clayton, D. (1968) *Principles of Stellar Evolution and Nucleosynthesis*, Pergamon Press, New York.

28 Kovetz, A., Lamb, D.Q. and Van Horn, H.M. (1972) Exchange contribution to the thermodynamic potential of a partially degenerate semirelativistic electron gas. *Astrophysical Journal*, **174**, 109.

29 Copeland, H., Jensen, J.O. and Jørgensen, H.E. (1970) Homogeneous models for population i and population ii compositions. *Astronomy & Astrophysics*, **5**, 12.

30 Vardya, M.S. (1961) Physical atmospheric parameters for late-type stars. *Astrophysical Journal*, **133**, 107.

31 Jiang, S.Y. and Huang, R.Q. (1997) The effect of turbulent pressure on the red giants and AGB stars. I. On the internal structure and evolution. *Astronomy & Astrophysics*, **317**, 114.

32 Arnett, D. (1996) *Supernovae and Nucleosynthesis*, Princeton University Press, Princeton.

33 von Weizsäcker, C.F. (1937) Physikalische Zeitschrift, **38**, 176.

34 Bethe, H.A. and Critchfield, C.L. (1938) Physical Review, **54**, 248.

35 Rolf, C.E. and Rodney, W.S. (1988) *Cauldrons in the Cosmos*, The University of Chicago Press, Chicago and London.

36 Formicola, A., Imbriani, G., Costantini, H. *et al.* (2004) Astrophysical S-factor of $^{14}N(p, \gamma)^{15}O$, *Physics Letters B*, **591**, 61.

37 Runkle, R.C., Champagne, A.E., Angulo, C., Fox, C., Iliadis, C., Longland, R. and Pollanen, J. (2005) Direct measurement of the $^{14}N(p, \gamma)^{15}O$ S Factor. *Physical Review Letters*, **94**, 2503.

38 Sugimoto, D. (1971) Mixing between stellar envelope and core in advanced phases of evolution. III Stellar core of initial mass $1.5 M_\odot$, Progress of Theoretical Physics, **45**, 761.

39 Arnould, M., Goriely, S. and Jorissen, A. (1999) Non-explosive hydrogen and helium burnings: abundance predictions from the NACRE reaction rate compilation. *Astronomy & Astrophysics*, **347**, 572.

40 Kunz, R., Fey, M., Jaeger, M., Mayer, A., Hammer, J.W., Staudt, G., Harissopulos, S. and Paradellis, T. (2002) Astrophysical reaction rate of $^{12}C(\alpha, \gamma)^{16}O$, *Astrophysical Journal*, **567**, 643.

41 Rogers, F.J. and Iglesias, C.A. (1992) Rosseland mean opacities for variable compositions. *Astrophysical Journal*, **401**, 361.

42 Salaris, M., Chieffi, A. and Straniero, O. (1993) The alpha-enhanced isochrones and their impact on the FITS to the

Galactic globular cluster system. *Astrophysical Journal*, **414**, 580.

43 Ferguson, J.W., Alexander, D.R., Allard, F., Barman, T., Bodnarik, J.G., Hauschildt, P.H., Heffner-Wong, A. and Tamanai, A. (2005) *Astrophysical Journal*, **623**, 585.

44 Marigo, P. (2002) Asymptotic Giant Branch evolution at varying surface C/O ratio: effects of changes in molecular opacities. *Astronomy & Astrophysics*, **387**, 507.

45 Lederer, M.T. and Aringer, B. (2009) Low temperature Rosseland opacities with varied abundances of carbon and nitrogen. *Astronomy & Astrophysics*, **494**, 403.

46 Weiss, A. and Ferguson, J.W. (2009) New asymptotic giant branch models for a range of metallicities. *Astronomy & Astrophysics*, **508**, 1343.

47 Borysow, A., Jørgensen, U.G. and Zheng, C. (1997) Model atmospheres of cool, low-metallicity stars: the importance of collision-induced absorption. *Astronomy & Astrophysics*, **324**, 185.

48 Lenzuni, P., Chernoff, D.F. and Salpeter, E.E. (1991) Rosseland and Planck mean opacities of a zero-metallicity gas. *Astrophysical Journal Supplement Series*, **76**, 759.

49 Baiko, D.A. and Yakovlev, D.G. (1995) Thermal and electrical conductivities of Coulomb crystals in neutron stars and white dwarfs. *Astronomy Letters*, **21**, 702.

50 Hubbard, W.B. and Lampe, M. (1969) Thermal conduction by electrons in stellar matter. *Astrophysical Journal Supplement Series*, **18**, 297.

51 Itoh, N., Hayashi, H. and Kohyama, Y. (1993) Electrical and thermal conductivities of dense matter in the crystalline lattice phase. III. Inclusion of lower densities. *Astrophysical Journal*, **418**, 405.

52 Cassisi, S., Potekhin, A.Y., Pietrinferni, A. and Salaris, M. (2007) Updated electron-conduction opacities: The impact on low-mass stellar models. *Astrophysical Journal*, **661**, 1094.

53 Potekhin, A.Y. (1999) Electron conduction in magnetized neutron star envelopes. *Astronomy & Astrophysics*, **351**, 787.

54 Potekhin A.Y., Baiko D.A., Haensel P. and Yakovlev, D.G. (1999) Transport properties of degenerate electrons in neutron star envelopes and white dwarf cores. *Astronomy & Astrophysics*, **346**, 345.

55 Ziman, J.M. (1960) *Electrons and Phonons*, Oxford University Press, Oxford.

56 Mihalas, D. (1967) Methods in Computational Physics, **7**, 1.

57 Mihalas, D. (1978) *Stellar Atmospheres*, 2nd edn, W.H. Freeman and Co., San Francisco, 650p.

58 Krishna Swamy, K.S. (1966) Profiles of strong lines in K-dwarfs. *Astrophysical Journal*, **145**, 174.

59 Van den Berg, D.A., Edvardsson, B., Eriksson, K. and Gustafsson, B. (2008) On the use of blanketed atmospheres as boundary conditions for stellar evolutionary models. *Astrophysical Journal*, **675**, 746.

60 Morel, P., van't Veer, C., Provost, J., Berthomieu, G., Castelli, F., Cayrel, R., Goupil, M.J. and Lebreton, Y. (1994) Incorporating the atmosphere in stellar structure models: The solar case. *Astronomy & Astrophysics*, **286**, 91.

61 Salaris, M. and Cassisi, S. (1996) New molecular opacities and effective temperature of RGB stellar models. *Astronomy & Astrophysics*, **305**, 858.

62 Gustafsson, B., Edvardsson, B., Eriksson, K., Mizuno-Wiedner, M., Jørgensen, U.G. and Plez, B. (2003) *Stellar Atmosphere Modelling*, (eds I. Hubeny, D. Mihalas and K. Werner), ASP Conf. Ser. 288, ASP, San Francisco, p. 331.

63 Reimers, D. (1975) *Problems in Stellar Atmospheres and Envelopes*, (eds B. Baschek, W.H. Kegel and G. Traving), Springer, Berlin, p. 229.

64 Catelan, M. (2009) Horizontal branch stars: The interplay between observations and theory and insights into the formation of the galaxy. *Astrophysics & Space Science*, **320**, 261.

65 Origlia, L., Ferraro, F.R., Fusi Pecci, F. and Rood, R.T. (2002) ISOCAM observations of galactic globular clusters: Mass loss along the red giant branch. *Astrophysical Journal*, **571**, 458.

66 Origlia, L., Rood, T.R., Fabbri, S., Ferraro, F.R., Fusi Pecci, F. and Rich, M. (2007) The first empirical mass-loss law for population II giants. *Astrophysical Journal*, **667**, L85.

67 Willson, L.A. and Bowen, G.H. (1984) Effects of pulsation and mass loss on stellar evolution. *Nature*, **312**, 429.

68 Koopmann, R.A., Lee, Y.-W., Demarque, P. and Howard, J.M. (1994) Mass loss during the RR Lyrae phase of the horizontal branch: Mass dispersion on the horizontal branch and RR Lyrae period changes. *Astrophysical Journal*, **423**, 380.

69 Unglaub, K. and Bues, I. (2001) The influence of diffusion and mass loss on the chemical composition of subdwarf B stars. *Astronomy & Astrophysics*, **374**, 570.

70 Vink, J.S. and Cassisi, S. (2002) Hot horizontal branch stars: Predictions for mass loss: winds, rotation and the low gravity problem. *Astronomy & Astrophysics*, **392**, 553.

71 Vink, J.S., de Koter, A. and Lamers, H.J.G.L.M. (2000) New theoretical mass-loss rates of O and B stars. *Astronomy & Astrophysics*, **362**, 295.

72 Wachter, A., Schröder, K.-P., Winters, J., Arndt, T. and Sedlmayr, E. (2002) An improved mass-loss description for dust-driven superwinds and tip-AGB evolution models. *Astronomy & Astrophysics*, **384**, 452.

73 Mattsson, L., Wahlin, R., Höfner, S. and Eriksson, K. (2008) Intense mass loss from C-rich AGB stars at low metallicity?. *Astronomy & Astrophysics*, **484**, L5.

74 van Loon, J., Cioni, M.-R., Zijlstra, A. and Loup, C. (2005) An empirical formula for the mass-loss rates of dust-enshrouded red supergiants and oxygen-rich Asymptotic Giant Branch stars. *Astronomy & Astrophysics*, **438**, 273.

75 Burgers, J.M. (1969) *Flow Equations for Composite Gases*, Academic Press, New York.

76 Bahcall, J.N. and Loeb, A. (1990) Element diffusion in stellar interiors. *Astrophysical Journal*, **360**, 267.

77 Cox, A.J., Guzik, J.A. and Kidman, P.B. (1989) Oscillations of solar models with internal element diffusion. *Astrophysical Journal*, **342**, 1187.

78 Thoul, A.A., Bahcall, J.N. and Loeb, A. (1994) Element diffusion in the solar interior. *Astrophysical Journal*, **421**, 828.

79 Aller, L.H. and Chapman, S. (1960) Diffusion in the Sun. *Astrophysical Journal*, **132**, 461.

80 Schlattl, H. and Salaris, M. (2003) Quantum corrections to microscopic diffusion constants. *Astronomy & Astrophysics*, **402**, 29.

81 Michaud, G., Charland, Y., Vauclair, S. and Vauclair, G. (1976) Diffusion in main-sequence stars – Radiation forces, time scales, anomalies. *Astrophysical Journal*, **210**, 447.

82 LeBlanc, F. and Michaud, G. (1995) Radiative accelerations on iron. *Astronomy & Astrophysics*, **303**, 166.

83 LeBlanc, F., Michaud, G. and Richer, J. (2000) Opacity sampling in radiative acceleration calculations. *Astrophysical Journal*, **538**, 876.

84 Baglin, A., Auvergne, M., Catala, C. and Michel, E. (2001) The CoRoT Team 1998, ESA Special Publications, **464**, 395.

85 Basri, G., Borucki, W.J. and Koch, D. (2005) The Kepler mission: A wide-field transit search for terrestrial planets. *New Astronomy Reviews*, **49**, 478.

86 Ostriker, J.P. and Mark, J.W.-K. (1968) Rapidly rotating stars. I. The self-consistent-field method. *Astrophysical Journal*, **151**, 1075.

87 Monaghan, J.J. and Roxburgh, I.W. (1965) The structure of rapidly rotating polytropes. *Monthly Notices of the Royal Astronomical Society*, **131**, 13.

88 Deupree, R.G. (1990) Stellar evolution with arbitrary rotation laws. I – Mathematical techniques and test cases. *Astrophysical Journal*, **357**, 175.

89 Maeder, A. (2009) *Physics, Formation and Evolution of Rotating Stars*, Springer, Berlin, Heidelberg.

90 Zahn, J.-P. (1992) Circulation and turbulence in rotating stars. *Astronomy & Astrophysics*, **265**, 115.

91 Endal, A.S. and Sofia, S. (1976) The evolution of rotating stars. I – Method and exploratory calculations for a 7-solar-

mass star. *Astrophysical Journal*, **210**, 184.

92 Kippenhahn, R. and Thomas, H.-C. (1970) *Stellar Rotation*, (ed. A. Slattebak), IAU Colloquium 4, Reidel, Dordrecht, p. 20.

93 Meynet, G. and Maeder, A. (1997) Stellar evolution with rotation. I. The computational method and the inhibiting effect of the μ-gradient. *Astronomy & Astrophysics*, **321**, 465.

94 Talon, S., Zahn, J.-P., Maeder, A. and Meynet, G. (1997) Rotational mixing in early-type stars: the main-sequence evolution of a $9 M_\odot$ star. *Astronomy & Astrophysics*, **322**, 209.

95 Herwig, F., Bloecker, T., Schoenberner, D. and El Eid, M. (1997) Stellar evolution of low and intermediate-mass stars. IV. Hydrodynamically-based overshoot and nucleosynthesis in AGB stars. *Astronomy & Astrophysics*, **324**, L81.

96 Langer, N., Fricke, K.J. and Sugimoto, D. (1983) Semiconvective diffusion and energy transport. *Astronomy & Astrophysics*, **126**, 207.

97 Brun, A.S. and Palacios, A. (2009) Numerical simulations of a rotating red giant star. I. Three-dimensional models of turbulent convection and associated mean flows. *Astrophysical Journal*, **702**, 1078.

98 Brown, D. (2007) Impact of rotation on the evolution of low-mass stars, Ph.D. thesis, John Moores University, Liverpool.

99 Bretherton, F.P. (1969) Quarterly Journal of the Royal Meteorological Society, **95**, 213.

100 Talon, S., Kumar, P. and Zahn, J.-P. (2002) Angular momentum extraction by gravity waves in the Sun. *Astrophysical Journal*, **574**, L175.

101 Zahn, J.-P., Talon, S. and Matias, J. (1997) Angular momentum transport by internal waves in the solar interior. *Astronomy & Astrophysics*, **322**, 320.

102 Press, W.H. (1981) Radiative and other effects from internal waves in solar and stellar interiors. *Astrophysical Journal*, **245**, 286.

103 Ando, H. (1986) Wave-rotation interaction and episodic mass-loss in Be stars. *Astronomy & Astrophysics*, **163**, 97.

104 Rogers, T.M. and Glatzmeier, G.A. (2005) Gravity waves in the Sun. *Monthly Notices of the Royal Astronomical Society*, **364**, 1135.

105 Talon, S. and Charbonnel, C. (2005) Hydrodynamical stellar models including rotation, internal gravity waves, and atomic diffusion. I. Formalism and tests on pop I dwarfs. *Astronomy & Astrophysics*, **440**, 981.

106 Talon, S. and Charbonnel, C. (2008) Angular momentum transport by internal gravity waves. IV. Wave generation by surface convection zone, from the pre-main sequence to the early-AGB in intermediate mass stars. *Astronomy & Astrophysics*, **482**, 597.

107 Parker, E.N. (1955) Hydromagnetic dynamo models. *Astrophysical Journal*, **122**, 293.

108 Hall, J.C. (2008) Stellar chromospheric activity. *Living Reviews in Solar Physics*, **5**, 2.

109 Kiraga, M. and Stepien, K. (2007) Age-rotation-activity relations for M dwarf stars. *Acta Astronomica*, **57**, 149.

110 Küker, M., Rüdiger, G. and Schultz, M. (2001) Circulation-dominated solar shell dynamo models with positive alpha-effect. *Astronomy & Astrophysics*, **374**, 301.

111 Chabrier, G. and Küker, M. (2006) Large-scale α^2-dynamo in low-mass stars and brown dwarfs. *Astronomy & Astrophysics*, **446**, 1027.

112 Braithwaite, J. and Spruit, H.C. (2004) A fossil origin for the magnetic field in A stars and white dwarfs. *Nature*, **431**, 819.

113 Denissenkov, P.A. and Pinsonneault, M. (2007) A revised prescription for the Tayler–Spruit dynamo: Magnetic angular momentum transport in stars. *Astrophysical Journal*, **655**, 1157.

114 Spruit, H.C. (1999) Differential rotation and magnetic fields in stellar interiors. *Astronomy & Astrophysics*, **349**, 189.

115 Eggenberger, P., Maeder, A. and Meynet, G. (2005) Stellar evolution with rotation and magnetic fields. IV. The solar rotation profile. *Astronomy & Astrophysics*, **440**, L9.

116 Couvidat, S., García, R.A., Turck-Chiéze, S. *et al.* (2003) The rotation of the deep solar layers. *Astrophysical Journal*, **597**, L77.

117 Gratton, R.G., Sneden, C., Carretta, E. and Bragaglia, A. (2000) Mixing along the red giant branch in metal-poor field stars. *Astronomy & Astrophysics*, **354**, 169.

118 Ulrich, R.K. (1972) Thermohaline convection in stellar interiors. *Astrophysical Journal*, **172**, 165.

119 Charbonnel, C. and Zahn, J.-P. (2007) Thermohaline mixing: a physical mechanism governing the photospheric composition of low-mass giants. *Astronomy & Astrophysics*, **467**, L15.

120 Eggleton, P.P., Dearborn, D.S.P. and Lattanzio, J.C. (2006) Deep mixing of ^{3}He: Reconciling Big Bang and stellar nucleosynthesis. *Science*, **314**, 1580.

121 Stern, M.E. (1960) The salt-fountain and thermohaline convection. *Tellus*, **12**, 172.

122 Kippenhahn, R., Ruschenplatt, G. and Thomas, H.C. (1980) The time scale of thermohaline mixing in stars. *Astronomy & Astrophysics*, **91**, 175.

123 Charbonnel, C. and Lagarde, N. (2010) Thermohaline instability and rotation-induced mixing. I. Low- and intermediate-mass solar metallicity stars up to the end of the AGB. *Astronomy & Astrophysics*, **522**, 10.

124 Théado, S. and Vauclair, S. (2012) Metal-rich accretion and thermohaline instabilities in exoplanet-host stars: Consequences on the light elements abundances. *Astrophysical Journal*, **744**, 123.

125 Traxler, A., Garaud, P. and Stellmach, S. (2011) Numerically determined transport laws for fingering ("thermohaline") convection in astrophysics. *Astrophysical Journal*, **728**, L29.

126 Cantiello, M. and Langer, N. (2010) Thermohaline mixing in evolved low-mass stars. *Astronomy & Astrophysics*, **521**, A9.

127 Kumar, S.S. (1963) Models for stars of very low mass. *Astrophysical Journal*, **137**, 1121.

128 Allard, F. and Hauschildt, P.H. (1995) Model atmospheres for M (sub)dwarf stars. 1: The base model grid. *Astrophysical Journal*, **445**, 433.

129 Allard, F., Hauschildt, P.H., Alexander, D.R. and Staarfield, S. (1997) Model atmospheres of very low mass stars and brown dwarfs. *Annual Review Astronomy & Astrophysics*, **35**, 137.

130 Auman J. (1969) *Low Luminosity Stars*, Gordon and Breach, New York.

131 Baraffe, I., Chabrier, G., Allard, F. and Hauschildt, P.H. (1995) New evolutionary tracks for very low mass stars. *Astrophysical Journal*, **446**, L35.

132 Chabrier, G. and Baraffe, I. (2000) Theory of low-mass stars and substellar objects. *Annual Review Astronomy & Astrophysics*, **38**, 337.

133 Brocato, E., Cassisi, S. and Castellani, V. (1998) Stellar models for very low-mass main-sequence stars – The role of model atmospheres. *Monthly Notices of the Royal Astronomical Society*, **295**, 711.

134 Chabrier, G. and Baraffe, I. (1997) Structure and evolution of low-mass stars. *Astronomy & Astrophysics*, **327**, 1039.

135 Saumon, D., Bergeron, P., Lunine, J.I., Hubbard, W.B. and Burrows, A. (1994) Cool zero-metallicity stellar atmospheres. *Astrophysical Journal*, **424**, 333.

136 Zoccali, M., Cassisi, S., Frogel, J.A., Gould, A., Ortolani, S., Renzini, A., Rich, R.M. and Stephens, A.W. (2000) The initial mass function of the galactic bulge down to $\sim 0.15 M_\odot$. *Astrophysical Journal*, **530**, 418.

137 Henry, T.J. and McCarthy Jr., D.W. (1993) The mass-luminosity relation for stars of mass 1.0 to 0.08 solar mass. *Astronomical Journal*, **106**, 773.

138 Kroupa, P. and Tout, C.A. (1997) The theoretical mass-magnitude relation of low mass stars and its metallicity dependence. *Monthly Notices of the Royal Astronomical Society*, **287**, 402.

139 Ribas, I., Morales, J.C., Jordi, C., Baraffe, I., Chabrier, G. and Gallardo, J. (2008) Fundamental properties of low-mass stars. *Memoire della Societa Astronomica Italiana*, **79**, 562.

140 Cassisi, S., Castellani, V., Ciarcelluti, P., Piotto, G. and Zoccali, M. (2000) Galactic globular clusters as a test for very-low-

mass stellar models. *Monthly Notices of the Royal Astronomical Society*, **315**, 679.

141 Ségransan, D., Kervella, P., Forveille, T. and Queloz, D. (2003) First radius measurements of very low mass stars with the VLTI. *Astronomy & Astrophysics*, **397**, L5.

142 Roxburgh, I.W. (1992) Limits on convective penetration from stellar cores. *Astronomy & Astrophysics*, **266**, 291.

143 Woo, J-.H. and Demarque, P. (2001) Empirical constraints on convective core overshoot. *Astronomical Journal*, **122**, 1602.

144 Borucki, W., Koch, D., Batalha, N., Caldwell, D., Christensen-Dalsgaard, J., Cochran, W.D. *et al.* (2009) *KEPLER: Search for Earth-Size Planets in the Habitable Zone*, Proceedings of the International Astronomical Union, IAU Symposium, vol. 253, Cambridge University Press, p. 289.

145 Noerdlinger, P.D. (1977) Diffusion of helium in the Sun. *Astronomy & Astrophysics*, **57**, 407.

146 Bahcall, J.N., Pinsonneault, M.H. and Wasserburg, G.J. (1995) Solar models with helium and heavy-element diffusion. *Reviews Modern Physics*, **67**, 781.

147 Grevesse, N. and Sauval, A.J. (1998) Standard solar composition. *Space Science Reviews*, **85**, 161.

148 Pietrinferni, A., Cassisi, S., Salaris, M. and Castelli, F. (2004) A large stellar evolution database for population synthesis studies. I. Scaled solar models and isochrones. *Astrophysical Journal*, **612**, 168.

149 Gough, D.O. and Thompson, M.J. (1991) The inversion problem, in *Solar Interior and Atmosphere*, University of Arizona Press, Tucson, p. 519.

150 Degl'Innocenti, S., Dziembowski, W.A., Fiorentini, G. and Ricci, B. (1997) Helioseismology and standard solar models. *Astroparticle Physics*, **7**, 77.

151 Rogers, F.J., Swenson, F.J. and Iglesias, C.A. (1996) OPAL equation-of-state tables for astrophysical application. *Astrophysical Journal*, **456**, 902.

152 Brun, A.S., Turck-Chiéze, S. and Morel, P. (1998) Standard solar models in the light of new helioseismic constraints. I. The solar core. *Astrophysical Journal*, **506**, 913.

153 Bahcall, J.N., Cleveland, B.T., Davis Jr., R. and Rowley, J.K. (1985) Chlorine and gallium solar neutrino experiments. *Astrophysical Journal*, **292**, L79.

154 Gavrin, V.N., Abdurashitov, J.N., Bowles, T.J. *et al.* (1999) Solar neutrino results from SAGE. *Nuclear Physics B*, **77**, 20.

155 Hampel, W., Heusser, G., Kiko, J. *et al.* (1998) Final results of the ^{51}Cr neutrino source experiments in GALLEX. *Physics Letters B*, **420**, 114.

156 Castellani, V., Degl'innocenti, S., Fiorentini, G., Lissia, M. and Ricci, B. (1994) Neutrinos from the Sun: Experimental results confronted with solar models. *Physical Review D*, **50**, 4749.

157 Fogli, G.L., Lisi, E., Montanino, D. and Palazzo, A. (2000) Three-flavor MSW solutions of the solar neutrino problem. *Physical Review D*, **62**, 3002.

158 McDonald, A.B., Ahmad, Q.R., Allen, R.C. *et al.* (2002) *Direct Evidence for Neutrino Flavor Transformation from Neutral-Current Interactions in SNO*, (eds V. Elias, R. Epp and R.C. Myers), Theoretical Physics, MRST 2002, American Institute of Physics Conference Series, vol. 646, p. 43.

159 Asplund, M., Grevesse, N., Sauval, A.J. and Scott, P. (2009) The chemical composition of the Sun. *Annual Review Astronomy & Astrophysics*, **47**, 481.

160 Baumann, P., Ramírez, I., Meléndez, J., Asplund, M. and Lind, K. (2010) Lithium depletion in solar-like stars: no planet connection. *Astronomy & Astrophysics*, **519**, 87.

161 Charbonnel, C. and Talon, S. (1995) Influence of gravity waves on the internal rotation and Li abundance of solar-type stars. *Science*, **309**, 2189.

162 Asplund, M., Grevesse, N. and Sauval, A.J. (2005) Cosmic abundances as records of stellar evolution and nucleosynthesis, in *ASP Conference Series*, **336**, Astronomical Society of the Pacific, 25.

163 Antia, H.M. and Basu, S. (2011) Are recent solar heavy element abundances

consistent with helioseismology? *Journal Physics Conference Series*, **271**, 2034.

164 Basu, S. and Antia, H.M. (2008) Helioseismology and solar abundances. *Physics Reports*, **457**, 217.

165 Serenelli, A. (2010) New results on standard solar models. *Astrophysics and Space Science*, **328**, 13.

166 Caffau, E., Ludwig, H.-G., Steffen, M., Freytag, B. and Bonifacio, P. (2011) Solar chemical abundances determined with a CO5BOLD 3D model atmosphere. *Solar Physics*, **268**, 255.

167 Grevesse, N., Asplund, M., Sauval, A.J. and Scott, P. (2010) The chemical composition of the Sun. *Astrophysics and Space Science*, **328**, 179.

168 Lodders, K., Palme, H., and Gail, H.P. (2009) Landolt–Börnstein – Group VI Astronomy and Astrophysics Numerical Data and Functional Relationships in Science and Technology, Solar System, vol. 4B, (ed. J.E. Trümper), Springer, p. 44.

169 Sandage, A. (1953) The color-magnitude diagram for the globular cluster M3. *Astronomical Journal*, **58**, 61.

170 Bailyn, C.D. (1995) Blue stragglers and other stellar anomalies: Implications for the dynamics of globular clusters. *Annual Review Astronomy & Astrophysics*, **33**, 133.

171 Ferraro, F.R., Fusi Pecci, F., Cacciari, C., Corsi, C., Buonanno, R., Fahlman, G.G. and Richer, H.B. (1993) Blue stragglers in the galactic globular clusters M3: Evidence for two populations. *Astronomical Journal*, **106**, 2324.

172 Shara, M.M., Saffer, R.A. and Livio, M. (1997) The first direct measurement of the mass of a blue straggler in the core of a globular cluster: BSS 19 in 47 Tucanae. *Astrophysical Journal*, **489**, L59.

173 Leonard, P.J.T. (1989) Stellar collisions in globular clusters and the blue straggler problem. *Astronomical Journal*, **98**, 217.

174 Preston, G.W. and Sneden, C. (2000) What are these blue metal-poor stars? *Astronomical Journal*, **120**, 1014.

175 Benz, W. and Hills, J.G. (1987) Three-dimensional hydrodynamical simulations of stellar collisions. I – Equal-mass

main-sequence stars. *Astrophysical Journal*, **323**, 614.

176 Davies, M.B. and Hansen, B.M.S. (1998) Neutron star retention and millisecond pulsar production in globular clusters. *Monthly Notices of the Royal Astronomical Society*, **301**, 15.

177 Benz, W. and Hills, J.G. (1992) Three-dimensional hydrodynamical simulations of colliding stars. III – Collisions and tidal captures of unequal-mass main-sequence stars. *Astrophysical Journal*, **389**, 546.

178 Lombardi Jr., J.C., Warren, J.S., Rasio, F.A., Sills, A. and Warren, A.R. (2002) Stellar collisions and the interior structure of blue stragglers. *Astrophysical Journal*, **568**, 939.

179 Ouellette, J.A. and Pritchet, C.J. (1998) The evolution of blue stragglers formed via stellar collisions. *Astrophysical Journal*, **115**, 2539.

180 Sandquist, E.L., Bolte, M. and Hernquist, L. (1997) Composition mixing during blue straggler formation and evolution. *Astrophysical Journal*, **477**, 335.

181 Leonard, P.J.T. and Livio, M. (1995) The rotational rates of blue stragglers produced by physical stellar collisions. *Astronomical Journal*, **447**, L121.

182 De Marco, O., Shara, M.M., Zurek, D., Ouellette, J.A., Lanz, T., Saffer, R.A. and Sepinsky, J.F. (2005) A spectroscopic analysis of blue stragglers, horizontal branch stars and turnoff stars in four globular clusters. *Astrophysical Journal*, **632**, 894.

183 Cameron, A.C., Campbell, C.G. and Quaintrell, H. (1995) Rotational evolution of magnetic T Tauri stars with accretion discs. II. Approach to the main sequence. *Astronomy & Astrophysics*, **298**, 133.

184 Fusi Pecci, F., Ferraro, F.R., Corsi, C.E., Cacciari, C. and Buonanno, R. (1992) On the blue stragglers and horizontal branch morphology in Galactic globular clusters – Some speculations and a new working scenario. *Astrophysical Journal*, **104**, 1831.

185 Ferraro, F.R., Sills, A., Rood, R.T., Paltrinieri, B. and Buonanno, R. (2003) Blue straggler stars: A direct comparison of

star counts and population ratios in six galactic globular clusters. *Astrophysical Journal*, **588**, 464.

186 Piotto, G., De Angeli, F., King, I.R., Djorgovski, S.G., Bono, G., Cassisi, S., Meylan, G. *et al.* (2004) Relative frequencies of blue stragglers in galactic globular clusters: Constraints for the formation mechanisms. *Astrophysical Journal*, **604**, L109.

187 Leigh, N., Sills, A. and Knigge, C. (2007) Where the blue stragglers roam: Searching for a link between formation and environment. *Astrophysical Journal*, **661**, 210.

188 Sigurdsson, S., Davies, M.B. and Bolte, M. (1994) Modeling the radial distribution of blue stragglers in M3. *Astrophysical Journal*, **431**, L115.

189 Mapelli, M., Sigurdsson, S., Ferraro, F.R., Colpi, M., Possenti, A. and Lanzoni, B. (2006) The radial distribution of blue straggler stars and the nature of their progenitors. *Monthly Notices of the Royal Astronomical Society*, **373**, 361.

190 Dalessandro, E., Lanzoni, B., Ferraro, F.R., Rood, R.T., Milone, A., Piotto, G. and Valenti, E. (2008) Blue straggler stars in the unusual globular cluster NGC 6388. *Astrophysical Journal*, **677**, 1069.

191 Gilliland, R.L., Bono, G., Edmonds, P.D., Caputo, F., Cassisi, S., Petro, L.D., Saha, A. and Shara, M.M. (1998) Oscillating blue stragglers in the core of 47 Tucanae. *Astrophysical Journal*, **507**, 818.

192 Santolamazza, P., Marconi, M., Bono, G., Caputo, F., Cassisi, S. and Gilliland, R.L. (2001) Linear nonadiabatic properties of SX phoenicis variables. *Astrophysical Journal*, **554**, 1124.

193 Petersen, J.O. and Christensen-Dalsgaard, J. (1996) Pulsation models of δ Scuti variables. I. The high-amplitude double-mode stars. *Astronomy & Astrophysics*, **312**, 463.

194 Sweigart, A.V., Greggio, L. and Renzini, A. (1990) The development of the red giant branch. II – Astrophysical properties. *Astrophysical Journal*, **364**, 527.

195 Sugimoto, D. and Fujimoto, M.Y. (2000) Why stars become red giants. *Astrophysical Journal*, **538**, 837.

196 Stancliffe, R.J., Chieffi, A., Lattanzio, J.C. and Church, R.P. (2009) Why do low-mass stars become red giants? *Publications of the Astronomical Society of Australia*, **26**, 203.

197 Renzini, A. and Buzzoni, A. (1986) Global properties of stellar populations and the spectral evolution of galaxies, in *Spectral Evolution of Galaxies*, Proceedings of the Fourth Workshop, Erice, Italy, 1985, Dordrecht, D. Reidel Publishing Co., p. 195.

198 Despain, K.H. (1981) low-mass evolution: Zero-age main sequence to asymptotic-giant branch. *Astrophysical Journal*, **251**, 639

199 Mengel, J.G., and Sweigart, A.V. (1981) The helium-core flash in globular cluster stars, in *Astrophysical Parameters for Globular Clusters*, IAU Colloquium 68, (eds A.G.D. Philip and D.S. Hayes), Davis Press, Schenectady, p. 277

200 Brown, T.M., Sweigart, A.V., Lanz, T., Landsman, W.B. and Hubeny, I. (2001) Flash mixing on the white dwarf cooling curve: Understanding hot horizontal branch anomalies in NGC 2808. *Astrophysical Journal*, **562**, 368.

201 Cassisi, S., Schlattl, H., Salaris, M. and Weiss, A. (2003) First full evolutionary computation of the helium flash-induced mixing in population II stars. *Astrophysical Journal*, **582**, L43.

202 Schlattl, H., Cassisi, S., Salaris, M. and Weiss, A. (2001) On the helium flash in low-mass population III red giant stars. *Astrophysical Journal*, **559**, 1082.

203 Bildsten, L., Paxton, B., Moore, K. and Macias, P.J. (2011) Acoustic signatures of the helium core flash. *Astrophysical Journal*, **744**, L6.

204 Mocák, M., Müller, E., Weiss, A. and Kifonidis, K. (2008) The core helium flash revisited. I. One and two-dimensional hydrodynamic simulations. *Astronomy & Astrophysics*, **490**, 265.

205 Mocák, M., Müller, E., Weiss, A. and Kifonidis, K. (2009) The core helium flash revisited. II. Two and three-dimensional

hydrodynamic simulations. *Astronomy & Astrophysics*, **501**, 659.

206 Mocák, M., Campbell, S.W., Müller, E. and Kifonidis, K. (2010) The core helium flash revisited. III. From population I to population III stars. *Astronomy & Astrophysics*, **520**, 114.

207 Cassisi, S., Salaris, M., Castelli, F. and Pietrinferni, A. (2004) Color transformations and bolometric corrections for galactic halo stars: α-enhanced versus scaled-solar results. *Astrophysical Journal*, **616**, 498.

208 Vandenberg, D.A., Bergbusch, P.A., Dotter, A., Ferguson, J.W., Michaud, G., Richer, J. and Proffitt, C. (2012) Stellar models with enhanced abundances of C, N, O, Ne, Na, Mg, Si, S, Ca and Ti, in turn, at constant helium and iron abundances, *Astrophysical Journal*, **755**, 15.

209 Salaris, M. and Cassisi, S. (2008) Stellar models with the ML2 theory of convection. *Astronomy & Astrophysics*, **487**, 1075.

210 Tassoul, M., Fontaine, G. and Winget, D.E. (1990) Evolutionary models for pulsation studies of white dwarfs. *Astrophysical Journal Supplement Series*, **72**, 335.

211 Bergeron, P., Wesemael, F., Lamontagne, R. *et al.* (1995) Optical and ultraviolet analyses of ZZ ceti stars and study of the atmospheric convective efficiency in DA white dwarfs. *Astrophysical Journal*, **449**, 258.

212 Thomas, H.C. (1967) Ph.D. thesis, L.M. University, München.

213 King, C.R., Da Costa, G.S. and Demarque, P. (1985) The luminosity function on the subgiant branch of 47 Tucanae A comparison of observation and theory. *Astrophysical Journal*, **299**, 674.

214 Cassisi, S. and Salaris, M. (1997) A critical investigation on the discrepancy between the observational and the theoretical red giant luminosity function bump. *Monthly Notices of the Royal Astronomical Society*, **285**, 593.

215 Di Cecco, A., Bono, G., Stetson, P.B. *et al.* (2010) On the ΔV_{HB}^{Bump} parameter in globular clusters. *Astrophysical Journal*, **712**, 527.

216 Fusi Pecci, F., Ferraro, F.R., Crocker, D.A., Rood, R.T. and Buonanno, R. (1990) The variation of the red giant luminosity function "Bump" with metallicity and the age of the globular clusters. *Astronomy & Astrophysics*, **238**, 95.

217 Monelli, M., Cassisi, S., Bernard, E.J., Hidalgo, S.L., Aparicio, A., Gallart, C. and Skillman, E.D. (2010) The ACS LCID project. IV. Detection of the red giant branch bump in isolated galaxies of the local group. *Astrophysical Journal*, **718**, 707.

218 Nataf, D.M., Udalski, A., Gould, A. and Pinsonneault, M.H. (2011) OGLE-III detection of the anomalous galactic bulge red giant branch bump: Evidence of enhanced helium enrichment. *Astrophysical Journal*, **730**, 118.

219 Cassisi, S., Marín-Franch, A., Salaris, M., Aparicio, A., Monelli, M. and Pietrinferni, A. (2011) The magnitude difference between the main sequence turn off and the red giant branch bump in galactic globular clusters. *Astronomy & Astrophysics*, **527**, 59.

220 Richard, O., Michaud, G., Richer, J., Turcotte, S., Turck-Chiéze, S. and Van den Berg, D. (2002) Models of metal-poor stars with gravitational settling and radiative accelerations. I. Evolution and abundance anomalies. *Astrophysical Journal*, **568**, 979.

221 Van den Berg, D.A., Richard, O., Michaud, G. and Richer, J. (2002) Models of metal-poor stars with gravitational settling and radiative accelerations. II. The age of the oldest stars. *Astrophysical Journal*, **571**, 487.

222 Cassisi, S., Degl'Innocenti, S. and Salaris, M. (1997) The effect of diffusion on the red giant luminosity function bump. *Monthly Notices of the Royal Astronomical Society*, **290**, 515.

223 Michaud, G., Richer, J. and Richard, O. (2010) Atomic diffusion during red giant evolution. *Astronomy & Astrophysics*, **510**, A104.

224 Cassisi, S., Castellani, V., Degl'Innocenti, S. and Weiss, A. (1998) An updated theoretical scenario for globular cluster stars. *Astronomy & Astrophysics Supplement Series*, **129**, 267.

225 Demarque, P., Mengel, J.G. and Sweigart, A.V. (1973) Rotating solar models with low neutrino flux. *Astrophysical Journal*, **183**, 997.

226 Mengel, J.G. and Gross, P.G. (1976) Possible effects of internal rotation in low-mass stars. *Astrophysics and Space Science*, **41**, 407.

227 Brown, D., Salaris, M., Cassisi, S. and Pietrinferni, A. (2008) Impact of rotation on the evolution of low-mass stars. *Memorie della Societá Astronomica Italiana*, **79**, 579.

228 Deliyannis, C.P., Demarque, P. and Pinsonneault, M.H. (1989) The ages of globular cluster stars – Effects of rotation on pre-main-sequence, main-sequence and turnoff evolution. *Astrophysical Journal*, **347**, L73.

229 Van den Berg, D.A., Larson, A.M. and De Propris, R. (1989) The luminosity function of M30: Evidence for rapidly rotating cores in the cluster giants? *Publications of the Astronomical Society of the Pacific*, **110**, 744.

230 Castellani, V., Giannone, P. and Renzini, A. (1971) Overshooting of convective cores in helium-burning horizontal-branch stars. *Astrophysics and Space Science*, **10**, 340.

231 Castellani, V., Giannone, P. and Renzini, A. (1971) Induced semi-convection in helium-burning horizontal-branch stars. *Astrophysics and Space Science*, **10**, 355.

232 Sweigart, A.V. and Demarque, P. (1972) Effects of semiconvection on the Horizontal-Branch. *Astronomy & Astrophysics*, **20**, 445.

233 Dorman, B. and Rood, R.T. (1993) On partial mixing zones in horizontal-branch stellar cores. *Astrophysical Journal*, **409**, 387.

234 Michaud, G., Richer, J. and Richard, O. (2007) Horizontal branch evolution and atomic diffusion. *Astrophysical Journal*, **670**, 1178.

235 Ledoux, W.P. (1947) Stellar models with convection and with discontinuity of the mean molecular weight. *Astrophysical Journal*, **105**, 305.

236 Straniero, O., Domínguez, I., Imbriani, G. and Piersanti, L. (2003) The chemical composition of white dwarfs as a test of convective efficiency during core helium burning. *Astrophysical Journal*, **583**, 878.

237 Castellani, V., Chieffi, A., Pulone, L. and Tornambé, A. (1985) Helium-burning evolutionary phases in population II stars. I. Breathing pulses in horizontal branch stars. *Astrophysical Journal*, **296**, 204.

238 Caputo, F., Chieffi, A., Tornambé, A., Castellani, V. and Pulone, L. (1989) The "Red Giant Clock" as an indicator for the efficiency of central mixing in horizontal-branch stars. *Astrophysical Journal*, **340**, 241.

239 Rood, R.T. (1973) Metal-poor stars. V. Horizontal-branch morphology. *Astrophysical Journal*, **184**, 815.

240 Sandage, A. and Wallerstein, G. (1960) Color-magnitude diagram for disk globular cluster NGC 6356 compared with halo clusters. *Astrophysical Journal*, **131**, 598.

241 Faulkner, J. (1966) On the nature of the horizontal branch. *Astrophysical Journal*, **144**, 978.

242 van den Bergh, S. (1967) UBV photometry of globular clusters. *Astronomical Journal*, **72**, 70.

243 Sandage, A. and Wildey, R. (1967) The anomalous color-magnitude diagram of the remote globular cluster NGC 7006. *Astrophysical Journal*, **150**, 469.

244 Gratton, R.G., Carretta, E., Bragaglia, A., Lucatello, S. and D'Orazi, V. (2010) The second and third parameters of the horizontal branch in globular clusters. *Astronomy & Astrophysics*, **517**, 81.

245 Marín-Franch, A., Aparicio, A., Piotto, G. *et al.* (2009) The ACS survey of galactic globular clusters. VII. Relative ages. *Astrophysical Journal*, **694**, 1498.

246 Renzini, A. (1981) *Evolutionary Effects of Mass Loss in Low-Mass Stars*, Proceedings of the Meeting Effects of Mass Loss on Stellar Evolution, vol. 89, (eds C. Chiosi and R. Stalio), D. Reidel Publishing Co., Dordrecht, p. 319.

247 Castellani, M. and Castellani, V. (1993) Mass loss in globular cluster red giants: an evolutionary investigations. *Astrophysical Journal*, **407**, 649.

248 Castellani, V., Luridiana, V. and Romaniello, M. (1994) Red giant stragglers and the He white dwarfs in galactic globulars. *Astrophysical Journal*, **428**, 633.

249 D'Cruz, N.L., Dorman, B., Root, R.T. and O'Connell, R.W. (1996) The origin of extreme horizontal branch stars. *Astrophysical Journal*, **466**, 359.

250 Sweigart, A.V. (1997) *Helium Mixing in Globular Cluster Stars*, Proceeding of The Third Conference on Faint Blue Stars (eds A.G.D. Philip, J. Liebert, R. Saffer and D.S. Hayes), Davis Press, p. 3.

251 Miller Bertolami, M.M., Althaus, L.G., Unglaub, K. and Weiss, A. (2008) Modeling He-rich subdwarfs through the hot-flasher scenario. *Astronomy & Astrophysics*, **491**, 253.

252 D'Cruz, N.L., Dorman, O'Connell, R.W., B., Root, R.T. *et al.* (2000) Hubble space telescope observations of new horizontal-branch structures in the globular cluster ω Centauri. *Astrophysical Journal*, **530**, 352.

253 Moehler, S., Dreizler, S., Lanz, T., Bono, G., Sweigart, A.V., Calamida, A. and Nonino, M. (2011) The hot horizontal-branch stars in ω Centauri. *Astronomy & Astrophysics*, **526**, 136.

254 Brown, T.M., Lanz, T., Sweigart, A.V., Cracraft, M., Hubeny, I. and Landsman, W.B. (2012) Flash mixing on the white dwarf cooling curve: spectroscopic confirmation in NGC 2808. *Astrophysical Journal*, **748**, 85.

255 Heber, U. (2009) Hot subdwarf stars, *Annual Review Astronomy & Astrophysics*, **47**, 211.

256 Cox, J.P. (1980) *Theory of Stellar Pulsation*, Princeton University Press.

257 Gautschy, A. and Saio, H. (1996) Stellar pulsations across the HR diagram: Part 2. *Annual Review of Astronomy and Astrophysics*, **34**, 551.

258 Sandage, A. and Tammann, G.A. (2006) Absolute magnitude calibrations of population I and II cepheids and other pulsating variables in the instability strip of the Hertzsprung–Russell Diagram. *Annual Review Astronomy and Astrophysics*, **44**, 93.

259 Eddington, A.S. (1917) The pulsation theory of cepheid variables. *The Observatory*, **40**, 290.

260 Eddington, A.S. (1918) On the radiative equilibrium of the stars: A correction. *Monthly Notices of the Royal Astronomical Society*, **79**, 2.

261 Bailey, S.I. (1902) A discussion of variable stars in the cluster ω Centauri, *Annals of Harvard College Observatory, Cambridge, The Observatory*, **38**, 13.

262 Bailey, S.I. and Pickering, E.C. (1913) Variable stars in the cluster Messier 3, *Annals of Harvard College Observatory, Cambridge, The Observatory*, **78**, 1.

263 van Albada, T.S. and Baker, N. (1971) On the masses, luminosities and compositions of horizontal-branch stars. *Astrophysical Journal*, **169**, 311.

264 Bono, G., Caputo, F., Castellani, V. and Marconi, M. (1997) Nonlinear investigation of the pulsational properties of RR Lyrae variables. *Astronomy & Astrophysics Supplement Series*, **121**, 327.

265 Di Criscienzo, M., Marconi, M. and Caputo, F. (2004) RR Lyrae stars in galactic globular clusters. III. Pulsational predictions for metal content $Z = 0.0001$ to $Z = 0.006$. *Astrophysical Journal*, **612**, 1092.

266 De Santis, R. and Cassisi, S. (1999) A pulsational approach to the luminosity of horizontal branch stellar structures. *Monthly Notices Royal Astronomical Society*, **308**, 97.

267 Oosterhoff, P.T. (1939) Some remarks on the variable stars in globular clusters. *The Observatory*, **62**, 104.

268 van Albada, T.S. and Baker, N. (1973) On the two Oosterhoff groups of globular clusters. *Astrophysical Journal*, **185**, 477.

269 Bono, H., Caputo, F. and Marconi, M. (1995) The topology of the RR Lyrae instability strip and the Oosterhoff dichotomy. *Astronomical Journal*, **110**, 2365.

270 Gingold, R.A. (1985) The evolutionary status of type II cepheids. *Memorie della Societá Astronomica Italiana*, **56**, 169.

271 Di Criscienzo, M., Caputo, F., Marconi, M. and Cassisi, S. (2007) Synthetic properties of bright metal-poor variables

II. BL Herculis stars. *Astronomy & Astrophysics*, **471**, 893.

272 Fiorentino, G., Limongi, M., Caputo, F. and Marconi, M. (2006) Synthetic properties of bright metal-poor variables I. "Anomalous Cepheids". *Astronomy & Astrophysics*, **460**, 155.

273 Zinn, R. and King, C.R. (1982) The mass of the Anomalous Cepheid in the globular cluster NGC 5466. *Astrophysical Journal*, **262**, 700.

274 Castellani, V. and Degl'innocenti, S. (1995) Dwarf spheroidals and the evolution of not-too-old population II stars. *Astronomy & Astrophysics*, **298**, 827.

275 Demarque, P. and Hirshfeld, A. (1975) On the nature of the bright variables in dwarf spheroidal galaxies. *Astrophysical Journal*, **202**, 346.

276 Hirshfeld, A. (1980) The stellar content of dwarf spheroidal galaxies. *Astrophysical Journal*, **241**, 111.

277 Sills, A., Karakas, A. and Lattanzio, J. (2009) Blue stragglers after the main sequence. *Astrophysical Journal*, **692**, 1411.

278 Renzini, A., Mengel, J.G. and Sweigart, A.V. (1977) The anomalous Cepheids in dwarf spheroidal galaxies as binary systems. *Astronomy & Astrophysics*, **56**, 369.

279 Kilkenny, D., Koen, C., O'Donoghue, D. and Stobie, R.S. (1997) A new class of rapidly pulsating star – I. EC 14026-2647, the class prototype. *Monthly Notices of the Royal Astronomical Society*, **285**, 640.

280 Charpinet, S., Fontaine, G., Brassard, P. and Dorman, B. (1996) The potential of asteroseismology for hot, subdwarf B stars: A new class of pulsating stars? *Astrophysical Journal*, **471**, L103.

281 Charpinet, S., Fontaine, G., Brassard, P., Chayer, P., Rogers, F.J., Iglesias, C.A. and Dorman, B. (1997) A driving mechanism for the newly discovered class of pulsating subdwarf B stars. *Astrophysical Journal*, **483**, L123.

282 Randall, S.K., Calamida, A., Fontaine, G., Bono, G. and Brassard, P. (2011) Rapidly pulsating hot subdwarfs in ω Centauri: A new instability strip on the extreme horizontal branch? *Astrophysical Journal*, **737**, L27.

283 Salaris, M and Girardi, L. (2002) Population effects on the red giant clump absolute magnitude: The *K*-band. *Monthly Notices of the Royal Astronomical Society*, **337**, 332.

284 Girardi, L. Gronewegen, M.A.T., Weiss, A. and Salaris, M. (1998) Fine structure of the red giant clump from Hipparcos data and distance determinations based on its mean magnitude. *Monthly Notices of the Royal Astronomical Society*, **301**, 149.

285 Girardi, L. and Salaris, M. (2001) Population effects on the red giant clump absolute magnitude and distance determinations to nearby galaxies. *Monthly Notices of the Royal Astronomical Society*, **323**, 109.

286 Cordier, S., Pietrinferni, A., Cassisi, S. and Salaris, M. (2007) A large stellar evolution database for population synthesis studies. III. Inclusion of the full asymptotic giant branch phase and web tools for stellar population analyses. *Astronomical Journal*, **133**, 468.

287 Grundahl, F., Van den Berg, D.A. and Andersen, M.I. (1998) Stroemgren photometry of globular clusters: The distance and age of M13, evidence for two populations of horizontal-branch stars. *Astrophysical Journal*, **500**, L179.

288 Grundhal, F., Catelan, M., Landsman, W.B., Stetson, P.B. and Andersen, M.I. (1999) Hot horizontal-branch stars: The ubiquitous nature of the "jump" in Strömgren *u*, low gravities and the role of radiative levitation on metals. *Astrophysical Journal*, **524**, 242.

289 Bedin, L.R., Piotto, G., Zoccali, M., Stetson, P.B., Saviane, I., Cassisi, S. and Bono, G. (2000) The anomalous Galactic globular cluster NGC 2808. Mosaic wide-field multi-band photometry. *Astronomy & Astrophysics*, **363**, 159.

290 Momany, Y., Piotto, G., Recio-Blanco, A., Bedin, L.R., Cassisi, S. and Bono, G. (2002) A new feature along the extended blue horizontal branch of NGC 6752. *Astrophysical Journal*, **576**, L65.

291 Moehler, S. (2001) Hot stars in globular clusters: A spectroscopist's view. *Publication of the Astronomical Society of Pacific*, **113**, 1162.

292 Hui-Bon-Hoa, A., LeBlanc, F. and Hauschildt, P.H. (2000) Diffusion in the atmospheres of blue horizontal-branch stars. *Astrophysical Journal*, **535**, L43.

293 Michaud, G., Vauclair, G. and Vauclair, S. (1983) Chemical separation in horizontal-branch stars. *Astrophysical Journal*, **267**, 256.

294 Glaspey, J.W., Michaud, G., Moffat, A.F.J. and Demers, S. (1989) Abundance anomalies in hot horizontal-branch stars of the globular cluster NGC 6752. *Astrophysical Journal*, **339**, 926.

295 Behr, B.B., Cohen, J.G., McCarthy, J.K. and Djorgovski, G.S. (1999) Striking photospheric anomalies in blue horizontal branch stars in globular cluster M13. *Astrophysical Journal*, **517**, L135.

296 Behr, B.B., Cohen, J.G. and McCarthy, J.K. (2000) Rotations and abundances of blue horizontal-branch stars in globular cluster M15. *Astrophysical Journal*, **531**, L37.

297 Behr, B.B. (2003) Chemical abundances and rotation velocities of blue horizontal-branch stars in six globular clusters. *Astrophysical Journal Supplement Series*, **149**, 67.

298 Pace, G., Recio-Blanco, A., Piotto, G. and Momany, Y. (2006) Abundance anomalies in hot horizontal branch stars of the galactic globular cluster NGC 2808. *Astronomy & Astrophysics*, **452**, 493.

299 Peterson, R.C., Rood, R.T. and Crocker, D.A. (1995) Rotation and oxygen line strengths in blue horizontal-branch stars. *Astrophysical Journal*, **453**, 214.

300 Behr, B.B. (2003) Rotation velocities of red and blue field horizontal-branch stars. *Astrophysical Journal Supplement Series*, **149**, 101.

301 Ferraro, F., Paltrinieri, B., Fusi Pecci, F., Rood, R.T. and Dorman, B. (1998) Multimodal distributions along the horizontal branch. *Astrophysical Journal*, **500**, 311.

302 Caloi, V. (1999) On the gap in horizontal branch at $(B - V)$ about zero. *Astronomy & Astrophysics*, **343**, 904.

303 Greenstein, G.S. (1967) Helium deficiency in old halo B stars. *Nature*, **213**, 871.

304 Michaud, G., Richer, J. and Richard, O. (2008) Abundance anomalies in horizontal branch stars and atomic diffusion. *Astrophysical Journal*, **675**, 1223.

305 Michaud, G., Richer, J. and Richard, O. (2007) Horizontal branch evolution, metallicity and sdB stars. *Astronomy & Astrophysics*, **529**, A60.

306 Quievy, D., Charbonneau, P., Michaud, G. and Richer, J. (2009) Abundances anomalies and meridional circulation in horizontal branch stars. *Astronomy & Astrophysics*, **500**, 1163.

307 Sills, A. and Pinsonneault, M.H. (2000) Rotation of horizontal-branch stars in globular clusters. *Astrophysical Journal*, **540**, 489.

308 Michaud, G. and Charland, Y. (1986) Mass loss in A and F stars – the λ bootis stars. *Astrophysical Journal*, **311**, 326.

309 Soker, N. (1998) Can planets influence the horizontal branch morphology? *Astronomical Journal*, **116**, 1308.

310 Montegriffo, P., Ferraro, F. R., Fusi Pecci, F. and Origlia, L. (1995) IR-array photometry of galactic globular clusters – II. JK photometry of 47 Tuc. *Monthly Notices of the Royal Astronomical Society*, **276**, 739.

311 Castellani, V., Chieffi, A. and Pulone, L. (1991) The evolution of He-burning stars – horizontal and asymptotic branches in galactic globulars. *Astrophysical Journal Supplement Series*, **76**, 911.

312 Renzini, A. and Fusi Pecci, F. (1988) Tests of evolutionary sequences using color-magnitude diagrams of globular clusters. *Annual Review Astronomy & Astrophysics*, **26**, 199.

313 Ferraro, F.R., Messineo, M., Fisi Pecci, F., De Palo, M.A., Straniero, O., Chieffi, A. and Limongi, M. (1999) The giant, horizontal and asymptotic branches of galactic globular clusters. I. The catalog, photometric observables and features. *Astronomical Journal*, **118**, 1738.

314 Schwarzschild, M. and Härm, R. (1965) Thermal instability in non-degenerate stars. *Astrophysical Journal*, **142**, 855.

315 Paczyński, B. (1970) Evolution of single stars. I. Stellar evolution from main sequence to white dwarf or carbon ignition. *Acta Astronomica*, **20**, 47.

316 Paczyński, B. (1970) Evolution of single stars. III. Stationary shell sources. *Acta Astronomica*, **20**, 287.

317 Iben Jr., I. and Renzini, A. (1983) Asymptotic giant branch evolution and beyond. *Annual Review Astronomy & Astrophysics*, **21**, 271.

318 Cristallo, S., Piersanti, L., Straniero, O., Gallino, R., Domínguez, I., Abia, C., Di Rico, G., Quintini, M. and Bisterzo, S. (2011) Evolution, nucleosynthesis and yields of low-mass, asymptotic giant branch stars at different metallicities. II The FRUITY database. *Astrophysical Journal Supplement Series*, **197**, 17.

319 Straniero, O., Chieffi, A., Limongi, M., Busso, M., Gallino, R. and Arlandini, C. (1997) Evolution and nucleosynthesis in low-mass asymptotic giant branch stars. I. Formation of population I carbon stars. *Astrophysical Journal*, **478**, 332.

320 Cameron, A.G.W. (1955) Origin of anomalous abundances of the elements in giant stars. *Astrophysical Journal*, **121**, 144.

321 Herwig, F. (2005) Evolution of asymptotic giant branch stars. *Annual Review Astronomy & Astrophysics*, **43**, 435.

322 Karacas, A.I. (2010) Updated stellar yields from asymptotic giant branch models. *Monthly Notices of the Royal Astronomical Society*, **403**, 1413.

323 Ulrich, R.K. (1973) *Explosive Nucleosynthesis*, (eds D.N. Schramm and W.D. Arnett), University of Texas Press, Austin, p. 139.

324 Sackmann, I.-J., Smith, R.L. and Despain, K.H. (1974) Carbon and eruptive stars: surface enrichment of lithium, carbon, nitrogen and ^{13}C by deep mixing. *Astrophysical Journal*, **187**, 555.

325 Suda, T., Aikawa, M., Machida, M.N., Fujimoto, M.Y. and Iben Jr., I. (2004) Is HE 0107-5240 a primordial star? The characteristics of extremely metal-poor carbon-rich stars. *Astrophysical Journal*, **611**, 476.

326 Straniero, O., Gallino, R., Busso, M., Chieffi, A., Raiteri, C.M., Limongi, M. and Salaris, M. (1995) Radiative ^{13}C burning in asymptotic giant branch stars

and s-processing. *Astrophysical Journal*, **440**, L85.

327 Denissenkov, P.A. and Tout, C.A. (2003) Partial mixing and formation of the ^{13}C pocket by internal gravity waves in asymptotic giant branch stars. *Monthly Notices of the Royal Astronomical Society*, **340**, 722.

328 Herwig, F. (2000) The evolution of AGB stars with convective overshoot. *Astronomy & Astrophysics*, **360**, 952.

329 Langer, N., Heger, A., Wellstein, S. and Herwig, F. (1999) Mixing and nucleosynthesis in rotating TP-AGB stars. *Astronomy & Astrophysics*, **346**, L37.

330 Longland, R., Iliadis, C. and Karakas, A.I. (2012) Reaction rates for the s-process neutron source ^{22}Ne $+ \alpha$, Physics Review C, **85**, 065809.

331 Cameron, A.G.W. (1955) Nuclear reactions in stars and nucleogenesis. *Publications of the Astronomical Society of the Pacific*, **69**, 201.

332 Busso, M., Gallino, R., and Wasserburg, G.J. (1999) Nucleosynthesis in asymptotic giant branch stars: relevance for galactic enrichment and solar system formation. *Annual Review Astronomy & Astrophysics*, **37**, 239.

333 Lugaro, M., Karakas, A.I., Stancliffe, R.J. and Rijs, C. (2012) The s-process in asymptotic giant branch stars of low metallicity and the composition of carbon-enhanced metal-poor stars, *Astrophysical Journal*, **747**, 2.

334 Salaris, M., Serenelli, A., Weiss, A. and Miller Bertolami, M. (2009) Semi-empirical white dwarf initial-final mass relationships: A thorough analysis of systematic uncertainties due to stellar evolution models. *Astrophysical Journal*, **692**, 1013.

335 Marigo, P. and Girardi, L. (2007) Evolution of asymptotic giant branch stars. I. Updated synthetic TP-AGB models and their basic calibration. *Astronomy & Astrophysics*, **469**, 239.

336 Wagenhuber, J. and Groenewegen, M.A.T. (1998) New input data for synthetic AGB evolution. *Astronomy & Astrophysics*, **340**, 183.

337 Izzard, R.G., Tout, C.A., Karakas, A.I. and Pols, O.R. (2004) A new synthetic model for asymptotic giant branch stars. *Monthly Notices of the Royal Astronomical Society*, **350**, 407.

338 Marigo, P., Bressan, A., and Chiosi, C. (1998) TGP-AGP stars with envelope burning. *Astronomy & Astrophysics*, **331**, 564.

339 Lattanzio, J.C. and Wood, P.R. (2003) Evolution, nucleosynthesis and pulsation of AGB stars, in *Asymptotic Giant Branch Stars*, (eds H.J. Habing and H. Olofsson), Springer, p. 23.

340 Ya'Ari, A. and Tuchman, Y. (1999) On the pulsation mode of Mira variables: nonlinear calculations compared with radii observations. *Astrophysical Journal*, **514**, L35.

341 Fox, M.W. and Wood, P.R. (1982) Theoretical growth rates, periods and pulsation constants for long-period variables. *Astrophysical Journal*, **259**, 198.

342 Ya'Ari, A. and Tuchman, Y. (1996) Long-term non-linear thermal effects in the pulsation of Mira variables. *Astrophysical Journal*, **456**, 350.

343 Wood, P.R. (2000) Variable red giants in the LMC: Pulsating stars and binaries?. *Publication of the Astronomical Society of Pacific*, **17**, 18.

344 Hughes, S.M.G. and Wood, P.R. (1990) Long-period variables in the Large Magellanic Cloud. II – Infrared photometry, spectral classification, AGB evolution and spatial distribution. *Astronomical Journal*, **99**, 784.

345 Whitelock, P. (2012) Asymptotic Giant Branch variables in the Galaxy and the Local Group, *Astrophysics and Space Science*, **341**, 123.

346 Feast, M. (1999) Pulsation modes in Mira and Semiregular Variables, in *Asymptotic Giant Branch stars*, (eds T. Le Bertre, A. Lébre and C. Waelkens), IAU Symposium 191, ASP Series, Astronoical Society of the Pacific, p. 109.

347 Zijlstra, A.A. and Bedding, T.R. (2002) Period evolution in Mira variables. *Journal of the American Association of Variable Star Observers*, **31**, 2.

348 Uttenthaler, S., Van Stiphout, K., Voet, K., Van Winckel, H., Van Eck, S., Jorissen, A., Kerschbaum, F. *et al.* (2011) The evolutionary state of Miras with changing pulsation period. *Astronomy & Astrophysics*, **531**, A88.

349 Gustafsson, B. and Höfner, S. (2003) Evolution, nucleosynthesis and pulsation of AGB stars, in *Asymptotic Giant Branch Stars*, (eds H.J. Habing and H. Olofsson), Springer, p. 149.

350 Fowler, W.A. and Hoyle, F. (1960) Nuclear cosmochronology. *Annals of Physics*, **10**, 280.

351 Reeves, H. (1991) Nucleochronology revised. *Astronomy & Astrophysics*, **244**, 294.

352 Meyer, B.S. and Truran, J.W. (2000) Nucleocosmochronology. *Physics Reports*, **333/334**, 1.

353 Verbena, J.L., Schröder, K.-P. and Wachter, A. (2011) Outflow dynamics of dust-driven wind models and implications for cool envelopes of PNe. *Monthly Notices of the Royal Astronomical Society*, **415**, 2270.

354 Fleischer, A.J., Gauger, A. and Sedlmayr, E. (1992) Circumstellar dust shells around long-period variables. I. Dynamical models of C-stars including dust formation, growth and evaporation. *Astronomy & Astrophysics*, **266**, 321.

355 Bloecker, T. (1995) Stellar evolution of low and intermediate-mass stars. I. Mass loss on the AGB and its consequences for stellar evolution. *Astronomy & Astrophysics*, **297**, 727.

356 Althaus, L.G., Miller Bertolami, M.M., García-Berro, E., Córsico, A.H. and Gil-Pons, P. (2005) The formation of DA white dwarfs with thin hydrogen envelopes. *Astronomy & Astrophysics Letters*, **440**, L1.

357 Miller Bertolami, M.M., Althaus, L.G., Serenelli, A.M. and Panei, J.A. (2006) New evolutionary calculations for the born again scenario. *Astronomy & Astrophysics*, **449**, 313.

358 Althaus, L.G., Córsico, A.H., Miller Bertolami, M.M., García-Berro, E. and Kepler, S.O. (2008) Evidence of thin helium envelopes in PG1159 stars. *Astrophysical Journal*, **677**, L35.

359 Miller Bertolami, M.M. and Althaus, L.G. (2007) The born-again (very late thermal pulse) scenario revisited: the mass of the remnants and implications for V4334 Sgr. *Monthly Notices of the Royal Astronomical Society*, **380**, 763.

360   Iben Jr., I. (1984) On the frequency of planetary nebula nuclei powered by helium burning and on the frequency of white dwarfs with hydrogen-deficient atmospheres. *Astrophysical Journal*, **277**, 333.

361   Schönberner, D. (1986) Late stages of stellar evolution. III. The observed evolution of central stars of planetary nebulae. *Astronomy & Astrophysics*, **169**, 189.

362   Vassiliadis, E. and Wood, P.R. (1994) Post-asymptotic giant branch evolution of low- to intermediate-mass stars. *Astrophysical Journal Supplement Series*, **92**, 125.

363   Herwig, F., Blöcker, T., Langer, N. and Driebe, T. (1999) On the formation of hydrogen-deficient post-AGB stars. *Astronomy & Astrophysics*, **349**, L5.

364   Asplund, M. (1999) *Sakurai's Object – Stellar Evolution in Real Time*, (eds T. Le Bertre, A. Lébre and C. Waelkens), IAU Symposium 191, ASP Series, Astronomical Society of the Pacific, p. 481.

365   Jeffery, C.S. and Schönberner, D. (2006) Stellar archaeology: the evolving spectrum of FG Sagittae, *Astronomy & Astrophysics*, **459**, 885.

366   Gautschy, A. (1997) A further look into the pulsating PG1159 stars. *Astronomy & Astrophysics*, **320**, 811.

367   Castellani, M., Limongi, M. and Tornambe, A. (1992) The contribution of advanced post-Hayashi track evolutionary phases to the ultraviolet light of elliptical galaxies. *Astrophysical Journal*, **389**, 227.

368   Catalán, S., Isern, J., García-Berro, E., Ribas, I., Allende Prieto, C. and Bonanos, A.Z. (2008) The initial-final mass relationship from white dwarfs in common proper motion pairs. *Astronomy & Astrophysics*, **477**, 213.

369   Kalirai, J.S., Saul Davis, D., Richer, H.B., Bergeron, P., Catelan, M., Hansen, B.M.S. and Rich, R.M. (2009) The masses of population II white dwarfs. *Astrophysical Journal*, **705**, 408.

370   Althaus, L.G., Córsico, A.H., Isern, J. and García-Berro, E. (2010) Evolutionary and pulsational properties of white dwarf stars. *Astronomy & Astrophysics Review*, **18**, 417.

371   Winget, D.E. and Kepler, S.O. (2008) Pulsating white dwarf stars and precision asteroseismology. *Annual Review of Astronomy and Astrophysics*, **46**, 157.

372   Renedo, I., Althaus, L.G., Miller Bertolami, M.M., Romero, A.D., Córsico, A.H., Rohrmann, R.D. and García-Berro, E. (2010) New cooling sequences for old white dwarfs. *Astrophysical Journal*, **717**, 183.

373   Chandrasekhar, S. (1939) *An Introduction to the Study of Stellar Structure*, The University of Chicago Press.

374   Mestel, L. (1952) On the theory of white dwarf stars. I. The energy sources of white dwarfs. *Monthly Notices of the Royal Astronomical Society*, **112**, 583.

375   Frogel, J.A., Stephens, A., Ramírez, S. and DePoy, D.L. (2001) An accurate, easy-to-use abundance scale for globular clusters based on 2.2 micron spectra of giant stars. *Astronomical Journal*, **122**, 1896.

376   Hansen, B.M.S. (1999) Cooling models for old white dwarfs. *Astrophysical Journal*, **520**, 680.

377   Salaris, M., Cassisi, S., Pietrinferni, A., Kowalski, P.M. and Isern, J. (2010) A large stellar evolution database for population synthesis studies. VI. White dwarf cooling sequences. *Astrophysical Journal*, **716**, 1241.

378   Salaris, M., García-Berro, E., Hernanz, M., Isern, J. and Saumon, D. (2000) The ages of very cool hydrogen-rich white dwarfs. *Astrophysical Journal*, **544**, 1036.

379   Fontaine, G., Brassard, P. and Bergeron, P. (2001) The potential of white dwarf cosmochronology. *Publications of the Astronomical Society of the Pacific*, **113**, 409.

380   Medin, Z. and Cumming, A. (2010) Crystallization of classical multicomponent plasmas, *Physical Review E*, **81**(3), 036107.

381   Segretain, L. and Chabrier, G. (1993) Crystallization of binary ionic mixtures in dense stellar plasmas. *Astronomy & Astrophysics*, **271**, L13.

382 García-Berro, E., Hernanz, M., Isern, J. and Mochkovitch, R. (1988) Properties of high-density binary mixtures and the age of the Universe from white dwarf stars. *Nature*, **333**, 642.

383 Isern, J., García-Berro, E., Hernanz, M. and Chabrier, G. (2000) The energetics of crystallizing white dwarfs revisited again. *Astrophysical Journal*, **528**, 397.

384 Segretain, L. (1996) Three-body crystallization diagrams and the cooling of white dwarfs. *Astronomy & Astrophysics*, **310**, 485.

385 Deloye, C.J. and Bildsten, L. (2002) Gravitational settling of ^{22}Ne in liquid white dwarf interiors: Cooling and seismological effects. *Astrophysical Journal*, **580**, 1077.

386 Althaus, L.G., García-Berro, E., Renedo, I., Isern, J., Córsico, A.H. and Rohrmann, R.D. (2010) Evolution of white dwarf stars with high-metallicity progenitors: The role of ^{22}Ne diffusion. *Astrophysical Journal*, **719**, 612.

387 Hughto, J., Schneider, A.S., Horowitz, C.J. and Berry, D.K. (2010) Diffusion of neon in white dwarf stars. *Physical Review E*, **82**, 066401.

388 Saumon, D. and Jacobson, S.B. (1999) Pure hydrogen model atmospheres for very cool white dwarfs. *Astrophysical Journal*, **511**, L107.

389 Salaris, M., Dominguez, I., García-Berro, E., Hernanz, M., Isern, J. and Mochkovitch, R. (1997) The cooling of CO white dwarfs: Influence of the internal chemical distribution. *Astrophysical Journal*, **486**, 413.

390 Bergeron, P., Wesemael, F. and Beauchamp, A. (1995) Photometric calibration of hydrogen- and helium-rich white dwarf models. *Publications of the Astronomical Society of the Pacific*, **107**, 1047.

391 Holberg, J.B. and Bergeron, P. (2006) Calibration of synthetic photometry using DA white dwarfs. *Astronomical Journal*, **132**, 1221.

392 Rohrmann, R.D. (2001) Hydrogen-model atmospheres for white dwarf stars. *Monthly Notices of the Royal Astronomical Society*, **323**, 699.

393 Kowalski, P.M. and Saumon, D. (2006) Found: The missing blue opacity in atmosphere models of cool hydrogen white dwarfs. *Astronomy & Astrophysics*, **651**, L137.

394 Rohrmann, R.D., Althaus, L.G. and Kepler, S.O. (2011) Lyman α wing absorption in cool white dwarf stars. *Monthly Notices of the Royal Astronomical Society*, **411**, 781.

395 Eisenstein, D.J. *et al.* (2006) A catalog of spectroscopically confirmed white dwarfs from the Sloan Digital Sky Survey Data Release 4. *Astrophysical Journal Supplement*, **167**, 40.

396 McCook, G.P. and Sion, E.M. (1987) A catalog of spectroscopically identified white dwarfs. *Astrophysical Journal Supplement*, **65**, 603.

397 Koester, D. and Chanmugam, G. (1990) Physics of white dwarf stars. *Reports of Progress in Physics*, **53**, 837.

398 Davis, D.S., Richer, H.B., Rich, R.M., Reitzel, D.R. and Kalirai, J.S. (2009) The spectral types of white dwarfs in Messier 4. *Astrophysical Journal*, **705**, 398.

399 Dufour, P., Liebert, J., Fontaine, G. and Behara, N. (2007) White dwarf stars with carbon atmospheres. *Nature*, **450**, 522.

400 Dufour, P., Fontaine, G., Liebert, J., Schmidt, G.D. and Behara, N. (2008) Hot DQ white dwarfs: Something different. *Astrophysical Journal*, **683**, 978.

401 Chen, E.Y. and Hansen, B.M.S. (2011) Cooling curves and chemical evolution curves of convective mixing white dwarf stars. *Monthly Notices of the Royal Astronomical Society*, **413**, 2827.

402 Prada Moroni, P.G. and Straniero, O. (2009) Very low-mass white dwarfs with a C–O core. *Astronomy & Astrophysics*, **507**, 1575.

403 Panei, J.A., Althaus, L.G., Chen, X. and Han, Z. (2007) Full evolution of low-mass white dwarfs with helium and oxygen cores. *Monthly Notices of the Royal Astronomical Society*, **382**, 779.

404 Iben Jr., I. and Tutukov, A.V. (1986) On the formation and evolution of a helium degenerate dwarf in a close binary. *Astrophysical Journal*, **311**, 742.

405 Webbink, R.F. (1975) Evolution of helium white dwarfs in close binaries.

Monthly Notices of the Royal Astronomical Society, **171**, 555.

406 An, D., Terndrup, D.M., Pinsonneault, M.H., Paulson, D.B., Hanson, R.B. and Stauffer, J.R. (2007) The distances to open clusters from main-sequence fitting. III. Improved accuracy with empirically calibrated isochrones. *Astrophysical Journal*, **655**, 233.

407 Carretta, E., Gratton, R.G., Clementini, G. and Fusi Pecci, F. (2000) Distances, ages and epoch of formation of globular clusters. *Astrophysical Journal*, **533**, 215.

408 Percival, S.M., Salaris, M. and Kilkenny, D. (2003) The open cluster distance scale. A new empirical approach. *Astronomy & Astrophysics*, **400**, 541.

409 Sandage, A. (1970) Main-sequence photometry, color-magnitude diagrams and ages for the globular clusters M3, M13, M15 and M92. *Astrophysical Journal*, **162**, 841.

410 Gratton, R.G., Bragaglia, A., Carretta, E., Clementini, G., Desidera, S., Grundahl, F. and Lucatello, S. (2003) Distances and ages of NGC 6397, NGC 6752 and 47 Tuc. *Astronomy & Astrophysics*, **408**, 529.

411 Sandage, A. and Saha, A. (2002) Bias properties of extragalactic distance indicators. XI. Methods to correct for observational selection bias for RR Lyrae absolute magnitudes from trigonometric parallaxes expected from the full-sky astrometric mapping explorer satellite. *Astronomical Journal*, **123**, 2047.

412 Percival, S.M., Salaris, M. and Groenewegen, M.A.T. (2005) The distance to the Pleiades. Main sequence fitting in the near infrared. *Astronomy & Astrophysics*, **429**, 887.

413 Straniero, O., Chieffi, A. and Limongi, M. (1997) Isochrones for hydrogen-burning globular cluster stars. III. From the Sun to the globular clusters. *Astrophysical Journal*, **490**, 425.

414 Korn, A.J., Grundahl, F., Richard, O., Barklem, P.S., Mashonkina, L., Collet, R., Piskunov, N. and Gustafsson, B. (2006) A probable stellar solution to the cosmological lithium discrepancy, *Nature*, **442**, 657.

415 Mucciarelli, A., Salaris, M., Lovisi, L., Ferraro, F.R., Lanzoni, B., Lucatello, S. and Gratton, R.G. (2011) Lithium abundance in the globular cluster M4: from the turn-off to the red giant branch bump. *Monthly Notices of the Royal Astronomical Society*, **412**, 8.

416 Salaris, M., Groenewegen, M.A.T. and Weiss, A. (2000) Atomic diffusion in metal poor stars. The influence on the Main Sequence fitting distance scale, subdwarfs ages and the value of $\Delta Y/\Delta Z$. *Astronomy & Astrophysics*, **355**, 299.

417 Sbordone, L., Salaris, M., Weiss, A. and Cassisi, S. (2011) Photometric signatures of multiple stellar populations in galactic globular clusters. *Astronomy & Astrophysics*, **534**, 9.

418 Lee, M.G., Freedman, W.L. and Madore, B.F. (1993) The tip of the red giant branch as a distance indicator for resolved galaxies. *Astrophysical Journal*, **417**, 553.

419 Madore, B.F. and Freedman, W.L. (1995) The tip of the red giant branch as a distance indicator for resolved galaxies. 2: Computer simulations. *Astronomical Journal*, **109**, 1645.

420 Cioni, M.-R.L., van der Marel, R.P., Loup, C. and Habing, H.J. (2000) The tip of the red giant branch and distance of the Magellanic Clouds: results from the DENIS survey. *Astronomy & Astrophysics*, **359**, 601.

421 Conn, A.R. *et al.* (2011) A Bayesian approach to locating the red giant branch tip magnitude (Part I). *Astrophysical Journal*, **740**, 69.

422 McConnachie, A.W., Irwin, M.J., Ferguson, A.M.N., Ibata, R.A., Lewis, G.F. and Tanvir, N. (2004) Determining the location of the tip of the red giant branch in old stellar populations: M33, Andromeda I and II. *Monthly Notices of the Royal Astronomical Society*, **350**, 243.

423 Makarov, D., Makarova, L., Rizzi, L., Tully, R.B., Dolphin, A.E., Sakai, S. and Shaya, E.J. (2006) Tip of the red giant branch distances. I. Optimization of a maximum likelihood algorithm. *Astronomical Journal*, **132**, 2729.

424 Bellazzini, M., Ferraro, F.R. and Pancino, E. (2001) A step toward the calibration of the red giant branch tip as a standard candle. *Astrophysical Journal,* **556**, 635.

425 Bellazzini, M., Ferraro, F.R., Sollima, A., Pancino, E. and Origlia, L. (2004) The calibration of the RGB tip as a standard candle. extension to near infrared colors and higher metallicity. *Astronomy & Astrophysics,* **424**, 199.

426 Thompson, I.B., Kaluzny, J., Rucinski, S.M., Krzeminski, W., Pych, W., Dotter, A. and Burley, G.S. (2010) The Cluster AgeS Experiment (CASE). IV. Analysis of the eclipsing binary V69 in the globular cluster 47 Tuc. *Astronomical Journal,* **139**, 329.

427 Rizzi, L., Tully, R.B., Makarov, D., Makarova, L., Dolphin, A.E., Sakai, S. and Shaya, E.J. (2007) Tip of the red giant branch distances. II. Zero-point calibration. *Astrophysical Journal,* **661**, 815.

428 Bellazzini, M. (2008) The tip of the red giant branch. *Memorie della Società Astronomica Italiana,* **79**, 440.

429 Barker, M.K., Sarajedini, A. and Harris, J. (2004) Variations in star formation history and the red giant branch tip, *Astrophysical Journal,* **606**, 869.

430 Salaris, M. and Girardi, L. (2005) Tip of the red giant branch distances to galaxies with composite stellar populations. *Monthly Notices of the Royal Astronomical Society,* **357**, 669.

431 Cole, A.A., Smecker-Hane, T.A. and Gallagher III, J.S. (2000) The metallicity distribution function of red giants in the Large Magellanic Cloud. *Astronomical Journal,* **120**, 1808.

432 Pietrinferni, A., Cassisi, S., Salaris, M. and Castelli, F. (2006) A large stellar evolution database for population synthesis studies. II. Stellar models and isochrones for an α-enhanced metal distribution. *Astrophysical Journal,* **642**, 797.

433 Groenewegen, M.A.T. and Salaris, M. (1999) The absolute magnitudes of RR Lyrae stars from HIPPARCOS parallaxes. *Astronomy & Astrophysics,* **348**, L33.

434 Cassisi, S., Castellani, M., Caputo, F. and Castellani, V. (2004) RR Lyrae variables in galactic globular clusters. IV. Synthetic HB and RR Lyrae predictions. *Astronomy & Astrophysics,* **426**, 641.

435 Corwin, T.M. and Carney, B.W. (2001) BV photometry of the RR Lyrae variables of the globular cluster M3. *Astronomical Journal,* **122**, 3183.

436 Marconi, M., Caputo, F., Di Criscienzo, M. and Castellani, M. (2003) RR Lyrae Stars in galactic globular clusters. II. A theoretical approach to variables in M3. *Astrophysical Journal,* **596**, 299.

437 Carretta, E., Bragaglia, A., Gratton, R., D'Orazi, V. and Lucatello, S. (2009) Intrinsic iron spread and a new metallicity scale for globular clusters. *Astronomy & Astrophysics,* **508**, 695.

438 Clementini, G., Gratton, R., Bragaglia, A., Carretta, E., Di Fabrizio, L. and Maio, M. (2003) Distance to the Large Magellanic Cloud: The RR Lyrae stars. *Astronomical Journal,* **125**, 1309.

439 McNamara, D. (1997) Luminosities of SX Phoenicis, large-amplitude delta Scuti and RR Lyrae stars. *Publication of the Astronomical Society of the Pacific,* **109**, 1221.

440 Cacciari, C. and Clementini, G. (2003) Globular cluster distances from RR Lyrae stars, in *Stellar Candles for the Extragalactic Distance Scale,* Lecture Notes in Physics, vol. 635, (eds D. Alloin and W. Gieren), Springer, p. 105.

441 Caputo, F. (1997) The period-magnitude diagram of RR Lyrae stars – I. The controversy about the distance scale. *Monthly Notices of the Royal Astronomical Society,* **284**, 994.

442 Caputo, F., Castellani, V., Marconi, M. and Ripepi, V. (2000) Pulsational M_V versus [Fe/H] relation(s) for globular cluster RR Lyrae variables. *Monthly Notices of the Royal Astronomical Society,* **316**, 819.

443 Iglesias, C.A. and Rogers, F.J. (1996) Updated opal opacities. *Astrophysical Journal,* **464**, 943.

444 Caughlan, G.R. and Fowler, W.A. (1988) Thermonuclear reaction rates V, in *Atomic Data and Nuclear Data Tables,* vol. 40, Elsevier, p. 283.

445 Haft, M., Raffelt, G. and Weiss, A. (1994) Standard and nonstandard plasma neutrino emission revisited. *Astrophysical Journal*, **425**, 222.

446 Castellani, V. and Degl'innocenti, S. (1999) Microscopic diffusion and the calibration of globular cluster ages. *Astronomy & Astrophysics*, **344**, 97.

447 Longmore, A.J., Dixon, R., Skillen, I., Jameson, R.F. and Fernley, J.A. (1990) Globular cluster distances from the RR Lyrae log(period)-infrared magnitude relation. *Monthly Notices of the Royal Astronomical Society*, **247**, 684.

448 Longmore, A.J., Fernley, J.A. and Jameson, R.F. (1986) RR Lyrae stars in globular clusters – Better distances from infrared measurements?. *Monthly Notices of the Royal Astronomical Society*, **220**, 279.

449 Jones, R.V., Carney, B.W., Storm, J. and Latham, D.W. (1992) The Baade–Wesselink method and the distances to RR Lyrae stars. VII – The field stars SW Andromedae and DX Delphini and a comparison of recent Baade–Wesselink analyses. *Astrophysical Journal*, **386**, 646.

450 Liu, T. and Janes, K.A. (1990) The luminosity scale of RR Lyrae stars with the Baade–Wesselink method. II – The absolute magnitudes of 13 field RR Lyrae stars. *Astrophysical Journal*, **354**, 273.

451 Bono, G., Caputo, F., Castellani, V., Marconi, M. and Storm, J. (2001) Theoretical insights into the RR Lyrae K-band period-luminosity relation. *Monthly Notices of the Royal Astronomical Society*, **326**, 1183.

452 Bono, G., Caputo, F., Castellani, V., Marconi, M., Storm, J. and Degl'Innocenti, S. (2003) A pulsational approach to near-infrared and visual magnitudes of RR Lyr stars. *Monthly Notices of the Royal Astronomical Society*, **34**, 1097.

453 Catelan, M., Pritzl, B.J. and Smith, H.A. (2004) The RR Lyrae period-luminosity relation. I. Theoretical calibration. *Astrophysical Journal Supplement Series*, **154**, 633.

454 Del Principe, M., Piersimoni, A.M., Storm, J., Caputo, F., Bono, G., Stet-son, P.B., Castellani, M. *et al.* (2006) A pulsational distance to ω Centauri based on near-infrared period-luminosity relations of RR Lyrae Stars. *Astrophysical Journal*, **652**, 362.

455 Del Principe, M., Piersimoni, A.M., Bono, G., Di Paola, A., Dolci, M. and Marconi, M. (2005) Near-infrared observations of RR Lyrae variables in galactic globular clusters. I. The case of M92. *Astronomical Journal*, **129**, 2714.

456 Cannon, R.D. (1970) Red giants in old open clusters. *Monthly Notices of the Royal Astronomical Society*, **150**, 111.

457 Stanek, K.Z. and Garnavich, P.M. (1998) Distance to M31 with the Hubble Space Telescope and HIPPARCOS red clump stars. *Astrophysical Journal*, **503**, L131.

458 Alves, D.R., Rejkuba, M., Minniti, D. and Cook, K.H. (2002), K-band red clump distance to the Large Magellanic Cloud. *Astrophysical Journal*, **573**, L51.

459 Groenewegen, M.A.T. (2008) The red clump absolute magnitude based on revised Hipparcos parallaxes. *Astronomy & Astrophysics*, **488**, 935.

460 Salaris, M., Percival, S. and Girardi, L. (2003) A theoretical analysis of the systematic errors in the red clump distance to the Large Magellanic Cloud (LMC). *Monthly Notices of the Royal Astronomical Society*, **345**, 1030.

461 Percival, S.M. and Salaris, M. (2003) An empirical test of the theoretical population corrections to the red clump absolute magnitude. *Monthly Notices of the Royal Astronomical Society*, **343**, 539.

462 Udalski, A. (2000) The Optical gravitational lensing experiment: Red clump stars as a distance indicator. *Astrophysical Journal Letters*, **531**, L25.

463 Pietrzyński, G., Górski, M., Gieren, W., Laney, D., Udalski, A. and Ciechanowska, A. (2010) The Araucaria Project. Population effects on the V- and I-band magnitudes of red clump stars. *Astronomical Journal*, **140**, 1038.

464 Renzini, A., Bragaglia, A., Ferraro, F.R., Gilmozzi, R., Ortolani, S., Holberg, J.B., Liebert, J., Wesemael, F. and Bohlin, R.C. (1996) The white dwarf distance to the globular cluster NGC 6752

(and its age) with the Hubble Space Telescope. *Astrophysical Journal*, **465**, L23.

465 Salaris, M., Cassisi, S., García-Berro, E., Isern, J. and Torres, S. (2001) On the white dwarf distances to galactic globular clusters. *Astronomy & Astrophysics*, **371**, 921.

466 Hut, P., McMillan, S. and Romani, R.W. (1992) The evolution of a primordial binary population in a globular cluster. *Astrophysical Journal*, **389**, 527.

467 Renzini, A. (1991) *Three Steps on the Age Ladder*, (eds R.T. Rood and A. Renzini), Proceeding of Advances in Stellar Evolution, Cambridge Contemporary Astrophysics, p. 46.

468 Van den Berg, D.A., Bolte, M. and Stetson, P. (1996) The age of the galactic globular cluster system. *Annual Review Astronomy & Astrophysics*, **34**, 461.

469 Di Cecco, A., Becucci, R., Bono, G., Monelli, M., Stetson, P.B., Degl'Innocenti, S., Prada Moroni, P.G. *et al.* (2010) On the absolute age of the globular cluster M92. *Publication of Astronomical Society of the Pacific*, **122**, 991.

470 Sarajedini, A. (2008) *Relative and Absolute Ages of Galactic Globular Clusters*, (eds E.E. Mamajek, D.R. Soderblom and R.F.G. Wyse), Proceedings of the Age of Stars, IAU Symposium 258, Cambridge University Press, p. 221.

471 Chaboyer, B. and Kim, Y.-C. (1995) The OPAL equation of state and low-metallicity isochrones. *Astrophysical Journal*, **454**, 767.

472 Rogers, F.J. (1994) *The Equation of State in Astrophysics*, Proceedings of IAU Colloquium No. 147, (eds G. Chabrier, and E. Schatzman), Cambridge University Press, p. 16.

473 Rogers, F.J. and Nayfonov, A. (2002) Updated and expanded OPAL equation-of-state tables: Implications for helioseismology. *Astrophysical Journal*, **576**, 1064.

474 Cassisi, S., Salaris, M. and Irwin, A.W. (2003) The initial helium content of galactic globular cluster stars from the *R*-parameter: Comparison with the cosmic microwave background constraint. *Astrophysical Journal*, **588**, 862.

475 Seaton, M.J. and Badnell, N.R. (2004) A comparison of Rosseland-mean opacities from OP and OPAL. *Monthly Notices of the Royal Astronomical Society*, **354**, 475.

476 Chaboyer, B., Demarque, P., Kernan, P.J. and Krauss, L.M. (1998) The age of globular clusters in light of Hipparcos: resolving the age problem?. *Astrophysical Journal*, **494**, 96.

477 Turcotte, S., Richer, J., Michaud, G., Iglesias, C. and Rogers, F. (1998) Consistent solar evolution model including diffusion and radiative acceleration effects. *Astrophysical Journal*, **504**, 539.

478 Richard, O., Vauclair, S., Charbonnel, C. and Dziembowski, W.A. (1996) New solar models including helioseismological constraints and light-element depletion. *Astronomy & Astrophysics*, **312**, 1000.

479 Brun, A.S., Turck-Chiéze, S. and Zahn, J.P. (1999) Standard solar models in the light of new helioseismic constraints. II. Mixing below the convective zone. *Astrophysical Journal*, **525**, 1032.

480 Salaris, M. and Weiss, A. (2002) Homogeneous age dating of 55 galactic globular clusters. Clues to the galaxy formation mechanisms. *Astronomy & Astrophysics*, **388**, 492.

481 De Angeli, F., Piotto, G., Cassisi, S., Busso, G., Recio-Blanco, A., Salaris, M., Aparicio, A. and Rosenberg, A. (2005) Galactic globular cluster relative ages. *Astronomical Journal*, **130**, 116.

482 Rosenberg, A., Saviane, I., Piotto, G. and Aparicio, A. (1999) Galactic globular cluster relative ages. *Astronomical Journal*, **118**, 2306.

483 Stetson, P.B., Van den Berg, D.A. and Bolte, M. (1996) The relative ages of galactic globular clusters. *Publications of the Astronomical Society of the Pacific*, **108**, 560.

484 Freeman, K.C. (1993) The globular clusters-galaxy connection, in *Astronomical Society of the Pacific Conference Series*, vol. 48, edited by G.H. Smith and J.P. Brodie), Proceedings of the 11th Santa Cruz Summer Workshop in Astronomy and Astrophysics, held 19–29 July 1992, at the University of California, Astro-

nomical Society of the Pacific (ASP), p. 608.

485 Eggen, O.J., Lynden-Bell, D. and Sandage, A.R. (1962) Evidence from the motions of old stars that the galaxy collapsed. *Astrophysical Journal*, **136**, 748.

486 Salaris, M., Weiss, A. and Percival, S.M. (2004) The age of the oldest open clusters. *Astronomy & Astrophysics*, **414**, 163.

487 Montgomery, K.A., Marschall, L.A. and Janes, K.A. (1993) CCD photometry of the old open cluster M67. *Astronomical Journal*, **106**, 181.

488 Bedin, L.R., Salaris, M., Piotto, G., Cassisi, S., Milone, A.P., Anderson, J. and King, I.R. (2008b) The puzzling white dwarf cooling sequence in NGC 6791: A simple solution. *Astrophysical Journal*, **679**, L29.

489 Bedin, L.R., Salaris, M., Piotto, G., Anderson, J., King, I.R. and Cassisi, S. (2009) The end of the white dwarf cooling sequence in M4: An efficient approach. *Astrophysical Journal*, **697**, 965.

490 Bedin, L.R., King, I.R., Anderson, J., Piotto, G., Salaris, M., Cassisi, S. and Serenelli, A. (2008a) Reaching the end of the white dwarf cooling sequence in NGC 6791. *Astrophysical Journal*, **678**, 1279.

491 García-Berro, E., Torres, S., Althaus, L.G., Renedo, I., Lorén-Aguilar, P., Córsico, A.H., Rohrmann, R.D., Salaris, M. and Isern, J. (2010) A white dwarf cooling age of 8 Gyr for NGC 6791 from physical separation processes. *Nature*, **465**, 194.

492 Brogaard, K., Bruntt, H., Grundahl, F., Clausen, J.V., Frandsen, S., Van den Berg, D.A. and Bedin, L.R. (2011) Age and helium content of the open cluster NGC 6791 from multiple eclipsing binary members. I. Measurements, methods and first results. *Astronomy & Astrophysics*, **525**, A2.

493 Basu, S. *et al.* (2011) Sounding open clusters: Asteroseismic constraints from Kepler on the properties of NGC 6791 and NGC 6819. *Astrophysical Journal*, **729**, L10.

494 Hilker, M. (2000) Revised Strömgren metallicity calibration for red giants. *Astronomy & Astrophysics*, **355**, 994.

495 Schuster, W.J. and Nissen, P.E. (1989) Ubvy-beta photometry of high-velocity and metal-poor stars. III. Metallicities and ages of the halo stars. *Astronomy & Astrophysics*, **222**, 69.

496 Nissen, P.E. and Schuster, W.J. (1991) Uvby-beta photometry of high-velocity and metal-poor stars. V. Distances, kinematics and ages of halo and disk stars. *Astronomy & Astrophysics*, **251**, 457.

497 Jofré, P. and Weiss, A. (2011) The age of the Milky Way halo stars from the Sloan Digital Sky Survey. *Astronomy & Astrophysics*, **533**, A59.

498 Salaris, M. and Weiss, A. (2001) Atomic diffusion in stellar interiors and field halo subdwarfs ages, in *Astronomical Society of the Pacific Conference Series*, vol. 245, (eds T. von Hippel, C. Simpson and N. Manset), Astronomical Society of the Pacific, p. 367.

499 Catalán, S., Isern, J., García-Berro, E. and Ribas, I. (2008) The initial-final mass relationship of white dwarfs revisited: Effect on the luminosity function and mass distribution. *Monthly Notices of the Royal Astronomical Society*, **387**, 1693.

500 Skumanich, A. (1972) Time scales for CA II emission decay, rotational braking and lithium depletion. *Astrophysical Journal*, **171**, 565.

501 Radick, R.R., Thompson, D.T., Lockwood, G.W., Duncan, D.K. and Baggett, W.E. (1987) The activity, variability, and rotation of lower mainsequence Hyades stars. *Astrophysical Journal*, **321**, 459.

502 Meibom, S., Mathieu, R.D. and Stassun, K.G. (2009) Stellar rotation in M35: Mass-period relations, spin-down rates and gyrochronology. *Astrophysical Journal*, **695**, 679.

503 Barnes, S.A. (2007) Ages for illustrative field stars using gyrochronology: Viability, limitations and errors. *Astrophysical Journal*, **669**, 1167.

504 Epstein, C.R. and Pinsonneault, M.H. (2012) How good of a clock is rotation? The stellar rotation–mass–age relationship for old field stars, arXiv:1203.1618.

505 Mamajek, E.E. and Hillenbrand, L.A. (2008) Improved age estimation for solar-type dwarfs using activity-rotation diagnostics. *Astrophysical Journal*, **687**, 1264.

506 Rocha-Pinto, H.J. and Maciel, W.J. (1998) Metallicity effects on the chromospheric activity-age relation for late-type dwarfs. *Monthly Notices of the Royal Astronomical Society*, **298**, 332.

507 Rocha-Pinto, H.J., Maciel, W.J., Scalo, J. and Flynn, C. (2000) Chemical enrichment and star formation in the Milky Way disk. I. Sample description and chromospheric age-metallicity relation. *Astronomy & Astrophysics*, **358**, 850.

508 Noyes, R.W., Hartmann, L.W., Baliunas, S.L., Duncan, D.K. and Vaughan, A.H. (1984) Rotation, convection and magnetic activity in lower main-sequence stars. *Astrophysical Journal*, **279**, 763.

509 Izotov, Y.I. and Thuan, T.X. (1998) The primordial abundance of ^{4}He revisited. *Astrophysical Journal*, **500**, 188.

510 Olive, K.A., Steigman, G. and Skillman, E.D. (1997) The primordial abundance of ^{4}He: an update. *Astrophysical Journal*, **483**, 788.

511 Burles, S., Nollett, K.M. and Turner, M.S. (2001) Big Bang nucleosynthesis predictions for precision cosmology. *Astrophysical Journal*, **552**, L1.

512 Sievers, J.L., Bond, J.R., Cartwright, J.K. et al. (2003) Cosmological parameters from cosmic background imager observations and comparisons with BOOMERANG, DASI and MAXIMA. *Astrophysical Journal*, **591**, 599.

513 Villanova, S., Piotto, G. and Gratton, R.G. (2009) The helium content of globular clusters: Light element abundance correlations and HB morphology. I. NGC 6752. *Astronomy & Astrophysics*, **499**, 755.

514 Villanova, S., Geisler, D., Piotto, G. and Gratton, R.G. (2012) The helium content of globular clusters: NGC 6121 (M4). *Astrophysical Journal*, **748**, 62.

515 Iben Jr., I. (1968) Age and initial helium abundance of stars in the globular cluster M15. *Nature*, **220**, 143.

516 Salaris, M., Riello, M., Cassisi, S. and Piotto, G. (2004) The initial helium abundance of the galactic globular cluster

system. *Astronomy & Astrophysics*, **420**, 911.

517 Caputo, F., Cayrel, R. and Cayrel de Strobel, G. (1983) The galactic globular cluster system – Helium content versus metallicity. *Astronomy & Astrophysics*, **123**, 135.

518 Sandquist, E.L. (2000) A catalogue of helium abundance indicators from globular cluster photometry. *Monthly Notices of the Royal Astronomical Society*, **313**, 571.

519 Troisi, F., Bono, G., Stetson, P.B. et al. (2011) On a new parameter to estimate the helium content in old stellar systems. *Publications of the Astronomical Society of the Pacific*, **123**, 879.

520 Spite, F. and Spite, M. (1982) Abundance of lithium in unevolved halo stars and old disk stars – Interpretation and consequences. *Astronomy & Astrophysics*, **115**, 357.

521 Thorburn, J.A. (1994) The primordial lithium abundance from extreme subdwarfs: New observations. *Astrophysical Journal*, **421**, 318.

522 Cyburt, R.H., Fields, B.D. and Olive, K.A. (2008) An update on the big bang nucleosynthesis prediction for ^{7}Li: the problem worsens. *Journal of Cosmology and Astroparticle Physics*, **11**, 12.

523 Salaris, M. and Weiss, A. (2001) Atomic diffusion in metal-poor stars. II. Predictions for the Spite plateau. *Astronomy & Astrophysics*, **376**, 955.

524 Richard, O., Michaud, G. and Richer, J. (2005) Implications of WMAP observations on Li abundance and stellar evolution models. *Astrophysical Journal*, **619**, 538.

525 Piau, L. (2008) Lithium isotopes in population II dwarfs. *Astrophysical Journal*, **689**, 1279.

526 Swenson, F.J. (1995) Lithium in halo dwarfs: The undoing of diffusion by mass loss. *Astrophysical Journal*, **438**, L87.

527 Vauclair, S. and Charbonnel, C. (1995) Influence of a stellar wind on the lithium depletion in halo stars: a new step towards the lithium primordial abundance. *Astronomy & Astrophysics*, **295**, 715.

528 Lind, K., Primas, F., Charbonnel, C., Grundahl, F. and Asplund, M. (2009) Signatures of intrinsic Li depletion and Li-Na anti-correlation in the metal-poor globular cluster NGC 6397. *Astronomy & Astrophysics*, **503**, 545.

529 Mucciarelli, A., Salaris, M. and Bonifacio, P. (2012) Giants reveal what dwarfs conceal: Li abundance in lower red giant branch stars as diagnostic of the primordial Li. *Monthly Notices of the Royal Astronomical Society*, **419**, 2195.

530 Fields, B.D. (2011) The primordial lithium problem. *Annual Review of Nuclear and Particle Science*, **61**, 47.

531 Cohen, J.G. (1978) Abundances in globular cluster red giants. I. M3 and M13. *Astrophysical Journal*, **223**, 487.

532 Osborn, W. (1971) Two new CN-strong globular cluster stars. *The Observatory*, **91**, 223.

533 Gratton, R., Sneden, C. and Carretta, E. (2004) Abundance variations within globular clusters. *Annual Review Astronomy & Astrophysics*, **42**, 385.

534 Gratton, R.G., Carretta, E. and Bragaglia, A. (2012) Multiple populations in globular clusters. Lessons learned from the Milky Way globular clusters, *The Astronomy and Astrophysics Review*, **20**, 50.

535 Salaris, M., Cassisi, S. and Weiss, A. (2002) Red giant branch stars: The theoretical framework. *Publications of the Astronomical Society of the Pacific*, **114**, 375.

536 Villanova, S., Geisler, D. and Piotto, G. (2009) Detailed abundances of red giants in the globular cluster NGC 1851: C + N + O and the origin of multiple populations. *Astrophysical Journal*, **722**, L18.

537 Yong, D., Grundahl, F., D'Antona, F., Karakas, A.I., Lattanzio, J.C. and Norris, J.E. (2009) A large C + N + O abundance spread in giant stars of the globular cluster NGC 1851. *Astrophysical Journal Letters*, **695**, L62.

538 Marino, A.F., Milone, A.P., Sneden, C. *et al.* (2012) The double sub-giant branch of NGC 6656 (M22): a chemical charac-

terization, *Astronomy & Astrophysics*, **541**, 15.

539 Gratton, R.G., Bonifacio, P., Bragaglia, A. *et al.* (2001) The O–Na and Mg–Al anticorrelations in turn-off and early subgiants in globular clusters. *Astronomy & Astrophysics*, **369**, 87.

540 Carretta, E., Bragaglia, A., Gratton, R.G. *et al.* (2009) Na-O anticorrelation and HB. VII. The chemical composition of first and second-generation stars in 15 globular clusters from GIRAFFE spectra. *Astronomy & Astrophysics*, **505**, 117.

541 Cottrell, P.L. and Da Costa, G.S. (1981) Correlated cyanogen and sodium anomalies in the globular clusters 47 Tuc and NGC 6752. *Astrophysical Journal*, **245**, L79.

542 Carretta, E., Bragaglia, A., Gratton, R.G., Recio-Blanco, A., Lucatello, S., D'Orazi, V. and Cassisi, S. (2010) Properties of stellar generations in globular clusters and relations with global parameters. *Astronomy & Astrophysics*, **516**, 55.

543 Carretta, E., Bragaglia, A., Gratton, R.G., Lucatello, S., Bellazzini, M., Catanzaro, G., Leone, F., Momany, Y., Piotto, G. and D'Orazi, V. (2010) Detailed abundances of a large sample of giant stars in M54 and in the Sagittarius nucleus. *Astronomy & Astrophysics*, **520**, 95.

544 Marino, A.F., Milone, A.P., Piotto, G., Villanova, S., Gratton, R., D'Antona, F., Anderson, J. *et al.* (2011) Sodium-oxygen anticorrelation and neutron-capture elements in Omega Centauri stellar populations. *Astrophysical Journal*, **731**, 64.

545 Piotto, G. (2010) Observational evidence of multiple stellar populations in star clusters. *Publications of the Korean Astronomical Society*, **25**, 91.

546 Anderson, J. (1998) Ph.D. Thesis, University of California, Berkeley.

547 Bedin, L.R., Piotto, G., Anderson, J., Cassisi, S., King, I.R., Momany, Y. and Carraro, G. (2004) ω Centauri: The population puzzle goes deeper. *Astrophysical Journal*, **605**, L125.

548 Norris, J.E. (2004) The helium abundances of ω Centauri. *Astrophysical Journal*, **612**, L25.

549 D'Antona, F., Bellazzini, M., Caloi, V., Fusi Pecci, F., Galleti, S. and Rood, R.T. (2005) A helium spread among the main-sequence stars in NGC 2808. *Astrophysical Journal*, **631**, 868.

550 Piotto, G., Bedin, L.R., Anderson, J., King, I.R., Cassisi, S., Milone, A.P., Villanova, S., Pietrinferni, A. and Renzini, A. (2007) A triple main sequence in the globular cluster NGC 2808. *Astrophysical Journal*, **661**, L53.

551 Milone, A.P., Piotto, G., Bedin, L.R., King, I.R., Anderson, J., Marino, A.F., Bellini, A. *et al.* (2012) Multiple stellar populations in 47 Tucanae. *Astrophysical Journal*, **744**, 58.

552 Milone, A.P., Marino, A.F., Piotto, G., Bedin, L.R., Anderson, J., Aparicio, A., Cassisi, S. *et al.* (2012) A double main sequence in the globular cluster NGC 6397. *Astrophysical Journal*, **745**, 27.

553 Bragaglia, A., Carretta, E., Gratton, R.G., Lucatello, S., Milone, A., Piotto, G., D'Orazi, V. *et al.* (2011) X-shooter observations of main-sequence stars in the globular cluster NGC 2808: First chemical tagging of a He-normal and a He-rich dwarf. *Astrophysical Journal*, **720**, L41.

554 Milone, A.P., Bedin, L.R., Piotto, G., Anderson, J., King, I.R., Sarajedini, A., Dotter, A. *et al.* (2008) The ACS survey of galactic globular clusters. III. The double subgiant branch of NGC 1851. *Astrophysical Journal*, **673**, 241.

555 Cassisi, S., Salaris, M., Pietrinferni, A., Piotto, G., Milone, A.P., Bedin, L.R. and Anderson, J. (2008) The double subgiant branch of NGC 1851: The role of the CNO abundance. *Astrophysical Journal*, **672**, L115.

556 D'Antona, F. and Caloi, V. (2008) The fraction of second generation stars in globular clusters from the analysis of the horizontal branch. *Monthly Notices of the Royal Astronomical Society*, **390**, 693.

557 Yoon, S.-J., Joo, S.-J., Ree, C.H., Han, S.-I., Kim, D.-G. and Lee, Y.-W. (2008) On the origin of bimodal horizontal branches in massive globular clusters: The case of NGC 6388 and NGC 6441. *Astrophysical Journal*, **677**, 1080.

558 Dalessandro, E., Salaris, M., Ferraro, F.R., Cassisi, S., Lanzoni, B., Rood, R.T., Fusi Pecci, F. and Sabbi, E. (2011) The peculiar horizontal branch of NGC 2808. *Monthly Notices of the Royal Astronomical Society*, **410**, 694.

559 Marino, A.F., Villanova, S., Milone, A.P., Piotto, G., Lind, K., Geisler, D. and Stetson, P.B. (2011) Sodium-oxygen anticorrelation among horizontal branch stars in the globular cluster M4. *Astrophysical Journal*, **730**, L16.

560 Gratton, R.G., Lucatello, S., Carretta, E., Bragaglia, A., D'Orazi, V. and Momany, Y. (2011) The Na-O anticorrelation in horizontal branch stars. I. NGC 2808. *Astronomy & Astrophysics*, **534**, 123.

561 Salaris, M., Weiss, A., Ferguson, J.W. and Fusilier, D.J. (2006) On the primordial scenario for abundance variations within globular clusters: the isochrone test. *Astrophysical Journal*, **645**, 1131.

562 Pietrinferni, A., Cassisi, S., Salaris, M., Percival, S. and Ferguson, J.W. (2009) A large stellar evolution database for population synthesis studies. V. Stellar models and isochrones with CNONa abundance anticorrelations. *Astrophysical Journal*, **697**, 275.

563 Marín-Franch, A., Cassisi, S., Aparicio, A. and Pietrinferni, A. (2010) The impact of enhanced He and CNONa abundances on globular cluster relative age-dating methods. *Astrophysical Journal*, **714**, 1072.

564 de Mink, S.E., Pols, O.R., Langer, N. and Izzard, R.G. (2009) Massive binaries as the source of abundance anomalies in globular clusters. *Astronomy & Astrophysics*, **507**, L1.

565 Yi, S.K. (2009) Multiple population theory: the extreme helium population problem, in *The Ages of Stars*, Proceedings of IAU Symposium 258, (eds E.E. Mamajek, D.R. Soderblom and R.F.G. Wyse), vol. 4, Cambridge University Press, p. 253.

566 D'Antona, F., Gratton, R.G. and Chieffi, A. (1983) CNO self-pollution in globular clusters – A model and its possible observational tests. *Memorie della Società Astronomica Italiana*, **54**, 173.

567 Ventura, P., Carini, R. and D'Antona, F. (2011) A deep insight into the Mg-Al-Si nucleosynthesis in massive asymptotic giant branch and super-asymptotic giant branch stars. *Monthly Notices of the Royal Astronomical Society*, **415**, 3865.

568 Decressin, T., Meynet, G., Charbonnel, C., Prantzos, N. and Ekström, S. (2007) Fast rotating massive stars and the origin of the abundance patterns in galactic globular clusters. *Astronomy & Astrophysics*, **464**, 1029.

569 Renzini, A. (2008) Origin of multiple stellar populations in globular clusters and their helium enrichment. *Monthly Notices of the Royal Astronomical Society*, **391**, 354.

570 Valcarce, A.A.R. and Catelan, M. (2011) Formation of multiple populations in globular clusters: another possible scenario. *Astronomy & Astrophysics*, **533**, 120.

571 Cameron, A.G.W. and Fowler, W.A. (1971) Lithium and the *s*-process in red-giant stars. *Astrophysical Journal*, **164**, 111.

572 Bekki, K., Campbell, S.W., Lattanzio, J.C. and Norris, J.E. (2007) Origin of abundance inhomogeneity in globular clusters. *Monthly Notices of the Royal Astronomical Society*, **377**, 335.

573 D'Ercole, A., Vesperini, E., D'Antona, F., McMillan, S.L.W. and Recchi, S. (2008) Formation and dynamical evolution of multiple stellar generations in globular clusters. *Monthly Notices of the Royal Astronomical Society*, **391**, 825.

574 D'Ercole, A., D'Antona, F. and Vesperini, E. (2011) Formation of multiple populations in globular clusters: constraints on the dilution by pristine gas. *Monthly Notices of the Royal Astronomical Society*, **415**, 1304.

575 Conroy, C. and Spergel, D.N. (2011) On the formation of multiple stellar populations in globular clusters. *Astrophysical Journal*, **726**, 36.

576 Bekki, K. (2011) Secondary star formation within massive star clusters: origin of multiple stellar populations in globular clusters. *Monthly Notices of the Royal Astronomical Society*, **412**, 224.

577 Aparicio, A. and Hidalgo, S.L. (2009) IAC-pop: Finding the star formation history of resolved galaxies. *Astronomical Journal*, **138**, 558.

578 de Boer, T.J.L., Tolstoy, E., Hill, V., Saha, A., Olsen, K., Starkenburg, E., Lemasle, B., Irwin, M.J. and Battaglia, G. (2012) The star formation and chemical evolution history of the sculptor dwarf spheroidal galaxy. *Astronomy & Astrophysics*, **539**, A103.

579 Dolphin, A.E. (2002) Numerical methods of star formation history measurement and applications to seven dwarf spheroidals. *Monthly Notices of the Royal Astronomical Society*, **332**, 91.

580 Harris, J. and Zaritsky, D. (2001) A method for determining the star formation history of a mixed stellar population. *Astrophysical Journal Supplement*, **136**, 25.

581 Gallart, C., Aparicio, A. and Vilchez, J.M. (1996) The local group dwarf irregular galaxy NGC 6822. I. The stellar content. *Astronomical Journal*, **112**, 1928.

582 Hidalgo, S.L., Aparicio, A., Skillman, E. et al. (2011) The ACS LCID project. V. The star formation history of the dwarf galaxy LGS-3: Clues to cosmic reionization and feedback. *Astrophysical Journal*, **730**, 14.

583 Monelli, M. et al. (2010) The ACS LCID project. III. The Star Formation History of the Cetus dSph Galaxy: A post-reionization fossil. *Astrophysical Journal*, **720**, 1225.

584 Monelli, M. et al. (2012) The ACS LCID project. VII. The blue stragglers population in the isolated dSph galaxies Cetus and Tucana. *Astrophysical Journal*, **744**, 157.

585 Mighell, K.J. (1999) Parameter Estimation in astronomy with poisson-distributed data. I. The χ^2_{gamma} statistic. *Astrophysical Journal*, **518**, 380.

586 Monelli, M. et al. (2010) The ACS LCID project. VI. The star formation history of the Tucana dSph and the relative ages of the isolated dSph galaxies. *Astrophysical Journal*, **722**, 1864.

587 Dolphin, A.E. (2012) On the estimation of systematic uncertainties of star formation histories, *Astrophysical Journal*, **751**, 60.

588 Hernandez, X., Valls-Gabaud, D. and Gilmore, G. (1999) Deriving star formation histories: Inverting Hertzsprung–Russell diagrams through a variational calculus maximum likelihood method. *Monthly Notices of the Royal Astronomical Society*, **304**, 705.

589 Hernandez, X. and Valls-Gabaud, D. (2008) A robust statistical estimation of the basic parameters of single stellar populations – I. Method. *Monthly Notices of the Royal Astronomical Society*, **383**, 1603.

590 Small, E. (2011) Ph.D. thesis, Liverpool John Moores University.

591 Tinsley, B.M. (1968) Evolution of the stars and gas in galaxies. *Astrophysical Journal*, **151**, 547.

592 Percival, S.M., Salaris, M., Cassisi, S. and Pietrinferni, A. (2009) A large stellar evolution database for population synthesis studies. IV. Integrated properties and spectra. *Astrophysical Journal*, **690**, 427.

593 Schiavon, R.P. (2007) Population synthesis in the blue. IV. Accurate model predictions for lick indices and UBV colors in single stellar populations. *Astrophysical Journal Supplement*, **171**, 146.

594 Buzzoni, A. (1989) Evolutionary population synthesis in stellar systems. I. A global approach. *Astrophysical Journal Supplement*, **71**, 817.

595 González Delgado, R.M., Cerviño, M., Martins, L.P., Leitherer, C. and Hauschildt, P.H. (2005) Evolutionary stellar population synthesis at high spectral resolution: optical wavelengths. *Monthly Notices of the Royal Astronomical Society*, **357**, 945.

596 Bruzual, G. and Charlot, S. (2003) Stellar population synthesis at the resolution of 2003. *Monthly Notices of the Royal Astronomical Society*, **344**, 1000.

597 Jimenez, R., MacDonald, J., Dunlop, J.S., Padoan, P. and Peacock, J.A. (2004) Synthetic stellar populations: Single stellar populations, stellar interior models and primordial protogalaxies. *Monthly Notices of the Royal Astronomical Society*, **112**, 583.

598 Conroy, C., Gunn, J.E. and White, M. (2009) The Propagation of uncertainties in stellar population synthesis modeling. I. The relevance of uncertain aspects of stellar evolution and the initial mass function to the derived physical properties of galaxies. *Astrophysical Journal*, **699**, 486.

599 Conroy, C. and Gunn, J.E. (2010) The propagation of uncertainties in stellar population synthesis modeling. III. Model calibration, comparison and evaluation. *Astrophysical Journal*, **712**, 833.

600 Raimondo, G. (2009) Joint analysis of near-infrared properties and surface brightness fluctuations of Large Magellanic Cloud star clusters. *Astrophysical Journal*, **700**, 1247.

601 Tantalo, R. and Chiosi, C. (2004) Star formation history in early-type galaxies – I. The line absorption indices diagnostics. *Monthly Notices of the Royal Astronomical Society*, **353**, 405.

602 Leitherer, C., Schaerer, D., Goldader, J.D., González Delgado, R.M., Robert, C., Kune, D.F., de Mello, D.F., Devost, D. and Heckman, T.M. (1999) Starburst99: Synthesis models for galaxies with active star formation. *Astronomy & Astrophysics Supplement*, **123**, 3.

603 Schulz, J., Fritze-v. Alvensleben, U., Möller, C.S. and Fricke, K.J. (2002) Spectral and photometric evolution of simple stellar populations at various metallicities. *Astronomy & Astrophysics*, **392**, 1.

604 Thomas, D., Maraston, C. and Johansson, J. (2011) Flux-calibrated stellar population models of Lick absorption-line indices with variable element abundance ratios. *Monthly Notices of the Royal Astronomical Society*, **412**, 2183.

605 Silva, L., Granato, G.L., Bressan, A. and Danese, L. (1998) Modeling the effects of dust on galactic spectral energy distributions from the ultraviolet to the millimeter band. *Astrophysical Journal*, **509**, 103.

606 Vazdekis, A., Sánchez-Blázquez, P., Falcón-Barroso, J., Cenarro, A.J., Beasley, M.A., Cardiel, N., Gorgas, J. and Peletier, R.F. (2010) Evolutionary stellar population synthesis with MILES –

I. The base models and a new line index system. *Monthly Notices of the Royal Astronomical Society*, **404**, 1639.

607 Maraston, C. and Strömbäck, G. (2011) Stellar population models at high spectral resolution. *Monthly Notices of the Royal Astronomical Society*, **418**, 2785.

608 Worthey, G. (1994) Comprehensive stellar population models and the disentanglement of age and metallicity effects. *Astrophysical Journal Supplement*, **95**, 107.

609 Le Borgne, D., Rocca-Volmerange, B., Prugniel, P., Lançon, A., Fioc, M. and Soubiran, C. (2004) Evolutionary synthesis of galaxies at high spectral resolution with the code PEGASE-HR. Metallicity and age tracers. *Astronomy & Astrophysics*, **425**, 881.

610 Prugniel, P., Vauglin, I. and Koleva, M. (2011) The atmospheric parameters and spectral interpolator for the stars of MILES. *Astronomy & Astrophysics*, **531**, A165.

611 Westera, P., Lejeune, T., Buser, R., Cuisinier, F. and Bruzual, G. (2002) A standard stellar library for evolutionary synthesis. III. Metallicity calibration. *Astronomy & Astrophysics*, **381**, 524.

612 Prugniel, P. and Soubiran, C. (2001) A database of high and medium-resolution stellar spectra. *Astronomy & Astrophysics*, **369**, 1048.

613 Bertone, E., Buzzoni, A., Chávez, M. and Rodríguez-Merino, L.H. (2008) Probing Atlas model atmospheres at high spectral resolution. Stellar synthesis and reference template validation. *Astronomy & Astrophysics*, **485**, 823.

614 Dotter, A., Chaboyer, B., Jevremović, D., Kostov, V., Baron, E. and Ferguson, J.W. (2008) The Dartmouth stellar evolution database. *Astrophysical Journal Supplement*, **178**, 89.

615 Valdes, F., Gupta, R., Rose, J.A., Singh, H.P. and Bell, D.J. (2004) The Indo-US library of coudé feed stellar spectra. *Astrophysical Journal Supplement*, **152**, 251.

616 Coelho, P., Barbuy, B., Meléndez, J., Schiavon, R.P. and Castilho, B.V. (2005) A library of high resolution synthetic stellar spectra from 300 nm to 1.8 μm

with solar and α-enhanced composition. *Astronomy & Astrophysics*, **443**, 735.

617 Lejeune, T. and Schaerer, D. (2001) Database of Geneva stellar evolution tracks and isochrones for $(UBV)_J(RI)_C$JHKLL′M, HST-WFPC2, Geneva and Washington photometric systems. *Astronomy & Astrophysics*, **366**, 538.

618 Rayner, J.T., Cushing, M.C. and Vacca, W.D. (2009) The Infrared Telescope Facility (IRTF) spectral library: Cool stars. *Astrophysical Journal Supplement*, **185**, 289.

619 Aringer, B., Girardi, L., Nowotny, W., Marigo, P. and Lederer, M.T. (2009) Synthetic photometry for carbon rich giants. I. Hydrostatic dust-free models. *Astronomy & Astrophysics*, **503**, 913.

620 Bertelli, G., Girardi, L., Marigo, P. and Nasi, E. (2008) Scaled solar tracks and isochrones in a large region of the Z-Y plane. I. From the ZAMS to the TP-AGB end for $0.15-2.5\,M_\odot$ stars. *Astronomy & Astrophysics*, **484**, 815.

621 Girardi, L., Bressan, A., Bertelli, G. and Chiosi, C. (2000) Evolutionary tracks and isochrones for low- and intermediate-mass stars: From 0.15 to $7\,M_\odot$ and from $Z = 0.0004$ to 0.03. *Astronomy & Astrophysics Supplement*, **141** 371.

622 Lançon, A. and Mouhcine, M. (2002) The modelling of intermediate-age stellar populations. II. Average spectra for upper AGB stars and their use. *Astronomy & Astrophysics*, **393**, 167.

623 Gustafsson, B., Edvardsson, B., Eriksson, K., Jørgensen, U.G., Nordlund, Å. and Plez, B. (2008) A grid of MARCS model atmospheres for late-type stars. I. Methods and general properties. *Astronomy & Astrophysics*, **486**, 951.

624 Yi, S.K., Kim, Y.-C. and Demarque, P. (2003) The Y^2 stellar evolutionary tracks. *Astrophysical Journal Supplement*, **144**, 259.

625 Sánchez-Blázquez, P., Peletier, R.F., Jiménez-Vicente, J., Cardiel, N., Cenarro, A.J., Falcón-Barroso, J., Gorgas, J., Selam, S. and Vazdekis, A. (2006) Medium-resolution Isaac Newton Telescope library of empirical spectra. *Monthly No-*

tices of the *Royal Astronomical Society*, **371**, 703.

626 Martins, L.P., González Delgado, R.M., Leitherer, C., Cerviño, M. and Hauschildt, P. (2005) A high-resolution stellar library for evolutionary population synthesis. *Monthly Notices of the Royal Astronomical Society*, **358**, 49.

627 Van den Berg, D.A., Bergbusch, P.A. and Dowler, P.D. (2006) The Victoria–Regina stellar models: Evolutionary tracks and isochrones for a wide range in mass and metallicity that allow for empirically constrained amounts of convective core overshooting. *Astrophysical Journal Supplement*, **162**, 375.

628 Gregg, M.D., Silva, D., Rayner, J., Valdes, F., Worthey, G., Pickles, A., Rose, J.A., Vacca, W. and Carney, B. (2004) The HST/STIS next generation spectral library. *Bulletin of the American Astronomical Society*, **36**, 1496.

629 Munari, U., Sordo, R., Castelli, F. and Zwitter, T. (2005) An extensive library of 2500–10 500 Å synthetic spectra. *Astronomy & Astrophysics*, **442**, 1127.

630 Pickles, A.J. (1998) A stellar spectral flux library: 1150–25 000 Å. *Publications of the Astronomical Society of the Pacific*, **110**. 863.

631 Brott, I. and Hauschildt, P.H. (2005) A PHOENIX model atmosphere grid for gaia in Turon C., in *The Three-Dimensional Universe with Gaia: A PHOENIX Model Atmosphere Grid for Gaia*, (eds K.S. O'Flaherty and M.A.C. Perryman), ESA SP-576. ESA, Noordwijk, p. 565.

632 Le Borgne, J.-F., Bruzual, G., Pelló, R., Lançon, A., Rocca-Volmerange, B., Sanahuja, B., Schaerer, D., Soubiran, C. and Vílchez-Gómez, R. (2003) STELIB: A library of stellar spectra at $R = 2000$. *Astronomy & Astrophysics*, **402**, 433.

633 Rauch, T. (2003) A grid of synthetic ionizing spectra for very hot compact stars from NLTE model atmospheres. *Astronomy & Astrophysics*, **403**, 709.

634 Jehin, E., Bagnulo, S., Melo, C., Ledoux, C. and Cabanac, R. (2005) The UVES paranal observatory project: a public library of high resolution stellar spectra, (eds V. Hill, P. Francois, F. Pri-

mas), IAU Symposium 228, Cambridge University Press, Cambridge, p. 261.

635 Smith, L.J., Norris, R.P.F. and Crowther, P.A. (2002) Realistic ionizing fluxes for young stellar populations from 0.05 to $2Z_\odot$. *Monthly Notices of the Royal Astronomical Society*, **337**, 1309.

636 Chen, Y., Trager, S., Peletier, R. and Lançon, A. (2011) XSL: The X-Shooter Spectral Library. *Journal of Physics Conference Series*, **328**, 2023.

637 Lanz, T. and Hubeny, I. (2003), A grid of non-LTE line-blanketed model atmospheres of O-type stars. *Astrophysical Journal Supplement*, **146**, 417.

638 Lanz, T. and Hubeny, I. (2007) A grid of NLTE line-blanketed model atmospheres of early B-type stars. *Astrophysical Journal Supplement*, **169**, 83.

639 Rodríguez-Merino, L.H., Chavez, M., Bertone, E. and Buzzoni, A. (2005) UVBLUE: A new high-resolution theoretical library of ultraviolet stellar spectra. *Astrophysical Journal*, **626**, 411.

640 Kroupa, P. (2002) The initial mass function of stars: Evidence for uniformity in variable systems. *Science*, **295**, 82.

641 Bessell, M.S. and Brett, J.M. (1988) JHKLM photometry – Standard systems, passbands and intrinsic colors. *Publications of the Astronomical Society of the Pacific*, **100**, 1134.

642 Dotter, A., Chaboyer, B., Ferguson, J.W., Lee, H.-C., Worthey, G., Jevremović, D. and Baron, E. (2007) Stellar population models and individual element abundances. I. Sensitivity of stellar evolution models. *Astrophysical Journal*, **666**, 403.

643 Lee, H.-C., Worthey, G., Dotter, A., Chaboyer, B., Jevremović, D., Baron, E., Briley, M.M., Ferguson, J.W., Coelho, P. and Trager, S.C. (2009) Stellar population models and individual element abundances. II. Stellar spectra and integrated light models. *Astrophysical Journal*, **694**, 902.

644 Bothun, G.D., Romanishin, W., Strom, S.E. and Strom, K.M. (1984) A possible relationship between metal abundance and luminosity for disk galaxies. *Astronomical Journal*, **89**, 1300.

645 Hempel, M. and Kissler-Patig, M. (2004) Extragalactic globular clusters in the

near infrared. IV. Quantifying the age structure using Monte-Carlo simulations. *Astronomy & Astrophysics*, **419**, 863.

646 Hempel, M., Hilker, M., Kissler-Patig, M., Puzia, T.H., Minniti, D. and Goudfrooij, P. (2003) Extragalactic globular clusters in the near infrared III. NGC 5846 and NGC 7192. Quantifying the age distribution of sub-populations. *Astronomy & Astrophysics*, **405**, 487.

647 Elson, R.A.W. and Fall, S.M. (1985) Age calibration and age distribution for rich star clusters in the Large Magellanic Cloud. *Astrophysical Journal*, **299**, 211.

648 James, P.A., Salaris, M., Davies, J.I., Phillipps, S. and Cassisi, S. (2006) Optical/near-infrared colours of early-type galaxies and constraints on their star formation histories. *Monthly Notices of the Royal Astronomical Society*, **367**, 339.

649 Yi, S.K. (2004) Uncertainties of synthetic integrated colors as age indicators. *Astrophysical Journal*, **582**, 202.

650 Li, Z. and Han, Z. (2008) Colour pairs for constraining the age and metallicity of stellar populations. *Monthly Notices of the Royal Astronomical Society*, **385**, 1270.

651 Salaris, M. and Cassisi, S. (2007) Colour–colour diagrams and extragalactic globular cluster ages. Systematic uncertainties using the $(V-K)-(V-I)$ diagram. *Astronomy & Astrophysics*, **461**, 493.

652 Kaviraj, S., Rey, S.-C., Rich, R.M., Yoon, S.-J. and Yi, S.K. (2007) Better age estimation using ultraviolet-optical colours: breaking the age-metallicity degeneracy. *Monthly Notices of the Royal Astronomical Society*, **381**, L74.

653 Yi, S.K., Peng, E., Ford, H., Kaviraj, S. and Yoon, S.-J. (2004) Globular clusters as probes of galaxy evolution: NGC 5128. *Monthly Notices of the Royal Astronomical Society*, **349**, 1493.

654 Leblanc, F., Hui-Bon-Hoa, A. and Khalack, V.R. (2010) Stratification of the elements in the atmospheres of blue horizontal-branch stars. *Monthly Notices of the Royal Astronomical Society*, **409**, 1606.

655 Brodie, J.P. and Strader, J. (2005) Extragalactic globular clusters and galaxy

formation. *Annual Review of Astronomy and Astrophysics*, **44**, 193.

656 Blakeslee, J.P., Cantiello, M. and Peng, E.W. (2010) The mass-metallicity relation of globular clusters in the context of nonlinear color-metallicty relations. *Astrophysical Journal*, **710**, 51.

657 Yoon, S.-J., Yi, S.K. and Lee, Y.-W. (2006) Explaining the color distributions of globular cluster systems in elliptical galaxies. *Science*, **311**, 1129.

658 Peng, E.W., Jordán, A., Côté, P., Blakeslee, J.P., Ferrarese, L., Mei, S., West, M.J., Merritt, D., Milosavljević, M. and Tonry, J.L. (2006) The ACS Virgo cluster survey. IX. The color distributions of globular cluster systems in early-type galaxies. *Astrophysical Journal*, **639**, 95.

659 Mieske, S., Jordán, A., Côté, P., Peng, E.W., Ferrarese, L., Blakeslee, J.P., Mei, S., Baumgardt, H., Tonry, J.L., Infante, L. and West, M.J. (2010) The ACS Fornax cluster survey. IX. The color-magnitude relation of globular cluster systems. *Astrophysical Journal*, **710**, 1672.

660 Alves-Brito, A., Hau, G.K.T., Forbes, D.A., Spitler, L.R., Strader, J., Brodie, J.P. and Rhode, K.L. (2011) Spectra of globular clusters in the Sombrero galaxy: Evidence for spectroscopic metallicity bimodality. *Monthly Notices of the Royal Astronomical Society*, **417**, 1823.

661 Woodley, K.A., Harris, W.E., Puzia, T.H., Gómez, M., Harris, G.L.H. and Geisler, D. (2010) The ages, metallicities and alpha element enhancements of globular clusters in the elliptical NGC 5128: A homogeneous spectroscopic study with Gemini/Gemini multi-object spectrograph. *Astrophysical Journal*, **708**, 1335.

662 Yoon, S.-J., Sohn, S.T., Lee, S.-Y., Kim, H.-S., Cho, J., Chung, C. and Blakeslee, J.P. (2011) Nonlinear color-metallicity relations of globular clusters. II. A test on the nonlinearity scenario for color bimodality using the *u*-band colors: The Case of M87 (NGC 4486). *Astrophysical Journal*, **743**, 149.

663 Anders, P., Bissantz, N., Fritze-v. Alvensleben, U. and de Grijs, R. (2004) Analysing observed star cluster SEDs with evolutionary synthesis models: systematic uncertainties. *Monthly Notices of the Royal Astronomical Society*, **347**, 196.

664 de Grijs, R., Anders, P., Lamers, H.J.G.L.M., Bastian, N., Fritze-v. Alvensleben, U., Parmentier, G., Sharina, M.E. and Yi, S. (2005) Systematic uncertainties in the analysis of star cluster parameters based on broad-band imaging observations. *Monthly Notices of the Royal Astronomical Society*, **359**, 874.

665 Calzetti, D., Armus, L., Bohlin, R.C., Kinney, A.L., Koornneef, J. and Storchi-Bergmann, T. (2000) The dust content and opacity of actively star-forming galaxies. *Astrophysical Journal*, **533**, 682.

666 Groenewegen, M.A.T. (2006) The mid- and far-infrared colours of AGB and post-AGB stars. *Astronomy & Astrophysics*, **448**, 181.

667 Piovan, L., Tantalo, R., and Chiosi, C. (2006) Modelling galaxy spectra in presence of interstellar dust – I. The model of interstellar medium and the library of dusty single stellar populations. *Monthly Notices of the Royal Astronomical Society*, **366**, 923.

668 Marigo, P., Girardi, L., Bressan, A., Groenewegen, M.A.T., Silva, L. and Granato, G.L. (2008) Evolution of asymptotic giant branch stars. II. Optical to far-infrared isochrones with improved TP-AGB models. *Astronomy & Astrophysics*, **482**, 883.

669 Carter, D., Smith, D.J.B., Percival, S.M., Baldry, I.K., Collins, C.A., James, P.A., Salaris, M., Simpson, C., Stott, J.P. and Mobasher, B. (2009) Optical and near-infrared colours as a discriminant of the age and metallicity of stellar populations. *Monthly Notices of the Royal Astronomical Society*, **397**, 695.

670 Anders, P., Lamers, H.J.G.L.M. and Baumgardt, H. (2009) The photometric evolution of dissolving star clusters. II. Realistic models. Colours and M/L ratios. *Astronomy & Astrophysics*, **502**, 817.

671 Baumgardt, H. and Makino, J. (2003) Dynamical evolution of star clusters in tidal fields. *Monthly Notices of the Royal Astronomical Society*, **340**, 227.

672 Cerviño, M. and Valls-Gabaud, D. (2003) On biases in the predictions of stellar population synthesis models. *Monthly Notices of the Royal Astronomical Society*, **338**, 481.

673 Chiosi, C., Bertelli, G. and Bressan, A. (1988) Integrated colours and ages of LMC clusters – The nature of the bi-modal distribution of the (*B–V*) colours. *Astronomy & Astrophysics*, **196**, 84.

674 Fagiolini, M., Raimondo, G. and Degl'Innocenti, S. (2007) Monte Carlo simulations of metal-poor star clusters. *Astronomy & Astrophysics*, **462**, 107.

675 Popescu, B. and Hanson, M.M. (2010) MASSCLEANage. Stellar cluster ages from integrated colors. *Astrophysical Journal*, **724**, 296.

676 McLaughlin, D.E. (1994) An analytical study of the globular-cluster luminosity function. *Publications of the Astronomical Society of the Pacific*, **106**, 47.

677 Li, Z. and Han, Z. (2008) How binary interactions affect spectral stellar population synthesis. *Astrophysical Journal*, **685**, 225.

678 Li, Z.-M. and Han, Z.-W. (2009) Fitting formulae for the effects of binary interactions on lick indices and colors of stellar populations. *Research in Astronomy and Astrophysics*, **9**, 191.

679 Zhang, F., Han, Z., Li, L. and Hurley, J.R. (2004) Evolutionary population synthesis for binary stellar populations. *Astronomy & Astrophysics*, **415**, 117.

680 Zhang, F., Han, Z., Li, L. and Hurley, J.R. (2005) Inclusion of binaries in evolutionary population synthesis. *Monthly Notices of the Royal Astronomical Society*, **357**, 1088.

681 Hurley, J.R., Tout, C.A. and Pols, O.R. (2002) Evolution of binary stars and the effect of tides on binary populations. *Monthly Notices of the Royal Astronomical Society*, **329**, 897.

682 Chabrier, G. (2003) The galactic disk mass function: Reconciliation of the Hubble Space Telescope and nearby determinations. *Astrophysical Journal*, **586**, L133.

683 Bell, E.F. and de Jong, R.S. (2001) Stellar mass-to-light ratios and the Tully–Fisher relation. *Astrophysical Journal*, **550**, 212.

684 Taylor, E.N. *et al.* (2011) Galaxy And Mass Assembly (GAMA): Stellar mass estimates. *Monthly Notices of the Royal Astronomical Society*, **418**, 1587.

685 Fryer, C.L. and Kalogera, V. (2001) Theoretical black hole mass distributions. *Astrophysical Journal*, **554**, 548.

686 McLaughlin, D.E. and van der Marel, R.P. (2005) Resolved massive star clusters in the Milky Way and its satellites: Brightness profiles and a catalog of fundamental parameters. *Astrophysical Journal Supplement Series*, **161**, 304.

687 Tonry, J. and Schneider, D.P. (1988) A new technique for measuring extragalactic distances. *Astrophysical Journal*, **96**, 807.

688 Blakeslee, J.P., Ajhar, E.A. and Tonry, J.L. (1999) Distances from surface brightness fluctuations, in *Post-Hipparcos Cosmic Candles*, (eds A. Heck and F. Caputo), Kluwer Academic Publishers, p. 181.

689 Blakeslee, J.P., Vazdekis, A. and Ajhar, E.A. (2001) Stellar populations and surface brightness fluctuations: new observations and models. *Monthly Notices of the Royal Astronomical Society*, **320**, 193.

690 Cantiello, M., Raimondo, G., Brocato, E. and Capaccioli, M. (2003) New optical and near-infrared surface brightness fluctuation models: A primary distance indicator ranging from globular clusters to distant galaxies? *Astronomical Journal*, **125**, 2783.

691 Liu, M.C., Charlot, S. and Graham, J.R. (2000) Theoretical predictions for surface brightness fluctuations and implications for stellar populations of elliptical galaxies. *Astrophysical Journal*, **543**, 644.

692 Marín-Franch, A. and Aparicio, A. (2006) Surface-brightness fluctuations in stellar populations. IAC-star models for the optical and near-IR wavelengths. *Astronomy & Astrophysics*, **450**, 979.

693 Mouhcine, M., González, R.A. and Liu, M.C. (2005) New near-infrared surface brightness fluctuation models. *Monthly Notices of the Royal Astronomical Society*, **362**, 1208.

694 Lee, H.-C., Worthey, G. and Blakeslee, J.P. (2010) Effects of α-element enhancement and the thermally pulsing-asymptotic giant branch on surface brightness fluctuation magnitudes and broadband colors. *Astrophysical Journal*, **710**, 421.

695 Ferrarese, L., Ford, H.C., Huchra, J., Kennicutt, Jr., R.C., Mould, J.R., Sakai, S., Freedman, W.L., Stetson, P.B., Madore, B.F., Gibson, B.K., Graham, J.A., Hughes, S.M., Illingworth, G.D., Kelson, D.D., Macri, L., Sebo, K. and Silbermann, N.A. (2000) A database of Cepheid distance moduli and tip of the red giant branch, globular cluster luminosity function, planetary nebula luminosity function and surface brightness fluctuation data useful for distance determinations. *Astrophysical Journal Supplement*, **128**, 431.

696 Tonry, J.L., Dressler, A., Blakeslee, J.P., Ajhar, E.A., Fletcher, A.B., Luppino, G.A., Metzger, M.R. and Moore, C.B. (2001) The SBF survey of galaxy distances. IV. SBF magnitudes, colors and distances. *Astrophysical Journal*, **546**, 681.

697 Liu, M.C., Graham, J.R. and Charlot, S. (2002) Surface brightness fluctuations of Fornax cluster galaxies: Calibration of infrared surface brightness fluctuations and evidence for recent star formation. *Astrophysical Journal*, **564**, 216.

698 Raimondo, G., Cantiello, M., Brocato, E. and Capaccioli, M. (2004) Surface brightness fluctuations: A powerful tool for investigating unresolved stellar populations. *Memorie della Societá Astronomica Italiana*, **75**, 198.

699 Raimondo, G., Brocato, E., Cantiello, M. and Capaccioli, M. (2005) New optical and near-infrared surface brightness fluctuation models. II. Young and intermediate-age stellar populations. *Astronomical Journal*, **130**, 2625.

700 Chavez, M., Bertone, E., Buzzoni, A., Franchini, M., Malagnini, M.L., Morossi, C. and Rodriguez-Merino, L.H. (2007) Synthetic mid-UV spectroscopic indices of stars. *Astrophysical Journal*, **657**, 1046.

701 Fanelli, M.N., O'Connell, R.W., Burstein, D. and Wu, C.-C. (1992) Spectral synthesis in the ultraviolet. IV –

A library of mean stellar groups. *Astrophysical Journal Supplement*, **82**, 197.

702 Maraston, C., Nieves Colmenárez, L., Bender, R. and Thomas, D. (2009) Absorption line indices in the UV. I. Empirical and theoretical stellar population models. *Astronomy & Astrophysics*, **493**, 425.

703 Trager, S.C., Worthey, G., Faber, S.M., Burstein, D. and Gonzalez, J.J. (1998) Old stellar populations. VI. Absorption-line spectra of galaxy nuclei and globular clusters. *Astrophysical Journal Supplement*, **116**, 1.

704 Tripicco, M.J. and Bell, R.A. (1995) Modeling the LICK/IDS spectral feature indices using synthetic spectra. *Astronomical Journal*, **110**, 3035.

705 Silva, D.R., Kuntschner, H. and Lyubenova, M. (2008) A new approach to the study of stellar populations in early-type galaxies: *K*-band spectral indices and an application to the fornax cluster. *Astrophysical Journal*, **674**, 194.

706 Schiavon, R.P., Barbuy, B. and Bruzual, A.G. (2000) Near-infrared spectral features in single-aged stellar populations. *Astrophysical Journal*, **532**, 453.

707 Cenarro, A.J., Cardiel, N., Gorgas, J., Peletier, R.F., Vazdekis, A. and Prada, F. (2001) Empirical calibration of the near-infrared Ca II triplet – I. The stellar library and index definition. *Monthly Notices of the Royal Astronomical Society*, **326**, 959.

708 Rogers, B., Ferreras, I., Peletier, R. and Silk, J. (2010) Exploring the star formation history of elliptical galaxies: beyond simple stellar populations with a new line strength estimator. *Monthly Notices of the Royal Astronomical Society*, **402**, 447.

709 Rose, J.A. (1984) Spectral anomalies in the Hyades and Pleiades and in field stars with active chromospheres. *Astronomical Journal*, **89**, 1238.

710 Worthey, G., Faber, S.M., Gonzalez, J.J. and Burstein, D. (1994) Old stellar populations. 5: Absorption feature indices for the complete LICK/IDS sample of stars. *Astrophysical Journal Supplement*, **94**, 687.

711 Lee, H.-C., Worthey, G. and Dotter, A. (2009) Comparison of alpha-element-enhanced simple stellar population models with Milky Way globular clusters. *Astronomical Journal*, **138**, 1442.

712 Coelho, P., Bruzual, G., Charlot, S., Weiss, A., Barbuy, B. and Ferguson, J.W. (2007) Spectral models for solar-scaled and α-enhanced stellar populations. *Monthly Notices of the Royal Astronomical Society*, **382**, 498.

713 Schiavon, R.P., Faber, S.M., Rose, J.A. and Castilho, B.V. (2002) Population synthesis in the blue. II. The spectroscopic age of 47 Tucanae. *Astrophysical Journal*, **580**, 873.

714 González J. (1993) Ph.D. thesis, University of California.

715 van Dokkum, P.G. and Conroy, C. (2010) A substantial population of low-mass stars in luminous elliptical galaxies. *Nature*, **468**, 940.

716 Percival, S.M. and Salaris, M. (2011) Modelling realistic horizontal branch morphologies and their impact on spectroscopic ages of unresolved stellar systems. *Monthly Notices of the Royal Astronomical Society*, **412**, 2445.

717 Schiavon, R.P., Rose, J.A., Courteau, S. and MacArthur, L.A. (2005) A library of integrated spectra of galactic globular clusters. *Astrophysical Journal Supplement*, **160**, 163.

718 Schiavon, R.P., Rose, J.A., Courteau, S. and MacArthur, L.A. (2004) The identification of blue horizontal-branch stars in the integrated spectra of globular clusters. *Astrophysical Journal*, **608**, L33.

719 O'Connell, R.W. (1999) Far-ultraviolet radiation from elliptical galaxies. *Annual Review of Astronomy & Astrophysics*, **37**, 603.

720 Bureau, M. *et al.* (2011) The SAURON project – XVIII. The integrated UV-line-strength relations of early-type galaxies. *Monthly Notices of the Royal Astronomical Society*, **414**, 1887.

721 Burstein, D., Bertola, F., Buson, L.M., Faber, S.M. and Lauer, T.R. (1988) The far-ultraviolet spectra of early-type galaxies. *Astrophysical Journal*, **328**, 440.

722 Rich, R. *et al.* (2005) Systematics of the ultraviolet rising flux in a GALEX/SDSS sample of early-type galaxies. *Astrophysical Journal*, **619**, L107.

723 Yi, S.K., Lee, J., Sheen, Y.-K., Jeong, H., Suh, H. and Oh, K. (2011) The ultraviolet upturn in elliptical galaxies and environmental effects. *Astrophysical Journal Supplement Series*, **195**, 22.

724 Chung, C., Yoon, S.-J. and Lee, Y.-W. (2011) The effect of helium-enhanced stellar populations on the ultraviolet-upturn phenomenon of early-type galaxies. *Astrophysical Journal*, **740**, L45.

725 Han, Z., Podsiadlowski, P. and Lynas-Gray, A. (2010) The formation of hot subdwarf stars and its implications for the UV-upturn phenomenon of elliptical galaxies. *Astrophysics & Space Science*, **329**, 41.

726 Ree, C.H. *et al.* (2007) The look-back time evolution of far-ultraviolet flux from the brightest cluster elliptical galaxies at $z < 0.2$. *Astrophysical Journal Supplement*, **173**, 607.

727 Korn, A.J., Maraston, C. and Thomas, D. (2005) The sensitivity of Lick indices to abundance variations. *Astronomy & Astrophysics*, **438**, 685.

728 Thomas, D., Maraston, C. and Bender, R. (2003) Stellar population models of Lick indices with variable element abundance ratios. *Monthly Notices of the Royal Astronomical Society*, **339**, 897.

729 Tantalo, R. and Chiosi, C. (2004b) Measuring age, metallicity and abundance ratios from absorption-line indices. *Monthly Notices of the Royal Astronomical Society*, **353**, 917.

730 Graves, G.J. and Schiavon, R.P. (2008) Measuring ages and elemental abundances from unresolved stellar populations: Fe, Mg, C, N and Ca. *Astrophysical Journal Supplement*, **177**, 446.

731 Thomas, D., Johansson, J. and Maraston, C. (2011) Chemical abundance ratios of galactic globular clusters from modelling integrated light spectroscopy. *Monthly Notices of the Royal Astronomical Society*, **412**, 2199.

732 Coelho, P., Percival, S. and Salaris, M. (2011) Chemical abundance anticorrelations in globular cluster stars: The effect on cluster integrated spectra. *Astrophysical Journal*, **734**, 72.

733 Li, Z. and Han, Z. (2007) How young stellar populations affect the ages and metallicities of galaxies. *Astronomy & Astrophysics*, **471**, 795.

734 Serra, P. and Trager, S.C. (2007) On the interpretation of the age and chemical composition of composite stellar populations determined with line-strength indices. *Monthly Notices of the Royal Astronomical Society*, **374**, 769.

735 Cardelli, J.A., Clayton, G.C. and Mathis, J.S. (1989) The relationship between infrared, optical and ultraviolet extinction. *Astrophysical Journal*, **345**, 245.

736 Cid Fernandes, R. and González Delgado, R.M. (2010) Testing spectral models for stellar populations with star clusters – I. Methodology. *Monthly Notices of the Royal Astronomical Society*, **403**, 780.

737 Koleva, M., Prugniel, P., Bouchard, A. and Wu, Y. (2009) ULySS: a full spectrum fitting package. *Astronomy & Astrophysics*, **501**, 1269.

738 Ocvirk, P., Pichon, C., Lançon, A. and Thiébaut, E. (2006) STECKMAP: STEllar content and kinematics from high resolution galactic spectra via maximum a posteriori. *Monthly Notices of the Royal Astronomical Society*, **365**, 740.

739 Walcher, J., Groves, B., Budavári, T. and Dale, D. (2011) Fitting the integrated spectral energy distributions of galaxies. *Astrophysics and Space Science*, **331**, 1.

740 Heavens, A.F., Jimenez, R. and Lahav, O. (2000) Massive lossless data compression and multiple parameter estimation from galaxy spectra. *Monthly Notices of the Royal Astronomical Society*, **317**, 965.

741 Martins, L.P. and Coelho, P. (2007) Testing the accuracy of synthetic stellar libraries. *Monthly Notices of the Royal Astronomical Society*, **381**, 1329.

742 Percival, S.M. and Salaris, M. (2009) The impact of systematic uncertainties in stellar parameters on integrated spectra of stellar populations. *Astrophysical Journal*, **703**, 1123.

743 Dias, B., Coelho, P., Barbuy, B., Kerber, L. and Idiart, T. (2010) Age and metallicity of star clusters in the Small Magellanic Cloud from integrated spectroscopy. *Astronomy & Astrophysics*, **520**, 85.

744 González Delgado, R.M. and Cid Fernandes, R. (2010) Testing spectral models for stellar populations with star clusters – II. Results. *Monthly Notices of the Royal Astronomical Society*, **403**, 797.

745 González, R.A., Liu, M.C. and Bruzual A.G. (2004) Infrared surface brightness fluctuations of magellanic star clusters. *Astrophysical Journal*, **611**, 270.

746 Koleva, M., Prugniel, P., Ocvirk, P., Le Borgne, D. and Soubiran, C. (2008) Spectroscopic ages and metallicities of stellar populations: Validation of full spectrum fitting. *Monthly Notices of the Royal Astronomical Society*, **385**, 1998.

747 Zhang, Y., Han, Z., Liu, J., Zhang, F. and Kang, X. (2012) Testing three derivative methods of stellar population synthesis models. *Monthly Notices of the Royal Astronomical Society*, **421**, 1678.

748 Harris, W.E. (1996) A catalog of parameters for globular clusters in the Milky Way. *Astronomical Journal*, **112**, 1487.

Index

Old Stellar Populations, First Edition. S. Cassisi and M. Salaris.
© 2013 WILEY-VCH Verlag GmbH & Co. KGaA. Published 2013 by WILEY-VCH Verlag GmbH & Co. KGaA.